SOLUTIONS MANUAL

RAMAMURTHY MANI

Electrical and Computer Engineering Department

Boston University

SIGNALS & SYSTEMS

SECOND EDITION

ALAN V. OPPENHEIM

ALAN S. WILLSKY

Massachusetts Institute of Technology

WITH S. HAMID NAWAB

Boston University

PRENTICE HALL

Upper Saddle River, NJ 07458

Acquisitions Editor: *Tom Robbins*
Project Editor: *Kimberly Dellas*
Special Projects Manager: *Barbara Murray*
Supplement Acquisitions Editor: *Nancy Garcia*
Production Coordinator: *Donna Sullivan*

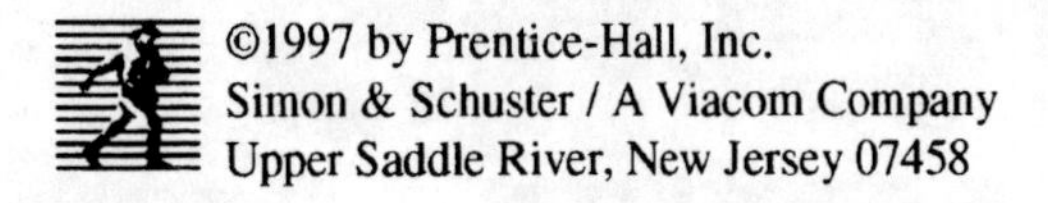

Simon & Schuster / A Viacom Company
Upper Saddle River, New Jersey 07458

Printed in the United States of America

10 9 8 7 6 5 4 3 2

ISBN 0-13-616939-2

Prentice-Hall International (UK) Limited, *London*
Prentice-Hall of Australia Pty. Limited, *Sydney*
Prentice-Hall Canada Inc., *Toronto*
Prentice-Hall Hispanoamericana, S.A., *Mexico*
Prentice-Hall of India Private Limited, *New Delhi*
Prentice-Hall of Japan, Inc., *Tokyo*
Simon & Schuster Asia Pte. Ltd., *Singapore*
Editora Prentice-Hall do Brasil, Ltda., *Rio de Janeiro*

Preface

More than half of the 600+ problems in the second edition of *Signals & Systems* are new, while the remainder are the same as in the first edition. This manual contains solutions to the new problems, as well as updated solutions for the problems from the first edition. These problems and their solutions were mostly generated during courses taught by Alan V. Oppenheim and Alan S. Willsky at Massachusetts Institute of Technology and by S. Hamid Nawab at Boston University. Numerous teaching assistants at MIT and BU have thus significantly contributed to this material over the years.

In preparing the manual, previously written solutions (including those from the solutions manual for the first edition) were reviewed and revised, new solutions were written wherever necessary, and the entire document was typed in LaTeXformat. We intend to update this document periodically by removing errors and improving solutions.

Contents

Chapter 1 Answers

1.1. Converting from polar to Cartesian coordinates:

$\frac{1}{2}e^{j\pi} = \frac{1}{2}\cos\pi = -\frac{1}{2}$, $\quad \frac{1}{2}e^{-j\pi} = \frac{1}{2}\cos(-\pi) = -\frac{1}{2}$

$e^{j\frac{\pi}{2}} = \cos\left(\frac{\pi}{2}\right) + j\sin\left(\frac{\pi}{2}\right) = j$, $\quad e^{-j\frac{\pi}{2}} = \cos\left(\frac{\pi}{2}\right) - j\sin\left(\frac{\pi}{2}\right) = -j$

$e^{j5\frac{\pi}{2}} = e^{j\frac{\pi}{2}} = j$, $\quad \sqrt{2}e^{j\frac{\pi}{4}} = \sqrt{2}\left(\cos\left(\frac{\pi}{4}\right) + j\sin\left(\frac{\pi}{4}\right)\right) = 1+j$

$\sqrt{2}e^{\frac{9j\pi}{4}} = \sqrt{2}e^{\frac{j\pi}{4}} = 1+j$, $\quad \sqrt{2}e^{\frac{-9j\pi}{4}} = \sqrt{2}e^{\frac{-j\pi}{4}} = 1-j$

$\sqrt{2}e^{\frac{-j\pi}{4}} = 1-j$

1.2. Converting from Cartesian to polar coordinates:

$5 = 5e^{j0}$, $\quad -2 = 2e^{j\pi}$, $\quad -3j = 3e^{-j\frac{\pi}{2}}$

$\frac{1}{2} - j\frac{\sqrt{3}}{2} = e^{-j\frac{\pi}{3}}$, $\quad 1+j = \sqrt{2}e^{j\frac{\pi}{4}}$, $\quad (1-j)^2 = 2e^{-j\frac{\pi}{2}}$

$j(1-j) = e^{j\frac{\pi}{4}}$, $\quad \frac{1+j}{1-j} = e^{j\frac{\pi}{2}}$, $\quad \frac{\sqrt{2}+j\sqrt{2}}{1+j\sqrt{3}} = e^{-j\frac{\pi}{12}}$

1.3. **(a)** $E_\infty = \int_0^\infty e^{-4t}dt = \frac{1}{4}$, $P_\infty = 0$, because $E_\infty < \infty$

(b) $x_2(t) = e^{j(2t+\frac{\pi}{4})}$, $|x_2(t)| = 1$. Therefore, $E_\infty = \int_{-\infty}^{\infty} |x_2(t)|^2 dt = \int_{-\infty}^{\infty} dt = \infty$, $P_\infty = \lim_{T\to\infty} \frac{1}{2T}\int_{-T}^{T} |x_2(t)|^2 dt = \lim_{T\to\infty} \frac{1}{2T}\int_{-T}^{T} dt = \lim_{T\to\infty} 1 = 1$

(c) $x_3(t) = \cos(t)$. Therefore, $E_\infty = \int_{-\infty}^{\infty} |x_3(t)|^2 dt = \int_{-\infty}^{\infty} \cos^2(t)dt = \infty$,

$$P_\infty = \lim_{T\to\infty} \frac{1}{2T}\int_{-T}^{T} \cos^2(t)dt = \lim_{T\to\infty} \frac{1}{2T}\int_{-T}^{T} \left(\frac{1+\cos(2t)}{2}\right) dt = \frac{1}{2}$$

(d) $x_1[n] = \left(\frac{1}{2}\right)^n u[n]$, $|x_1[n]|^2 = \left(\frac{1}{4}\right)^n u[n]$. Therefore, $E_\infty = \sum_{n=-\infty}^{\infty} |x_1[n]|^2 = \sum_{n=0}^{\infty} \left(\frac{1}{4}\right)^n = \frac{4}{3}$,

$P_\infty = 0$, because $E_\infty < \infty$.

(e) $x_2[n] = e^{j(\frac{\pi n}{2}+\frac{\pi}{8})}$, $|x_2[n]|^2 = 1$. Therefore, $E_\infty = \sum_{n=-\infty}^{\infty} |x_2[n]|^2 = \infty$,

$$P_\infty = \lim_{N\to\infty} \frac{1}{2N+1}\sum_{n=-N}^{N} |x_2[n]|^2 = \lim_{N\to\infty} \frac{1}{2N+1}\sum_{n=-N}^{N} 1 = 1.$$

(f) $x_3[n] = \cos(\frac{\pi}{4}n)$. Therefore, $E_\infty = \sum_{n=-\infty}^{\infty} |x_3[n]|^2 = \sum_{n=-\infty}^{\infty} \cos^2(\frac{\pi}{4}n) = \infty$,

$$P_\infty = \lim_{N\to\infty} \frac{1}{2N+1}\sum_{n=-N}^{N} \cos^2(\frac{\pi}{4}n) = \lim_{N\to\infty} \frac{1}{2N+1}\sum_{n=-N}^{N} \left(\frac{1+\cos(\frac{\pi}{2}n)}{2}\right) = \frac{1}{2}$$

1.4. **(a)** The signal $x[n]$ is shifted by 3 to the right. The shifted signal will be zero for $n < 1$ and $n > 7$.

(b) The signal $x[n]$ is shifted by 4 to the left. The shifted signal will be zero for $n < -6$ and $n > 0$.

(c) The signal $x[n]$ is flipped. The flipped signal will be zero for $n < -4$ and $n > 2$.

(d) The signal $x[n]$ is flipped and the flipped signal is shifted by 2 to the right. This new signal will be zero for $n < -2$ and $n > 4$.

(e) The signal $x[n]$ is flipped and the flipped signal is shifted by 2 to the left. This new signal will be zero for $n < -6$ and $n > 0$.

1.5. **(a)** $x(1-t)$ is obtained by flipping $x(t)$ and shifting the flipped signal by 1 to the right. Therefore, $x(1-t)$ will be zero for $t > -2$.

(b) From (a), we know that $x(1-t)$ is zero for $t > -2$. Similarly, $x(2-t)$ is zero for $t > -1$. Therefore, $x(1-t) + x(2-t)$ will be zero for $t > -2$.

(c) $x(3t)$ is obtained by linearly compressing $x(t)$ by a factor of 3. Therefore, $x(3t)$ will be zero for $t < 1$.

(d) $x(t/3)$ is obtained by linearly stretching $x(t)$ by a factor of 3. Therefore, $x(t/3)$ will be zero for $t < 9$.

1.6. **(a)** $x_1(t)$ is not periodic because it is zero for $t < 0$.

(b) $x_2[n] = 1$ for all n. Therefore, it is periodic with a fundamental period of 1.

(c) $x_3[n]$ is as shown in the Figure S1.6.

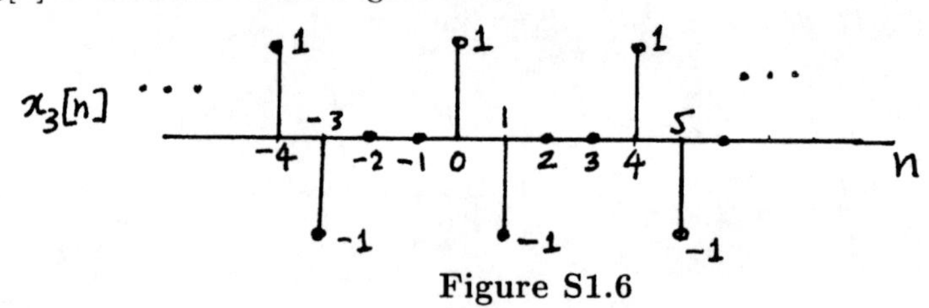

Figure S1.6

Therefore, it is periodic with a fundamental period of 4.

1.7. **(a)**

$$\mathcal{E}v\{x_1[n]\} = \frac{1}{2}(x_1[n] + x_1[-n]) = \frac{1}{2}(u[n] - u[n-4] + u[-n] - u[-n-4])$$

Therefore, $\mathcal{E}v\{x_1[n]\}$ is zero for $|n| > 3$.

(b) Since $x_2(t)$ is an odd signal, $\mathcal{E}v\{x_2(t)\}$ is zero for all values of t.

(c)

$$\mathcal{E}v\{x_3[n]\} = \frac{1}{2}(x_1[n] + x_1[-n]) = \frac{1}{2}[(\frac{1}{2})^n u[n-3] - (\frac{1}{2})^{-n} u[-n-3]]$$

Therefore, $\mathcal{E}v\{x_3[n]\}$ is zero when $|n| < 3$ and when $|n| \to \infty$.

(d)

$$\mathcal{E}v\{x_4(t)\} = \frac{1}{2}(x_4(t) + x_4(-t)) = \frac{1}{2}[e^{-5t}u(t+2) - e^{5t}u(-t+2)$$

Therefore, $\mathcal{E}v\{x_4(t)\}$ is zero only when $|t| \to \infty$.

1.8. (a) $\mathcal{R}e\{x_1(t)\} = -2 = 2e^{0t}\cos(0t+\pi)$

(b) $\mathcal{R}e\{x_2(t)\} = \sqrt{2}\cos(\frac{\pi}{4})\cos(3t+2\pi) = \cos(3t) = e^{0t}\cos(3t+0)$

(c) $\mathcal{R}e\{x_3(t)\} = e^{-t}\sin(3t+\pi) = e^{-t}\cos(3t+\frac{\pi}{2})$

(d) $\mathcal{R}e\{x_4(t)\} = -e^{-2t}\sin(100t) = e^{-2t}\sin(100t+\pi) = e^{-2t}\cos(100t+\frac{\pi}{2})$

1.9. (a) $x_1(t)$ is a periodic complex exponential.

$$x_1(t) = je^{j10t} = e^{j(10t+\frac{\pi}{2})}$$

The fundamental period of $x_1(t)$ is $\frac{2\pi}{10} = \frac{\pi}{5}$.

(b) $x_2(t)$ is a complex exponential multiplied by a decaying exponential. Therefore, $x_2(t)$ is not periodic.

(c) $x_3[n]$ is a periodic signal.

$$x_3[n] = e^{j7\pi n} = e^{j\pi n}$$

$x_3[n]$ is a complex exponential with a fundamental period of $\frac{2\pi}{\pi} = 2$.

(d) $x_4[n]$ is a periodic signal. The fundamental period is given by $N = m(\frac{2\pi}{3\pi/5}) = m(\frac{10}{3})$. By choosing $m = 3$, we obtain the fundamental period to be 10.

(e) $x_5[n]$ is not periodic. $x_5[n]$ is a complex exponential with $\omega_0 = 3/5$. We cannot find any integer m such that $m(\frac{2\pi}{\omega_0})$ is also an integer. Therefore, $x_5[n]$ is not periodic.

1.10.

$$x(t) = 2\cos(10t+1) - \sin(4t-1)$$

Period of first term in RHS $= \frac{2\pi}{10} = \frac{\pi}{5}$
Period of second term in RHS $= \frac{2\pi}{4} = \frac{\pi}{2}$
Therefore, the overall signal is periodic with a period which is the least common multiple of the periods of the first and second terms. This is equal to π.

1.11.

$$x[n] = 1 + e^{j\frac{4\pi}{7}n} - e^{j\frac{2\pi}{5}n}$$

Period of the first term in the RHS = 1
Period of the second term in the RHS $= m(\frac{2\pi}{4\pi/7}) = 7$ (when $m = 2$)
Period of the third term in the RHS $= m(\frac{2\pi}{2\pi/5}) = 5$ (when $m = 1$)
Therefore, the overall signal $x[n]$ is periodic with a period which is the least common multiple of the periods of the three terms in $x[n]$. This is equal to 35.

1.12. The signal $x[n]$ is as shown in Figure S1.12. $x[n]$ can be obtained by flipping $u[n]$ and then shifting the flipped signal by 3 to the right. Therefore, $x[n] = u[-n+3]$. This implies that $M = -1$ and $n_0 = -3$.

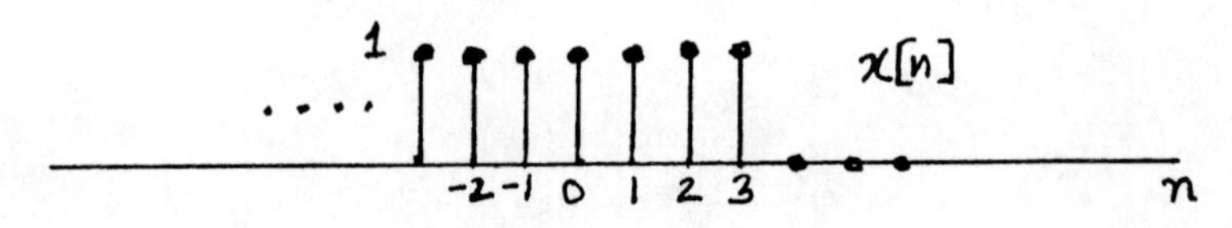

Figure S1.12

1.13.

$$y(t) = \int_{-\infty}^{t} x(\tau)d\tau = \int_{-\infty}^{t} (\delta(\tau+2) - \delta(\tau-2))dt = \begin{cases} 0, & t < -2 \\ 1, & -2 \le t \le 2 \\ 0, & t > 2 \end{cases}$$

Therefore,

$$E_\infty = \int_{-2}^{2} dt = 4$$

1.14. The signal $x(t)$ and its derivative $g(t)$ are shown in Figure S1.14.

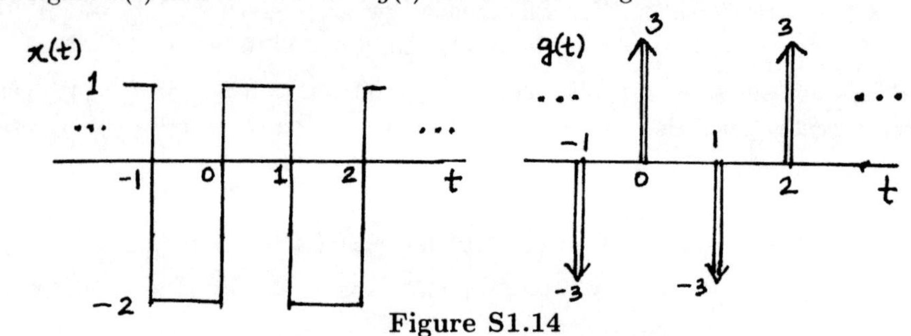

Figure S1.14

Therefore,

$$g(t) = 3\sum_{k=-\infty}^{\infty} \delta(t-2k) - 3\sum_{k=-\infty}^{\infty} \delta(t-2k-1)$$

This implies that $A_1 = 3$, $t_1 = 0$, $A_2 = -3$, and $t_2 = 1$.

1.15. **(a)** The signal $x_2[n]$, which is the input to S_2, is the same as $y_1[n]$. Therefore,

$$\begin{aligned} y_2[n] &= x_2[n-2] + \frac{1}{2}x_2[n-3] \\ &= y_1[n-2] + \frac{1}{2}y_1[n-3] \\ &= 2x_1[n-2] + 4x_1[n-3] + \frac{1}{2}(2x_1[n-3] + 4x_1[n-4]) \\ &= 2x_1[n-2] + 5x_1[n-3] + 2x_1[n-4] \end{aligned}$$

The input-output relationship for S is

$$y[n] = 2x[n-2] + 5x[n-3] + 2x[n-4]$$

(b) The input-output relationship does not change if the order in which S_1 and S_2 are connected in series is reversed. We can easily prove this by assuming that S_1 follows S_2. In this case, the signal $x_1[n]$, which is the input to S_1, is the same as $y_2[n]$. Therefore,

$$\begin{aligned} y_1[n] &= 2x_1[n] + 4x_1[n-1] \\ &= 2y_2[n] + 4y_2[n-1] \\ &= 2(x_2[n-2] + \frac{1}{2}x_2[n-3]) + 4(x_2[n-3] + \frac{1}{2}x_2[n-4]) \\ &= 2x_2[n-2] + 5x_2[n-3] + 2x_2[n-4] \end{aligned}$$

The input-output relationship for S is once again

$$y[n] = 2x[n-2] + 5x[n-3] + 2x[n-4]$$

1.16. (a) The system is not memoryless because $y[n]$ depends on past values of $x[n]$.

(b) The output of the system will be $y[n] = \delta[n]\delta[n-2] = 0$.

(c) From the result of part (b), we may conclude that the system output is always zero for inputs of the form $\delta[n-k]$, $k \in \mathcal{I}$. Therefore, the system is not invertible.

1.17. (a) The system is not causal because the output $y(t)$ at some time may depend on future values of $x(t)$. For instance, $y(-\pi) = x(0)$.

(b) Consider two arbitrary inputs $x_1(t)$ and $x_2(t)$.

$$x_1(t) \longrightarrow y_1(t) = x_1(\sin(t))$$

$$x_2(t) \longrightarrow y_2(t) = x_2(\sin(t))$$

Let $x_3(t)$ be a linear combination of $x_1(t)$ and $x_2(t)$. That is,

$$x_3(t) = ax_1(t) + bx_2(t)$$

where a and b are arbitrary scalars. If $x_3(t)$ is the input to the given system, then the corresponding output $y_3(t)$ is

$$\begin{aligned} y_3(t) &= x_3(\sin(t)) \\ &= ax_1(\sin(t)) + bx_2(\sin(t)) \\ &= ay_1(t) + by_2(t) \end{aligned}$$

Therefore, the system is linear.

1.18. (a) Consider two arbitrary inputs $x_1[n]$ and $x_2[n]$.

$$x_1[n] \longrightarrow y_1[n] = \sum_{k=n-n_0}^{n+n_0} x_1[k]$$

$$x_2[n] \longrightarrow y_2[n] = \sum_{k=n-n_0}^{n+n_0} x_2[k]$$

Let $x_3[n]$ be a linear combination of $x_1[n]$ and $x_2[n]$. That is,

$$x_3[n] = ax_1[n] + bx_2[n]$$

where a and b are arbitrary scalars. If $x_3[n]$ is the input to the given system, then the corresponding output $y_3[n]$ is

$$\begin{aligned} y_3[n] &= \sum_{k=n-n_0}^{n+n_0} x_3[k] \\ &= \sum_{k=n-n_0}^{n+n_0} (ax_1[k] + bx_2[k]) = a \sum_{k=n-n_0}^{n+n_0} x_1[k] + b \sum_{k=n-n_0}^{n+n_0} x_2[k] \\ &= ay_1[n] + by_2[n] \end{aligned}$$

Therefore, the system is linear.

(b) Consider an arbitrary input $x_1[n]$. Let

$$y_1[n] = \sum_{k=n-n_0}^{n+n_0} x_1[k]$$

be the corresponding output. Consider a second input $x_2[n]$ obtained by shifting $x_1[n]$ in time:

$$x_2[n] = x_1[n - n_1]$$

The output corresponding to this input is

$$y_2[n] = \sum_{k=n-n_0}^{n+n_0} x_2[k] = \sum_{k=n-n_0}^{n+n_0} x_1[k - n_1] = \sum_{k=n-n_1-n_0}^{n-n_1+n_0} x_1[k]$$

Also note that

$$y_1[n - n_1] = \sum_{k=n-n_1-n_0}^{n-n_1+n_0} x_1[k].$$

Therefore,

$$y_2[n] = y_1[n - n_1]$$

This implies that the system is time-invariant.

(c) If $|x[n]| < B$, then

$$y[n] \le (2n_0 + 1)B$$

Therefore, $C \le (2n_0 + 1)B$.

1.19. **(a)** (i) Consider two arbitrary inputs $x_1(t)$ and $x_2(t)$.

$$x_1(t) \longrightarrow y_1(t) = t^2 x_1(t-1)$$

$$x_2(t) \longrightarrow y_2(t) = t^2 x_2(t-1)$$

Let $x_3(t)$ be a linear combination of $x_1(t)$ and $x_2(t)$. That is,

$$x_3(t) = a x_1(t) + b x_2(t)$$

where a and b are arbitrary scalars. If $x_3(t)$ is the input to the given system, then the corresponding output $y_3(t)$ is

$$\begin{aligned} y_3(t) &= t^2 x_3(t-1) \\ &= t^2 (a x_1(t-1) + b x_2(t-1)) \\ &= a y_1(t) + b y_2(t) \end{aligned}$$

Therefore, the system is **linear**.

(ii) Consider an arbitrary input $x_1(t)$. Let

$$y_1(t) = t^2 x_1(t-1)$$

be the corresponding output. Consider a second input $x_2(t)$ obtained by shifting $x_1(t)$ in time:

$$x_2(t) = x_1(t - t_0)$$

The output corresponding to this input is

$$y_2(t) = t^2 x_2(t-1) = t^2 x_1(t - 1 - t_0)$$

Also note that

$$y_1(t - t_0) = (t - t_0)^2 x_1(t - 1 - t_0) \neq y_2(t)$$

Therefore, the system is **not time-invariant**.

(b) (i) Consider two arbitrary inputs $x_1[n]$ and $x_2[n]$.

$$x_1[n] \longrightarrow y_1[n] = x_1^2[n-2]$$

$$x_2[n] \longrightarrow y_2[n] = x_2^2[n-2]$$

Let $x_3[n]$ be a linear combination of $x_1[n]$ and $x_2[n]$. That is,

$$x_3[n] = a x_1[n] + b x_2[n]$$

where a and b are arbitrary scalars. If $x_3[n]$ is the input to the given system, then the corresponding output $y_3[n]$ is

$$\begin{aligned} y_3[n] &= x_3^2[n-2] \\ &= (a x_1[n-2] + b x_2[n-2])^2 \\ &= a^2 x_1^2[n-2] + b^2 x_2^2[n-2] + 2ab x_1[n-2] x_2[n-2] \\ &\neq a y_1[n] + b y_2[n] \end{aligned}$$

Therefore, the system is **not linear**.

(ii) Consider an arbitrary input $x_1[n]$. Let

$$y_1[n] = x_1^2[n-2]$$

be the corresponding output. Consider a second input $x_2[n]$ obtained by shifting $x_1[n]$ in time:

$$x_2[n] = x_1[n - n_0]$$

The output corresponding to this input is

$$y_2[n] = x_2^2[n-2] = x_1^2[n-2-n_0]$$

Also note that

$$y_1[n-n_0] = x_1^2[n-2-n_0]$$

Therefore,

$$y_2[n] = y_1[n-n_0]$$

This implies that the system is **time-invariant**.

(c) (i) Consider two arbitrary inputs $x_1[n]$ and $x_2[n]$.

$$x_1[n] \longrightarrow y_1[n] = x_1[n+1] - x_1[n-1]$$

$$x_2[n] \longrightarrow y_2[n] = x_2[n+1] - x_2[n-1]$$

Let $x_3[n]$ be a linear combination of $x_1[n]$ and $x_2[n]$. That is,

$$x_3[n] = ax_1[n] + bx_2[n]$$

where a and b are arbitrary scalars. If $x_3[n]$ is the input to the given system, then the corresponding output $y_3[n]$ is

$$\begin{aligned} y_3[n] &= x_3[n+1] - x_3[n-1] \\ &= ax_1[n+1] + bx_1[n+1] - ax_1[n-1] - bx_2[n-1] \\ &= a(x_1[n+1] - x_1[n-1]) + b(x_2[n+1] - x_2[n-1]) \\ &= ay_1[n] + by_2[n] \end{aligned}$$

Therefore, the system is **linear**.

(ii) Consider an arbitrary input $x_1[n]$. Let

$$y_1[n] = x_1[n+1] - x_1[n-1]$$

be the corresponding output. Consider a second input $x_2[n]$ obtained by shifting $x_1[n]$ in time:

$$x_2[n] = x_1[n-n_0]$$

The output corresponding to this input is

$$y_2[n] = x_2[n+1] - x_2[n-1] = x_1[n+1-n_0] - x_1[n-1-n_0]$$

Also note that

$$y_1[n-n_0] = x_1[n+1-n_0] - x_1[n-1-n_0]$$

Therefore,

$$y_2[n] = y_1[n-n_0]$$

This implies that the system is **time-invariant**.

(d) (i) Consider two arbitrary inputs $x_1(t)$ and $x_2(t)$.

$$x_1(t) \longrightarrow y_1(t) = \mathcal{O}d\{x_1(t)\}$$

$$x_2(t) \longrightarrow y_2(t) = \mathcal{O}d\{x_2(t)\}$$

Let $x_3(t)$ be a linear combination of $x_1(t)$ and $x_2(t)$. That is,

$$x_3(t) = ax_1(t) + bx_2(t)$$

where a and b are arbitrary scalars. If $x_3(t)$ is the input to the given system, then the corresponding output $y_3(t)$ is

$$\begin{aligned} y_3(t) &= \mathcal{O}d\{x_3(t)\} \\ &= \mathcal{O}d\{ax_1(t) + bx_2(t)\} \\ &= a\mathcal{O}d\{x_1(t)\} + b\mathcal{O}d\{x_2(t)\} = ay_1(t) + by_2(t) \end{aligned}$$

Therefore, the system is **linear**.

(ii) Consider an arbitrary input $x_1(t)$. Let

$$y_1[t) = \mathcal{O}d\{x_1(t)\} = \frac{x_1(t) - x_1(-t)}{2}$$

be the corresponding output. Consider a second input $x_2(t)$ obtained by shifting $x_1[n]$ in time:

$$x_2(t) = x_1(t-t_0)$$

The output corresponding to this input is

$$\begin{aligned} y_2(t) &= \mathcal{O}d\{x_2(t)\} = \frac{x_2(t) - x_2(-t)}{2} \\ &= \frac{x_1(t-t_0) - x_1(-t-t_0)}{2} \end{aligned}$$

Also note that

$$y_1(t-t_0) = \frac{x_1(t-t_0) - x_1(-t+t_0)}{2} \neq y_2(t)$$

Therefore, the system is **not time-invariant**.

1.20. (a) Given

$$x(t) = e^{j2t} \longrightarrow y(t) = e^{j3t}$$
$$x(t) = e^{-j2t} \longrightarrow y(t) = e^{-j3t}$$

Since the system is linear,

$$x_1(t) = \frac{1}{2}(e^{j2t} + e^{-j2t}) \longrightarrow y_1(t) = \frac{1}{2}(e^{j3t} + e^{-j3t})$$

Therefore,

$$x_1(t) = \cos(2t) \longrightarrow y_1(t) = \cos(3t)$$

(b) We know that

$$x_2(t) = \cos\left(2(t - \frac{1}{2})\right) = \frac{e^{-j}e^{j2t} + e^{j}e^{-j2t}}{2}$$

Using the linearity property, we may once again write

$$x_1(t) = \frac{1}{2}(e^{-j}e^{j2t} + eje^{-j2t}) \longrightarrow y_1(t) = \frac{1}{2}(e^{-j}e^{j3t} + e^{j}e^{-j3t}) = \cos(3t - 1)$$

Therefore,

$$x_1(t) = \cos(2(t - 1/2)) \longrightarrow y_1(t) = \cos(3t - 1)$$

1.21. The signals are sketched in Figure S1.21.

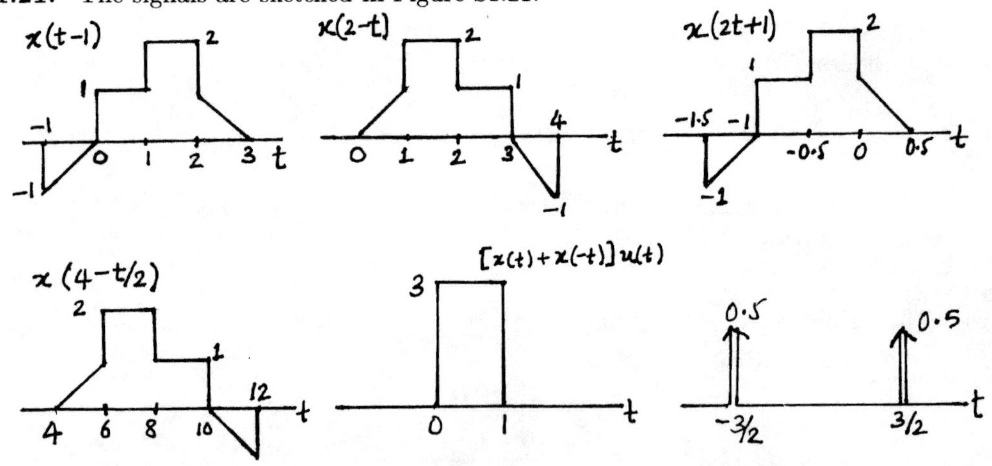

Figure S1.21

1.22. The signals are sketched in Figure S1.22.

1.23. The even and odd parts are sketched in Figure S1.23.

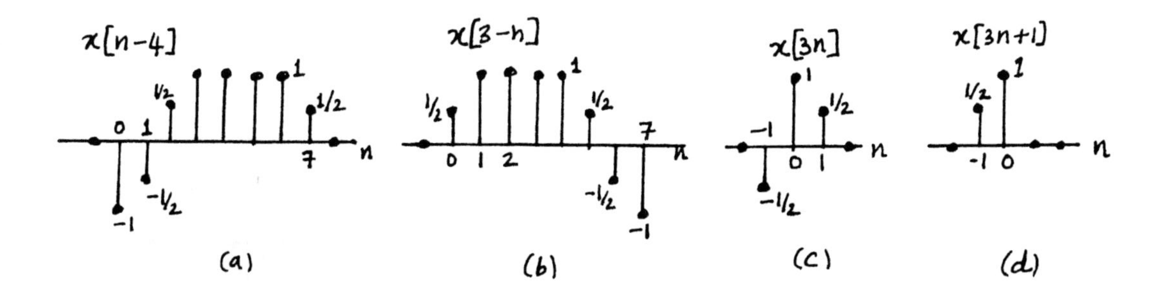

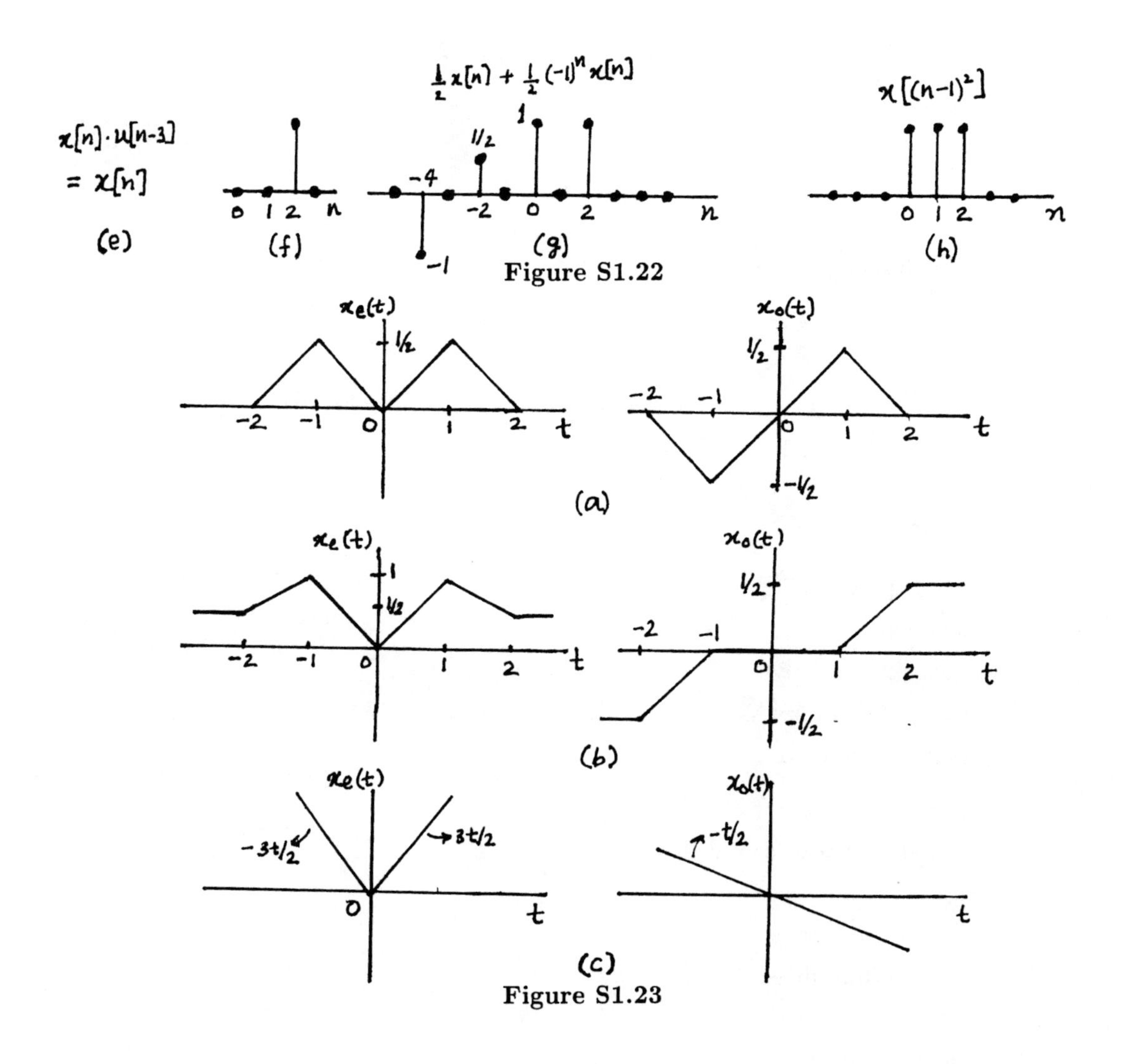

Figure S1.22

Figure S1.23

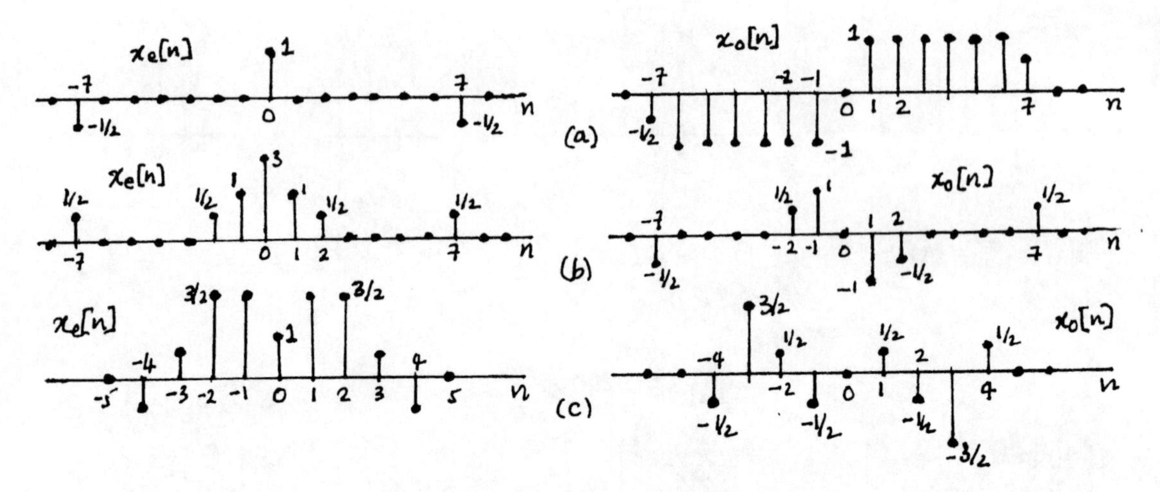

Figure S1.24

1.24. The even and odd parts are sketched in Figure S1.24.

1.25. (a) Periodic, period $= 2\pi/(4) = \pi/2$.

(b) Periodic, period $= 2\pi/(\pi) = 2$.

(c) $x(t) = [1 + \cos(4t - 2\pi/3)]/2$. Periodic, period $= 2\pi/(4) = \pi/2$.

(d) $x(t) = \cos(4\pi t)/2$. Periodic, period $= 2\pi/(4\pi) = 1/2$.

(e) $x(t) = [\sin(4\pi t)u(t) - \sin(4\pi t)u(-t)]/2$. Not periodic.

(f) Not periodic.

1.26. (a) Periodic, period $= 7$.

(b) Not periodic.

(c) Periodic, period $= 8$.

(d) $x[n] = (1/2)[\cos(3\pi n/4) + \cos(\pi n/4)]$. Periodic, period $= 8$.

(e) Periodic, period $= 16$.

1.27. (a) Linear, stable.

(b) Memoryless, linear, causal, stable.

(c) Linear

(d) Linear, causal, stable.

(e) Time invariant, linear, causal, stable.

(f) Linear, stable.

(g) Time invariant, linear, causal.

1.28. **(a)** Linear, stable.

(b) Time invariant, linear, causal, stable.

(c) Memoryless, linear, causal.

(d) Linear, stable.

(e) Linear, stable.

(f) Memoryless, linear, causal, stable.

(g) Linear, stable.

1.29. **(a)** Consider two inputs to the system such that

$$x_1[n] \xrightarrow{S} y_1[n] = \mathcal{R}e\{x_1[n]\} \quad \text{and} \quad x_2[n] \xrightarrow{S} y_2[n] = \mathcal{R}e\{x_2[n]\}.$$

Now consider a third input $x_3[n] = x_1[n] + x_2[n]$. The corresponding system output will be

$$\begin{aligned} y_3[n] &= \mathcal{R}e\{x_3[n]\} \\ &= \mathcal{R}e\{x_1[n] + x_2[n]\} \\ &= \mathcal{R}e\{x_1[n]\} + \mathcal{R}e\{x_2[n] \\ &= y_1[n] + y_2[n] \end{aligned}$$

Therefore, we may conclude that the system is additive.

Let us now assume that the input-output relationship is changed to $y[n] = \mathcal{R}e\{e^{j\pi/4}x[n]\}$. Also, consider two inputs to the system such that

$$x_1[n] \xrightarrow{S} y_1[n] = \mathcal{R}e\{e^{j\pi/4}x_1[n]\}$$

and

$$x_2[n] \xrightarrow{S} y_2[n] = \mathcal{R}e\{e^{j\pi/4}x_2[n]\}.$$

Now consider a third input $x_3[n] = x_1[n] + x_2[n]$. The corresponding system output will be

$$\begin{aligned} y_3[n] &= \mathcal{R}e\{e^{j\pi/4}x_3[n]\} \\ &= \cos(\pi n/4)\mathcal{R}e\{x_3[n]\} - \sin(\pi n/4)\mathcal{I}m\{x_3[n]\} \\ &\quad + \cos(\pi n/4)\mathcal{R}e\{x_1[n]\} - \sin(\pi n/4)\mathcal{I}m\{x_1[n]\} \\ &+ \cos(\pi n/4)\mathcal{R}e\{x_2[n]\} - \sin(\pi n/4)\mathcal{I}m\{x_2[n]\} \\ &= \mathcal{R}e\{e^{j\pi/4}x_1[n]\} + \mathcal{R}e\{e^{j\pi/4}x_2[n]\} \\ &= y_1[n] + y_2[n] \end{aligned}$$

Therefore, we may conclude that the system is additive.

(b) (i) Consider two inputs to the system such that

$$x_1(t) \overset{S}{\rightarrow} y_1(t) = \frac{1}{x_1(t)}\left[\frac{dx_1(t)}{dt}\right]^2 \quad \text{and} \quad x_2(t) \overset{S}{\rightarrow} y_2(t) = \frac{1}{x_1(t)}\left[\frac{dx_2(t)}{dt}\right]^2.$$

Now consider a third input $x_3(t) = x_1(t) + x_2(t)$. The corresponding system output will be

$$\begin{aligned} y_3(t) &= \frac{1}{x_3(t)}\left[\frac{dx_3(t)}{dt}\right]^2 \\ &= \frac{1}{x_1(t)+x_2(t)}\left[\frac{d[x_1(t)+x_2(t)]}{dt}\right]^2 \\ &\neq y_1(t) + y_2(t) \end{aligned}$$

Therefore, we may conclude that the system is not additive.

Now consider a fourth input $x_4(t) = ax_1(t)$. The corresponding output will be

$$\begin{aligned} y_4(t) &= \frac{1}{x_4(t)}\left[\frac{dx_4(t)}{dt}\right]^2 \\ &= \frac{1}{ax_1(t)}\left[\frac{d[ax_1(t)]}{dt}\right]^2 \\ &= \frac{a}{x_1(t)}\left[\frac{dx_1(t)}{dt}\right]^2 \\ &= ay_1(t) \end{aligned}$$

Therefore, the system is homogeneous.

(ii) This system is not additive. Consider the following example. Let $x_1[n] = 2\delta[n+2] + 2\delta[n+1] + 2\delta[n]$ and $x_2[n] = \delta[n+2] + 2\delta[n+1] + 3\delta[n]$. The corresponding outputs evaluated at $n = 0$ are

$$y_1[0] = 2 \quad \text{and} \quad y_2[0] = 3/2.$$

Now consider a third input $x_3[n] = x_1[n] + x_2[n] = 3\delta[n+2] + 4\delta[n+1] + 5\delta[n]$. The corresponding output evaluated at $n = 0$ is $y_3[0] = 15/4$. Clearly, $y_3[0] \neq y_1[0] + y_2[0]$. This implies that the system in not additive.

No consider an input $x_4[n]$ which leads to the output $y_4[n]$. We know that

$$y_4[n] = \begin{cases} \frac{x_4[n]x_4[n-2]}{x_4[n-1]}, & x_4[n-1] \neq 0 \\ 0, & \text{otherwise} \end{cases}.$$

Let us now consider another input $x_5[n] = ax_4[n]$. The corresponding output is

$$y_5[n] = \begin{cases} a\frac{x_4[n]x_4[n-2]}{x_4[n-1]}, & x_4[n-1] \neq 0 \\ 0, & \text{otherwise} \end{cases} = ay_4[n].$$

Therefore, the system is homogeneous.

1.30. **(a)** Invertible. Inverse system: $y(t) = x(t+4)$.

(b) Non invertible. The signals $x(t)$ and $x_1(t) = x(t) + 2\pi$ give the same output.

(c) Non invertible. $\delta[n]$ and $2\delta[n]$ give the same output.

(d) Invertible. Inverse system: $y(t) = dx(t)/dt$.

(e) Invertible. Inverse system: $y[n] = x[n+1]$ for $n \geq 0$ and $y[n] = x[n]$ for $n < 0$.

(f) Non invertible. $x[n]$ and $-x[n]$ give the same result.

(g) Invertible. Inverse system: $y[n] = x[1-n]$.

(h) Invertible. Inverse system: $y(t) = x(t) + dx(t)/dt$.

(i) Invertible. Inverse system: $y[n] = x[n] - (1/2)x[n-1]$.

(j) Non invertible. If $x(t)$ is any constant, then $y(t) = 0$.

(k) Non invertible. $\delta[n]$ and $2\delta[n]$ result in $y[n] = 0$.

(l) Invertible. Inverse system: $y(t) = x(t/2)$.

(m) Non invertible. $x_1[n] = \delta[n] + \delta[n-1]$ and $x_2[n] = \delta[n]$ give $y[n] = \delta[n]$.

(n) Invertible. Inverse system: $y[n] = x[2n]$.

1.31. **(a)** Note that $x_2(t) = x_1(t) - x_1(t-2)$. Therefore, using linearity we get $y_2(t) = y_1(t) - y_1(t-2)$. This is as shown in Figure S1.31.

(b) Note that $x_3(t) = x_1(t) + x_1(t+1)$. Therefore, using linearity we get $y_3(t) = y_1(t) + y_1(t+1)$. This is as shown in Figure S1.31.

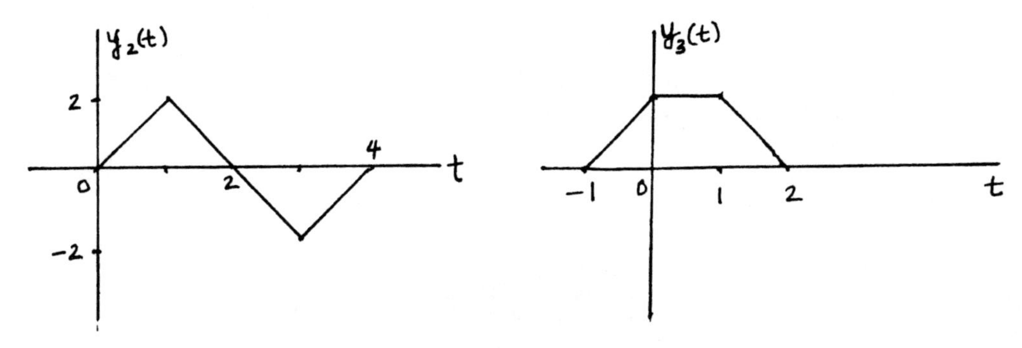

Figure S1.31

1.32. All statements are true.

(1) $x(t)$ periodic with period T; $y_1(t)$ periodic, period $T/2$.

(2) $y_1(t)$ periodic, period T; $x(t)$ periodic, period $2T$.

(3) $x(t)$ periodic, period T; $y_2(t)$ periodic, period $2T$.

(4) $y_2(t)$ periodic, period T; $x(t)$ periodic, period $T/2$.

1.33. (1) True. $x[n] = x[n+N]; y_1[n] = y_1[n+N_0]$. i.e. periodic with $N_0 = N/2$ if N is even, and with period $N_0 = N$ if N is odd.

(2) False. $y_1[n]$ periodic does no imply $x[n]$ is periodic. i.e. let $x[n] = g[n] + h[n]$ where

$$g[n] = \begin{cases} 1, & n \text{ even} \\ 0, & n \text{ odd} \end{cases} \quad \text{and} \quad h[n] = \begin{cases} 0, & n \text{ even} \\ (1/2)^n, & n \text{ odd} \end{cases}.$$

Then $y_1[n] = x[2n]$ is periodic but $x[n]$ is clearly not periodic.

(3) True. $x[n+N] = x[n]$; $y_2[n+N_0] = y_2[n]$ where $N_0 = 2N$

(4) True. $y_2[n+N] = y_2[n]$; $x[n+N_0] = x[n]$ where $N_0 = N/2$

1.34. (a) Consider

$$\sum_{n=-\infty}^{\infty} x[n] = x[0] + \sum_{n=1}^{\infty} \{x[n] + x[-n]\}.$$

If $x[n]$ is odd, $x[n] + x[-n] = 0$. Therefore, the given summation evaluates to zero.

(b) Let $y[n] = x_1[n]x_2[n]$. Then

$$y[-n] = x_1[-n]x_2[-n] = -x_1[n]x_2[n] = -y[n].$$

This implies that $y[n]$ is odd.

(c) Consider

$$\begin{aligned} \sum_{n=-\infty}^{\infty} x^2[n] &= \sum_{n=-\infty}^{\infty} \{x_e[n] + x_o[n]\}^2 \\ &= \sum_{n=-\infty}^{\infty} x_e^2[n] + \sum_{n=-\infty}^{\infty} x_o^2[n] + 2\sum_{n=-\infty}^{\infty} x_e[n]x_o[n]. \end{aligned}$$

Using the result of part (b), we know that $x_e[n]x_o[n]$ is an odd signal. Therefore, using the result of part (a) we may conclude that

$$2\sum_{n=-\infty}^{\infty} x_e[n]x_o[n] = 0.$$

Therefore,

$$\sum_{n=-\infty}^{\infty} x^2[n] == \sum_{n=-\infty}^{\infty} x_e^2[n] + \sum_{n=-\infty}^{\infty} x_o^2[n].$$

(d) Consider

$$\int_{-\infty}^{\infty} x^2(t)dt = \int_{-\infty}^{\infty} x_e^2(t)dt + \int_{-\infty}^{\infty} x_o^2(t)dt + 2\int_{-\infty}^{\infty} x_e(t)x_o(t)dt.$$

Again, since $x_e(t)x_o(t)$ is odd,

$$\int_{-\infty}^{\infty} x_e^2(t)x_o(t)dt = 0.$$

Therefore,

$$\int_{-\infty}^{\infty} x^2(t)dt = \int_{-\infty}^{\infty} x_e^2(t)dt + \int_{-\infty}^{\infty} x_o^2(t)dt.$$

1.35. We want to find the smallest N_0 such that $m(2\pi/N)N_0 = 2\pi k$ or $N_0 = kN/m$, where k is an integer. If N_0 has to be an integer, then N must be a multiple of m/k and m/k must be an integer. This implies that m/k is a divisor of both m and N. Also, if we want the smallest possible N_0, then m/k should be the GCD of m and N. Therefore, $N_0 = N/\text{gcd}(m, N)$.

1.36. **(a)** If $x[n]$ is periodic $e^{j\omega_0(n+N)T} = e^{j\omega_0 nT}$, where $\omega_0 = 2\pi/T_0$. This implies that

$$\frac{2\pi}{T_0}NT = 2\pi k \quad \Rightarrow \quad \frac{T}{T_0} = \frac{k}{N} = \text{a rational number.}$$

(b) If $T/T_0 = p/q$ then $x[n] = e^{j2\pi n(p/q)}$. The fundamental period is $q/\text{gcd}(p, q)$ and the fundamental frequency is

$$\frac{2\pi}{q}\text{gcd}(p, q) = \frac{2\pi}{p}\frac{p}{q}\text{gcd}(p, q) = \frac{\omega_0}{p}\text{gcd}(p, q) = \frac{\omega_0 T}{p}\text{gcd}(p, q).$$

(c) $p/\text{gcd}(p, q)$ periods of $x(t)$ are needed.

1.37. **(a)** From the definition of $\phi_{xy}(t)$, we have

$$\begin{aligned}\phi_{xy}(t) &= \int_{-\infty}^{\infty} x(t+\tau)y(\tau)d\tau \\ &= \int_{-\infty}^{\infty} y(-t+\tau)x(\tau)d\tau \\ &= \phi_{yx}(-t).\end{aligned}$$

(b) Note from part (a) that $\phi_{xx}(t) = \phi_{xx}(-t)$. This implies that $\phi_{xx}(t)$ is even. Therefore, the odd part of $\phi_{xx}(t)$ is zero.

(c) Here, $\phi_{xy}(t) = \phi_{xx}(t-T)$ and $\phi_{yy}(t) = \phi_{xx}(t)$.

1.38. **(a)** We know that $2\delta_\Delta(2t) = \delta_{\Delta/2}(t)$. Therefore,

$$\lim_{\Delta\to 0}\delta_\Delta(2t) = \lim_{\Delta\to 0}\frac{1}{2}\delta_{\Delta/2}(t).$$

This implies that

$$\delta(2t) = \frac{1}{2}\delta(t).$$

(b) The plots are as shown in Figure S1.38.

1.39. We have

$$\lim_{\Delta\to 0} u_\Delta(t)\delta(t) = \lim_{\Delta\to 0} u_\Delta(0)\delta(t) = 0.$$

Also,

$$\lim_{\Delta\to 0} u_\Delta(t)\delta_\Delta(t) = \frac{1}{2}\delta(t).$$

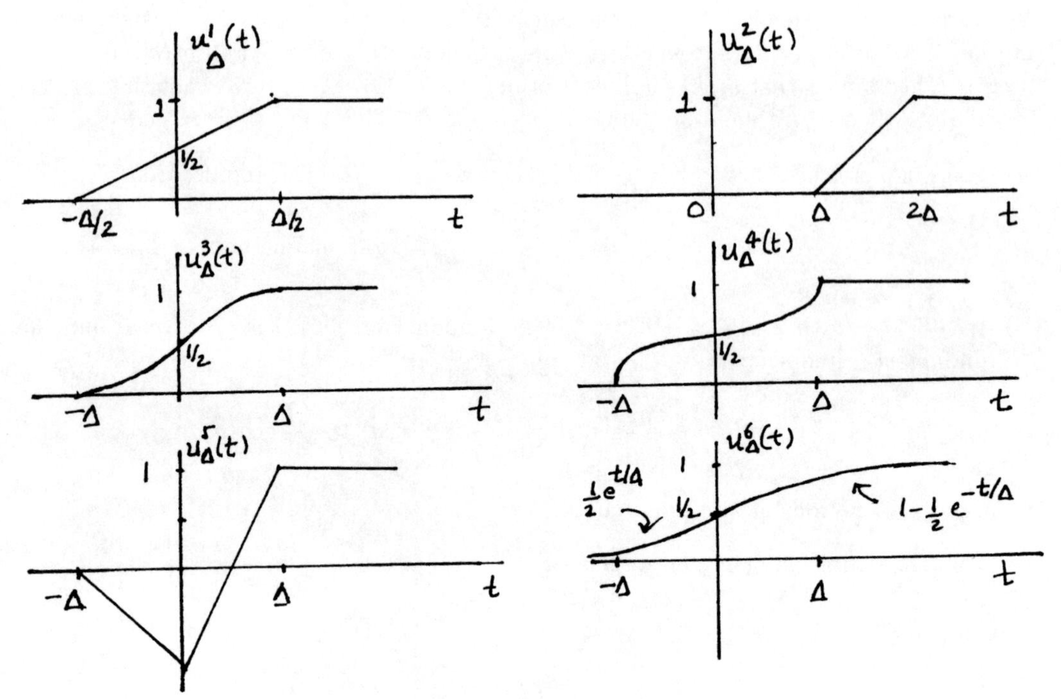

Figure S1.38

We have

$$g(t) = \int_{-\infty}^{\infty} u(\tau)\delta(t-\tau)d\tau = \int_{0}^{\infty} u(\tau)\delta(t-\tau)d\tau.$$

Therefore,

$$g(t) = \begin{cases} 0, & t < 0 & \because \delta(t-\tau) = 0 \\ 1, & t > 0 & \because u(\tau)\delta(t-\tau) = \delta(t-\tau) \\ \text{undefined} & \text{for } t = 0 & \end{cases}.$$

1.40. **(a)** If a system is additive, then

$$0 = x(t) - x(t) \longrightarrow y(t) - y(t) = 0.$$

Also, if a system is homogeneous, then

$$0 = 0.x(t) \longrightarrow y(t).0 = 0.$$

(b) $y(t) = x^2(t)$ is such a system.

(c) No. For example, consider $y(t) = \int_{-\infty}^{t} x(\tau)d\tau$ with $x(t) = u(t) - u(t-1)$. Then $x(t) = 0$ for $t > 1$, but $y(t) = 1$ for $t > 1$.

1.41. (a) $y[n] = 2x[n]$. Therefore, the system is time invariant.

(b) $y[n] = (2n-1)x[n]$. This is not time-invariant because $y[n-N_0] \neq (2n-1)x[n-N_0]$.

(c) $y[n] = x[n]\{1+(-1)^n+1+(-1)^{n-1}\} = 2x[n]$. Therefore, the system is time invariant.

1.42. (a) Consider two systems S_1 and S_2 connected in series. Assume that if $x_1(t)$ and $x_2(t)$ are the inputs to S_1, then $y_1(t)$ and $y_2(t)$ are the outputs, respectively. Also, assume that if $y_1(t)$ and $y_2(t)$ are the inputs to S_2, then $z_1(t)$ and $z_2(t)$ are the outputs, respectively. Since S_1 is linear, we may write

$$ax_1(t) + bx_2(t) \xrightarrow{S_1} ay_1(t) + by_2(t),$$

where a and b are constants. Since S_2 is also linear, we may write

$$ay_1(t) + by_2(t) \xrightarrow{S_2} az_1(t) + bz_2(t),$$

We may therefore conclude that

$$ax_1(t) + bx_2(t) \xrightarrow{S_1,S_2} az_1(t) + bz_2(t).$$

Therefore, the series combination of S_1 and S_2 is linear.

Since S_1 is time invariant, we may write

$$x_1(t-T_0) \xrightarrow{S_1} y_1(t-T_0)$$

and

$$y_1(t-T_0) \xrightarrow{S_2} z_1(t-T_0).$$

Therefore,

$$x_1(t-T_0) \xrightarrow{S_1,S_2} z_1(t-T_0).$$

Therefore, the series combination of S_1 and S_2 is time invariant.

(b) False. Let $y(t) = x(t) + 1$ and $z(t) = y(t) - 1$. These correspond to two nonlinear systems. If these systems are connected in series, then $z(t) = x(t)$ which is a linear system.

(c) Let us name the output of system 1 as $w[n]$ and the output of system 2 as $z[n]$. Then,

$$\begin{aligned} y[n] &= z[2n] = w[2n] + \frac{1}{2}w[2n-1] + \frac{1}{4}w[2n-2] \\ &= x[n] + \frac{1}{2}x[n-1] + \frac{1}{4}x[n-2] \end{aligned}$$

The overall system is linear and time-invariant.

1.43. (a) We have

$$x(t) \xrightarrow{S} y(t).$$

Since S is time-invariant,

$$x(t-T) \xrightarrow{S} y(t-T).$$

Now, if $x(t)$ is periodic with period T, $x(t) = x(t-T)$. Therefore, we may conclude that $y(t) = y(t-T)$. This implies that $y(t)$ is also periodic with period T. A similar argument may be made in discrete time.

(b)

1.44. **(a)** Assumption: If $x(t) = 0$ for $t < t_0$, then $y(t) = 0$ for $t < t_0$. To prove that: The system is causal.
Let us consider an arbitrary signal $x_1(t)$. Let us consider another signal $x_2(t)$ which is the same as $x_1(t)$ for $t < t_0$. But for $t > t_0$, $x_2(t) \neq x_1(t)$. Since the system is linear,

$$x_1(t) - x_2(t) \longrightarrow y_1(t) - y_2(t).$$

Since $x_1(t) - x_2(t) = 0$ for $t < t_0$, by our assumption $y_1(t) - y_2(t) = 0$ for $t < t_0$. This implies that $y_1(t) = y_2(t)$ for $t < t_0$. In other words, the output is not affected by input values for $t \geq t_0$. Therefore, the system is causal.

Assumption: The system is causal. To prove that: If $x(t) = 0$ for $t < t_0$, then $y(t) = 0$ for $t < t_0$.
Let us assume that the signal $x(t) = 0$ for $t < t_0$. Then we may express $x(t)$ as $x(t) = x_1(t) - x_2(t)$, where $x_1(t) = x_2(t)$ for $t < t_0$. Since the system is linear, the output to $x(t)$ will be $y(t) = y_1(t) - y_2(t)$. Now, since the system is causal, $y_1(t) = y_2(t)$ for $t < t_0$ implies that $y_1(t) = y_2(t)$ for $t < t_0$. Therefore, $y(t) = 0$ for $t < t_0$.

(b) Consider $y(t) = x(t)x(t+1)$. Now, $x(t) = 0$ for $t < t_0$ implies that $y(t) = 0$ for $t < t_0$. Note that the system is nonlinear and non-causal.

(c) Consider $y(t) = x(t) + 1$. This system is nonlinear and causal. This does not satisfy the condition of part (a).

(d) Assumption: The system is invertible. To prove that: $y[n] = 0$ for all n only if $x[n] = 0$ for all n.
Consider

$$x[n] = 0 \longrightarrow y[n].$$

Since the system is linear,

$$2x[n] = 0 \longrightarrow 2y[n].$$

Since the input has not changed in the two above equations, we require that $y[n] = 2y[n]$. This implies that $y[n] = 0$. Since we have assumed that the system is invertible, only one input could have led to this particular output. That input must be $x[n] = 0$.

Assumption: $y[n] = 0$ for all n if $x[n] = 0$ for all n. To prove that: The system is invertible.
Suppose that

$$x_1[n] \longrightarrow y_1[n]$$

and

$$x_2[n] \longrightarrow y_1[n].$$

Since the system is linear,

$$x_1[n] - x_2[n] \longrightarrow y_1[n] - y_1[n] = 0.$$

By the original assumption, we must conclude that $x_1[n] = x_2[n]$. That is, any particular $y_1[n]$ can be produced by only one distinct input $x_1[n]$. Therefore, the system is invertible.

(e) $y[n] = x^2[n]$.

1.45. **(a)** Consider

$$x_1(t) \xrightarrow{S} y_1(t) = \phi_{hx_1}(t)$$

and

$$x_2(t) \xrightarrow{S} y_2(t) = \phi_{hx_2}(t).$$

Now, consider $x_3(t) = ax_1(t) + bx_2(t)$. The corresponding system output will be

$$\begin{aligned} y_3(t) &= \int_{-\infty}^{\infty} x_3(\tau)h(t+\tau)d\tau \\ &= a\int_{-\infty}^{\infty} x_1(\tau)h(t+\tau)d\tau + b\int_{-\infty}^{\infty} x_2(\tau)h(t+\tau)d\tau \\ &= a\phi_{hx_1}(t) + b\phi_{hx_2}(t) \\ &= ay_1(t) + by_2(t) \end{aligned}$$

Therefore, S is linear.

Now, consider $x_4(t) = x_1(t-T)$. The corresponding system output will be

$$\begin{aligned} y_4(t) &= \int_{-\infty}^{\infty} x_4(\tau)h(t+\tau)d\tau \\ &= \int_{-\infty}^{\infty} x_1(\tau - T)h(t+\tau)d\tau \\ &= \int_{-\infty}^{\infty} x_1(\tau)h(t+\tau+T)d\tau \\ &= \phi_{hx_1}(t+T) \end{aligned}$$

Clearly, $y_4(t) \neq y_1(t-T)$. Therefore, the system is not time-invariant.

The system is definitely not causal because the output at any time depends on future values of the input signal $x(t)$.

(b) The system will then be linear, time invariant and non-causal.

1.46. The plots are as in Figure S1.46.

1.47. **(a)** The overall response of the system of Figure P1.47(a) = (the response of the system to $x[n] + x_1[n]$) − the response of the system to $x_1[n]$ = (Response of a linear system L to $x[n] + x_1[n]$+ zero input response of S) − (Response of a linear system L to $x_1[n]$+ zero input response of S) = (Response of a linear system L to $x[n]$).

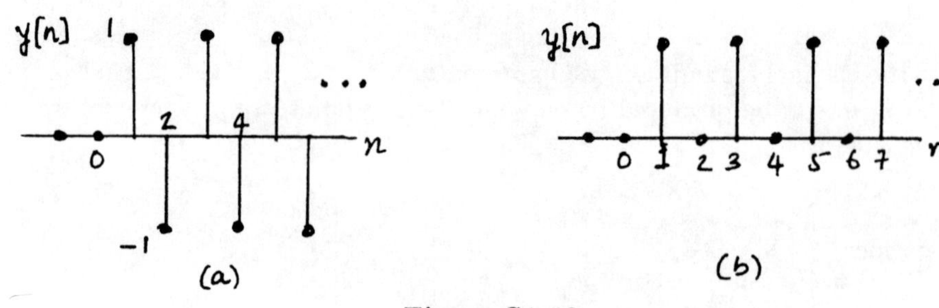

Figure S1.46

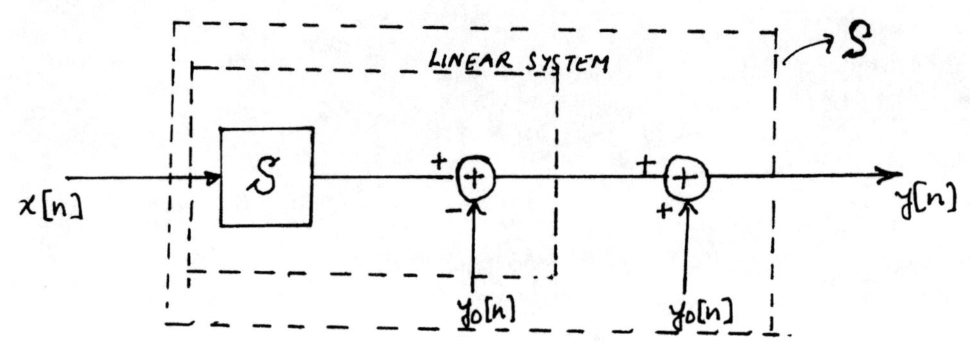

Figure S1.47

(b) If $x_1[n] = 0$ for all n, then $y_1[n]$ will be the zero-input response $y_0[n]$. S may then be redrawn as shown in Figure S1.47. This is the same as Figure 1.48.

(c) (i) Incrementally linear.

$$x[n] \longrightarrow x[n] + 2x[n+1] \qquad \text{and} \qquad y_0[n] = n$$

(ii) Incrementally linear.

$$x[n] \longrightarrow \begin{cases} 0, & n \text{ even} \\ \displaystyle\sum_{k=-\infty}^{(n-1)/2} x[k], & n \text{ odd} \end{cases}.$$

and

$$y_0[n] = \begin{cases} n/2, & n \text{ even} \\ (n-1)/2, & n \text{ odd}. \end{cases}$$

(iii) Not incrementally linear. Eg. choose $y_0[n] = 3$. Then

$$y[n] - y_0[n] = \begin{cases} x[n] - x[n-1], & x[0] \geq 0 \\ x[n] - x[n-1] - 6, & x[0] < 0. \end{cases}$$

Still non-linear: eg.: If $x_1[n] = -\delta[n]$ and $x_2[n] = -2\delta[n]$, then $y_1[n] = -\delta[n] + \delta[n-1] - 6$ and $y_2[n] = -2\delta[n] + 2\delta[n-1] - 6 \neq 2y_1[n]$.

(iv) Incrementally linear.

$$x(t) \longrightarrow x(t) + t dx(t)/dt - 1 \qquad \text{and} \qquad y_0(t) = 1.$$

(v) Incrementally linear

$$x[n] \longrightarrow 2\cos(\pi n)x[n] \qquad \text{and} \qquad y_0[n] = \cos^2(\pi n)$$

(d) Let $x[n] \xrightarrow{S} y[n]$ and $x[n] \xrightarrow{L} z[n]$. Then, $y[n] = z[n] + c$. For time invariance, we require that when the input is $x[n-n_0]$, the output be

$$y[n-n_0] = z[n-n_0] + c.$$

This implies that we require

$$x[n-n_0] \xrightarrow{L} z[n-n_0]$$

which in turn implies that L should be time invariant. We also require that $y_0[n] = c$ =constant independent of n.

1.48. We have

$$z_0 = r_0 e^{j\theta_0} = r_0 \cos\theta_0 + j r_0 \sin\theta_0 = x_0 + j y_0$$

(a) $z_1 = x_0 - j y_0$

(b) $z_2 = \sqrt{x_0^2 + y_0^2}$

(c) $z_3 = -x_0 - j y_0 = -z_0$

(d) $z_4 = -x_0 + j y_0$

(e) $z_5 = x_0 + j y_0$

The plots for the points are as shown in the Figure S1.48.

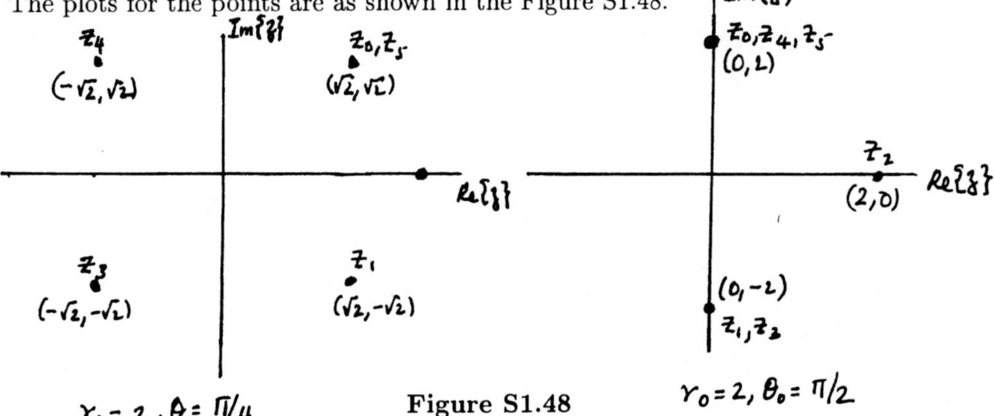

Figure S1.48

1.49. (a) Here, $r = \sqrt{1+3} = 2$. Also, $\cos\theta = 1/2$, $\sin\theta = \sqrt{3}/2$. This implies that $\theta = \pi/3$. Therefore, $1 + j\sqrt{3} = 2e^{j\pi/3}$.

(b) $5e^{j\pi}$

(c) $5\sqrt{2}e^{j5\pi/4}$

(d) $5e^{j\tan^{-1}(4/3)} = 5e^{j(53.13°)}$

(e) $8e^{-j\pi}$

(f) $4\sqrt{2}e^{j5\pi/4}$

(g) $2\sqrt{2}e^{-j5\pi/12}$

(h) $e^{-j2\pi/3}$

(i) $e^{j\pi/6}$

(j) $\sqrt{2}e^{j11\pi/12}$

(k) $4\sqrt{2}e^{-j\pi/12}$

(l) $\frac{1}{2}e^{j\pi/3}$

Plot depicting these points is as shown in Figure S1.49.

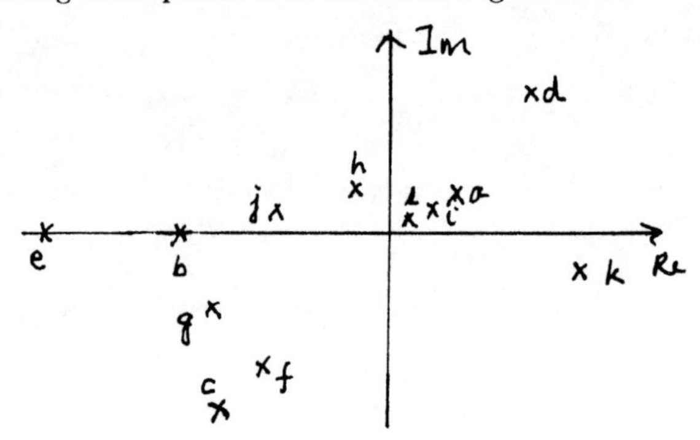

Figure S1.49

1.50. (a) $x = r\cos\theta, y = r\sin\theta$

(b) We have

$$r = \sqrt{x^2 + y^2}$$

and

$$\theta = \sin^{-1}\left[\frac{y}{\sqrt{x^2+y^2}}\right] = \cos^{-1}\left[\frac{x}{\sqrt{x^2+y^2}}\right] = \tan^{-1}\left[\frac{y}{x}\right].$$

θ is undefined if $r = 0$ and also irrelevant. θ is not unique since θ and $\theta + 2m\pi$ ($m \in$ integer) give the same results.

(c) θ and $\theta + \pi$ have the same value of tangent. We only know that the complex number is either $z_1 re^{j\theta}$ or $z_2 = re^{j(\theta+\pi)} = -z_1$.

1.51. **(a)** We have

$$e^{j\theta} = \cos\theta + j\sin\theta. \qquad \text{(S1.51-1)}$$

and

$$e^{-j\theta} = \cos\theta - j\sin\theta. \qquad \text{(S1.51-2)}$$

Summing eqs. (S1.51-1) and (S1.51-2) we get

$$\cos\theta = \frac{1}{2}(e^{j\theta} + e^{-j\theta}).$$

(b) Subtracting eq. (S1.51-2) from (S1.51-1) we get

$$\sin\theta = \frac{1}{2j}(e^{j\theta} - e^{-j\theta}).$$

(c) We now have $e^{j(\theta+\phi)} = e^{j\theta}e^{j\phi}$. Therefore,

$$\begin{aligned}\cos(\theta+\phi) + j\sin(\theta+\phi) &= (\cos\theta\cos\phi - \sin\theta\sin\phi)\\ &+ j(\sin\theta\cos\phi + \cos\theta\sin\phi)\end{aligned} \qquad \text{(S1.51-3)}$$

Putting $\theta = \phi$ in eq. (S1.51-3), we get

$$\cos 2\theta = \cos^2\theta - \sin^2\theta.$$

Putting $\theta = -\phi$ in eq. (S1.51-3), we get

$$1 = \cos^2\theta + \sin^2\theta.$$

Adding the two above equations and simplifying

$$\cos^2\theta = \frac{1}{2}(1 + \cos 2\theta).$$

(d) Equating the real parts in eq. (S1.51-3) with arguments $(\theta+\phi)$ and $(\theta-\phi)$ we get

$$\cos(\theta+\phi) = \cos\theta\cos\phi - \sin\theta\sin\phi$$

and

$$\cos(\theta-\phi) = \cos\theta\cos\phi + \sin\theta\sin\phi.$$

Subtracting the two above equations, we obtain

$$\sin\theta\sin\phi = \frac{1}{2}[\cos(\theta-\phi) - \cos(\theta+\phi)].$$

(e) Equating imaginary parts in in eq. (S1.51-3), we get

$$\sin(\theta+\phi) = \sin\theta\cos\phi + \cos\theta\sin\phi.$$

1.52. **(a)** $zz^* = re^{j\theta}re^{-j\theta} = r^2$

(b) $z/z^* = re^{j\theta}r^{-1}e^{j\theta} = e^{j2\theta}$

(c) $z + z^* = x + jy + x - jy = 2x = 2\mathcal{R}e\{z\}$

(d) $z - z^* = x + jy - x + jy = 2jy = 2\mathcal{I}m\{z\}$

(e) $(z_1 + z_2)^* = ((x_1 + x_2) + j(y_1 + y_2))^* = x_1 - jy_1 + x_2 - jy_2 = z_1^* + z_2^*$

(f) Consider $(az_1z_2)^*$ for $a > 0$.

$$(az_1z_2)^* = (ar_1r_2e^{j(\theta_1+\theta_2)})^* = ar_1e^{-j\theta_1}r_2e^{-j\theta_2} = az_1^*z_2^*.$$

For $a < 0$, $a = |a|e^{j\pi}$. Therefore,

$$(az_1z_2)^* = (|a|r_1r_2e^{j(\theta_1+\theta_2+\pi)})^* = |a|e^{-j\pi}r_1e^{-j\theta_1}r_2e^{-j\theta_2} = az_1^*z_2^*.$$

(g) For $|z_2| \neq 0$,

$$\left(\frac{z_1}{z_2}\right)^* = \frac{r_1}{r_2}e^{-j\theta_1}e^{j\theta_2} = \frac{r_1e^{-j\theta_1}}{r_2e^{-j\theta_2}} = \frac{z_1^*}{z_2^*}.$$

(h) From (c), we get

$$\mathcal{R}e\{\frac{z_1}{z_2}\} = \frac{1}{2}\left[\left(\frac{z_1}{z_2}\right) + \left(\frac{z_1}{z_2}\right)^*\right].$$

Using (g) on this, we get

$$\mathcal{R}e\{\frac{z_1}{z_2}\} = \frac{1}{2}\left[\left(\frac{z_1}{z_2}\right) + \left(\frac{z_1^*}{z_2^*}\right)\right] = \frac{1}{2}\left[\frac{z_1z_2^* + z_1^*z_2}{z_2z_2^*}\right].$$

1.53. **(a)** $(e^z)^* = (e^x e^{jy})^* = e^x e^{-jy} = e^{x-jy} = e^{z^*}$.

(b) Let $z_3 = z_1z_2^*$ and $z_4 = z_1^*z_2$. Then,

$$\begin{aligned} z_1z_2^* + z_1^*z_2 &= z_3 + z_3^* = 2\mathcal{R}e\{z_3\} = 2\mathcal{R}e\{z_1z_2^*\} \\ &= z_4^* + z_4 = 2\mathcal{R}e\{z_4\} = 2\mathcal{R}e\{z_1^*z_2\} \end{aligned}$$

(c) $|z| = |re^{j\theta}| = r = |re^{-j\theta}| = |z^*|$

(d) $|z_1z_2| = |r_1r_2e^{j(\theta_1+\theta_2)}| = |r_1r_2| = |r_1||r_2| = |z_1||z_2$

(e) Since $z = x + jy$, $|z| = \sqrt{x^2 + y^2}$. By the triangle inequality,

$$\mathcal{R}e\{z\} = x \leq \sqrt{x^2 + y^2} = |z|$$

and

$$\mathcal{I}m\{z\} = y \leq \sqrt{x^2 + y^2} = |z|.$$

(f) $|z_1z_2^* + z_1^*z_2| = |2\mathcal{R}e\{z_1z_2^*\}| = |2r_1r_2\cos(\theta_1 - \theta_2)| \leq 2r_1r_2 = 2|z_1z_2|$.

(**g**) Since $r_1 > 0, r_2 > 0$ and $-1 \leq \cos(\theta_1 - \theta_2) \leq 1$,

$$\begin{aligned}(|z_1| - |z_2|)^2 &= r_1^2 + r_2^2 - 2r_1r_2 \\ &\leq r_1^2 + r_2^2 + 2r_1r_2\cos(\theta_1 - \theta_2) \\ &= |z_1 + z_2|^2\end{aligned}$$

and

$$(|z_1| + |z_2|)^2 = r_1^2 + r_2^2 + 2r_1r_2 \geq |z_1 + z_2|^2.$$

1.54. (**a**) For $\alpha = 1$, it is fairly obvious that

$$\sum_{n=0}^{N-1} \alpha^n = N.$$

For $\alpha \neq 1$, we may write

$$(1-\alpha)\sum_{n=0}^{N-1} \alpha^n = \sum_{n=0}^{N-1} \alpha^n - \sum_{n=0}^{N-1} \alpha^{n+1} = 1 - \alpha^N.$$

Therefore,

$$\sum_{n=0}^{N-1} \alpha^n = \frac{1-\alpha^N}{1-\alpha}.$$

(**b**) For $|\alpha| < 1$,

$$\lim_{N\to\infty} \alpha^N = 0.$$

Therefore, from the result of the previous part,

$$\lim_{N\to\infty} \sum_{n=0}^{N-1} \alpha^n = \sum_{n=0}^{\infty} \alpha^n = \frac{1}{1-\alpha}.$$

(**c**) Differentiating both sides of the result of part (b) wrt α, we get

$$\begin{aligned}\frac{d}{d\alpha}\left(\sum_{n=0}^{\infty} \alpha^N\right) &= \frac{d}{d\alpha}\left(\frac{1}{1-\alpha}\right) \\ \sum_{n=0}^{\infty} n\alpha^{n-1} &= \frac{1}{(1-\alpha)^2}\end{aligned}$$

(**d**) We may write

$$\sum_{n=k}^{\infty} \alpha^n = \alpha^k \sum_{n=0}^{\infty} \alpha^n = \frac{\alpha^k}{1-\alpha} \text{ for } |\alpha| < 1.$$

1.55. (**a**) The desired sum is

$$\sum_{n=0}^{9} e^{j\pi n/2} = \frac{1 - e^{j\pi 10/2}}{1 - e^{j\pi/2}} = 1 + j.$$

(b) The desired sum is

$$\sum_{n=-2}^{7} e^{j\pi n/2} = e^{-j2\pi/2}\sum_{n=0}^{9} e^{j\pi n/2} = -(1+j).$$

(c) The desired sum is

$$\sum_{n=0}^{\infty}(1/2)^n e^{j\pi n/2} = \frac{1}{1-(1/2)e^{j\pi/2}} = \frac{4}{5}+j\frac{2}{5}.$$

(d) The desired sum is

$$\sum_{n=2}^{\infty}(1/2)^n e^{j\pi n/2} = (1/2)^2 e^{j\pi 2/2}\sum_{n=0}^{\infty}(1/2)^n e^{j\pi n/2} = -\frac{1}{4}\left[\frac{4}{5}+j\frac{2}{5}\right].$$

(e) The desired sum is

$$\sum_{n=0}^{9}\cos(\pi n/2) = \frac{1}{2}\sum_{n=0}^{9} e^{j\pi n/2} + \frac{1}{2}\sum_{n=0}^{9} e^{-j\pi n/2} = \frac{1}{2}(1+j)+\frac{1}{2}(1-j) = 1.$$

(f) The desired sum is

$$\begin{aligned}\sum_{n=0}^{\infty}(1/2)^n\cos(\pi n/2) &= \frac{1}{2}\sum_{n=0}^{\infty}(1/2)^n e^{j\pi n/2} + \frac{1}{2}\sum_{n=0}^{\infty}(1/2)^n e^{-j\pi n/2} \\ &= \frac{4}{10}+j\frac{2}{10}+\frac{4}{10}-j\frac{2}{10} = \frac{4}{5}.\end{aligned}$$

1.56. (a) The desired integral is

$$\int_0^4 e^{j\pi t/2}dt = \left.\frac{e^{\pi t/2}}{j\pi/2}\right|_0^4 = 0.$$

(b) The desired integral is

$$\int_0^6 e^{j\pi t/2}dt = \left.\frac{e^{\pi t/2}}{j\pi/2}\right|_0^6 = (2/j\pi)[e^{j3\pi}-1] = \frac{4j}{\pi}.$$

(c) The desired integral is

$$\int_2^8 e^{j\pi t/2}dt = \left.\frac{e^{\pi t/2}}{j\pi/2}\right|_2^8 = (2/j\pi)[e^{j4\pi}-e^{j\pi}] = -\frac{4j}{\pi}.$$

(d) The desired integral is

$$\int_0^{\infty} e^{-(1+j)t}dt = \left.\frac{e^{-(1+j)t}}{-(1+j)}\right|_0^{\infty} = \frac{1}{1+j} = \frac{1-j}{2}.$$

(e) The desired integral is

$$\int_0^\infty e^{-t}\cos(t)dt = \int_0^\infty \left[\frac{e^{-(1+j)t} + e^{-(1-j)t}}{2}\right] dt = \frac{1/2}{1+j} + \frac{1/2}{1-j} = \frac{1}{2}.$$

(f) The desired integral is

$$\int_0^\infty e^{-2t}\sin(3t)dt = \int_0^\infty \left[\frac{e^{-(2-3j)t} - e^{-(2+3j)t}}{2j}\right] dt = \frac{1/2j}{2-3j} + \frac{1/2j}{2+3j} = \frac{3}{13}.$$

Chapter 2 Answers

2.1. (a) We know that

$$y_1[n] = x[n] * h[n] = \sum_{k=-\infty}^{\infty} h[k]x[n-k] \tag{S2.1-1}$$

The signals $x[n]$ and $h[n]$ are as shown in Figure S2.1.

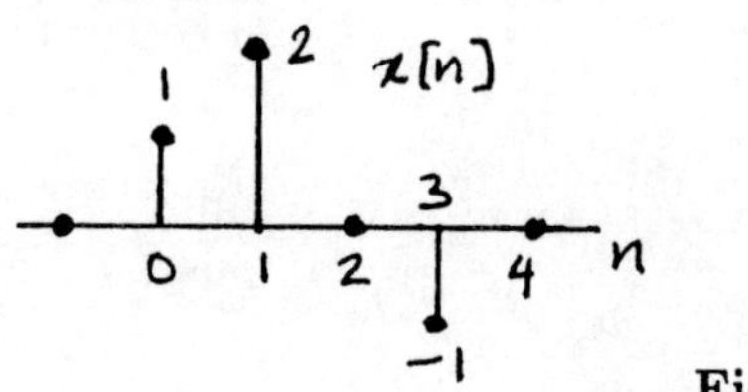

Figure S2.1

From this figure, we can easily see that the above convolution sum reduces to

$$\begin{aligned} y_1[n] &= h[-1]x[n+1] + h[1]x[n-1] \\ &= 2x[n+1] + 2x[n-1] \end{aligned}$$

This gives

$$y_1[n] = 2\delta[n+1] + 4\delta[n] + 2\delta[n-1] + 2\delta[n-2] - 2\delta[n-4]$$

(b) We know that

$$y_2[n] = x[n+2] * h[n] = \sum_{k=-\infty}^{\infty} h[k]x[n+2-k]$$

Comparing with eq. (S2.1-1), we see that

$$y_2[n] = y_1[n+2]$$

(c) We may rewrite eq. (S2.1-1) as

$$y_1[n] = x[n] * h[n] = \sum_{k=-\infty}^{\infty} x[k]h[n-k]$$

Similarly, we may write

$$y_3[n] = x[n] * h[n+2] = \sum_{k=-\infty}^{\infty} x[k]h[n+2-k]$$

Comparing this with eq. (S2.1), we see that

$$y_3[n] = y_1[n+2]$$

2.2. Using the given definition for the signal $h[n]$, we may write

$$h[k] = \left(\frac{1}{2}\right)^{k-1} \{u[k+3] - u[k-10]\}$$

The signal $h[k]$ is non zero only in the range $-3 \leq k \leq 9$. From this we know that the signal $h[-k]$ is non zero only in the range $-9 \leq k \leq 3$. If we now shift the signal $h[-k]$ by n to the right, then the resultant signal $h[n-k]$ will be non zero in the range $(n-9) \leq k \leq (n+3)$. Therefore,

$$A = n - 9, \qquad B = n + 3$$

2.3. Let us define the signals

$$x_1[n] = \left(\frac{1}{2}\right)^n u[n]$$

and

$$h_1[n] = u[n].$$

We note that

$$x[n] = x_1[n-2] \qquad \text{and} \qquad h[n] = h_1[n+2]$$

Now,

$$\begin{aligned} y[n] &= x[n] * h[n] = x_1[n-2] * h_1[n+2] \\ &= \sum_{k=-\infty}^{\infty} x_1[k-2]h_1[n-k+2] \end{aligned}$$

By replacing k with $m+2$ in the abovr summation, we obtain

$$y[n] = \sum_{m=-\infty}^{\infty} x_1[m]h_1[n-m] = x_1[n] * h_1[n]$$

Using the results of Example 2.1 in the text book, we may write

$$y[n] = 2\left[1 - \left(\frac{1}{2}\right)^{n+1}\right] u[n]$$

2.4. We know that

$$y[n] = x[n] * h[n] = \sum_{k=\infty}^{\infty} x[k]h[n-k]$$

The signals $x[n]$ and $y[n]$ are as shown in Figure S2.4. From this figure, we see that the above summation reduces to

$$y[n] = x[3]h[n-3] + x[4]h[n-4] + x[5]h[n-5] + x[6]h[n-6] + x[7]h[n-7] + x[8]h[n-8]$$

This gives

$$y[n] = \begin{cases} n-6, & 7 \leq n \leq 11 \\ 6, & 12 \leq n \leq 18 \\ 24-n, & 19 \leq n \leq 23 \\ 0, & \text{otherwise} \end{cases}$$

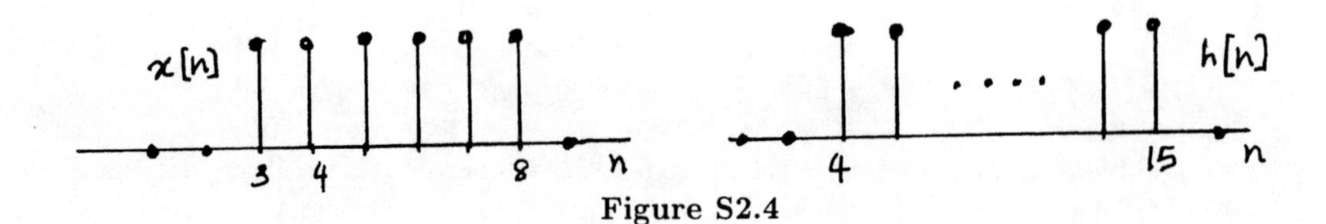

Figure S2.4

2.5. The signal $y[n]$ is

$$y[n] = x[n] * h[n] = \sum_{k=-\infty}^{\infty} x[k]h[n-k]$$

In this case, this summation reduces to

$$y[n] = \sum_{k=0}^{9} x[k]h[n-k] = \sum_{k=0}^{9} h[n-k]$$

From this it is clear that $y[n]$ is a summation of shifted replicas of $h[n]$. Since the last replica will begin at $n = 9$ and $h[n]$ is zero for $n > N$, $y[n]$ is zero for $n > N + 9$. Using this and the fact that $y[14] = 0$, we may conclude that N can *at most* be 4. Furthermore, since $y[4] = 5$, we can conclude that $h[n]$ has *at least* 5 non-zero points. The only value of N which satifies both these conditions is 4.

2.6. From the given information, we have:

$$\begin{aligned} y[n] &= x[n] * h[n] = \sum_{k=-\infty}^{\infty} x[k]h[n-k] \\ &= \sum_{k=-\infty}^{\infty} (\frac{1}{3})^{-k} u[-k-1]u[n-k-1] \\ &= \sum_{k=-\infty}^{-1} (\frac{1}{3})^{-k} u[n-k-1] \\ &= \sum_{k=1}^{\infty} (\frac{1}{3})^{k} u[n+k-1] \end{aligned}$$

Replacing k by $p-1$,

$$y[n] = \sum_{p=0}^{\infty} (\frac{1}{3})^{p+1} u[n+p] \qquad \text{(S2.6-1)}$$

For $n \geq 0$ the above equation reduces to,

$$y[n] = \sum_{p=0}^{\infty} (\frac{1}{3})^{p+1} = \frac{1}{3}\frac{1}{1-\frac{1}{3}} = \frac{1}{2}.$$

For $n < 0$ eq. (S2.6-1) reduces to,

$$\begin{aligned} y[n] &= \sum_{p=-n}^{\infty} (\frac{1}{3})^{p+1} = (\frac{1}{3})^{-n+1} \sum_{p=0}^{\infty} (\frac{1}{3})^{p} \\ &= (\frac{1}{3})^{-n+1} \frac{1}{1-\frac{1}{3}} = (\frac{1}{3})^{-n} \frac{1}{2} = \frac{3^n}{2} \end{aligned}$$

Therefore,

$$y[n] = \begin{cases} (3^n/2), & n < 0 \\ (1/2), & n \geq 0 \end{cases}$$

2.7. **(a)** Given that

$$x[n] = \delta[n-1],$$

we see that

$$y[n] = \sum_{k=-\infty}^{\infty} x[k]g[n-2k] = g[n-2] = u[n-2] - u[n-6]$$

(b) Given that

$$x[n] = \delta[n-2],$$

we see that

$$y[n] = \sum_{k=-\infty}^{\infty} x[k]g[n-2k] = g[n-4] = u[n-4] - u[n-8]$$

(c) The input to the system in part (b) is the same as the input in part (a) shifted by 1 to the right. If S is time invariant then the system output obtained in part (b) has to the be the same as the system output obtained in part (a) shifted by 1 to the right. Clealry, this is not the case. Therefore, the system is **not** LTI.

(d) If $x[n] = u[n]$, then

$$\begin{aligned} y[n] &= \sum_{k=-\infty}^{\infty} x[k]g[n-2k] \\ &= \sum_{k=0}^{\infty} g[n-2k] \end{aligned}$$

The signal $g[n-2k]$ is plotted for $k = 0, 1, 2$ in Figure S2.7. From this figure it is clear that

$$y[n] = \begin{cases} 1, & n = 0, 1 \\ 2, & n > 1 \\ 0, & \text{otherwise} \end{cases} = 2u[n] - \delta[n] - \delta[n-1]$$

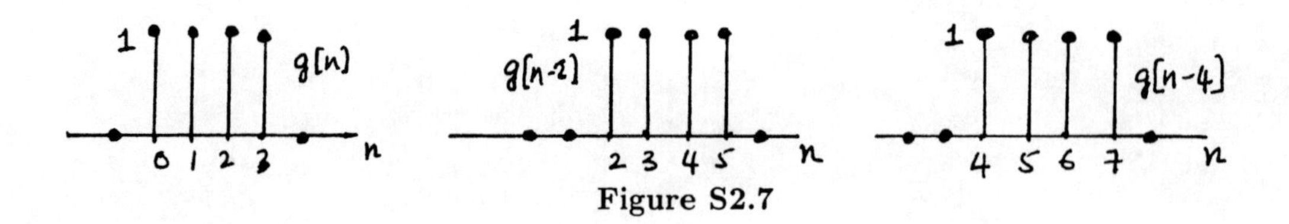

Figure S2.7

2.8. Using the convolution integral,

$$x(t) * h(t) = \int_{-\infty}^{\infty} x(\tau)h(t-\tau)d\tau = \int_{-\infty}^{\infty} h(\tau)x(t-\tau)d\tau.$$

Given that $h(t) = \delta(t+2) + 2\delta(t+1)$, the above integral reduces to

$$x(t) * y(t) = x(t+2) + 2x(t+1)$$

The signals $x(t+2)$ and $2x(t+1)$ are plotted in Figure S2.8.

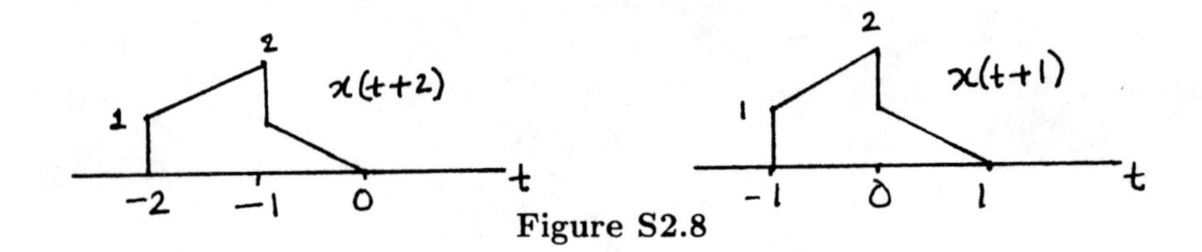

Figure S2.8

Using these plots, we can easily show that

$$y(t) = \begin{cases} t+3, & -2 < t \leq -1 \\ t+4, & -1 < t \leq 0 \\ 2-2t, & 0 < t \leq 1 \\ 0, & \text{otherwise} \end{cases}$$

2.9. Using the given definition for the signal $h(t)$, we may write

$$h(\tau) = e^{2\tau}u(-\tau+4) + e^{-2\tau}u(\tau-5) = \begin{cases} e^{-2\tau}, & \tau > 5 \\ e^{2\tau}, & \tau < 4 \\ 0, & 4 < \tau < 5 \end{cases}$$

Therefore,

$$h(-\tau) = \begin{cases} e^{2\tau}, & \tau < -5 \\ e^{-2\tau}, & \tau > -4 \\ 0, & -5 < \tau < -4 \end{cases}$$

If we now shift the signal $h(-\tau)$ by t to the right, then the resultant signal $h(t-\tau)$ will be

$$h(t-\tau) = \begin{cases} e^{-2(t-\tau)}, & \tau < t-5 \\ e^{2(t-\tau)}, & \tau > t-4 \\ 0, & (t-5) < \tau < (t-4) \end{cases}$$

Therefore,

$$A = t - 5, \qquad B = t - 4.$$

2.10. From the given information, we may sketch $x(t)$ and $h(t)$ as shown in Figure S2.10.

(a) With the aid of the plots in Figure S2.10, we can show that $y(t) = x(t) * h(t)$ is as shown in Figure S2.10.

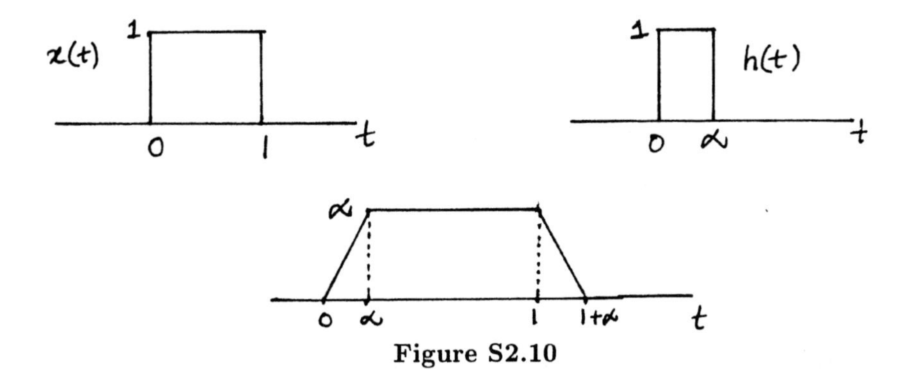

Figure S2.10

Therefore,

$$y(t) = \begin{cases} t, & 0 \le t \le \alpha \\ \alpha, & \alpha \le t \le 1 \\ 1 + \alpha - t, & 1 \le t \le (1+\alpha) \\ 0, & \text{otherwise} \end{cases}$$

(b) From the plot of $y(t)$, it is clear that $\frac{dy(t)}{dt}$ has discontinuities at 0, α, 1, and $1+\alpha$. If we want $\frac{dy(t)}{dt}$ to have only three discontinuities, then we need to ensure that $\alpha = 1$.

2.11. **(a)** From the given information, we see that $h(t)$ is non zero only for $0 \le t \le \infty$. Therefore,

$$\begin{aligned} y(t) &= x(t) * h(t) = \int_{-\infty}^{\infty} h(\tau) x(t-\tau) d\tau \\ &= \int_{0}^{\infty} e^{-3\tau} (u(t-\tau-3) - u(t-\tau-5)) d\tau \end{aligned}$$

We can easily show that $(u(t-\tau-3) - u(t-\tau-5))$ is non zero only in the range $(t-5) < \tau < (t-3)$. Therefore, for $t \le 3$, the above integral evaluates to zero. For $3 < t \le 5$, the above integral is

$$y(t) = \int_{0}^{t-3} e^{-3\tau} d\tau = \frac{1 - e^{-3(t-3)}}{3}$$

For $t > 5$, the integral is

$$y(t) = \int_{t-5}^{t-3} e^{-3\tau} d\tau = \frac{(1 - e^{-6}) e^{-3(t-5)}}{3}$$

Therefore, the result of this convolution may be expressed as

$$y(t) = \begin{cases} 0, & -\infty < t \le 3 \\ \frac{1-e^{-3(t-3)}}{3}, & 3 < t \le 5 \\ \frac{(1-e^{-6})e^{-3(t-5)}}{3}, & 5 < t \le \infty \end{cases}$$

(b) By differentiating $x(t)$ with respect to time we get

$$\frac{dx(t)}{dt} = \delta(t-3) - \delta(t-5)$$

Therefore,

$$g(t) = \frac{dx(t)}{dt} * h(t) = e^{-3(t-3)}u(t-3) - e^{-3(t-5)}u(t-5).$$

(c) From the result of part (a), we may compute the derivative of $y(t)$ to be

$$\frac{dy(t)}{dt} = \begin{cases} 0, & -\infty < t \le 3 \\ e^{-3(t-3)}, & 3 < t \le 5 \\ (e^{-6}-1)e^{-3(t-5)}, & 5 < t \le \infty \end{cases}$$

This is exactly equal to $g(t)$. Therefore, $g(t) = \frac{dy(t)}{dt}$.

2.12. The signal $y(t)$ may be written as

$$y(t) = \cdots + e^{-(t+6)}u(t+6) + e^{-(t+3)}u(t+3) + e^{-t}u(t) + e^{-(t-3)}u(t-3) + e^{-(t-6)}u(t-6) + \cdots$$

In the range $0 \le t < 3$, we may write $y(t)$ as

$$\begin{aligned} y(t) &= \cdots + e^{-(t+6)}u(t+6) + e^{-(t+3)}u(t+3) + e^{-t}u(t) \\ &= e^{-t} + e^{-(t+3)} + e^{-(t+6)} + \cdots \\ &= e^{-t}(1 + e^{-3} + e^{-6} + \cdots) \\ &= e^{-t}\frac{1}{1-e^{-3}} \end{aligned}$$

Therefore, $A = \frac{1}{1-e^{-3}}$.

2.13. **(a)** We require that

$$\left(\frac{1}{5}\right)^n u[n] - A\left(\frac{1}{5}\right)^{(n-1)} u[n-1] = \delta[n]$$

Putting $n = 1$ and solving for A gives $A = \frac{1}{5}$.

(b) From part (a), we know that

$$\begin{aligned} h[n] - \frac{1}{5}h[n-1] &= \delta[n] \\ h[n] * (\delta[n] - \frac{1}{5}\delta[n-1]) &= \delta[n] \end{aligned}$$

From the definition of an inverse system, we may argue that

$$g[n] = \delta[n] - \frac{1}{5}\delta[n-1].$$

2.14. (a) We first determine if $h_1(t)$ is absolutely integrable as follows

$$\int_{-\infty}^{\infty} |h_1(\tau)| d\tau = \int_{0}^{\infty} e^{-t} d\tau = 1$$

Therefore, $h_1(t)$ is the impulse response of a stable LTI system.

(b) We determine if $h_2(t)$ is absolutely integrable as follows

$$\int_{-\infty}^{\infty} |h_2(\tau)| d\tau = \int_{0}^{\infty} e^{-t} |\cos(2t)| d\tau$$

This integral is clearly finite-valued because $e^{-t}|\cos(2t)|$ is an exponentially decaying function in the range $0 \leq t \leq \infty$. Therefore, $h_2(t)$ is the impulse response of a stable LTI system.

2.15. (a) We determine if $h_1[n]$ is absolutely summable as follows

$$\sum_{k=-\infty}^{\infty} |h_1[k]| = \sum_{k=0}^{\infty} k|\cos(\frac{\pi}{4}k)|$$

This sum does not have a finite value because the function $k|\cos(\frac{\pi}{4}k)|$ increases as the value of k increases. Therefore, $h_1[n]$ cannot be the impulse response of a stable LTI system.

(b) We determine if $h_2[n]$ is absolutely summable as follows

$$\sum_{k=-\infty}^{\infty} |h_2[k]| = \sum_{k=-\infty}^{10} 3^k \approx 3^{11}/2$$

Therefore, $h_2[n]$ is the impulse response of a stable LTI system.

2.16. (a) **True**. This may be easily argued by noting that convolution may be viewed as the process of carrying out the superposition of a number of echos of $h[n]$. The first such echo will occur at the location of the first non zero sample of $x[n]$. In this case, the first echo will occur at N_1. The echo of $h[n]$ which occurs at $n = N_1$ will have its first non zero sample at the time location $N_1 + N_2$. Therefore, for all values of n which are lesser that $N_1 + N_2$, the output $y[n]$ is zero.

(b) **False**. Consider

$$\begin{aligned} y[n] &= x[n] * h[n] \\ &= \sum_{k=-\infty}^{\infty} x[k]h[n-k] \end{aligned}$$

From this,

$$\begin{aligned} y[n-1] &= \sum_{k=-\infty}^{\infty} x[k]h[n-1-k] \\ &= x[n] * h[n-1] \end{aligned}$$

This shows that the given statement is false.

(c) True. Consider

$$y(t) = x(t) * h(t) = \int_{-\infty}^{\infty} x(\tau)h(t-\tau)d\tau.$$

From this,

$$\begin{aligned} y(-t) &= \int_{-\infty}^{\infty} x(\tau)h(-t-\tau)d\tau \\ &= \int_{-\infty}^{\infty} x(-\tau)h(-t+\tau)d\tau \\ &= x(-t) * h(-t) \end{aligned}$$

This shows that the given statement is true.

(d) True. This may be argued by considering

$$y(t) = x(t) * h(t) = \int_{-\infty}^{\infty} x(\tau)h(t-\tau)d\tau.$$

In Figure S2.16, we plot $x(\tau)$ and $h(t-\tau)$ under the assumptions that (1) $x(t) = 0$ for $t > T_1$ and (2) $h(t) = 0$ for $t > T_2$. Clearly, the product $x(\tau)h(t-\tau)$ is zero if

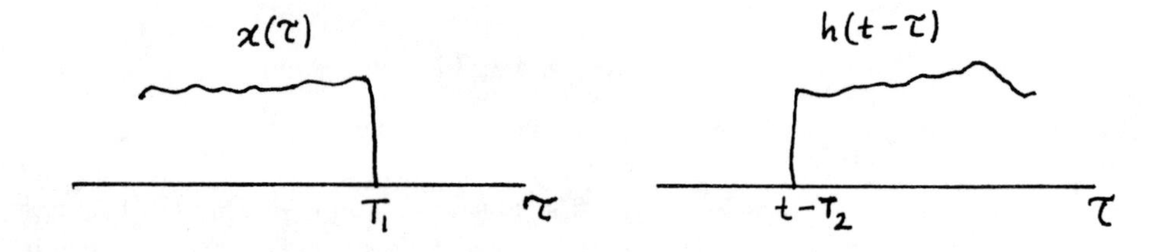

Figure S2.16

$t - T_2 > T_1$. Therefore, $y(t) = 0$ for $t > T_1 + T_2$.

2.17. **(a)** We know that $y(t)$ is the sum of the particular and homogeneous solutions to the given differential equation. We first determine the particular solution $y_p(t)$ by using the method specified in Example 2.14. Since we are given that the input is $x(t) = e^{(-1+3j)t}u(t)$ for $t > 0$, we hypothesize that for $t > 0$

$$y_p(t) = Ke^{(-1+3j)t}.$$

Substituting for $x(t)$ and $y(t)$ in the given differential equation,

$$(-1+3j)Ke^{(-1+3j)t} + 4Ke^{(-1+3j)t} = e^{(-1+3j)t}$$

This gives

$$(-1+3j)K+4K=1, \qquad \Rightarrow K=\frac{1}{3(1+j)}$$

Therefore,

$$y_p(t)=\frac{1}{3(1+j)}e^{(-1+3j)t}, \qquad t>0$$

In order to determine the homogeneous solution, we hypothesize that

$$y_h(t)=Ae^{st}$$

Since the homogeneous solution has to satisfy the following differential equation

$$\frac{dy_h(t)}{dt}+4y_h(t)=0,$$

we obtain

$$Ase^{st}+4Ae^{st}=Ae^{st}(s+4)=0.$$

This implies that $s=-4$ for any A. The overall solution to the differential equation now becomes

$$y(t)=Ae^{-4t}+\frac{1}{3(1+j)}e^{(-1+3j)t}, \qquad t>0$$

Now in order to determine the constant A, we use the fact that the system satisfies the condition of initial rest. Given that $y(0)=0$, we may conclude that

$$A+\frac{1}{3(1+j)}=0, \qquad A=\frac{-1}{3(1+j)}$$

Therefore for $t>0$,

$$y(t)=\frac{1}{3(1+j)}\left[-e^{-4t}+e^{(-1+3j)t}\right], \qquad t>0$$

Since the system satisfies the condition of initial rest, $y(t)=0$ for $t<0$. Therefore,

$$y(t)=\frac{1-j}{6}\left[-e^{-4t}+e^{(-1+3j)t}\right]u(t)$$

(b) The output will now be the real part of the answer obtained in part (a).

$$y(t)=\frac{1}{6}\left[e^{-t}\cos 3t+e^{-t}\sin 3t-e^{-4t}\right]u(t).$$

2.18. Since the system is causal, $y[n] = 0$ for $n < 1$. Now,

$$\begin{aligned} y[1] &= \frac{1}{4}y[0] + x[1] = 0 + 1 = 1 \\ y[2] &= \frac{1}{4}y[1] + x[2] = \frac{1}{4} + 0 = \frac{1}{4} \\ y[3] &= \frac{1}{4}y[2] + x[3] = \frac{1}{16} + 0 = \frac{1}{16} \\ &\vdots \\ y[m] &= (\frac{1}{4})^{m-1} \\ &\vdots \end{aligned}$$

Therefore,

$$y[n] = (\frac{1}{4})^{n-1}u[n-1]$$

2.19. **(a)** Consider the difference equation relating $y[n]$ and $w[n]$ for S_2:

$$y[n] = \alpha y[n-1] + \beta w[n]$$

From this we may write

$$w[n] = \frac{1}{\beta}y[n] - \frac{\alpha}{\beta}y[n-1]$$

and

$$w[n-1] = \frac{1}{\beta}y[n-1] - \frac{\alpha}{\beta}y[n-2]$$

Weighting the previous equation by $1/2$ and subtracting from the one before, we obtain

$$w[n] - \frac{1}{2}w[n-1] = \frac{1}{\beta}y[n] - \frac{\alpha}{\beta}y[n-1] - \frac{1}{2\beta}y[n-1] + \frac{\alpha}{2\beta}y[n-2]$$

Substituting this in the difference equation relating $w[n]$ and $x[n]$ for S_1,

$$\frac{1}{\beta}y[n] - \frac{\alpha}{\beta}y[n-1] - \frac{1}{2\beta}y[n-1] + \frac{\alpha}{2\beta}y[n-2] = x[n]$$

That is,

$$y[n] = (\alpha + \frac{1}{2})y[n-1] - \frac{\alpha}{2}y[n-2] + \beta x[n]$$

Comparing with the given equation relating $y[n]$ and $x[n]$, we obtain

$$\alpha = \frac{1}{4}, \qquad \beta = 1$$

(b) The difference equations relating the input and output of the systems S_1 and S_2 are

$$w[n] = \frac{1}{2}w[n-1] + x[n] \quad \text{and} \quad y[n] = \frac{1}{4}y[n-1] + w[n]$$

From these, we can use the method specifed in Example 2.15 to show that the impulse responses of S_1 and S_2 are

$$h_1[n] = \left(\frac{1}{2}\right)^n u[n]$$

and

$$h_2[n] = \left(\frac{1}{4}\right)^n u[n],$$

respectively. The overall impulse response of the system made up of a cascade of $S_1$and S_2 will be

$$\begin{aligned} h[n] &= h_1[n] * h_2[n] = \sum_{k=-\infty}^{\infty} h_1[k]h_2[n-k] \\ &= \sum_{k=0}^{\infty}(\frac{1}{2})^k(\frac{1}{4})^{n-k}u[n-k] \\ &= \sum_{k=0}^{n}(\frac{1}{2})^k(\frac{1}{4})^{n-k} = \sum_{k=0}^{n}(\frac{1}{2})^{2(n-k)} \\ &= [2(\frac{1}{2})^n - (\frac{1}{4})^n]u[n] \end{aligned}$$

2.20. **(a)**

$$\int_{-\infty}^{\infty} u_0(t)\cos(t)dt = \int_{-\infty}^{\infty} \delta(t)dt = 1$$

(b)

$$\int_0^5 \sin(2\pi t)\delta(t+3)dt = \sin(6\pi) = 0$$

(c) In order to evaluate the integral

$$\int_{-5}^{5} u_1(1-\tau)\cos(2\pi\tau)d\tau,$$

consider the signal

$$x(t) = \cos(2\pi t)[u(t+5) - u(t-5)].$$

We know that

$$\begin{aligned} \frac{dx(t)}{dt} &= u_1(t) * x(t) = \int_{-\infty}^{\infty} u_1(t-\tau)x(\tau)d\tau \\ &= \int_{-5}^{5} u_1(t-\tau)\cos(2\pi\tau)d\tau \end{aligned}$$

Now,

$$\left.\frac{dx(t)}{dt}\right|_{t=1} = \int_{-5}^{5} u_1(1-\tau)\cos(2\pi\tau)d\tau$$

which is the desired integral. We now evaluate the value of the integral as

$$\left.\frac{dx(t)}{dt}\right|_{t=1} = \sin(2\pi t)|_{t=1} = 0.$$

2.21. **(a)** The desired convolution is

$$\begin{aligned} y[n] &= x[n] * h[n] \\ &= \sum_{k=-\infty}^{\infty} x[k]h[n-k] \\ &= \beta^n \sum_{k=0}^{n} (\alpha/\beta)^k \text{ for } n \geq 0 \\ &= [\frac{\beta^{n+1} - \alpha^{n+1}}{\beta - \alpha}]u[n] \text{ for } \alpha \neq \beta. \end{aligned}$$

(b) From (a),

$$y[n] = \alpha^n \left[\sum_{k=0}^{n} 1\right] u[n] = (n+1)\alpha^n u[n].$$

(c) For $n \leq 6$,

$$y[n] = 4^n \left\{\sum_{k=0}^{\infty}(-\frac{1}{8})^k - \sum_{k=0}^{3}(-\frac{1}{8})^k\right\}.$$

For $n > 6$,

$$y[n] = 4^n \left\{\sum_{k=0}^{\infty}(-\frac{1}{8})^k - \sum_{k=0}^{n-1}(-\frac{1}{8})^k\right\}.$$

Therefore,

$$y[n] = \begin{cases} (8/9)(-1/8)^4 4^n, & n \leq 6 \\ (8/9)(-1/2)^n, & n > 6 \end{cases}$$

(d) The desired convolution is

$$\begin{aligned} y[n] &= \sum_{k=-\infty}^{\infty} x[k]h[n-k] \\ &= x[0]h[n] + x[1]h[n-1] + x[2]h[n-2] + x[3]h[n-3] + x[4]h[n-4] \\ &= h[n] + h[n-1] + h[n-2] + h[n-3] + h[n-4]. \end{aligned}$$

This is as shown in Figure S2.21.

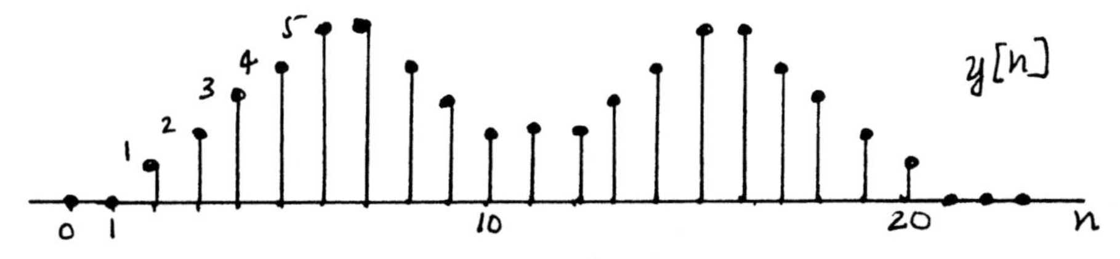

Figure S2.21

2.22. **(a)** The desired convolution is

$$\begin{aligned} y(t) &= \int_{-\infty}^{\infty} x(\tau)h(t-\tau)d\tau \\ &= \int_{0}^{t} e^{-\alpha t} e^{-\beta(t-\tau)} d\tau, \ t \geq 0 \end{aligned}$$

Then

$$y(t) = \begin{cases} \frac{e^{-\beta t}\{e^{-(\alpha-\beta)t}-1\}}{\beta-\alpha} u(t) & \alpha \neq \beta \\ te^{-\beta t}u(t) & \alpha = \beta \end{cases}.$$

(b) The desired convolution is

$$\begin{aligned} y(t) &= \int_{-\infty}^{\infty} x(\tau)h(t-\tau)d\tau \\ &= \int_{0}^{2} h(t-\tau)d\tau - \int_{2}^{5} h(t-\tau)d\tau. \end{aligned}$$

This may be written as

$$y(t) = \begin{cases} \int_{0}^{2} e^{2(t-\tau)}d\tau - \int_{2}^{5} e^{2(t-\tau)}d\tau, & t \leq 1 \\ \int_{t-1}^{2} e^{2(t-\tau)}d\tau - \int_{2}^{5} e^{2(t-\tau)}d\tau, & 1 \leq t \leq 3 \\ -\int_{t-1}^{5} e^{2(t-\tau)}d\tau, & 3 \leq t \leq 6 \\ 0, & 6 < t \end{cases}$$

Therefore,

$$y(t) = \begin{cases} (1/2)[e^{2t} - 2e^{2(t-2)} + e^{2(t-5)}], & t \leq 1 \\ (1/2)[e^{2} + e^{2(t-5)} - 2e^{2(t-2)}], & 1 \leq t \leq 3 \\ (1/2)[e^{2(t-5)} - e^{2}], & 3 \leq t \leq 6 \\ 0, & 6 < t \end{cases}$$

(c) The desired convolution is

$$\begin{aligned} y(t) &= \int_{-\infty}^{\infty} x(\tau)h(t-\tau)d\tau \\ &= \int_{0}^{2} \sin(\pi\tau)h(t-\tau)d\tau. \end{aligned}$$

This gives us

$$y(t) = \begin{cases} 0, & t < 1 \\ (2/\pi)[1 - \cos\{\pi(t-1)\}], & 1 < t < 3 \\ (2/\pi)[\cos\{\pi(t-3)\} - 1], & 3 < t < 5 \\ 0, & 5 < t \end{cases}$$

(d) Let

$$h(t) = h_1(t) - \frac{1}{3}\delta(t-2),$$

where

$$h_1(t) = \begin{cases} 4/3, & 0 \le t \le 1 \\ 0, & \text{otherwise} \end{cases}.$$

Now,

$$y(t) = h(t) * x(t) = [h_1(t) * x(t)] - \frac{1}{3}x(t-2).$$

We have

$$h_1(t) * x(t) = \int_{t-1}^{t} \frac{4}{3}(a\tau + b)d\tau = \frac{4}{3}[\frac{1}{2}at^2 - \frac{1}{2}a(t-1)^2 + bt - b(t-1)].$$

Therefore,

$$y(t) = \frac{4}{3}[\frac{1}{2}at^2 - \frac{1}{2}a(t-1)^2 + bt - b(t-1)] - \frac{1}{3}[a(t-2) + b] = at + b = x(t).$$

(e) $x(t)$ periodic implies $y(t)$ periodic. $\therefore$ determine 1 period only. We have

$$y(t) = \begin{cases} \int_{t-1}^{-\frac{1}{2}} (t - \tau - 1)d\tau + \int_{-\frac{1}{2}}^{t} (1 - t + \tau)d\tau = \frac{1}{4} + t - t^2, & -\frac{1}{2} < t < \frac{1}{2} \\ \int_{t-1}^{\frac{1}{2}} (1 - t + \tau)d\tau + \int_{\frac{1}{2}}^{t} (t - 1 - \tau)d\tau = t^2 - 3t + 7/4, & \frac{1}{2} < t < \frac{3}{2} \end{cases}.$$

The period of $y(t)$ is 2.

2.23. $y(t)$ is sketched in Figure S2.23 for the different values of T.

2.24. **(a)** We are given that $h_2[n] = \delta[n] + \delta[n-1]$. Therefore,

$$h_2[n] * h_2[n] = \delta[n] + 2\delta[n-1] + \delta[n-2].$$

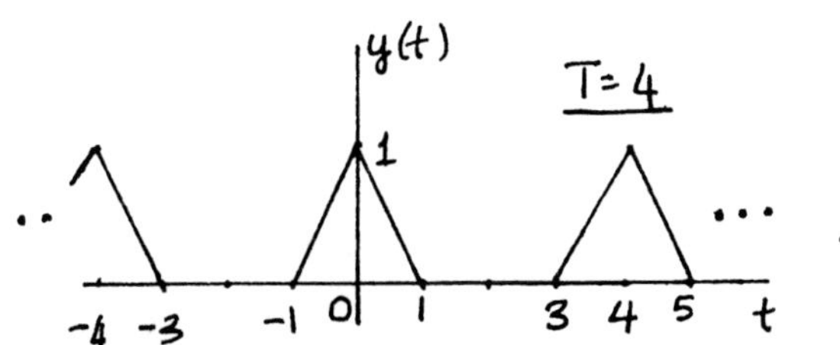

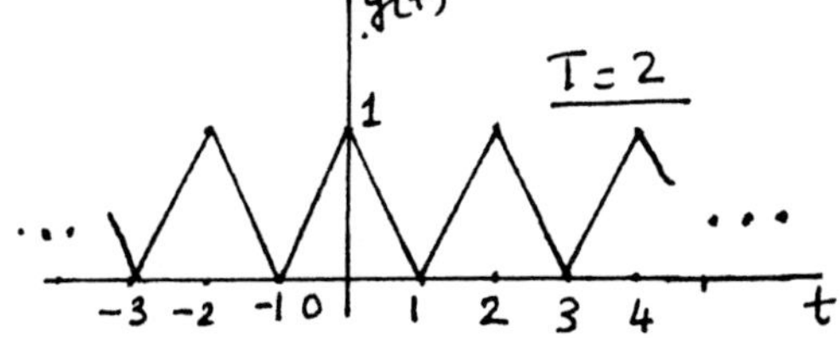

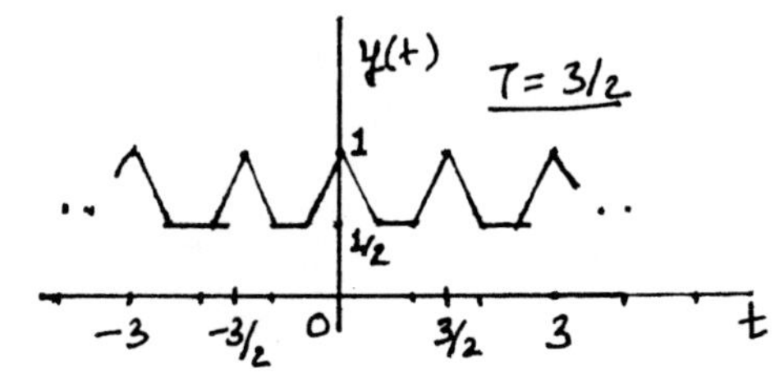

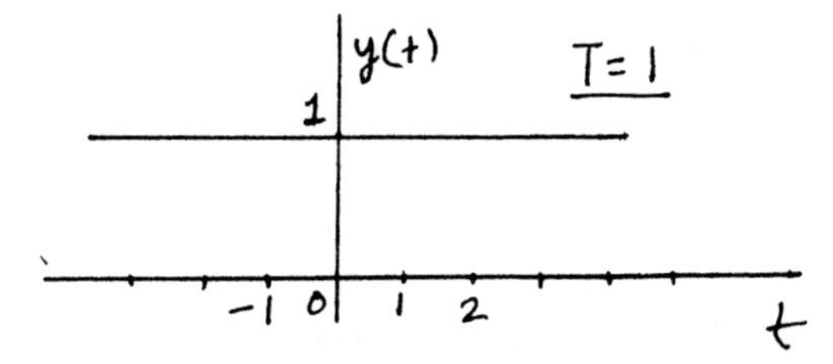

Figure S2.23

Since

$$h[n] = h_1[n] * [h_2[n] * h_2[n]],$$

we get

$$h[n] = h_1[n] + 2h_1[n-1] + h_1[n-2].$$

Therefore,

$$\begin{aligned}
h[0] &= h_1[0] & \Rightarrow \quad & h_1[0] = 1, \\
h[1] &= h_1[1] + 2h_1[0] & \Rightarrow \quad & h_1[1] = 3, \\
h[2] &= h_1[2] + 2h_1[1] + h_1[0] & \Rightarrow \quad & h_1[2] = 3, \\
h[3] &= h_1[3] + 2h_1[2] + h_1[1] & \Rightarrow \quad & h_1[3] = 2, \\
h[4] &= h_1[4] + 2h_1[3] + h_1[2] & \Rightarrow \quad & h_1[4] = 1 \\
h[5] &= h_1[5] + 2h_1[4] + h_1[3] & \Rightarrow \quad & h_1[5] = 0.
\end{aligned}$$

$h_1[n] = 0$ for $n < 0$ and $n \geq 5$.

(b) In this case,

$$y[n] = x[n] * h[n] = h[n] - h[n-1].$$

2.25. **(a)** We may write $x[n]$ as

$$x[n] = \left(\frac{1}{3}\right)^{|n|}.$$

Now, the desired convolution is

$$
\begin{aligned}
y[n] &= h[n] * x[n] \\
&= \sum_{k=-\infty}^{-1} (1/3)^{-k}(1/4)^{n-k}u[n-k+3] + \sum_{k=0}^{\infty}(1/3)^k(1/4)^{n-k}u[n-k+3] \\
&= (1/12)\sum_{k=0}^{\infty}(1/3)^k(1/4)^{n+k}u[n+k+4] + \sum_{k=0}^{\infty}(1/3)^k(1/4)^{n-k}u[n-k+3]
\end{aligned}
$$

By consider each summation in the above equation separately, we may show that

$$
y[n] = \begin{cases} (12^4/11)3^n, & n < -4 \\ (1/11)4^4, & n = -4 \\ (1/4)^n(1/11) + -3(1/4)^n + 3(256)(1/3)^n, & n \geq -3 \end{cases}.
$$

(b) Now consider the convolution

$$
y_1[n] = [(1/3)^n u[n]] * [(1/4)^n u[n+3]].
$$

We may show that

$$
y_1[n] = \begin{cases} 0, & n < -3 \\ -3(1/4)^n + 3(256)(1/3)^n, & n \geq -3 \end{cases}.
$$

Also, consider the convolution

$$
y_2[n] = [(3)^n u[-n-1]] * [(1/4)^n u[n+3]].
$$

We may show that

$$
y_2[n] = \begin{cases} (12^4/11)3^n, & n < -4 \\ (1/4)^n(1/11), & n \geq -3 \end{cases}.
$$

Clearly, $y_1[n] + y_2[n] = y[n]$ obtained in the previous part.

2.26. **(a)** We have

$$
\begin{aligned}
y_1[n] = x_1[n] * x_2[n] &= \sum_{k=-\infty}^{\infty} x_1[k]x_2[n-k] \\
&= \sum_{k=0}^{\infty}(0.5)^k u[n+3-k].
\end{aligned}
$$

This evaluates to

$$
y_1[n] = x_1[n] * x_2[n] = \begin{cases} 2\left\{1-(1/2)^{n+4}\right\}, & n \geq -3 \\ 0, & \text{otherwise} \end{cases}.
$$

(b) Now,

$$y[n] = x_3[n] * y_1[n] = y_1[n] - y_1[n-1].$$

Therefore,

$$y[n] = \begin{cases} 2\left\{1 - (1/2)^{n+3}\right\} + 2\left\{1 - (1/2)^{n+4}\right\} = (1/2)^{n+3}, & n \geq -2 \\ 1, & n = -3 \\ 0, & \text{otherwise} \end{cases}.$$

Therefore, $y[n] = (1/2)^{n+3}u[n+3]$.

(c) We have

$$y_2[n] = x_2[n] * x_3[n] = u[n+3] - u[n+2] = \delta[n+3].$$

(d) From the result of part (c), we get

$$y[n] = y_2[n] * x_1[n] = x_1[n+3] = (1/2)^{n+3}u[n+3].$$

2.27. The proof is as follows.

$$\begin{aligned} A_y &= \int_{-\infty}^{\infty} y(t)dt \\ &= \int_{-\infty}^{\infty}\int_{-\infty}^{\infty} x(\tau)h(t-\tau)d\tau dt \\ &= \int_{-\infty}^{\infty} x(\tau)\int_{-\infty}^{\infty} h(t-\tau)dtd\tau \\ &= \int_{-\infty}^{\infty} x(\tau)A_h d\tau \\ &= A_x A_h \end{aligned}$$

2.28. **(a)** Causal because $h[n] = 0$ for $n < 0$. Stable because $\sum_{n=0}^{\infty}(\frac{1}{5})^n = 5/4 < \infty$.

(b) Not causal because $h[n] \neq 0$ for $n < 0$. Stable because $\sum_{n=-2}^{\infty}(0.8)^n = 5 < \infty$.

(c) Anti-causal because $h[n] = 0$ for $n > 0$. Unstable because $\sum_{n=-\infty}^{0}(1/2)^n = \infty$

(d) Not causal because $h[n] \neq 0$ for $n < 0$. Stable because $\sum_{n=-\infty}^{3} 5^n = \frac{625}{4} < \infty$

(e) Causal because $h[n] = 0$ for $n < 0$. Unstable because the second term becomes infinite as $n \to \infty$.

(f) Not causal because $h[n] \neq 0$ for $n < 0$. Stable because $\sum_{n-\infty}^{\infty}|h[n]| = 305/3 < \infty$

(g) Causal because $h[n] = 0$ for $n < 0$. Stable because $\sum_{n=-\infty}^{\infty} |h[n]| = 1 < \infty$.

2.29. (a) Causal because $h(t) = 0$ for $t < 0$. Stable because $\int_{-\infty}^{\infty} |h(t)|dt = e^{-8}/4 < \infty$.

(b) Not causal because $h(t) \neq 0$ for $t < 0$. Unstable because $\int_{-\infty}^{\infty} |h(t)| = \infty$.

(c) Not causal because $h(t) \neq 0$ for $t < 0$. a Stable because $\int_{-\infty}^{\infty} |h(t)|dt = e^{100}/2 < \infty$.

(d) Not causal because $h(t) \neq 0$ for $t < 0$. Stable because $\int_{-\infty}^{\infty} |h(t)|dt = e^{-2}/2 < \infty$.

(e) Not causal because $h(t) \neq 0$ for $t < 0$. Stable because $\int_{-\infty}^{\infty} |h(t)|dt = 1/3 < \infty$.

(f) Causal because $h(t) = 0$ for $t < 0$. Stable because $\int_{-\infty}^{\infty} |h(t)|dt = 1 < \infty$.

(g) Causal because $h(t) = 0$ for $t < 0$. Unstable because $\int_{-\infty}^{\infty} |h(t)|dt = \infty$.

2.30. We need to find the output of the system when the input is $x[n] = \delta[n]$. Since we are asked to assume initial rest, we may conclude that $y[n] = 0$ for $n < 0$. Now,

$$y[n] = x[n] - 2y[n-1].$$

Therefore,

$$y[0] = x[0] - 2y[-1] = 1, \qquad y[1] = x[1] - 2y[0] = -2, \qquad y[2] = x[2] + 2y[2] = -4$$

and so on. In closed form,

$$y[n] = (-2)^n u[n].$$

This is the impulse response of the system.

2.31. Initial rest implies that $y[n] = 0$ for $n < -2$. Now

$$y[n] = x[n] + 2x[n-2] - 2y[n-1].$$

Therefore,

$$y[-2] = 1, \quad y[-1] = 0, \quad y[0] = 5, \quad y[1] = -$$

$$y[4] = 56, y[5] = -110, \quad y[n] = -110(-2)^{n-5} \quad \text{for} n \geq 5.$$

2.32. (a) If $y_h[n] = A(1/2)^n$, then we need to verify

$$A\left(\frac{1}{2}\right)^n - \frac{1}{2}A\left(\frac{1}{2}\right)^{n-1} = 0.$$

Clearly this is true.

(b) We now require that for $n \geq 0$

$$B\left(\frac{1}{3}\right)^n - \frac{1}{2}B\left(\frac{1}{3}\right)^{n-1} = \left(\frac{1}{3}\right)^n.$$

Therefore, $B = -2$.

(c) From eq. (P2.32-1), we know that $y[0] = x[0] + (1/2)y[-1] = x[0] = 1$. Now we also have

$$y[0] = A + B \quad \Rightarrow \quad A = 1 - B = 3.$$

2.33. (a) (i) From Example 2.14, we know that

$$y_1(t) = \left[\frac{1}{5}e^{3t} - \frac{1}{5}e^{-2t}\right]u(t).$$

(ii) We solve this along the lines of Example 2.14. First assume that $y_p(t)$ is of the form Ke^{2t} for $t > 0$. Then using eq. (P2.33-1), we get for $t > 0$

$$2Ke^{2t} + 2Ke^{2t} = e^{2t} \quad \Rightarrow \quad K = \frac{1}{4}.$$

We now know that $y_p(t) = \frac{1}{4}e^{2t}$ for $t > 0$. We may hypothesize the homogeneous solution to be of the form

$$y_h(t) = Ae^{-2t}.$$

Therefore,

$$y_2(t) = Ae^{-2t} + \frac{1}{4}e^{2t}, \qquad \text{for } t > 0.$$

Assuming initial rest, we can conclude that $y_2(t) = 0$ for $t \leq 0$. Therefore,

$$y_2(0) = 0 = A + \frac{1}{4} \quad \Rightarrow \quad A = -\frac{1}{4}.$$

Then,

$$y_2(t) = \left[-\frac{1}{4}e^{2t} + \frac{1}{4}e^{-2t}\right]u(t).$$

(iii) Let the input be $x_3(t) = \alpha e^{3t}u(t) + \beta e^{2t}u(t)$. Assume that the particular solution $y_p(t)$ is of the form

$$y_p(t) = K_1\alpha e^{3t} + K_2\beta e^{2t}$$

for $t > 0$. Using eq. (P2.33-1), we get

$$3K_1\alpha e^{3t} + 2K_2\beta e^{2t} + 2K_1\alpha e^{3t} + 2K_2\beta e^{2t} = \alpha^{3t} + \beta e^{2t}.$$

Equating the coefficients of e^{3t} and e^{2t} on both sides, we get

$$K_1 = \frac{1}{5} \quad \text{and} \quad K_2 = \frac{1}{4}.$$

Now hypothesizing that $y_h(t) = Ae^{-2t}$, we get

$$y_3(t) = \frac{1}{5}\alpha e^{3t} + \frac{1}{4}\beta e^{2t} + Ae^{-2t}$$

for $t > 0$. Assuming initial rest,

$$y_3(0) = 0 = A + \alpha/5 + \beta/4 \quad \Rightarrow \quad A = -\left(\frac{\alpha}{5} + \frac{\beta}{4}\right).$$

Therefore,

$$y_3(t) = \left\{\frac{1}{5}\alpha e^{3t} + \frac{1}{4}\beta e^{2t} - \left(\frac{\alpha}{5} + \frac{\beta}{4}\right) e^{-2t}\right\} u(t).$$

Clearly, $y_3(t) = \alpha y_1(t) + \beta y_2(t)$.

(iv) For the input-output pair $x_1(t)$ and $y_1(t)$, we may use eq. (P2.33-1) and the initial rest condition to write

$$\frac{dy_1(t)}{dt} + 2y_1(t) = x_1(t), \qquad y_1(t) = 0 \text{ for } t < t_1. \tag{S2.33–1}$$

For the input-output pair $x_2(t)$ and $y_2(t)$, we may use eq. (P2.33-1) and the initial rest condition to write

$$\frac{dy_2(t)}{dt} + 2y_2(t) = x_2(t), \qquad y_2(t) = 0 \text{ for } t < t_2. \tag{S2.33–2}$$

Scaling eq. (S2.33-1) by α and eq. (S2.33-2) by β and summing, we get

$$\frac{d}{dt}\{\alpha y_1(t) + \beta y_2(t)\} + 2\{\alpha y_1(t) + \beta y_2(t)\} = \alpha x_1(t) + \beta x_2(t),$$

and

$$y_1(t) + y_2(t) = 0 \text{ for } t < \min(t_1, t_2).$$

By inspection, it is clear that the output is $y_3(t) = \alpha y_1(t) + \beta y_2(t)$ when the input is $x_3(t) = \alpha x_1(t) + \beta x_2(t)$. Furthermore, $y_3(t) = 0$ for $t < t_3$, where t_3 denotes the time until which $x_3(t) = 0$.

(b) (i) Using the result of (a-ii), we may write

$$y_1(t) = \frac{K}{4}\left[e^{2t} - e^{-2t}\right] u(t).$$

(ii) We solve this along the lines of Example 2.14. First assume that $y_p(t)$ is of the form $KYe^{2(t-T)}$ for $t > T$. Then using eq. (P2.33-1), we get for $t > T$

$$2Ke^{2(t-T)} + 2Ke^{2(t-T)} = e^{2t} \quad \Rightarrow \quad K = \frac{1}{4}.$$

We now know that $y_p(t) = \frac{K}{4}e^{2(t-T)}$ for $t > T$. We may hypothesize the homogeneous solution to be of the form

$$y_h(t) = Ae^{-2t}.$$

Therefore,

$$y_2(t) = Ae^{-2t} + \frac{K}{4}e^{2(t-T)}, \qquad \text{for } t > T.$$

Assuming initial rest, we can conclude that $y_2(t) = 0$ for $t \le T$. Therefore,

$$y_2(T) = 0 = Ae^{-2T} + \frac{K}{4} \qquad \Rightarrow \qquad A = -\frac{K}{4}e^{2T}.$$

Then,

$$y_2(t) = \left[-\frac{K}{4}e^{-2(t-T)} + \frac{K}{4}e^{2(t-T)}\right]u(t-T).$$

Clearly, $y_2(t) = y_1(t-T)$.

(iii) Consider the input-output pair $x_1(t) \to y_1(t)$ where $x_1(t) = 0$ for $t < t_0$. Note that

$$\frac{dy_1(t)}{dt} + 2y_1(t) = x_1(t), \qquad y_1(t) = 0, \text{ for } t < t_0.$$

Since the derivative is a time-invariant operation, we may now write

$$\frac{dy_1(t-T)}{dt} + 2y_1(t-T) = x_1(t-T), \qquad y_1(t) = 0, \text{ for } t < t_0.$$

This suggests that if the input is a signal of the form $x_2(t) = x_1(t-T)$, then the output is a signal of the form $y_2(t) = y_1(t-T)$. Also, note that the new output $y_2(t)$ will be zero for $t < t_0 + T$. This supports time-invariance since $x_2(t)$ is zero for $t < t_0 + T$. Therefore, we may conclude that the system is time-invariant.

2.34. (a) Consider $x_1(t) \xrightarrow{S} y_1(t)$ and $x_2(t) \xrightarrow{S} y_2(t)$. We know that $y_1(1) = y_2(1) = 1$. Now consider a third input to the system which is $x_3(t) = x_1(t) + x_2(t)$. Let the corresponding output be $y_3(t)$. Now, note that $y_3(1) = 1 \neq y_1(1) + y_2(1)$. Therefore, the system is not linear. A specific example follows.

Consider an input signal $x_1(t) = e^{2t}u(t)$. From Problem 2.33(a-ii), we know that the corresponding output for $t > 0$ is

$$y_1(t) = \frac{1}{4}e^{2t} + Ae^{-2t}.$$

Using the fact that $y_1(1) = 1$, we get for $t > 0$

$$y_1(t) = \frac{1}{4}e^{2t} + \left(1 - \frac{e}{4}\right)e^{-2(t-1)}.$$

Now, consider a second signal $x_2(t) = 0$. Then, the corresponding output is

$$y_2(t) = Be^{-2t}$$

for $t > 0$. Using the fact that $y_2(1) = 1$, we get for $t > 0$

$$y_2(t) = e^{-2(t-1)}.$$

Now consider a third signal $x_3(t) = x_1(t) + x_2(t) = x_1(t)$. Note that the output will still be $y_3(t) = y_1(t)$ for $t > 0$. Clearly, $y_3(t) \neq y_1(t) + y_2(t)$ for $t > 0$. Therefore, the system is not linear.

(b) Again consider an input signal $x_1(t) = e^{2t}u(t)$. From part (a), we know that the corresponding output for $t > 0$ with $y_1(1) = 1$ is

$$y_1(t) = \frac{1}{4}e^{2t} + \left(1 - \frac{e}{4}\right)e^{-2(t-1)}.$$

Now, consider an input signal of the form $x_2(t) = x_1(t-T) = e^{2(t-T)}u(t-T)$. Then for $t > T$,

$$y_2(t) = \frac{1}{4}e^{2(t-T)} + Ae^{-2t}.$$

Using the fact that $y_2(1) = 1$ and also assuming that $T < 1$, , we get for $t > T$

$$y_2(t) = \frac{1}{4}e^{2(t-T)} + \left(1 - \frac{1}{4}e^{2(1-T)}\right)e^{-2(t-1)}.$$

Now note that $y_2(t) \neq y_1(t-T)$ for $t > T$. Therefore, the system is not time invariant.

(c) In order to show that the system is incrementally linear with the auxiliary condition specified as $y(1) = 1$, we need to first show that the system is linear with the auxiliary condition specified as $y(1) = 0$.

For an input-output pair $x_1(t)$ and $y_1(t)$, we may use eq. (P2.33-1) and the fact that $y_1(1) = 0$ to write

$$\frac{dy_1(t)}{dt} + 2y_1(t) = x_1(t), \qquad y_1(1) = 0. \tag{S2.34–1}$$

For an input-output pair $x_2(t)$ and $y_2(t)$, we may use eq. (P2.33-1) and the initial rest condition to write

$$\frac{dy_2(t)}{dt} + 2y_2(t) = x_2(t), \qquad y_2(1) = 0. \tag{S2.34–2}$$

Scaling eq. (S2.34-1) by α and eq. (S2.34-2) by β and summing, we get

$$\frac{d}{dt}\{\alpha y_1(t) + \beta y_2(t)\} + 2\{\alpha y_1(t) + \beta y_2(t)\} = \alpha x_1(t) + \beta x_2(t)$$

and

$$y_3(1) = y_1(1) + y_2(1) = 0.$$

By inspection, it is clear that the output is $y_3(t) = \alpha y_1(t) + \beta y_2(t)$ when the input is $x_3(t) = \alpha x_1(t) + \beta x_2(t)$. Furthermore, $y_3(1) = 0 = y_1(1) + y_2(1)$. Therefore, the system is linear.

Therefore, the overall system may be treated as the cascade of a linear system with an adder which adds the response of the system to the auxiliary conditions alone.

(d) In the previous part, we showed that the system is linear when $y(1) = 0$. In order to show that the system is not time-invariant, consider an input of the form $x_1(t) = e^{2t}u(t)$. From part (a), we know that the corresponding output will be

$$y_1(t) = \frac{1}{4}e^{2t} + Ae^{-2t}.$$

Using the fact that $y_1(1) = 0$, we get for $t > 0$

$$y_1(t) = \frac{1}{4}e^{2t} - \frac{1}{4}e^{-2(t-2)}.$$

Now consider an input of the form $x_2(t) = x_1(t - 1/2)$. Note that $y_2(1) = 0$. Clearly, $y_2(1) \neq y_1(1 - 1/2) = (1/4)(e - e^3)$. Therefore, $y_2(t) \neq y_1(t - 1/2)$ for all t. This implies that the system is not time invariant.

(e) A proof which is very similar to the proof for linearity used in part (c) may be used here. We may show that the system is not time invariant by using the method outlined in part (d).

2.35. (a) Since the system is linear, the response $y_1(t) = 0$ for all t.

(b) Now let us find the output $y_2(t)$ when the input is $x_2(t)$. The particular solution is of the form

$$y_p(t) = Y, \qquad t > -1.$$

Substituting in eq. (P2.33-1), we get

$$2Y = 1.$$

Now, including the homogeneous solution which is of the form $y_h(t) = Ae^{-2t}$, we get the overall solution:

$$y_2(t) = Ae^{-2t} + \frac{1}{2}, \qquad t > -1.$$

Since $y(0) = 0$, we get

$$y_2(t) = -\frac{1}{2}e^{-2t} + \frac{1}{2}, \qquad t > -1. \qquad \text{(S2.35–1)}$$

For $t < -1$, we note that $x_2(t) = 0$. Thus the particular solution is zero in this range and

$$y_2(t) = Be^{-2t}, \qquad t < -1. \qquad \text{(S2.35–2)}$$

Since the two pieces of the solution for $y_2(t)$ in eqs. (S2.35-1) and (S2.35-2) must match at $t = -1$, we can determine B from the equation

$$\frac{1}{2} - \frac{1}{2}e^2 = Be^2$$

which yields

$$y_2(t) = \left(\frac{1}{2} - \frac{1}{2}e^2\right)e^{-2(t+1)}, \qquad t < -1.$$

Now note that since $x_1(t) = x_2(t)$ for $t < -1$, it must be true that for a causal system $y_1(t) = y_2(t)$ for $t < -1$. However the results of parts (a) and (b) show that this is not true. Therefore, the system is not causal.

2.36. **(a)** Consider an input $x_1[n]$ such that $x_1[n] = 0$ for $n < n_1$. The corresponding output will be

$$y_1[n] = \frac{1}{2}y_1[n-1] + x_1[n], \qquad y_1[n] = 0 \text{ for } n < n_1. \tag{S2.36-1}$$

Also, consider another input $x_2[n]$ such that $x_2[n] = 0$ for $n < n_2$. The corresponding output will be

$$y_2[n] = \frac{1}{2}y_2[n-1] + x_2[n], \qquad y_2[n] = 0 \text{ for } n < n_2. \tag{S2.36-2}$$

Scaling eq. (S2.36-1) by α and eq. (S2.36-2) by β and summing, we get

$$\alpha y_1[n] + \beta y_2[n] = \frac{\alpha}{2}y_1[n-1] + \frac{\beta}{2}y_2[n-1] + \alpha x_1[n] + \beta x_2[n].$$

By inspection, it is clear that the output is $y_3[n] = \alpha y_1[n] + \beta y_2[n]$ when the input is $x_3[n] = \alpha x_1[n] + \beta x_2[n]$. Furthermore, $y_3(1) = 0 = y_1(1) + y_2(1)$. Therefore, the system is linear.

(b) Let us consider two inputs

$$x_1[n] = 0, \qquad \text{for all } n,$$

and

$$x_2[n] = \begin{cases} 0, & n < -1 \\ 1, & n \geq -1 \end{cases}.$$

Since the system is linear, the response to $x_1[n]$ is $y_1[n] = 0$ for all n. Now let us find the output $y_2[n]$ when the input is $x_2[n]$. Since $y_2[0] = 0$,

$$y_2[1] = (1/2)0 + 0 = 0, \qquad y_2[2] = (1/2)0 + 0 = 0, \qquad \cdots.$$

Therefore, $y_2[n] = 0$ for $n \geq 0$. Now, for $n < 0$, note that

$$y_2[0] = (1/2)y_2[-1] + x[0].$$

Therefore, $y_2[-1] = -2$. Proceeding similarly, we get $y_2[-2] = -4$, $y_2[-3] = -8$, and so on. Therefore, $y_2[n] = -(1/2)^n u[-n-1]$.

Now note that since $x_1[n] = x_2[n]$ for $n < 0$, it must be true that for a causal system $y_1[n] = y_2[n]$ for $n < 0$. However, the results obtained above show that this is not true. Therefore, the system is not causal.

2.37. Let us consider two inputs

$$x_1(t) = 0, \qquad \text{for all } t$$

and

$$x_2(t) = e^t[u(t) - u(t-1)].$$

Since the system is linear, the response $y_1(t) = 0$ for all t.

Now let us find the output $y_2(t)$ when the input is $x_2(t)$. The particular solution is of the form

$$y_p(t) = Ye^t, \qquad 0 < t < 1.$$

Substituting in eq. (P2.33-1), we get

$$3Y = 1.$$

Now, including the homogeneous solution which is of the form $y_h(t) = Ae^{-2t}$, we get the overall solution:

$$y_2(t) = Ae^{-2t} + \frac{1}{3}e^t, \qquad 0 < t < 1.$$

Assuming final rest, we have $y(1) = 0$. Using this we get $A = -e^3/3$. Therefore,

$$y_2(t) = -\frac{1}{3}e^{-2t+3} + \frac{1}{3}e^t, \qquad 0 < t < 1. \tag{S2.37-1}$$

For $t < 0$, we note that $x_2(t) = 0$. Thus the particular solution is zero in this range and

$$y_2(t) = Be^{-2t}, \qquad t < 0. \tag{S2.37-2}$$

Since the two pieces of the solution for $y_2(t)$ in eqs. (S2.37-1) and (S2.37-2) must match at $t = 0$, we can determine B from the equation

$$\frac{1}{3} - \frac{1}{3}e^3 = B$$

which yields

$$y_2(t) = \left(\frac{1}{3} - \frac{1}{3}e^3\right)e^{-2t}, \qquad t < 0.$$

Now note that since $x_1(t) = x_2(t)$ for $t < 0$, it must be true that for a causal system $y_1(t) = y_2(t)$ for $t < 0$. However, the results of obtained above show that this is not true. Therefore, the system is not causal.

2.38. The block diagrams are as shown in Figure S2.38.

2.39. The block diagrams are as shown in Figure S2.39.

2.40. (a) Note that

$$y(t) = \int_{-\infty}^{t} e^{-(t-\tau)}x(\tau - 2)d\tau = \int_{-\infty}^{t-2} e^{-(t-2-\tau')}x(\tau')d\tau'.$$

Therefore,

$$h(t) = e^{-(t-2)}u(t-2).$$

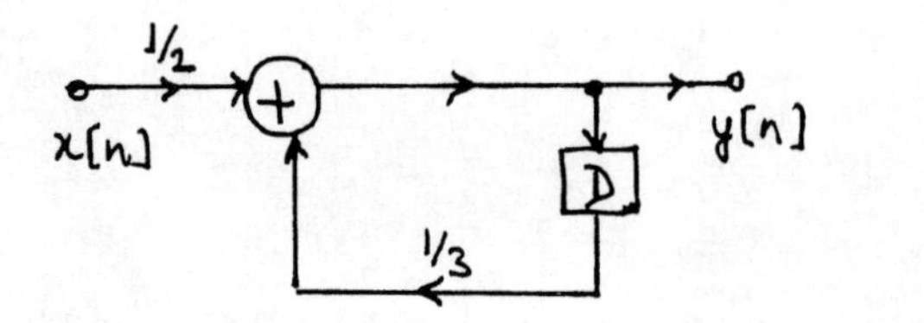

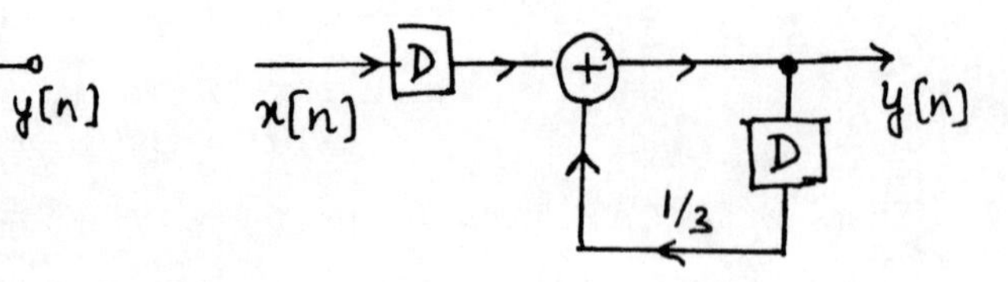

Figure S2.38

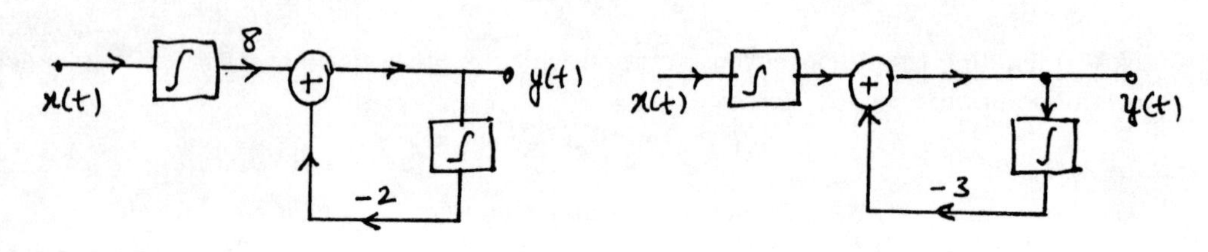

Figure S2.39

(b) We have

$$
\begin{aligned}
y(t) &= \int_{-\infty}^{\infty} h(\tau)x(t-\tau)d\tau \\
&= \int_{2}^{\infty} e^{-(\tau-2)}[u(t-\tau+1) - u(t-\tau-2)
\end{aligned}
$$

$h(\tau)$ and $x(t-\tau)$ are as shown in the figure below.

Using this figure, we may write

$$
y(t) = \begin{cases} 0, & t < 1 \\ \int_{2}^{t+1} e^{-(\tau-2)} d\tau = 1 - e^{-(t-1)}, & 1 < t < 4 \\ \int_{t-2}^{t+1} e^{-(\tau-2)} d\tau = e^{-(t-4)}[1 - e^{-3}], & t > 4 \end{cases}
$$

2.41. **(a)** We may write

$$
\begin{aligned}
g[n] &= x[n] - \alpha x[n-1] \\
&= \alpha^n u[n] - \alpha^n u[n-1] \\
&= \delta[n].
\end{aligned}
$$

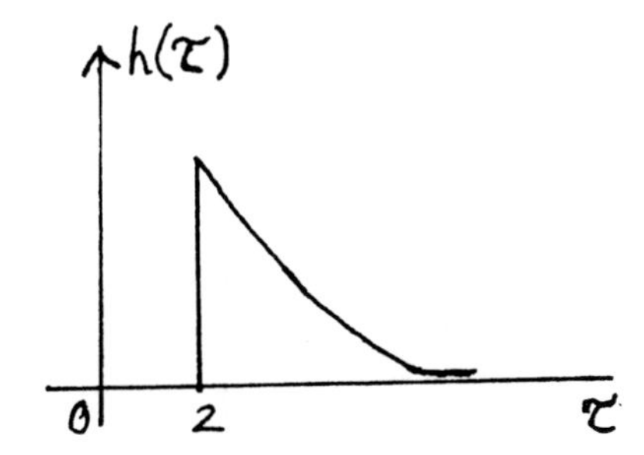

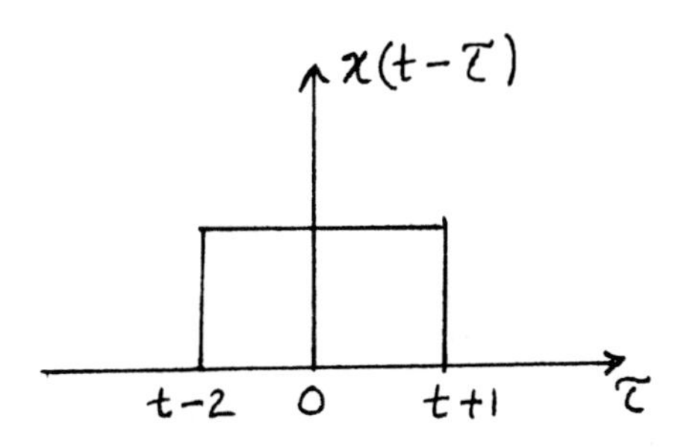

Figure S2.40

(b) Note that $g[n] = x[n] * \{\delta[n] - \alpha\delta[n-1]\}$. Therefore, from part (a), we know that $x[n] * \{\delta[n] - \alpha\delta[n-1]\} = \delta[n]$. Using this we may write

$$\begin{aligned} x[n] * \{\delta[n-1] - \alpha\delta[n-2]\} &= \delta[n-1], \\ x[n] * \{\delta[n+1] - \alpha\delta[n]\} &= \delta[n+1], \\ x[n] * \{\delta[n+2] - \alpha\delta[n+1]\} &= \delta[n+2]. \end{aligned}$$

Now note that

$$x[n] * h[n] = 4\delta[n+2] + 2\delta[n+1] + \delta[n] + \frac{1}{2}\delta[n-1].$$

Therefore,

$$\begin{aligned} x[n] * h[n] &= 4x[n] * \{\delta[n+2] - \alpha\delta[n+1]\} \\ &+ 2x[n] * \{\delta[n+1] - \alpha\delta[n]\} \\ &+ x[n] * \{\delta[n] - \alpha\delta[n-1]\} \\ &+ (1/2)x[n] * \{\delta[n-1] - \alpha\delta[n-2]\} \end{aligned}$$

This may be written as

$$\begin{aligned} x[n] * h[n] = x[n] &* \{4\delta[n+2] - 4\alpha\delta[n+1] + 2\delta[n+1] \\ &- 2\alpha\delta[n] + \delta[n] - \alpha\delta[n-1] \\ &+ (1/2)\delta[n-1] - (1/2)\delta[n-2] \end{aligned}$$

Therefore,

$$\begin{aligned} h[n] &= 4\delta[n+2] + (2-4\alpha)\delta[n+1] + (1-2\alpha)\delta[n] \\ &+ (1/2-\alpha)\delta[n-1] - (1/2)\delta[n-2] \end{aligned}$$

2.42. We have

$$y(t) = x(t) * h(t) = \int_{-0.5}^{0.5} e^{j\omega_0(t-\tau)} d\tau.$$

Therefore,

$$y(0) = \int_{-0.5}^{0.5} e^{-j\omega_0\tau} d\tau = \frac{2}{\omega_0}\sin(\omega_0/2).$$

(a) If $\omega_0 = 2\pi$, then $y(0) = 0$.

(b) Clearly, our answer to part (a) is not unique. Any $\omega_0 = 2k\pi$, $k \in \mathcal{I}$ and $k \neq 0$ will suffice.

2.43. **(a)** We first have

$$\begin{aligned}[x(t) * h(t)] * g(t) &= \int_{-\infty}^{\infty}\int_{-\infty}^{\infty} x(\tau)h(\sigma' - \tau)g(t - \sigma')d\tau d\sigma' \\ &= \int_{-\infty}^{\infty}\int_{-\infty}^{\infty} x(\tau)h(\sigma)g(t - \sigma - \tau)d\tau d\sigma\end{aligned}$$

Also,

$$\begin{aligned}x(t) * [h(t) * g(t)] &= \int_{-\infty}^{\infty}\int_{-\infty}^{\infty} x(t - \sigma')h(\tau)g(\sigma' - \tau)d\sigma' d\tau \\ &= \int_{-\infty}^{\infty}\int_{-\infty}^{\infty} x(\sigma)h(\tau)g(t - \tau - \sigma)d\tau d\sigma \\ &= \int_{-\infty}^{\infty}\int_{-\infty}^{\infty} x(\tau)h(\sigma)g(t - \sigma - \tau)d\tau d\sigma\end{aligned}$$

The equality is proved.

(b) (i) We first have

$$w[n] = u[n] * h_1[n] = \sum_{k=0}^{n}\left(-\frac{1}{2}\right)^k = \frac{2}{3}\left[1 - (-\frac{1}{2})^{n+1}\right]u[n].$$

Now,

$$y[n] = w[n] * h_2[n] = (n+1)u[n].$$

(ii) We first have

$$g[n] = h_1[n] * h_2[n] = \sum_{k=0}^{n}\left(-\frac{1}{2}\right)^k + \frac{1}{2}\sum_{k=0}^{n-1}(-\frac{1}{2})^k = u[n]$$

Now,

$$y[n] = u[n] * g[n] = u[n] * u[n] = (n+1)u[n].$$

The same result was obtained in both parts (i) and (ii).

(c) Note that

$$x[n] * (h_2[n] * h_1[n]) = (x[n] * h_2[n]) * h_1[n].$$

Also note that

$$x[n] * h_2[n] = \alpha^n u[n] - \alpha^n u[n-1] = \delta[n].$$

Therefore,

$$x[n] * h_1[n] * h_2[n] = \delta[n] * \sin 8n = \sin 8n.$$

2.44. **(a)** We have

$$x(t) * h(t) = \int_{-\infty}^{\infty} x(\tau)h(t-\tau)d\tau = \int_{-T_1}^{T_1} x(\tau)h(t-\tau)d\tau.$$

Note that $h(-\tau) = 0$ for $|\tau| > T_2$. Therefore, $h(t-\tau) = 0$ for $\tau > t+T_2$ and $\tau < -T_2+t$. Therefore, the above integral evaluates to zero either if $T_1 < -T_2 + t$ or $T_2 + t < -T_1$. This implies that the convolution integral is zero if $t > |T_1 + T_2|$.

(b) (i) We have

$$y[n] = h[n] * x[n] = \sum_{k=N_0}^{N_1} h[k]x[n-k].$$

Note that $x[-k] \neq 0$ for $-N_3 \leq n \leq -N_2$. Therefore, $x[-k+n] \neq 0$ for $-N_3+n \leq k \leq -N_2+n$. Clearly, the convolution sum is not zero if $-N_3+n \leq N_1$ **and** $-N_2+n \geq N_0$. Therefore, $y[n]$ is nonzero for $n \leq N_1 + N_3$ and $n \geq N_0 + N_2$.

(ii) We can easily show that $M_y = M_h + M_x - 1$.

(c) $h[n] = 0$ for $n > 5$.

(d) From the figure it is clear that

$$y(t) = h(t) * x(t) = \int_{-2}^{-1} x(t-\tau)d\tau + x(t-6).$$

Therefore,

$$y(0) = \int_{-2}^{-1} x(\tau)d\tau + x(-6).$$

This implies that $x(t)$ must be known for $1 \leq t \leq 2$ and for $t = -6$.

2.45. **(a)** (i) We have

$$\frac{x(t) - x(t-h)}{h} \overset{LTI}{\rightarrow} \frac{y(t) - y(t-h)}{h}.$$

Taking limit as $h \rightarrow 0$ on both sides of the above equation:

$$x'(t) \overset{LTI}{\rightarrow} y'(t)$$

(ii) Differentiating the convolution integral, we get

$$\begin{aligned} y'(t) &= \frac{d}{dt}\left[\int_{-\infty}^{\infty} x(t-\tau)h(\tau)d\tau\right] \\ &= \int_{-\infty}^{\infty} \frac{d}{dt}[x(t-\tau)]h(\tau)d\tau \\ &= \int_{-\infty}^{\infty} x'(t-\tau)h(\tau)d\tau \\ &= x'(t) * h(t). \end{aligned}$$

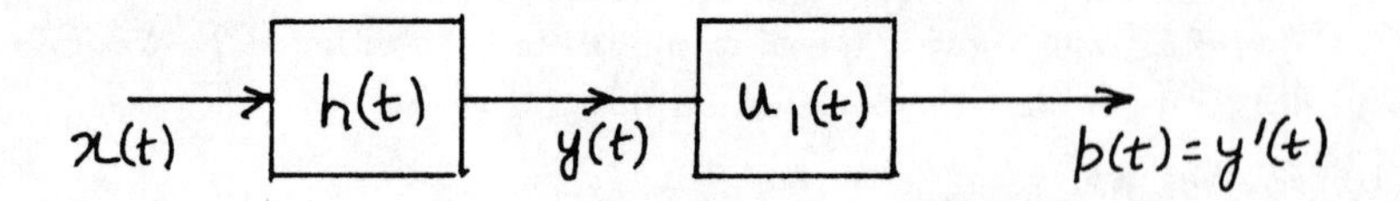

Figure S2.45

(iii) Let us name the output of the system with impulse response $u_1(t)$ as $w(t)$. Then, $w(t) = x(t) * u_1(t) = x'(t)$ and $z(t) = x'(t) * h(t)$.

Since both systems in the cascade are LTI, we may interchange their order as shown in Figure S2.45.
Then, $y(t) = x(t) * h(t)$ and $p(t) = y'(t)$. Since $z(t)$ and $p(t)$ have to be the same, we may conclude that $x'(t) * h(t) = y'(t)$.

(b) (i) We have already proved that $y'(t) = x'(t) * h(t)$. Now we may interchange $x(t)$ and $h(t)$ in the earlier proofs and they would all still hold. Therefore, we may argue that $y'(t) = x(t) * h'(t)$.

(ii) Consider

$$\begin{aligned} y(t) &= [x(t) * u(t)] * h'(t) \\ &= x(t) * [u(t) * u_1(t)] * h(t) \\ &= x(t) * h(t). \end{aligned}$$

This shows that $[x(t) * u(t)]h'(t)$ is equivalent to $x(t) * h(t)$. Now the same thing may be written as:

$$\begin{aligned} y(t) &= [x(t) * u(t)] * h'(t) \\ &= [[x(t) * u_1(t)] * h(t)] * u(t) \\ &= \int_{-\infty}^{t} x'(\tau)h(t-\tau)d\tau \\ &= x'(t) * [h(t) * u(t)] \\ &= x'(t) * \int_{-\infty}^{t} h(\tau)d\tau \end{aligned}$$

(c) Note that $x'(t) = \delta(t) - 5e^{-5t}u(t)$. Therefore, the output of the LTI system to $x'(t)$ will be $h(t) - 5\sin(\omega_0 t)$. Since this has to be equal to $y'(t) = \omega_0 \cos(\omega_0 t)$, we have

$$h(t) = \omega_0 \cos(\omega_0 t) + 5\sin(\omega_0 t).$$

(d) (i) We have

$$\begin{aligned} y(t) &= x(t) * [u_1(t) * u(t)] * h(t) \\ &= [x(t) * u_1(t)] * [u(t) * h(t)] \\ &= x'(t) * s(t) \\ &= \int_{-\infty}^{\infty} x'(\tau)s(t-\tau)d\tau \end{aligned}$$

(ii) Also,

$$\begin{aligned} x(t) &= x(t) * \delta(t) \\ &= [x(t) * u_1(t)] * u(t) \\ &= \int_{-\infty}^{\infty} x'(\tau)u(t-\tau)d\tau \end{aligned}$$

(e) In this case

$$x'(t) = e^t u(t) + \delta(t).$$

Therefore,

$$y(t) = s(t) + e^t u(t) * s(t).$$

This may be written as

$$\begin{aligned} y(t) &= [e^{-3t} - 2e^{-2t} + 1]u(t) \\ &+ [\frac{1}{4}(e^t - e^{-3t}) \\ &- \frac{2}{3}(e^t - e^{-2t}) - e^t - 1]u(t). \end{aligned}$$

(f) Using the fact that $[\delta[n] - \delta[n-1]] * u[n] = \delta[n]$ gives:

$$y[n] = [x[n] - x[n-1]] * s[n] = \sum_k [x[k] - x[k-1]]s[n-k]$$

and

$$x[n] = [x[n] - x[n-1]] * u[n] = \sum_{k=-\infty}^{\infty} [x[k] - x[k-1]]u[n-k].$$

2.46. Note that

$$\frac{dx(t)}{dt} = -6e^{-3t}u(t-1) + 2\delta(t-1) = -3x(t) + 2\delta(t-1).$$

Given that

$$x(t) = 2e^{-3t}u(t-1) \longrightarrow y(t)$$

we know that $\frac{dx(t)}{dt} = -3x(t) + 2\delta(t-1)$ must yield $-3y(t) + 2h(t-1)$ at the output. From the given information, we may conclude that $2h(t-1) = e^{-2t}u(t)$. Therefore,

$$h(t) = \frac{1}{2}e^{-2(t+1)}u(t+1).$$

2.47. **(a)** $y(t) = 2y_0(t)$.

(b) $y(t) = y_0(t) - y_0(t-2)$.

(c) $y(t) = y_0(t-1)$.

(d) Not enough information.

(e) $y(t) = y_0(-t)$.

(f) $y(t) = y_0''(t)$.

The signals for all parts of this problem are plotted in the Figure S2.47.

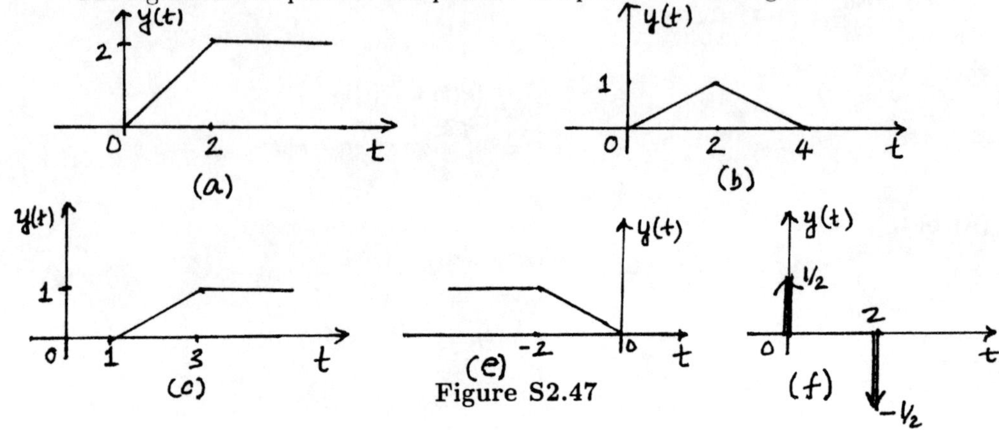

Figure S2.47

2.48. **(a)** True. If $h(t)$ periodic and nonzero, then

$$\int_{-\infty}^{\infty} |h(t)|dt = \infty.$$

Therefore, $h(t)$ is unstable.

(b) False. For example, inverse of $h[n] = \delta[n-k]$ is $g[n] = \delta[n+k]$ which is noncausal.

(c) False. For example $h[n] = u[n]$ implies that

$$\sum_{n=-\infty}^{\infty} |h[n]| = \infty.$$

This is an unstable system.

(d) True. Assuming that $h[n]$ is bounded and nonzero in the range $n_1 \le n \le n_2$,

$$\sum_{k=n_1} n_2 |h[k]| < \infty.$$

This implies that the system is stable.

(e) False. For example, $h(t) = e^t u(t)$ is causal but not stable.

(f) False. For example, the cascade of a causal system with impulse response $h_1[n] = \delta[n-1]$ and a non-causal system with impulse response $h_2[n] = \delta[n+1]$ leads to a system with overall impulse response given by $h[n] = h_1[n] * h_2[n] = \delta[n]$.

(g) False. For example, if $h(t) = e^{-t}u(t)$, then $s(t) = (1 - e^{-t})u(t)$ and

$$\int_0^{\infty} |1 - e^{-t}| dt = t + e^{-t}\Big|_0^{\infty} = \infty.$$

Although the system is stable, the step response is not absolutely integrable.

(h) True. We may write $u[n] = \sum_{k=0}^{\infty} \delta[n-k]$. Therefore,

$$s[n] = \sum_{k=0}^{\infty} h[n-k].$$

If $s[n] = 0$ for $n < 0$, then $h[n] = 0$ for $n < 0$ and the system is causal.

2.49. **(a)** It is a bounded input. $|x[n]| \le 1 = B_x$ for all n.

(b) Consider

$$\begin{aligned} y[0] &= \sum_{k=-\infty}^{\infty} x[-k]h[k] \\ &= \sum_{k=-\infty}^{\infty} \frac{h^2[k]}{|h[k]|} \\ &= \sum_{k=-\infty}^{\infty} |h[k]| \to \infty \end{aligned}$$

Therefore, the output is not bounded. Thus, the system is not stable and absolute summability is necessary.

(c) Let

$$x(t) = \begin{cases} 0, & \text{if } h(-t) = 0 \\ \frac{h(-t)}{|h(-t)|}, & \text{if } h(-t) \neq 0 \end{cases}.$$

Now, $|x(t)| \le 1$ for all t. Therefore, $x(t)$ is a bounded input Now,

$$\begin{aligned} y(0) &= \int_{-\infty}^{\infty} x(-\tau)h(\tau)d\tau \\ &= \int_{-\infty}^{\infty} \frac{h^2(\tau)}{|h(\tau)|} d\tau \\ &= \int_{-\infty}^{\infty} |h(t)| dt = \infty \end{aligned}$$

Therefore, the system is unstable if the impulse response is not absolutely integrable.

2.50. **(a)** The output will be $ax_1(t) + bx_2(t)$.

(b) The output will be $x_1(t - \tau)$.

2.51. **(a)** For the system of Figure P2.51(a) the response to an unit impulse is

$$y_1[n] = n(\frac{1}{2})^n u[n].$$

For the system of Figure P2.51(b) the response to an unit impulse is

$$y_2[n] = 0.$$

Clearly, $y_1[n] \neq y_2[n]$.

(b) For the system of Figure P2.51(a) the response to an unit impulse is

$$y[n] = (\frac{1}{2})^n u[n] + 2.$$

For the system of Figure P2.51(b) the response to an unit impulse is

$$y[n] = (\frac{1}{2})^n u[n] + 4.$$

Clearly, $y_1[n] \neq y_2[n]$.

2.52. We get

$$s[n] = h[n] * u[n] = \begin{cases} \sum_{k=0}^{n}(k+1)\alpha^k, & n \geq 0 \\ 0, & \text{otherwise} \end{cases}.$$

Noting that

$$\sum_{k=0}^{n}(k+1)\alpha^k = \frac{d}{d\alpha}\sum_{k=0}^{n+1}\alpha^k = \frac{d}{d\alpha}\left[\frac{1-\alpha^{n+2}}{1-\alpha}\right],$$

we get

$$\begin{aligned} s[n] &= \left[\frac{1-(n+2)\alpha^{n+1}}{1-\alpha} + \frac{1-\alpha^{n+2}}{(1-\alpha^2)}\right] u[n] \\ &= \left[\frac{1}{(1-\alpha)^2} - \frac{\alpha}{(1-\alpha)^2}\alpha^n + \frac{\alpha}{1-\alpha}(n+1)\alpha^n\right] u[n]. \end{aligned}$$

2.53. **(a)** Let us assume that

$$\sum_{k=0}^{N} a_k s_0^k = 0.$$

Then,

$$\sum_{k=0}^{N} a_k \frac{d^k}{dt^k}(Ae^{s_0 t}) = \sum_{k=0}^{N} A a_k e^{s_0 t} s_0^k = 0.$$

Therefore, $Ae^{s_0^t}$ is a solution of eq. (P2.53-1).

(b) Consider

$$\begin{aligned}
\sum_{k=0}^{N} a_k \frac{d^k}{dt^k}(Ate^{st}) &= \sum_{k=0}^{N} Aa_k ts^k e^{st} + \sum_{k=0}^{N} Aa_k k e^{st} s^{k-1} \\
&= Ate^{st}\sum_{k=0}^{N} a_k s^k + Ae^{st}\sum_{k=0}^{N} a_k \frac{d}{ds}(s^k) \\
&= Ate^{st}\sum_{k=0}^{N} a_k s^k + Ae^{st}\frac{d}{ds}\sum_{k=0}^{N} a_k s^k.
\end{aligned}$$

If s_i is a solution, then $\sum_{k=0}^{N} a_k s_i^k = 0$. This implies that $te^{s_i t}$ is a solution.

(c) (i) Here,

$$s^2 + 3s + 2 = 0, \quad \Rightarrow \quad s = -2, s = -1.$$

Therefore,

$$y_h(t) = Ae^{-2t} + Be^{-t}.$$

Since $y_h(0) = 0$, $y'_h(0) = 2$, $A + B = 0$ and $2A + B = 2$. Therefore, $A = -2$, $B = 2$.

$$y(t) = 2e^{-t} - 2e^{-2t}.$$

(ii) Here,

$$s^2 + 3s + 2 = 0 \quad \Rightarrow \quad y(t) = Ae^{-2t} + Be^{-t}.$$

Since $y(0) = 1$, $y'(0) = -1$, we have $y(t) = e^{-t}$.

(iii) $y(t) = 0$ because of initial rest condition.

(iv) Here,

$$s^2 + 2s + 1 = 0 = (s+1)^2 \quad \Rightarrow \quad s = -1, \sigma = 2.$$

and

$$y(t) = Ae^{-t} + Bte^{-t}.$$

Since $y(0) = 1$, $y'(0) = 1$, $A = 1$, $B = 2$. Therefore,

$$y(t) = e^{-t} + 2te^{-t}.$$

(v) Here,

$$s^3 + s^2 - s - 1 = 0 = (s-1)(s+1)^2 \quad \Rightarrow \quad y(t) = Ae^t + Be^{-t} + Cte^{-t}.$$

Since $y(0) = 1$, $y'(0) = 1$, and $y''(0) = -2$, we get $A = 1/2$, $B = 3/4$, $C = 3/2$. Therefore,

$$y(t) = \frac{1}{2}e^t + \frac{3}{4}e^{-t} + \frac{3}{2}te^{-t}.$$

(vi) Here, $s = -1 \pm 2j$ and

$$y(t) = Ae^{-t}e^{2jt} + Be^{-t}e^{-2jt}.$$

Since $y(0) = 1$, $y'(0) = 1$,

$$A = \frac{1}{2}(1-j) = B^*.$$

Therefore,

$$y(t) = e^{-t}[\cos 2t + \sin 2t].$$

2.54. **(a)** Let us assume that

$$\sum_{k=0}^{N} a_k z_0^k = 0.$$

Then, if $y[n] = Az_0^n$,

$$\sum_{k=0}^{N} a_k y[n-k] = \sum_{k=0}^{N} a_k (Az_0^{n-k}) = Az_0^n \sum_{k=0}^{N} a_k z_0^{-k} = 0.$$

Therefore, Az_0^n is a solution of eq. (P2.54-1).

(b) If $y[n] = nz^{n-1}$, then

$$\sum_{k=0}^{N} a_k y[n-k] = \sum_{k=0}^{N} a_k (n-k) z^{n-k-1}. \qquad \text{(S2.54–1)}$$

Taking the right-hand side of the equation that we want to prove,

$$\begin{aligned} R.H.S &= z^{n-N} \sum_{k=0}^{N} a_k (N-k) z^{N-k-1} + (n-N) \sum_{k=0}^{N} a_k \\ &= \sum_{k=0}^{N} a_k (n-k) z^{n-k-1} \qquad \text{(S2.54–2)} \end{aligned}$$

Comparing eqs. (S2.54-1) and (S2.54-2), we conclude that the equation is proved.

(c) (i) Here,

$$1 + \frac{3}{4}z^{-1} + \frac{1}{8}z^{-2} = 0 \quad \Rightarrow \quad z = -\frac{1}{2},\ z = -\frac{1}{4}.$$

Therefore,

$$y[n] = A(-\frac{1}{2})^n + B(-\frac{1}{4})^n.$$

Since $y[0] = 1$, $y[-1] = -6$, we get $A = -1$, $B = 2$, and

$$y[n] = 2(-\frac{1}{4})^n - (-\frac{1}{2})^n.$$

(ii) Here,

$$z^2 - 2z + 1 = 0.$$

Therefore,

$$y[n] = A(1)^n + Bn(1)^n = A + Bn.$$

Since, $y[0] = 1$, $y[1] = 0$ we get $A = 1$, $B = -1$, and

$$y[n] = 1 - n.$$

(iii) Only difference from previous part is initial conditions. Since $y[0] = 1$, $y[10] = 21$, we get $A = 1$, $B = 2$, and

$$y[n] = 1 + 2n.$$

(iv) Here,

$$z = \frac{1}{2\sqrt{2}}(1 \pm j).$$

Therefore,

$$y[n] = A[\frac{1}{2\sqrt{2}}(1+j)]^n + B[\frac{1}{2\sqrt{2}}(1-j)]^n.$$

Since $y[0] = 0$, $y[-1] = 1$, we get $A = \frac{j}{2\sqrt{2}}$, $B = \frac{-j}{2\sqrt{2}}$, and

$$y[n] = -\frac{1}{\sqrt{2}}(\frac{1}{2})^n \sin(n\pi/4).$$

2.55. **(a)** $y[0] = x[0] = 1$. $h[n]$ satisfies the equation

$$h[n] = \frac{1}{2}h[n-1], \qquad n \geq 1.$$

The auxiliary condition is $h[0] = 1$. Using the method introduced in the previous problem, we have $z = 1/2$. Therefore, $h[n] = A(1/2)^n$. Using the auxiliary condition,

$$h[n] = \left(\frac{1}{2}\right)^n u[n].$$

(b) From Figure P2.55(b), we know that if $x[n] = \delta[n]$, then

$$w[n] = h_a[n] = (\frac{1}{2})^n u[n].$$

This implies that

$$y[n] = h[n] = \left(\frac{1}{2}\right)^n u[n] + 2(\frac{1}{2})^{n-1}u[n-1].$$

(c) Plugging eq. (P2.55-3) into eq. (P2.55-1) gives:

$$\begin{aligned}\sum_m h[n-m]x[m] - \frac{1}{2}\sum_m h[n-m-1]x[m] &= \sum_{m=-\infty}^{n}(\frac{1}{2})^{n-m}x[m] - \sum_{m=-\infty}^{n-1}(\frac{1}{2})^{n-m}x[m] \\ &= (\frac{1}{2})^{n-n}x[n] \\ &= x[n].\end{aligned}$$

This implies that eq. (P2.55-3) satisfies eq. P(2.55-1).

(d) (i) Given that $a_0 \neq 0$ and that the system obeys initial rest, we get

$$a_0 y[0] = 1 \quad \Rightarrow \quad y[0] = \frac{1}{a_0}.$$

The homogeneous equation is

$$\sum_{k=0}^{N} a_k h[n-k] = 0$$

with the initial conditions

$$h[0] = 1/a_0, \qquad h[-1] = \cdots = h[-N+1] = 0.$$

(ii) We have

$$h[n] = \sum_{k=0}^{M} b_k h_1[n-k] = 0,$$

where $h_1[n]$ is as above.

(e) For $n > M$,

$$\sum_{k=0}^{N} a_k h[n-k] = 0$$

with

$$h[0] = y[0], \cdots, h[M] = y[M].$$

(f) (i) We get

$$h[n] = \begin{cases} 1, & n \text{ even, } n \geq 0 \\ 0, & n \text{ odd or } n < 0 \end{cases}.$$

(ii) We get

$$h[n] = \begin{cases} 1, & n \text{ even and } n \geq 0 \\ 2, & n \text{ odd and } n > 0 \\ 0, & n < 0 \end{cases}.$$

(iii) We get

$$h[n] = \begin{cases} 2, & n = 0, 2 \\ -1, & n \text{ even } n \geq 4 \\ 0, & \text{else} \end{cases}.$$

(iv) We get

$$h[n] = \frac{1}{2}[\cos\frac{\pi n}{6} + \sqrt{3}\sin\frac{\pi n}{6}].$$

2.56. **(a)** In this case, $s+2=0$ which implies that

$$y(t) = h(t) = Ae^{-2t}.$$

Since $y(0+) = 1$, $A = 1$ and

$$h(t) = e^{-2t}u(t).$$

Now consider eq. (P2.56-1).

$$\begin{aligned} L.H.S. &= \frac{d}{dt}\int_{-\infty}^{\infty} h(t-\tau)x(\tau)d\tau + 2\int_{-\infty}^{\infty} h(t-\tau)x(\tau)d\tau \\ &= \int_{-\infty}^{\infty} e^{-2(t-\tau)}\delta(t-\tau)x(\tau)d\tau \\ &= x(t) = R.H.S. \end{aligned}$$

This implies that $y(t)$ does solve the differential equation.

(b) Take

$$y(t) = \sum_l \alpha_l u_l(t).$$

Then

$$\sum_{k=0}^{N} a_k \sum_l \alpha_l u_{k+l}(t) = \delta(t).$$

Integrating between $t = 0^-$ and $t = 0^+$ and matching coefficients, we get $\alpha_l = 0$ except $\alpha_{-N} = 1/a_N$. This implies that for $0^- \leq t \leq 0^+$

$$y(t) = \frac{1}{a_N}u_{-N}(t)$$

and

$$y(0^+) = y'(0^+) = \cdots = y^{N-2'}(0^+) = 0$$

and

$$\left.\frac{d^{N-1}y(t)}{dt^{N-1}}\right|_{0+} = \frac{1}{a_N}.$$

(c) The impulse response is

$$h(t) = \sum_{k=0}^{M} b_k \frac{d^k h_b(t)}{dt^k}.$$

(d) (i) Taking

$$y(t) = \sum_r \alpha u_r(t)$$

we get

$$\sum_r [\alpha_r u_{r+2}(t) + 3\alpha_r u_{r+1}(t) + 2\alpha_r u_r] = \delta(t)$$

This implies that $r_{max} = -2$ and $\alpha_{-2} = 1$. Therefore, $h(0+) = 0$ and $h'(0^+) = 1$ constitute the initial conditions. Now,

$$s^3 + 3s + 2 = 0 \qquad \Rightarrow \qquad s = -2, s = -1.$$

Therefore,

$$h(t) = Ae^{-2}t + Be^{-t}, \; t \geq 0.$$

Applying initial conditions, we get $A = -1$, $B = 1$. Therefore,

$$h(t) = (e^{-t} - e^{-2t})u_{-1}(t).$$

(ii) The initial conditions are $h(0^+) = 0$ and $h'(0^+)1$, Also, $s = -1 \pm j$. Therefore

$$h(t) = [e^{-t} \sin t]u_{-1}(t).$$

(e) From part (c), if $M \geq N$, then $\sum_{k=0}^{M} b_k \frac{d^k h_b(t)}{dt^k}$ will contain singularity terms at $t = 0$ This implies that

$$h(t) = \sum_r \alpha_r u_r(t) + \cdots .$$

(f) (i) Now,

$$\sum_r \alpha_r u_{r+1}(t) + 2\sum \alpha_r u_r = 3u_1(t) + u_0(t).$$

Therefore, $r_{max} = 0$. Also

$$\alpha_0 u_1(t) + \alpha_{-1} u_0(t) + 2\alpha_0 u_0(t) = 3u_1(t) + u_0(t).$$

This gives $\alpha_0 = 3$ and $\alpha_{-1} = -5$. The initial condition is $h(0^+) = -5$ and

$$h(t) = 3u_0(t) - 5e^{-2t}u_{-1}(t) = 3\delta(t) - 5e^{-2t}u(t).$$

(ii) Here, $\alpha_1 = 1$, $\alpha_0 = -3$, $\alpha_{-1} = 13$, $\alpha_{-2} = -44$. Therefore $h(0^+) = 13$ and $h'(0^+) = -44$ and

$$h(t) = u_1(t) - 3u_0(t) + 18e^{-3t}u_{-1}(t) - 5e^{-2t}u_{-1}(t).$$

2.57. **(a)** Realizing that $x_2[n] = y_1[n]$, we may eliminate these from the two given difference equations. This would give us

$$y_2[n] = -ay_2[n-1] + b_0 x_1[n] + b_1 x_1[n-1].$$

This is the same as the overall difference equation.

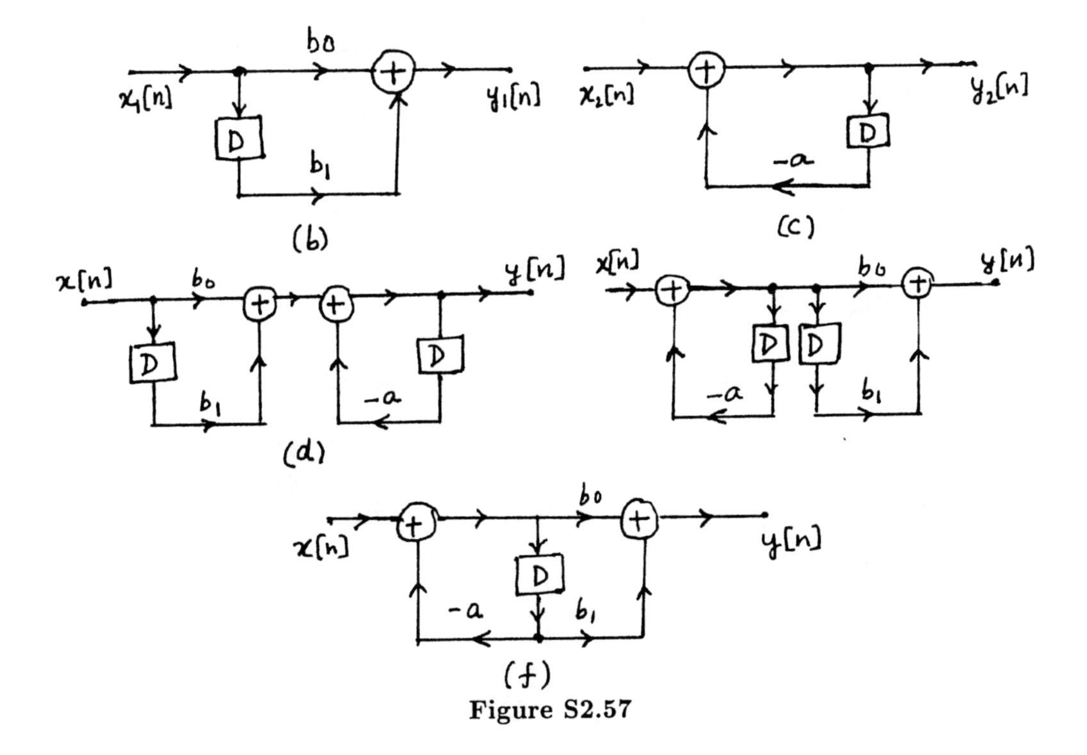

Figure S2.57

(b) The figures corresponding to the remaining parts of this problem are shown in the Figure S2.57.

2.58. **(a)** Realizing that $x_2[n] = y_1[n]$, we may eliminate these from the two given difference equations. This would give us

$$2y_2[n] - y_2[n-1] + y_2[n-3] = x_1[n] - 5x_1[n-4].$$

This is the same as the overall difference equation.

(b) The figures corresponding to the remaining parts of this problem are shown in Figure S2.58.

2.59. **(a)** Integrating the given differential equation once and simplifying, we get

$$y(t) = -\frac{a_0}{a_1}\int_{-\infty}^{t} y(\tau)d\tau + \frac{b_0}{a_1}\int_{-\infty}^{t} x(\tau)d\tau + \frac{b_1}{a_1}x(t).$$

Therefore, $A = -a_0/a_1$, $B = b_1/a_1$, $C = b_0/a_1$.

(b) Realizing that $x_2(t) = y_1(t)$, we may eliminate these from the two given integral equations. This would give us

$$y_2(t) = A\int_{-\infty}^{t} y_2(\tau)d\tau + B\int_{-\infty}^{t} x_1(\tau)d\tau + Cx_1(t).$$

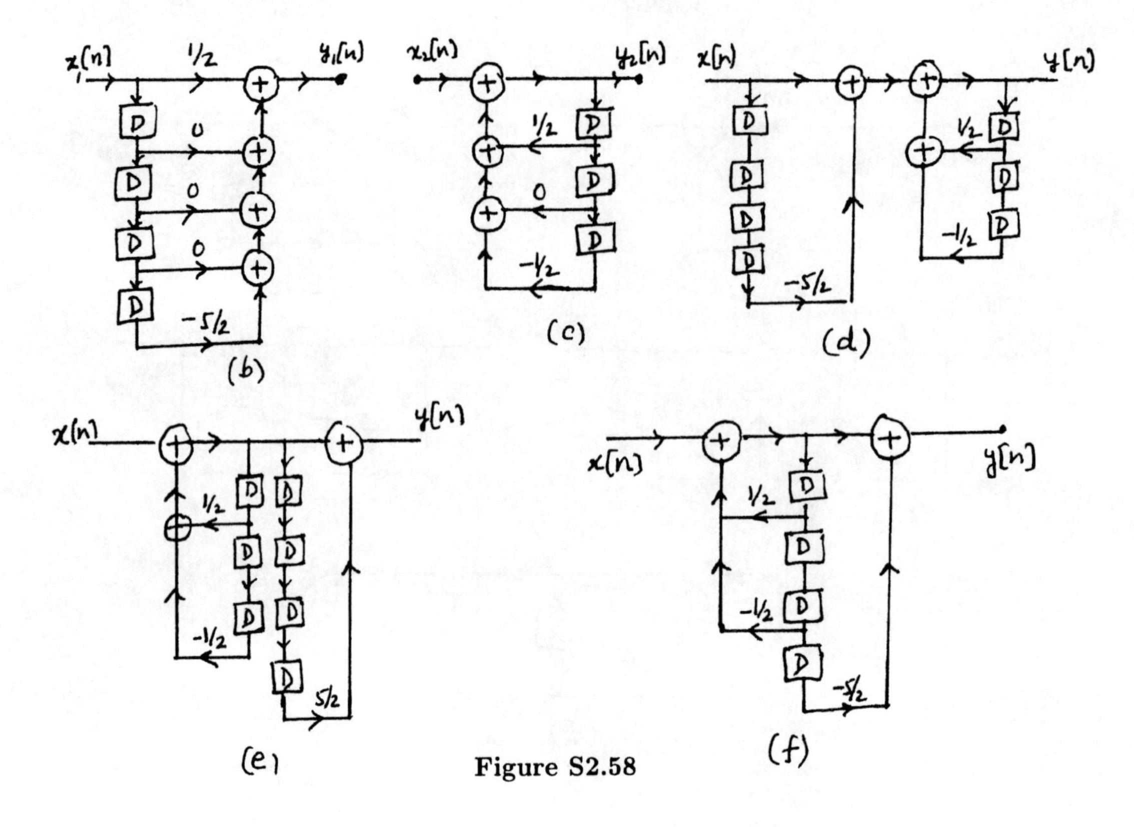

Figure S2.58

(c) The figures corresponding to the remaining parts of this problem are shown in Figure S2.59.

2.60. **(a)** Integrating the given differential equation once and simplifying, we get

$$\begin{aligned} y(t) &= -\frac{-a_1}{a_2}\int_{-\infty}^{t} y(\tau)d\tau - \frac{a_0}{a_2}\int_{-\infty}^{t}\int_{-\infty}^{\tau} y(\sigma)d\sigma d\tau \\ &+ \frac{b_0}{a_2}\int_{-\infty}^{t}\int_{-\infty}^{\tau} x(\sigma)d\sigma d\tau + \frac{b_1}{a_2}\int_{-\infty}^{t} x(\tau)d\tau + \frac{b_2}{a_1}x(t). \end{aligned}$$

Therefore, $A = -a_1/a_2$, $B = -a_0/a_2$, $C = b_2/a_1$, $D = b_1/a_2$, $E = b_0/a_2$.

(b) Realizing that $x_2(t) = y_1(t)$, we may eliminate these from the two given integral equations.

(c) The figures corresponding to the remaining parts of this problem are shown in Figure S2.60.

2.61. **(a)** (i) From Kirchoff's voltage law, we know that the input voltage must equal the sum of the voltages across the inductor and capacitor. Therefore,

$$x(t) = LC\frac{d^2y(t)}{dt^2} + y(t).$$

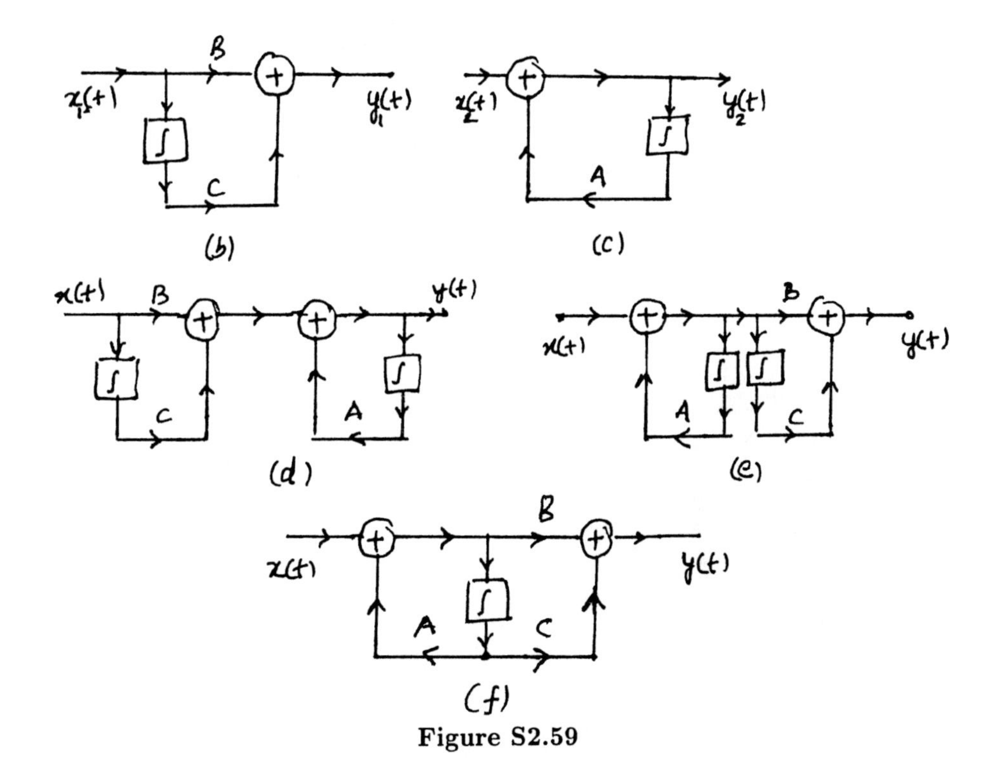

Figure S2.59

Using the values of L and C we get

$$\frac{d^2y(t)}{dt^2} + y(t) = x(t).$$

(ii) Using the results of Problem 2.53, we know that the homogeneous solution of the differential equation

$$\frac{d^2y(t)}{dt^2} + a_1\frac{dy(t)}{dt} + a_2y(t) = bx(t).$$

will have terms of the form $K_1e^{s_0t} + K_2e^{s_1t}$ where s_0 and s_1 are roots of the equation

$$s^2 + a_1s + a_2 = 0.$$

(It is assumed here that $s_0 \neq s_1$.) In this problem, $a_1 = 0$ and $a_2 = 1$. Therefore, the root of the equation are $s_0 = j$ and $s_1 = -j$.. The homogeneous solution is

$$y_h(t) = K_1e^{jt} + K_2e^{-jt}.$$

And, $\omega_1 = 1 = \omega_2$.

(iii) If the voltage and current are restricted to be real, then $K_1 = K_2 = K$. Therefore,

$$y_h(t) = 2K\cos(t) = 2K\sin(t + \pi/2).$$

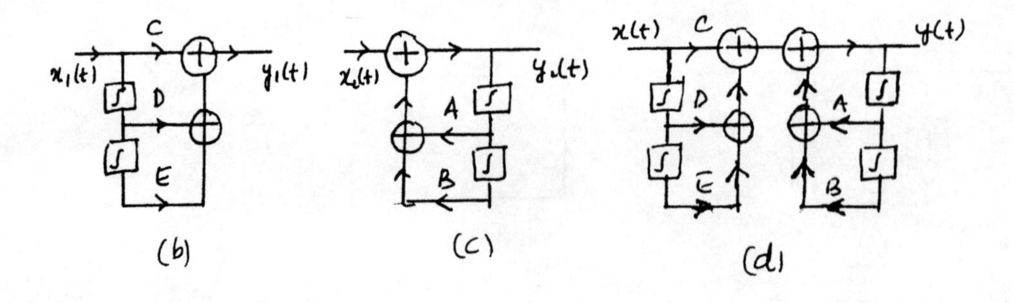

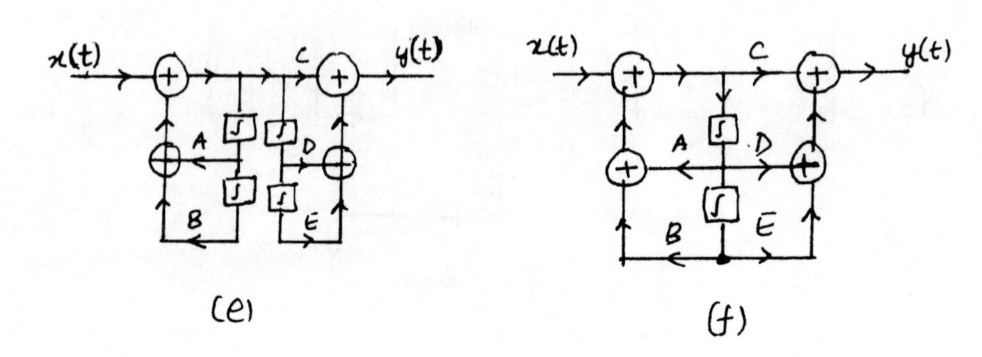

Figure S2.60

(b) (i) From Kirchoff's voltage law, we know that the input voltage must equal the sum of the voltages across the resistor and capacitor. Therefore,

$$x(t) = RC\frac{dy(t)}{dt} + y(t).$$

Using the values of R, L, and C we get

$$\frac{dy(t)}{dt} + y(t) = x(t).$$

(ii) The natural response of the system is the homogeneous solution of the above differential equation. Using the results of Problem 2.53, we know that the homogeneous solution of the differential equation

$$\frac{dy(t)}{dt} + a_1 y(t) = bx(t).$$

will have terms of the form $Ae^{s_0 t}$ where s_0 is the root of the equation

$$s + a_1 = 0.$$

In this problem, $a_1 = 1$. Therefore, the root of the equation are $s_0 = -1$. The homogeneous solution is

$$y_h(t) = Ke^{-t}.$$

And, $a = 1$.

(c) (i) From Kirchoff's voltage law, we know that the input voltage must equal the sum of the voltages across the resistor, inductor, and capacitor. Therefore,

$$x(t) = LC\frac{d^2y(t)}{dt^2} + RC\frac{dy(t)}{dt} + y(t).$$

Using the values of R, L, and C we get

$$\frac{d^2y(t)}{dt^2} + 2\frac{dy(t)}{dt} + 5y(t) = 5x(t).$$

(ii) Using the results of Problem 2.53, we know that the homogeneous solution of the differential equation

$$\frac{d^2y(t)}{dt^2} + a_1\frac{dy(t)}{dt} + a_2y(t) = bx(t).$$

will have terms of the form $K_1e^{s_0t} + K_2e^{s_1t}$ where s_0 and s_1 are roots of the equation

$$s^2 + a_1s + a_2 = 0.$$

(It is assumed here that $s_0 \neq s_1$.) In this problem, $a_1 = 2$ and $a_2 = 5$. Therefore, the root of the equation are $s_0 = -1 + 2j$ and $s_1 = -1 - 2j$. The homogeneous solution is

$$y_h(t) = K_1e^{-t}e^{2jt} + K_2e^{-t}e^{-2jt}.$$

And, $a = 1$.

(iii) If the voltage and current are restricted to be real, then $K_1 = K_2 = K$. Therefore,

$$y_h(t) = 2Ke^{-t}\cos(2t) = 2Ke^{-t}\sin(2t + \pi/2).$$

2.62. (a) The force $x(t)$ must equal the sum of the force required to displace the mass and the force required to stretch the spring. Therefore,

$$x(t) = m\frac{d^2y(t)}{dt^2} + Ky(t) = x(t).$$

Substituting the values of m and K, we get

$$\frac{d^2y(t)}{dt^2} + 4y(t) = 2x(t).$$

Using the results of Problem 2.53, we know that the homogeneous solution of the differential equation

$$\frac{d^2y(t)}{dt^2} + a_1\frac{dy(t)}{dt} + a_2y(t) = bx(t).$$

will have terms of the form $K_1e^{sot} + K_2e^{s_1t}$ where s_0 and s_1 are roots of the equation

$$s^2 + a_1s + a_2 = 0.$$

(It is assumed here that $s_0 \neq s_1$.) In this problem, $a_1 = 0$ and $a_2 = 4$. Therefore, the root of the equation are $s_0 = +2j$ and $s_1 = -2j$. The homogeneous solution is

$$y_h(t) = K_1 e^{2jt} + K_2 e^{-2jt}.$$

Assuming that $y(t)$ is real, we have $K_1 = K_2 = K$. Therefore,

$$y_h(t) = 2K\cos(2t).$$

Clearly, $y_h(t)$ is periodic.

(b) The force $x(t)$ must equal the sum of the force required to displace the mass and the force required to stretch the spring. Therefore,

$$x(t) = m\frac{dy(t)}{dt} + by(t).$$

Substituting the values of m and b, we get

$$\frac{dy(t)}{dt} + \frac{y(t)}{10000} = \frac{x(t)}{1000}.$$

Using the results of Problem 2.53, we know that the homogeneous solution of the differential equation

$$\frac{dy(t)}{dt} + a_1 y(t) = bx(t).$$

will have terms of the form $Ae^{s_0 t}$ where s_0 is the root of the equation

$$s + a_1 = 0.$$

In this problem, $a_1 = 1/10000$. Therefore, the root of the equation are $s_0 = -10^{-4}$. The homogeneous solution is

$$y_h(t) = Ke^{-10^{-4}t}.$$

Clearly, $y_h(t)$ decreases with increasing t.

(c) (i) We know that the input force $x(t)$ = (Force required to displace mass by $y(t)$) + (Force required to displace dashpot by $y(t)$) + (Force required to displace spring by $y(t)$). Therefore,

$$x(t) = m\frac{d^2y(t)}{dt^2} + b\frac{dy(t)}{dt} + Ky(t).$$

Using the values of m, b, and K we get

$$\frac{d^2y(t)}{dt^2} + 2\frac{dy(t)}{dt} + 2y(t) = x(t).$$

(ii) Using the results of Problem 2.53, we know that the homogeneous solution of the differential equation

$$\frac{d^2y(t)}{dt^2} + a_1\frac{dy(t)}{dt} + a_2y(t) = b_1x(t).$$

will have terms of the form $K_1e^{s_0t}+K_2e^{s_1t}$ where s_0 and s_1 are roots of the equation

$$s^2 + a_1s + a_2 = 0.$$

(It is assumed here that $s_0 \neq s_1$.) In this problem, $a_1 = 2$ and $a_2 = 2$. Therefore, the root of the equation are $s_0 = -1 + j$ and $s_1 = -1 - j$. The homogeneous solution is

$$y_h(t) = K_1e^{-t}e^{jt} + K_2e^{-t}e^{-jt}.$$

And, $a = 1$.

(iii) If the force is restricted to be real, then $K_1 = K_2 = K$. Therefore,

$$y_h(t) = 2Ke^{-t}\cos(t) = 2Ke^{-t}\sin(t + \pi/2).$$

2.63. **(a)** We have

$$\begin{aligned} y[n] &= \text{Amt. borrowed} - \text{Amt. paid} + \text{Compounded Amt from prev. month} \\ &= 100,000\delta[n] + 1.01y[n-1] - Du[n-1]. \end{aligned}$$

Therefore,

$$y[n] = 1.01y[n-1] - D, \qquad n > 0$$

and $y[0] = 100,000$ and $\gamma = 1.01$.

(b) We have

$$y_p[n] = 1.01y_p[n-1] - D.$$

This implies that $y_p[n] = 100D$. Also the homogeneous solution is of the form

$$y_h[n] = A(1.01)^n.$$

Therefore,

$$y[n] = y_h[n] + y_p[n] = A(1.01)^n + 100D$$

Using the initial condition $y[0] = 100000$, we have

$$A = 100000 - 100D.$$

Therefore,

$$y[n] = (100000 - 100D)(1.01)^n + 100D.$$

(c) We have

$$y[360] = 0 = (P - 100D)(1.01)^{360} + 100D.$$

Therefore, $D = \$1028.60$.

(d) Total payment = \$370, 296.

(e) The toughest question in this book!!

2.64. **(a)** We have $y(t) = x(t) * h(t)$ and $x(t) = y(t) * g(t)$. Therefore, $g(t) * h(t) = \delta(t)$. Now,

$$h(t) * g(t)|_{t=nT} = \sum_{k=0}^{n} h_k g_{n-k} \delta(t - nT).$$

Therefore we want

$$\sum_{k=0}^{n} h_k g_{n-k} = \begin{cases} 1, & n = 0 \\ 0, & n = 1, 2, 3, \cdots \end{cases}.$$

Therefore,

$$g_0 = \frac{1}{h_0}, \quad g_1 = -\frac{h_1}{h_0^2}, \quad g_2 = -\frac{1}{h_0}\left[\frac{-h_1^2}{h_0^2} + \frac{h_2}{h_0}\right], \cdots.$$

(b) In this case, $g_0 = 1$, $g_1 = -1/2$, $g_2 = (-1/2)^2$, $g_3 = (-1/2)^3$, and so on. This implies that

$$g(t) = \delta(t) + \sum_{k=1}^{\infty} \left(-\frac{1}{2}\right)^k \delta(t - kT).$$

(c) (i) Here, $h(t) = \sum_{k=0}^{\infty} \alpha^k \delta(t - T)$.

(ii) If $0 < \alpha < 1$, then $\alpha^k < 1$. Therefore, $h(t)$ is bounded and absolutely integrable and corresponds to a stable system. If $\alpha > 1$, then $h(t)$ is not absolutely integrable making the system unstable.

(iii) Here, $g(t) = 1 - \alpha\delta(t - T)$. The inverse system is as shown in the figure below.

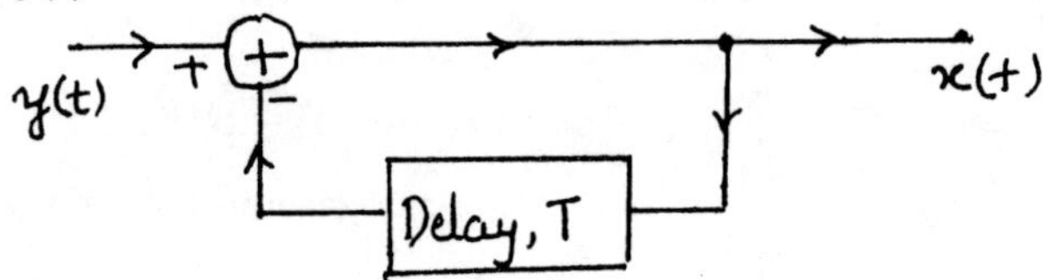

Figure S2.64

(d) If $x_1[n] = \delta[n]$, $y[n] = h[n]$. If $x_2[n] = \frac{1}{2}\delta[n] + \frac{1}{2}\delta[n - N]$, $y[n] = h[n]$.

2.65. **(a)** The autocorrelation sequences are as shown in Figure S2.65.

(b) The autocorrelation sequences are as shown in Figure S2.65.

(c) We get

$$\phi_{xy}[n] = \sum_{k=-\infty}^{\infty} h[-k]\phi_{xx}[n - k].$$

Therefore, $\phi_{xy}[n]$ may be viewed as

$$\phi_{xx}[n] \rightarrow \boxed{h[-n]} \rightarrow \phi_{xy}[n].$$

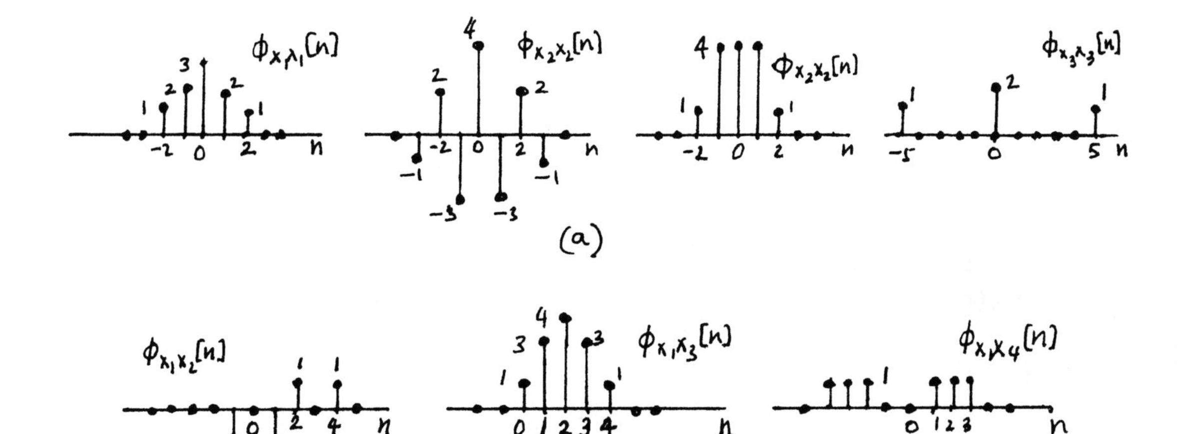

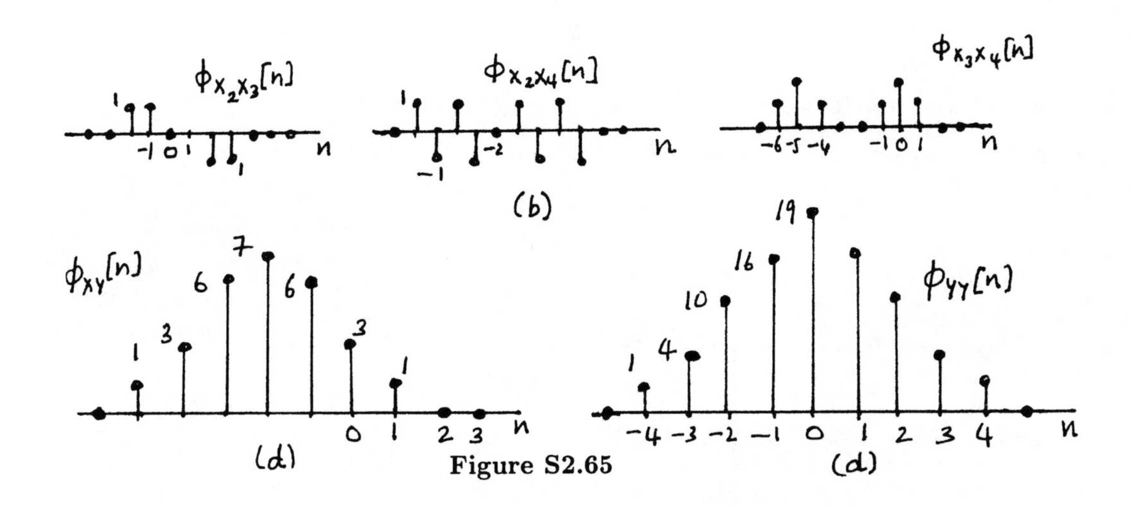

Figure S2.65

Also,

$$\phi_{yy}[n] = \sum_k \phi_{xx}[n-k]\phi_{hh}[k].$$

Therefore,$\phi_{yy}[n]$ may be viewed as

$$\phi_{xx}[n] \rightarrow \boxed{h[n] * h[-n]} \rightarrow \phi_{yy}[n].$$

(d) $\phi_{xy}[n]$ and ϕ_{yy} are as shown in Figure S2.65.

2.66. **(a)** The plot of $x_1(t)$ is as shown in Figure S2.66.

(b) The plots of $x_2(t)$ and $x_2(t)$ are as shown in Figure S2.66.

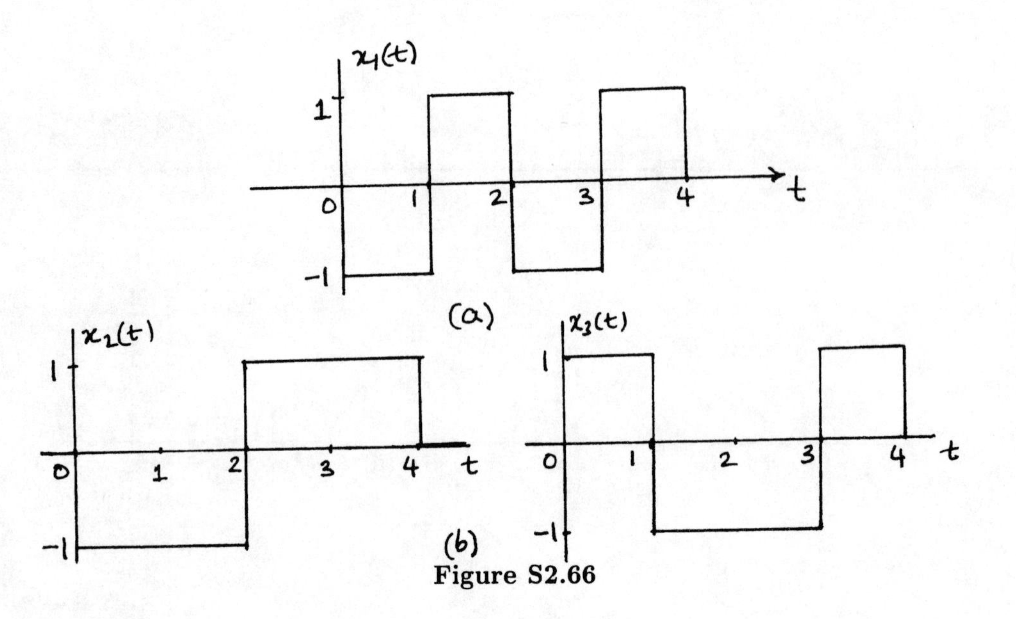

Figure S2.66

(c) $x_1(t) * h_2(t) = x_2(t) * h_3(t) = x_1(t) * h_3(t) = 0$ for $t = 4$.

2.67. (a) The autocorrelation functions are:

$$\phi_{x_1x_1}(t) = \begin{cases} \frac{1}{24}t^3 - \frac{1}{2}t + \frac{2}{3}, & 0 \le t \le 2 \\ 0, & t > 2 \end{cases} \quad \text{and} \quad \phi_{x_1x_1}(t) = \phi_{x_1x_1}(-t).$$

and

$$\phi_{x_2x_2}(t) = \begin{cases} 7(1-t), & 0 \le t \le 1 \\ 1-t, & 1 \le t \le 2 \\ t-3, & 2 \le t \le 3 \\ 3-t, & 3 \le t \le 4 \\ t-5, & 4 \le t \le 5 \\ 5-t, & 5 \le t \le 6 \\ t-7, & 6 \le t \le 7 \\ 0, & t > 7 \end{cases} \quad \text{and} \quad \phi_{x_2x_2}(t) = \phi_{x_2x_2}(-t).$$

(b) If the impulse response is $h(t) = x(T-t)$, then $y(t) = \phi_{xx}(t-T)$.

(c) We have

$$\begin{aligned} y(T) &= \int_0^T x(\tau)h(T-\tau)d\tau \\ &\le M^{1/2}\left[\int_0^T x^2(t)dt\right]^{1/2}. \end{aligned}$$

Therefore, $y(t)$ is at most $M^{1/2}\left[\int_0^T x^2(t)dt\right]^{1/2}$.

If we now choose

$$h(t) = \sqrt{\frac{M}{\int_0^T x^2(t)dt}} x(T-t).$$

then

$$y(T) = M^{1/2}[\int_0^T x^2(t)dt]^{1/2}.$$

Clearly, $y(T)$ is maximized for the above choice of $h(t)$.

(d) (i) The responses are as sketched in Figure S2.67.

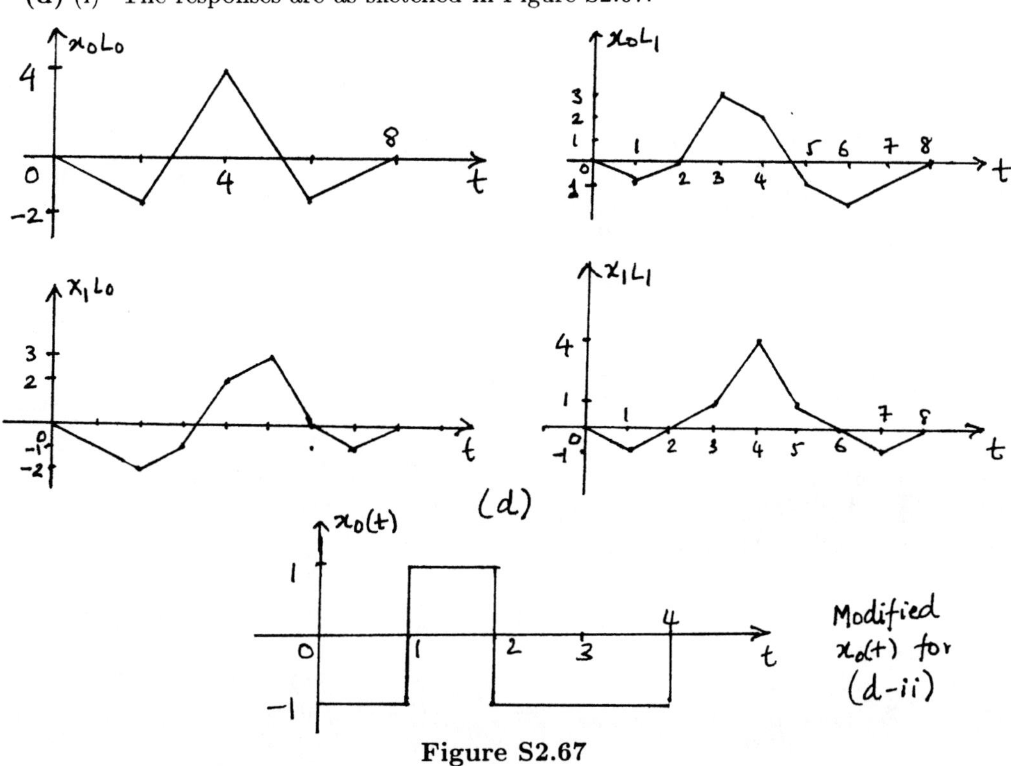

Figure S2.67

(ii) Let the impulse responses of L_0 and L_1 be $h_{L_0}(t)$ and $h_{L_1}(t)$. Then,

$$\begin{aligned} x_0(t) * h_{L_0}(t)|_{t=4} &= 4 \\ x_0(t) * h_{L_1}(t)|_{t=4} &= 2 \\ x_1(t) * h_{L_0}(t)|_{t=4} &= 2 \\ x_1(t) * h_{L_1}(t)|_{t=4} &= 4 \end{aligned}$$

To make the job of the receiver easier, modify $x_0(t)$ as shown in the figure below.

2.68. We have

$$\begin{aligned}\phi_{pp}(\tau) &= \int p(\tau)p(t+\tau)dt \\ &\leq [\int p^2(\tau)d\tau]^{1/2}[\int p^2(t+\tau)d\tau]^{1/2} \\ &\leq \int p^2(\tau)d\tau\end{aligned}$$

Therefore,

$$\phi_{pp}(\tau) \leq \phi_{pp}(0) \quad \Rightarrow \quad \phi_{pp}(0) = \max_t \phi_{pp}(t).$$

Also,

$$\phi_{xp}(t) = \phi_{pp}(t - t_0) \quad \Rightarrow \quad \phi_{xp}(t_0) = \phi_{pp}(0) = \max_t \phi_{xp}(t).$$

2.69. **(a)** Let $g(\tau) = x(t-\tau)$. Then

$$\int_{-\infty}^{\infty} g(\tau)u_1(\tau)d\tau = -g^{'}(0) = -x^{'}(t).$$

(b) Consider $r(t) = g(t)f(t)$. Then,

$$\int_{-\infty}^{\infty} r(t)u_1(t)dt = -r'(0) = -g'(0)f(0) - g(0)f'(0).$$

Also,

$$\int_{-\infty}^{\infty} g(t)f(0)u_1(t)dt - \int_{-\infty}^{\infty} g(t)f'(0)u_0(t)dt = -g'(0)f(0) - g(0)f'(0)$$

which is the same as above.

(c) $\int_{\infty}^{\infty} g(\tau)u_2(\tau)d\tau = g''(0).$

(d) We have

$$\begin{aligned}\int_{-\infty} g(\tau)f(\tau)u_2(\tau)d\tau &= \frac{d^2}{dt^2}[g(-t)f(-t)]|_{t=0} \\ &= -\frac{d}{dt}[g'(-t)f(-t) + g(-t)f'(-t)]_{t=0} \\ &= g''(0)f(0) - 2g'(0)f'(0) + g(0)f''(0)\end{aligned}$$

Therefore,

$$f(t)u_2(t) = f(0)u_2(t) - 2f'(0)u_1(t) + f''(0)u_0(t).$$

2.70. **(a)** We have

$$\begin{aligned}\sum_{m=-\infty}^{\infty} x[m]u_1[m] &= \sum_m x[m]\{\delta[m] - \delta[m-1]\} \\ &= x[0] - x[1].\end{aligned}$$

(b) We have

$$\begin{aligned} x[n]u_1[n] &= x[0]\delta[n] - x[1]\delta[n-1] + [x[0]\delta[n-1] - x[0]\delta[n-1]] \\ &= x[0]u_1[n] - \{x[1] - x[0]\}\delta[n-1] \\ &= x[0]\delta[n] - x[1]\delta[n-1] + x[1]\delta[n] - x[1]\delta[n] \\ &= x[1]u_1[n] - \{x[1] - x[0]\}\delta[n] \end{aligned}$$

(c) We have

$$u_2[n] = u_1[n] * u_1[n] = \delta[n] - 2\delta[n-1] + \delta[n-2]$$

and

$$u_3[n] = \delta[n] - 3\delta[n-1] + 3\delta[n-2] - \delta[n-3].$$

The plots for these signals are as shown in Figure S2.70.

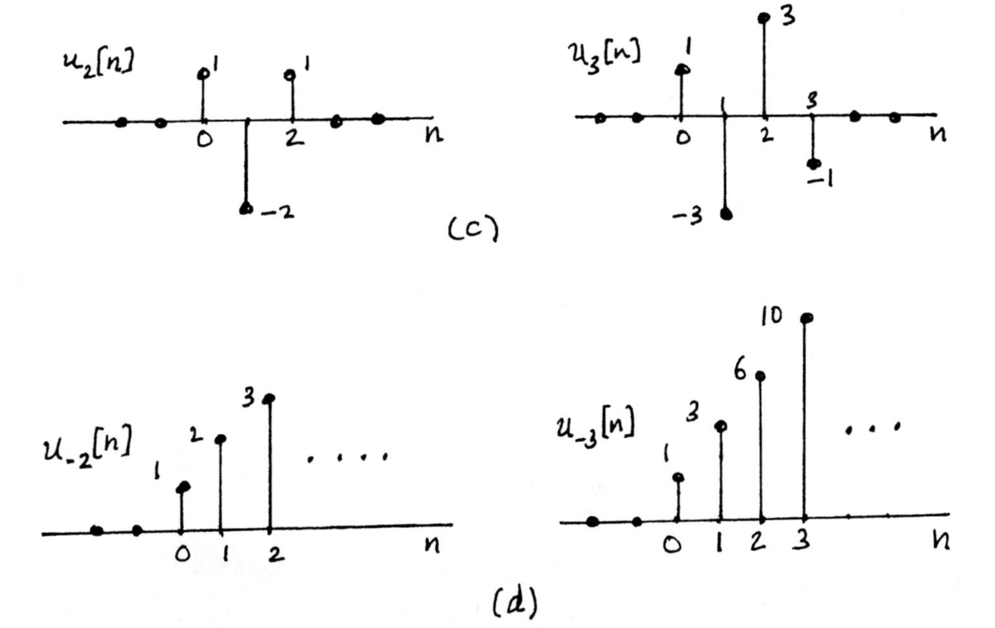

Figure S2.70

(d) We have

$$u_{-2}[n] = (n+1), \qquad n \geq 0$$

and

$$u_{-3}[n] = \frac{(n+1)(n+2)}{2}, \qquad n \geq 0.$$

The plots for these signals are as shown in the Figure S2.70.

(e) The statement is true for $k = 1, 2, 3$. Assume it is true for k. Then, for $k > 0$

$$u_{k+1}[n] = u_1[n] * u_k[n] = u_k[n] - u_k[n-1].$$

By induction, we may now claim that the statement is true for all $k > 0$.

(f) For $k = 1$, $u_{-1}[n] = u[n]$ which shows that the statement is true. For $k = 2$,

$$u_{-2}[n] = \frac{(n+1)!}{n!}u[n] = (n+1)u[n]$$

which again shows that the statement is true. Assume that it is true for $k - 1 > 0$. Then,

$$u_{-(k-1)}[n] = u_{-k}[n] - u_{-k}[n-1]. \qquad \text{(S2.70–1)}$$

Also,

$$\begin{aligned} u_{-(k-1)}[n] &= \frac{(n+k-2)!}{n!(k-2)!}u[n] \\ &= \frac{(n+k-1)!}{n!(k-1)!}u[n] - \frac{(n+k-2)!}{(n-1)!(k-2)!}u[n-2]. \end{aligned}$$

Using the above equation with eq. (S2.70-1), we get

$$u_{-k}[n] = \frac{(n+k-1)!}{n!(k-1)!}u[n].$$

By induction, we may now claim that the statement is true for all $k > 0$.

2.71. (a) We have

$$x(t) * [u_1(t) * u(t)] = x(t) = 1, \qquad \text{for all } t,$$

$$[x(t) * u_1(t)] * u(t) = 0 * u(t) = 0 \qquad \text{for all } t,$$

and

$$[x(t) * u(t)] * u_1(t) = \infty * u_1(t) = \text{undefined}.$$

(b) We have $x(t) = e^{-t}$, $h(t) = e^{-t}u(t)$, and $g(t) = u_1(t) + \delta(t)$. Therefore,

$$x(t) * [h(t) * g(t)] = x(t) = e^{-t},$$

$$[x(t) * g(t)] * h(t) = 0,$$

and

$$g(t) * [x(t) * h(t)] = g(t) * e^{-t}\int_0^\infty 1 d\tau = \text{undefined}.$$

(c) We have

$$x[n] * [h[n] * g[n]] = \left(\frac{1}{2}\right)^n * \delta[n] = \frac{1}{2}^n,$$

$$(x[n] * g[n]) * h[n] = 0 * h[n] = 0,$$

and

$$(x[n] * h[n]) * g[n] = \{(\frac{1}{2})^n \sum_{k=0}^{\infty} 1\} * g[n] = \infty.$$

(d) Let $h(t) = u_1(t)$. Then if the input is $x_1(t) = 0$, the output will be $y_1(t) = 0$. Now if $x_2(t) =$ constant, then $y_2(t) = 0$. Therefore, the system is not invertible.

Now note that

$$|\int_{-\infty}^{t} x_2(\tau)d\tau| = \begin{cases} 0 & \text{if } x_2(t) = 0 \forall t \\ \infty & \text{if } x_2(t) \neq 0 \end{cases}:$$

Therefore, if $\left|\int_{-\infty}^{t} cdt\right|_{t\to\infty} \neq \infty$, then only $x_2(t) = 0$ will yield $y_2(t) = 0$. Therefore, the system is invertible.

2.72. We have

$$\delta_\Delta(t) = \frac{1}{\Delta}u(t) * [\delta(t) - \delta(t-T)].$$

Differentiating both sides we get

$$\begin{aligned} \frac{d}{dt}\delta_\Delta t &= \frac{1}{\Delta}u'(t) * [\delta(t) - \delta(t-T)] \\ &= \frac{1}{\Delta}\delta(t) * [\delta(t) - \delta(t-T)] \\ &= \frac{1}{\Delta}[\delta(t) - \delta(t-T)] \end{aligned}$$

2.73. For $k = 1$, $u_{-1}(t) = u(t)$. Therefore, the given statement is true for $k = 1$. Now assume that it is true for some $k > 1$. Then,

$$\begin{aligned} u_{-(k+1)}(t) &= u(t) * u_{-k}(t) \\ &= \int_{-\infty}^{t} u_{-k}(t) = \int_{0}^{t} u_{-k}(\tau)d\tau \\ &= \int_{0}^{t} \frac{\tau^{k-1}}{(k-1)!}, \qquad t \geq 0 \\ &= \left.\frac{\tau^k}{k(k-1)!}\right|_{\tau=t\geq 0} = \frac{t^k}{k!}u(t). \end{aligned}$$

Chapter 3 Answers

3.1. Using the Fourier series synthesis eq. (3.38),

$$\begin{aligned} x(t) &= a_1 e^{j(2\pi/T)t} + a_{-1}e^{-j(2\pi/T)t} + a_3 e^{j3(2\pi/T)t} + a_{-3}e^{-j3(2\pi/T)t} \\ &= 2e^{j(2\pi/8)t} + 2e^{-j(2\pi/8)t} + 4je^{j3(2\pi/8)t} - 4je^{-j3(2\pi/8)t} \\ &= 4\cos(\frac{\pi}{4}t) - 8\sin(\frac{6\pi}{8}t) \\ &= 4\cos(\frac{\pi}{4}t) + 8\cos(\frac{3\pi}{4}t + \frac{\pi}{2}) \end{aligned}$$

3.2. Using the Fourier series synthesis eq. (3.95).

$$\begin{aligned} x[n] &= a_0 + a_2 e^{j2(2\pi/N)n} + a_{-2}e^{-j2(2\pi/N)n} + a_4 e^{j4(2\pi/N)n} + a_{-4}e^{-j4(2\pi/N)n} \\ &= 1 + e^{j(\pi/4)}e^{j2(2\pi/5)n} + e^{-j(\pi/4)}e^{-2j(2\pi/5)n} \\ &\quad +2e^{j(\pi/3)}e^{j4(2\pi/N)n} + 2e^{-j(\pi/3)}a_{-4}e^{-j4(2\pi/N)n} \\ &= 1 + 2\cos(\frac{4\pi}{5}n + \frac{\pi}{4}) + 4\cos(\frac{8\pi}{5}n + \frac{\pi}{3}) \\ &= 1 + 2\sin(\frac{4\pi}{5}n + \frac{3\pi}{4}) + 4\sin(\frac{8\pi}{5}n + \frac{5\pi}{6}) \end{aligned}$$

3.3. The given signal is

$$\begin{aligned} x(t) &= 2 + \frac{1}{2}e^{j(2\pi/3)t} + \frac{1}{2}e^{-j(2\pi/3)t} - 2je^{j(5\pi/3)t} + 2je^{-j(5\pi/3)t} \\ &= 2 + \frac{1}{2}e^{j2(2\pi/6)t} + \frac{1}{2}e^{-j2(2\pi/6)t} - 2je^{j5(2\pi/6)t} + 2je^{-j5(2\pi/6)t} \end{aligned}$$

From this, we may conclude that the fundamental frequency of $x(t)$ is $2\pi/6 = \pi/3$. The non-zero Fourier series coeffcients of $x(t)$ are:

$$a_0 = 2, \qquad a_2 = a_{-2} = \frac{1}{2}, \qquad a_5 = a_{-5}^* = -2j$$

3.4. Since $\omega_0 = \pi$, $T = 2\pi/\omega_0 = 2$. Therefore,

$$a_k = \frac{1}{2}\int_0^2 x(t)e^{-jk\pi t}dt$$

Now,

$$a_0 = \frac{1}{2}\int_0^1 1.5dt - \frac{1}{2}\int_1^2 1.5dt = 0$$

and for $k \neq 0$

$$\begin{aligned} a_k &= \frac{1}{2}\int_0^1 1.5e^{-jk\pi t}dt - \frac{1}{2}\int_1^2 1.5e^{-jk\pi t}dt \\ &= \frac{3}{2k\pi j}[1 - e{-}jk\pi] \\ &= \frac{3}{k\pi}e^{-jk(\pi/2)}\sin(\frac{k\pi}{2}) \end{aligned}$$

3.5. Both $x_1(1-t)$ and $x_1(t-1)$ are periodic with fundemental period $T_1 = \frac{2\pi}{\omega_1}$. Since $y(t)$ is a linear combination of $x_1(1-t)$ and $x_1(t-1)$, it is also periodic with fundemental period $T_2 = \frac{2\pi}{\omega_1}$. Therefore, $\omega_2 = \omega_1$.

Since $x_1(t) \stackrel{FS}{\longleftrightarrow} a_k$, using the results in Table 3.1 we have

$$x_1(t+1) \stackrel{FS}{\longleftrightarrow} a_k e^{jk(2\pi/T_1)}$$

$$x_1(t-1) \stackrel{FS}{\longleftrightarrow} a_k e^{-jk(2\pi/T_1)} \Rightarrow x_1(-t+1) \stackrel{FS}{\longleftrightarrow} a_{-k} e^{-jk(2\pi/T_1)}$$

Therefore,

$$x_1(t+1) + x_1(1-t) \stackrel{FS}{\longleftrightarrow} a_k e^{jk(2\pi/T_1)} + a_{-k} e^{-jk(2\pi/T_1)} = e^{-j\omega_1 k}(a_k + a_{-k})$$

3.6. **(a)** Comparing $x_1(t)$ with the Fourier series synthesis eq. (3.38), we obtain the Fourier series coefficients of $x_1(t)$ to be

$$a_k = \begin{cases} \left(\frac{1}{2}\right)^k, & 0 \le k \le 100 \\ 0, & \text{otherwise} \end{cases}$$

From Table 3.1 we know that if $x_1(t)$ is real, then a_k has to be conjugate-symmetric, i.e, $a_k = a^*_{-k}$. Since this is not true for $x_1(t)$, the signal is **not real valued**.

Similarly, the Fourier series coefficients of $x_2(t)$ are

$$a_k = \begin{cases} \cos(k\pi), & 100 \le k \le 100 \\ 0, & \text{otherwise} \end{cases}$$

From Table 3.1 we know that if $x_2(t)$ is real, then a_k has to be conjugate-symmetric, i.e, $a_k = a^*_{-k}$. Since this is true for $x_2(t)$, the signal is **real valued**.

Similarly, the Fourier series coefficients of $x_3(t)$ are

$$a_k = \begin{cases} j\sin(k\pi/2), & 100 \le k \le 100 \\ 0, & \text{otherwise} \end{cases}$$

From Table 3.1 we know that if $x_3(t)$ is real, then a_k has to be conjugate-symmetric, i.e, $a_k = a^*_{-k}$. Since this is true for $x_3(t)$, the signal is **real valued**.

(b) For a signal to be even, its Fourier series coefficients must be even. This is true only for $x_2(t)$.

3.7. Given that

$$x(t) \stackrel{FS}{\longleftrightarrow} a_k$$

we have

$$g(t) = \frac{dx(t)}{dt} \stackrel{FS}{\longleftrightarrow} b_k = jk\frac{2\pi}{T}a_k.$$

Therefore,

$$a_k = \frac{b_k}{j(2\pi/T)k}, \qquad k \neq 0$$

When $k = 0$,

$$a_k = \frac{1}{T}\int_{<T>} x(t)dt = \frac{2}{T} \quad \text{using given information}$$

Therefore,

$$a_k = \begin{cases} \frac{2}{T}, & k = 0 \\ \frac{b_k}{j(2\pi/T)k}, & k \neq 0 \end{cases}.$$

3.8. Since $x(t)$ is real and odd (clue 1), its Fourier series coefficients a_k are purely imaginary and odd (See Table 3.1). Therefore, $a_k = -a_{-k}$ and $a_0 = 0$. Also, since it is given that $a_k = 0$ for $|k| > 1$, the only unknown Fourier series coefficients are a_1 and a_{-1}. Using Parseval's relation,

$$\frac{1}{T}\int_{<T>} |x(t)|^2 dt = \sum_{k=-\infty}^{\infty} |a_k|^2,$$

for the given signal we have

$$\frac{1}{2}\int_0^2 |x(t)|^2 dt = \sum_{k=-1}^{1} |a_k|^2.$$

Using the information given in clue (4) along with the above equation,

$$|a_1|^2 + |a_{-1}|^2 = 1 \quad \Rightarrow \quad 2|a_1|^2 = 1$$

Therefore,

$$a_1 = -a_{-1} = \frac{1}{\sqrt{2}j} \quad \text{or} \quad a_1 = -a_{-1} = -\frac{1}{\sqrt{2}j}$$

The two possible signals which satisfy the given information are

$$x_1(t) = \frac{1}{\sqrt{2}j}e^{j(2\pi/2)t} - \frac{1}{\sqrt{2}j}e^{-j(2\pi/2)t} = -\sqrt{2}\sin(\pi t)$$

and

$$x_2(t) = -\frac{1}{\sqrt{2}j}e^{j(2\pi/2)t} + \frac{1}{\sqrt{2}j}e^{-j(2\pi/2)t} = \sqrt{2}\sin(\pi t)$$

3.9. The period of the given signal is 4. Therefore,

$$\begin{aligned} a_k &= \frac{1}{4}\sum_{n=0}^{3} x[n]e^{-j\frac{2\pi}{4}kn} \\ &= \frac{1}{4}[4 + 8e^{-j\frac{\pi}{2}k}] \end{aligned}$$

This gives

$$a_0 = 3, \quad a_1 = 1 - 2j, \quad a_2 = -1, \quad a_3 = 1 + 2j$$

3.10. Since the Fourier series coeffiecients repeat every N, we have

$$a_1 = a_{15}, \qquad a_2 = a_{16} \quad, \text{and} \qquad a_3 = a_{17}$$

Furthermore, since the signal is real and odd, the Fourier series coefficients a_k will be purely imaginary and odd. Therefore, $a_0 = 0$ and

$$a_1 = -a_{-1}, \qquad a_2 = -a_{-2} \qquad a_3 = -a_{-3}$$

Finally,

$$a_{-1} = -j, \qquad a_{-2} = -2j, \qquad a_{-3} = -3j$$

3.11. Since the Fourier series coefficients repeat every $N = 10$, we have $a_1 = a_{11} = 5$. Furthermore, since $x[n]$ is real and even, a_k is also real and even. Therefore, $a_1 = a_{-1} = 5$. We are also given that

$$\frac{1}{10}\sum_{n=0}^{9}|x[n]|^2 = 50.$$

Using Parseval's relation,

$$\begin{aligned}
\sum_{k=<N>} |a_k|^2 &= 50 \\
\sum_{k=-1}^{8} |a_k|^2 &= 50 \\
|a_{-1}|^2 + |a_1|^2 + a_0^2 + \sum_{k=2}^{8} |a_k|^2 &= 50 \\
a_0^2 + \sum_{k=2}^{8} |a_k|^2 &= 0
\end{aligned}$$

Therefore, $a_k = 0$ for $k = 2, \cdots, 8$. Now using the synthesis eq.(3.94), we have

$$\begin{aligned}
x[n] &= \sum_{k=<N>} a_k e^{j\frac{2\pi}{N}kn} = \sum_{k=-1}^{8} a_k e^{j\frac{2\pi}{10}kn} \\
&= 5e^{j\frac{2\pi}{10}n} + 5e^{-j\frac{2\pi}{10}n} \\
&= 10\cos(\frac{\pi}{5}n)
\end{aligned}$$

3.12. Using the multiplication property (see Table 3.2), we have

$$\begin{aligned}
x_1[n]x_2[n] &\overset{FS}{\longleftrightarrow} \sum_{l=<N>} a_l b_{k-l} = \sum_{k=0}^{3} a_l b_{k-l} \\
&\overset{FS}{\longleftrightarrow} a_0 b_k + a_1 b_{k-1} + a_2 b_{k-2} + a_3 b_{k-3} \\
&\overset{FS}{\longleftrightarrow} b_k + 2b_{k-1} + 2b_{k-2} + 2b_{k-3}
\end{aligned}$$

Since b_k is 1 for all values of k, it is clear that $b_k + 2b_{k-1} + 2b_{k-3} + 2b_{k-3}$ will be 6 for all values of k. Therefore,

$$x_1[n]x_2[n] \stackrel{FS}{\longleftrightarrow} 6, \qquad \text{for all } k.$$

3.13. Let us first evaluate the Fourier series coefficients of $x(t)$. Clearly, since $x(t)$ is real and odd, a_k is purely imaginary and odd. Therefore, $a_0 = 0$. Now,

$$\begin{aligned} a_k &= \frac{1}{8}\int_0^8 x(t)e^{-j(2\pi/8)kt}dt \\ &= \frac{1}{8}\int_0^4 e^{-j(2\pi/8)kt}dt - \frac{1}{8}\int_4^8 e^{-j(2\pi/8)kt}dt \\ &= \frac{1}{j\pi k}[1 - e^{-j\pi k}] \end{aligned}$$

Clearly, the above expression evaluates to zero for all even values of k. Therefore,

$$a_k = \begin{cases} 0, & k = 0, \pm 2, \pm 4, \cdots \\ \frac{2}{j\pi k}, & k = \pm 1, \pm 3, \pm 5, \cdots \end{cases}$$

When $x(t)$ is passed through an LTI system with frequency response $H(j\omega)$, the output $y(t)$ is given by (see Section 3.8)

$$y(t) = \sum_{k=-\infty}^{\infty} a_k H(jk\omega_0)e^{jk\omega_0 t}$$

where $\omega_0 = \frac{2\pi}{T} = \frac{\pi}{4}$. Since a_k is non zero only for odd values of k, we need to evaluate the above summation only for odd k. Furthermore, note that

$$H(jk\omega_0) = H(jk(\pi/4)) = \frac{\sin(k\pi)}{k(\pi/4)}$$

is always zero for odd values of k. Therefore,

$$y(t) = 0.$$

3.14. The signal $x[n]$ is periodic with period $N = 4$. Its Fourier series coefficients are

$$\begin{aligned} a_k &= \frac{1}{4}\sum_{n=0}^{3} x[n]e^{-j\frac{2\pi}{4}kn} \\ &= \frac{1}{4}, \qquad \text{for all } k \end{aligned}$$

From the results presented in Section 3.8, we know that the output $y[n]$ is given by

$$\begin{aligned} y[n] &= \sum_{k=0}^{3} a_k H(e^{j(2\pi/4)k})e^{jk(2\pi/4)n} \\ &= \tfrac{1}{4}H(e^{j0})e^{j0} + \tfrac{1}{4}H(e^{j(\pi/2)})e^{j(\pi/2)} \\ &\quad + \tfrac{1}{4}H(e^{j(3\pi/2)})e^{j(3\pi/2)} + \tfrac{1}{4}H(e^{j(\pi)})e^{j(\pi)} \end{aligned} \qquad \text{(S3.14–1)}$$

From the given information, we know that $y[n]$ is

$$\begin{aligned} y[n] &= \cos(\frac{5\pi}{2}n + \frac{\pi}{4}) \\ &= \cos(\frac{\pi}{2}n + \frac{\pi}{4}) \\ &= \frac{1}{2}e^{j(\frac{\pi}{2}n+\frac{\pi}{4})} + \frac{1}{2}e^{-j(\frac{\pi}{2}n+\frac{\pi}{4})} \\ &= \frac{1}{2}e^{j(\frac{\pi}{2}n+\frac{\pi}{4})} + \frac{1}{2}e^{j(3\frac{\pi}{2}n-\frac{\pi}{4})} \end{aligned}$$

Comparing this with eq. (S3.14-1), we have

$$H(e^{j0}) = H(e^{j\pi}) = 0$$

and

$$H(e^{j\frac{\pi}{2}}) = 2e^{j\frac{\pi}{4}}, \quad \text{and} \quad H(e^{3j\frac{\pi}{2}}) = 2e^{-j\frac{\pi}{4}}$$

3.15. From the results of Section 3.8,

$$y(t) = \sum_{k=-\infty}^{\infty} a_k H(jk\omega_0)e^{jk\omega_0 t}$$

where $\omega_0 = \frac{2\pi}{T} = 12$. Since $H(j\omega)$ is zero for $|\omega| > 100$, the largest value of $|k|$ for which a_k is nonzero should be such that

$$|k|\omega_0 \leq 100$$

This implies that $|k| \leq 8$. Therefore, for $|k| > 8$, a_k is guaranteed to be zero.

3.16. **(a)** The given signal $x_1[n]$ is

$$x_1[n] = (-1)^n = e^{j\pi n} = e^{j(2\pi/2)n}$$

Therefore, $x_1[n]$ is periodic with period $N = 2$ and it's Fourier series coefficients in the range $0 \leq k \leq 1$ are

$$a_0 = 0, \quad \text{and} \quad a_1 = 1$$

Using the results derived in Section 3.8, the output $y_1[n]$ is given by

$$\begin{aligned} y_1[n] &= \sum_{k=0}^{1} a_k H(e^{j2\pi k/2})e^{k(2\pi/2)} \\ &= 0 + a_1 H(e^{j\pi})e^{j\pi} \\ &= 0 \end{aligned}$$

(b) The signal $x_2[n]$ is periodic with period $N = 16$. The signal $x_2[n]$ may be written as

$$\begin{aligned} x_2[n] &= e^{j(2\pi/16)(0)n} - (j/2)e^{j(\pi/4)}e^{j(2\pi/16)(3)n} + (j/2)e^{-j(\pi/4)}e^{-j(2\pi/16)(3)n} \\ &= e^{j(2\pi/16)(0)n} - (j/2)e^{j(\pi/4)}e^{j(2\pi/16)(3)n} + (j/2)e^{-j(\pi/4)}e^{j(2\pi/16)(13)n} \end{aligned}$$

Therefore, the non-zero Fourier series coefficients of $x_2[n]$ in the range $0 \le k \le 15$ are

$$a_0 = 1, \qquad a_3 = -(j/2)e^{j(\pi/4)}, \qquad a_{13} = (j/2)e^{-j(\pi/4)}$$

Using the results derived in Section 3.8, the output $y_2[n]$ is given by

$$\begin{aligned} y_2[n] &= \sum_{k=0}^{15} a_k H(e^{j2\pi k/16}) e^{k(2\pi/16)} \\ &= 0 - (j/2)e^{j(\pi/4)} e^{j(2\pi/16)(3)n} + (j/2)e^{-j(\pi/4)} e^{j(2\pi/16)(13)n} \\ &= \sin(\frac{3\pi}{8}n + \frac{\pi}{4}) \end{aligned}$$

(c) The signal $x_3[n]$ may be written as

$$x_3[n] = \left[\left(\frac{1}{2}\right)^n u[n]\right] * \sum_{k=-\infty}^{\infty} \delta[n-4k] = g[n] * r[n]$$

where $g[n] = \left(\frac{1}{2}\right)^n u[n]$ and $r[n] = \sum_{k=-\infty}^{\infty} \delta[n-4k]$. Therefore, $y_3[n]$ may be obtained by passing the signal $r[n]$ through the filter with frequency response $H(e^{j\omega})$, and then convolving the result with $g[n]$.

The signal $r[n]$ is periodic with period 4 and its Fourier series coeffients are

$$a_k = \frac{1}{4}, \qquad \text{for all } k \text{ (See Problem 3.14)}$$

The output $q[n]$ obtained by passing $r[n]$ through the filter with frequency response $H(e^{j\omega})$ is

$$\begin{aligned} q[n] &= \sum_{k=0}^{3} a_k H(e^{j2\pi k/4}) e^{k(2\pi/4)} \\ &= (1/4)(H(e^{j0})e^{j0} + H(e^{j(\pi/2)})e^{j(\pi/2)} + H(e^{j\pi})e^{j\pi} + H(e^{j3(\pi/2)})e^{j3(\pi/2)}) \\ &= 0 \end{aligned}$$

Therefore, the final output $y_3[n] = q[n] * g[n] = 0$.

3.17. **(a)** Since complex exponentials are Eigen functions of LTI systems, the input $x_1(t) = e^{j5t}$ has to produce an output of the form Ae^{j5t}, where A is a complex constant. But clearly, in this case the output is not of this form. Therefore, system S_1 is definitely **not LTI**.

(b) This system may be LTI because it satisifies the Eigen function property of LTI systems.

(c) In this case, the output is of the form $y_3(t) = (1/2)e^{j5t} + (1/2)e^{-j5t}$. Clearly, the output contains a complex exponential with frequency -5 which was not present in the input $x_3(t)$. We know that an LTI system can never produce a complex exponential of frequency -5 unless there was complex exponential of the same frequency at its input. Since this is not the case in this problem, S_3 is definitely **not LTI**.

3.18. (a) By using an argument similar to the one used in part (a) of the previous problem, we conclude that S_1 is defintely not LTI.

(b) The output in this case is $y_2[n] = e^{j(3\pi/2)n} = e^{-j(\pi/2)n}$. Clearly this violates the eigen function property of LTI systems. Therefore, S_2 is definitely **not LTI**.

(c) The output in this case is $y_3[n] = 2e^{j(5\pi/2)n} = 2e^{j(\pi/2)n}$. This does not violate the eigen function property of LTI systems. Therefore, S_3 could possibly be an LTI system.

3.19. (a) Voltage across inductor $= L\frac{dy(t)}{dt}$.
Current through resistor $= \frac{L}{R}\frac{dy(t)}{dt}$.
Input current $x(t)$ = current through resistor + current through inductor
Therefore,

$$x(t) = \frac{L}{R}\frac{dy(t)}{dt} + y(t).$$

Substituting for R and L we obtain

$$\frac{dy(t)}{dt} + y(t) = x(t).$$

(b) Using the approach outlined in Section 3.10.1, we know that the output of this system will be $H(j\omega)e^{j\omega t}$ when the input is $e^{j\omega t}$. Substituting in the differential equation of part (a),

$$j\omega H(j\omega)e^{j\omega t} + H(j\omega)e^{j\omega t} = e^{j\omega t}$$

Therefore,

$$H(j\omega) = \frac{1}{1+j\omega}$$

(c) The signal $x(t)$ is periodic with period 2π. Since $x(t)$ can be expressed in the form

$$x(t) = \frac{1}{2}e^{j(2\pi/2\pi)t} + \frac{1}{2}e^{-j(2\pi/2\pi)t},$$

the non-zero Fourier series coefficients of $x(t)$ are

$$a_1 = a_{-1} = \frac{1}{2}.$$

Using the results derived in Section 3.8 (see eq.(3.124)), we have

$$\begin{aligned} y(t) &= a_1 H(j)e^{jt} + a_{-1}H(-j)e^{-jt} \\ &= (1/2)(\frac{1}{1+j}ejt + \frac{1}{1-j}e^{-jt}) \\ &= (1/2\sqrt{2})(e^{-j\pi/4}e^{jt} + e^{j\pi/4}e^{-jt}) \\ &= (1/\sqrt{2})\cos(t - \frac{\pi}{4}) \end{aligned}$$

3.20. (a) Current through the capacitor $= C\frac{dy(t)}{dt}$.
Voltage across resistor $= RC\frac{dy(t)}{dt}$.
Voltage across inductor $= LC\frac{d^2y(t)}{dt^2}$.
Input voltage = Voltage across resistor + Voltage across inductor + Voltage across capacitor.
Therefore,

$$x(t) = LC\frac{d^2y(t)}{dt^2} + RC\frac{dy(t)}{dt} + y(t)$$

Substituting for R, L and C, we have

$$\frac{d^2y(t)}{dt^2} + \frac{dy(t)}{dt} + y(t) = x(t)$$

(b) We will now use an approach similar to the one used in part (b) of the previous problem. If we assume that the input is of the form $e^{j\omega t}$, then the output will be of the form $H(j\omega)e^{j\omega t}$. Substituting in the above differential equation and simplifying, we obtain

$$H(j\omega) = \frac{1}{-\omega^2 + j\omega + 1}$$

(c) The signal $x(t)$ is periodic with period 2π. Since $x(t)$ can be expressed in the form

$$x(t) = \frac{1}{2j}e^{j(2\pi/2\pi)t} - \frac{1}{2j}e^{-j(2\pi/2\pi)t},$$

the non-zero Fourier series coefficients of $x(t)$ are

$$a_1 = a_{-1}^* = \frac{1}{2j}.$$

Using the results derived in Section 3.8 (see eq.(3.124)), we have

$$\begin{aligned} y(t) &= a_1 H(j)e^{jt} - a_{-1}H(-j)e^{-jt} \\ &= (1/2j)(\frac{1}{j}e^{jt} - \frac{1}{-j}e^{-jt}) \\ &= (-1/2)(e^{jt} + e^{-jt}) \\ &= -\cos(t) \end{aligned}$$

3.21. Using the Fourier series synthesis eq. (3.38),

$$\begin{aligned} x(t) &= a_1 e^{j(2\pi/T)t} + a_{-1}e^{-j(2\pi/T)t} + a_5 e^{j5(2\pi/T)t} + a_{-5}e^{-j5(2\pi/T)t} \\ &= je^{j(2\pi/8)t} - je^{-j(2\pi/8)t} + 2e^{j5(2\pi/8)t} + 2e^{-j5(2\pi/8)t} \\ &= -2\sin(\frac{\pi}{4}t) + 4\cos(\frac{5\pi}{4}t) \\ &= -2\cos(\frac{\pi}{4}t - \pi/2) + 4\cos(\frac{5\pi}{4}t). \end{aligned}$$

3.22. (a) (i) $T=1$, $a_0=0$, $a_k=\frac{j(-1)^k}{k\pi}$, $k \neq 0$.

(ii) Here,

$$x(t) = \begin{cases} t+2, & -2<t<-1 \\ 1, & -1<t<1 \\ 2-t, & 1<t<2 \end{cases}$$

$T=6$, $a_0 = 1/2$, and

$$a_k = \begin{cases} 0, & k \text{ even} \\ \frac{6}{\pi^2 k^2}\sin(\frac{\pi k}{2})\sin(\frac{\pi k}{6}), & k \text{ odd} \end{cases}.$$

(iii) $T=3$, $a_0=1$, and

$$a_k = \frac{3j}{2\pi^2 k^2}[e^{jk2\pi/3}\sin(k2\pi/3) + 2e^{jk\pi/3}\sin(k\pi/3)], \qquad k \neq 0.$$

(iv) $T=2$, $a_0 = -1/2$, $a_k = \frac{1}{2} - (-1)^k$, $k \neq 0$.

(v) $T=6$, $\omega_0 = \pi/3$, and

$$a_k = \frac{\cos(2k\pi/3) - \cos(k\pi/3)}{jk\pi/3}.$$

Note that $a_0 = 0$ and $a_{k\ even} = 0$.

(vi) $T=4$, $\omega_0=\pi/2$, $a_0 = 3/4$ and

$$a_k = \frac{e^{-jk\pi/2}\sin(k\pi/2) + e^{-jk\pi/4}\sin(k\pi/4)}{k\pi}, \qquad \forall k.$$

(b) $T=2$, $a_k = \frac{-1^k}{2(1+jk\pi)}[e - e^{-1}]$ for all k.

(c) $T=3$, $\omega_0 = 2\pi/3$, $a_0 = 1$ and

$$a_k = \frac{2e^{-j\pi k/3}}{\pi k}\sin(2\pi k/3) + \frac{e^{-j\pi k}}{\pi k}\sin(\pi k).$$

3.23. (a) First let us consider a signal $y(t)$ with FS coefficients

$$b_k = \frac{\sin(k\pi/4)}{k\pi}.$$

From Example 3.5, we know that $y(t)$ must be a periodic square wave which over one period is

$$y(t) = \begin{cases} 1, & |t| < 1/2 \\ 0, & 1/2 < |t| < 2 \end{cases}.$$

Now, note that $b_0 = 1/4$. Let us define another signal $z(t) = -1/4$ whose only nonzero FS coefficient is $c_0 = -1/4$. The signal $p(t) = y(t) + z(t)$ will have FS coefficients

$$d_k = a_k + c_k = \begin{cases} 0, & k=0 \\ \frac{\sin(k\pi/4)}{k\pi}, & \text{otherwise.} \end{cases}.$$

Now note that $a_k = d_k e^{j(\pi/2)k}$. Therefore, the signal $x(t) = p(t+1)$ which is as shown in Figure S2.23(a).

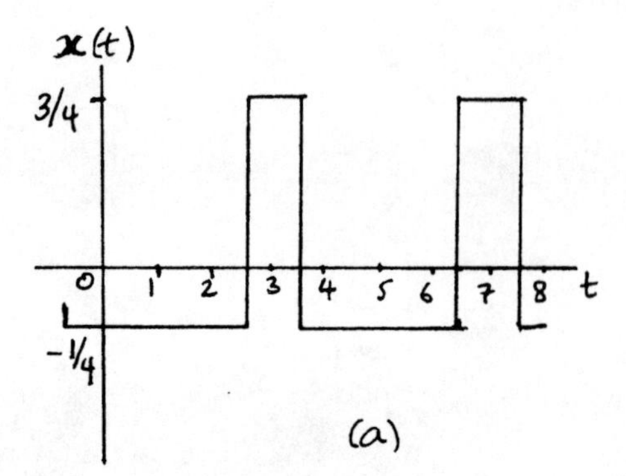

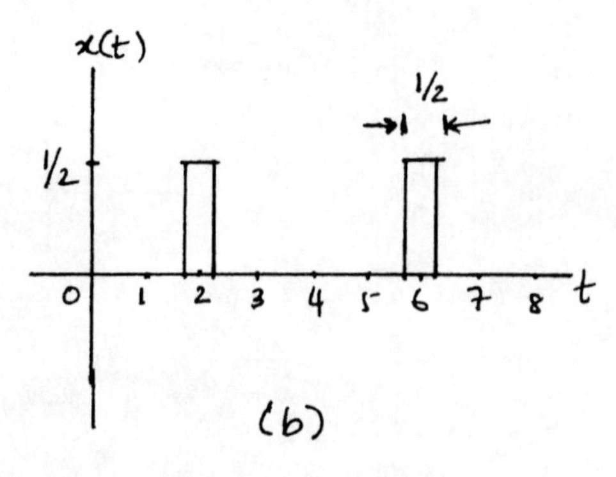

Figure S3.23

(b) First let us consider a signal $y(t)$ with FS coefficients

$$b_k = \frac{\sin(k\pi/8)}{2k\pi}.$$

From Example 3.5, we know that $y(t)$ must be a periodic square wave which over one period is

$$y(t) = \begin{cases} 1/2, & |t| < 1/4 \\ 0, & 1/4 < |t| < 2 \end{cases}.$$

Now note that $a_k = b_k e^{j\pi k}$. Therefore, the signal $x(t) = y(t+2)$ which is as shown in Figure S2.23(b).

(c) The only nonzero FS coefficients are $a_1 = a_{-1}^* = j$ and $a_2 = a_{-2}^* = 2j$. Using the FS synthesis equation, we get

$$\begin{aligned} x(t) &= a_1 e^{j(2\pi/T)t} + a_{-1} e^{-j(2\pi/T)t} + a_2 e^{j2(2\pi/T)t} + a_{-2} e^{-j2(2\pi/T)t} \\ &= j e^{j(2\pi/4)t} - j e^{-j(2\pi/4)t} + 2j e^{j2(2\pi/4)t} - 2j e^{-j2(2\pi/4)t} \\ &= -2\sin(\frac{\pi}{2}t) - 4\sin(\pi t) \end{aligned}$$

(d) The FS coefficients a_k may be written as the sum of two sets of FS coefficients b_k and c_k, where

$$b_k = 1, \qquad \text{for all } k$$

and

$$c_k = \begin{cases} 1, & k \text{ odd} \\ 0, & k \text{ even} \end{cases}.$$

The FS coefficients b_k correspond to the signal

$$y(t) = \sum_{k=-\infty}^{\infty} \delta(t - 4k)$$

and the FS coefficients c_k correspond to the signal

$$z(t) = \sum_{k=-\infty}^{\infty} e^{j(\pi/2)t} \delta(t - 2k).$$

Therefore,

$$x(t) = y(t) + p(t) = \sum_{k=-\infty}^{\infty} \delta(t - 4k) + \sum_{k=-\infty}^{\infty} e^{j(\pi/2)t}\delta(t - 2k).$$

3.24. **(a)** We have

$$a_0 = \frac{1}{2}\int_0^1 t dt + \frac{1}{2}\int_1^2 (2-t)dt = 1/2.$$

(b) The signal $g(t) = dx(t)/dt$ is as shown in Figure S3.24.

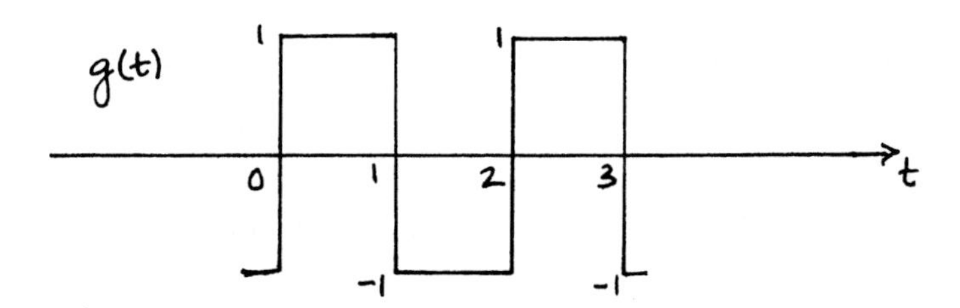

Figure S3.24

The FS coefficients b_k of $g(t)$ may be found as follows:

$$b_0 = \frac{1}{2}\int_0^1 dt - \frac{1}{2}\int_1^2 dt = 0$$

and

$$\begin{aligned} b_k &= \frac{1}{2}\int_0^1 e^{-j\pi kt}dt - \frac{1}{2}\int_1^2 e^{-j\pi kt}dt \\ &= \frac{1}{j\pi k}[1 - e^{-j\pi k}]. \end{aligned}$$

(c) Note that

$$g(t) = \frac{dx(t)}{dt} \stackrel{FS}{\longleftrightarrow} b_k = jk\pi a_k.$$

Therefore,

$$a_k = \frac{1}{jk\pi}b_k = -\frac{1}{\pi^2 k^2}\{1 - e^{-j\pi k}\}.$$

3.25. **(a)** The nonzero FS coefficients of $x(t)$ are $a_1 = a_{-1} = 1/2$.

(b) The nonzero FS coefficients of $x(t)$ are $b_1 = b_{-1}^* = 1/2j$.

(c) Using the multiplication property, we know that

$$z(t) = x(t)y(t) \stackrel{FS}{\longleftrightarrow} c_k = \sum_{l=-\infty}^{\infty} a_l b_{k-l}.$$

Therefore,

$$c_k = a_k * b_k = \frac{1}{4j}\delta[k-2] - \frac{1}{4j}\delta[k+2].$$

This implies that the nonzero Fourier series coefficients of $z(t)$ are $c_2 = c_{-2}^* = (1/4j)$.

(d) We have

$$z(t) = \sin(4t)\cos(4t) = \frac{1}{2}\sin(8t).$$

Therefore, the nonzero Fourier series coefficients of $z(t)$ are $c_2 = c_{-2} = (1/4j)$.

3.26. **(a)** If $x(t)$ is real, then $x(t) = x^*(t)$. This implies that for $x(t)$ real $a_k = a_{-k}^*$. Since this is not true in this case problem, $x(t)$ is not real.

(b) If $x(t)$ is even, then $x(t) = x(-t)$ and $a_k = a_{-k}$. Since this is true for this case, $x(t)$ is even.

(c) We have

$$g(t) = \frac{dx(t)}{dt} \stackrel{FS}{\longleftrightarrow} b_k = jk\frac{2\pi}{T_0}a_k.$$

Therefore,

$$b_k = \begin{cases} 0, & k = 0 \\ -k(1/2)^{|k|}(2\pi/T_0), & \text{otherwise} \end{cases}.$$

Since b_k is not even, $g(t)$ is not even.

3.27. Using the Fourier series synthesis eq. (3.38),

$$\begin{aligned} x[n] &= a_0 + a_2 e^{j2(2\pi/N)n} + a_{-2}e^{-j2(2\pi/N)n} + a_4 e^{j4(2\pi/N)n} + a_{-4}e^{-j4(2\pi/N)n} \\ &= 2 + 2e^{j\pi/6}e^{j(4\pi/5)n} + 2e^{-j\pi/6}e^{-j(4\pi/5)n} + e^{j\pi/3}e^{j(8\pi/5)n} + e^{-j\pi/3}e^{-j(8\pi/5)n} \\ &= 2 + 4\cos[(4\pi n/5) + \pi/6] + 2\cos[(8\pi n/5) + \pi/3] \\ &= 2 + 4\sin[(4\pi n/5) + 2\pi/3] + 2\sin[(8\pi n/5) + 5\pi/6] \end{aligned}$$

3.28. **(a)** $N = 7$,

$$a_k = \frac{1}{7}\frac{e^{-j4\pi k/7}\sin(5\pi k/7)}{\sin(\pi k/7)}.$$

(b) $N = 6$, a_k over one period ($0 \le k \le 5$) may be specified as: $a_0 = 4/6$,

$$a_k = \frac{1}{6}e^{-j\pi k/2}\frac{\sin(\frac{2\pi k}{3})}{\sin(\frac{\pi k}{6})}, \qquad 1 \le k \le 5.$$

(c) $N = 6$,

$$a_k = 1 + 4\cos(\pi k/3) - 2\cos(2\pi k/3).$$

(d) $N = 12$, a_k over one period $(0 \le k \le 11)$ may be specified as: $a_1 = \frac{1}{4j} = a_{11}^*$, $a_5 = -\frac{1}{4j} = a_7^*$, $a_k = 0$ otherwise.

(e) $N = 4$.

$$a_k = 1 + 2(-1)^k(1 - \frac{1}{\sqrt{2}})\cos(\frac{\pi k}{2}).$$

(f) $N = 12$,

$$\begin{aligned} a_k &= 1 + (1 - \frac{1}{\sqrt{2}})2\cos(\frac{\pi k}{6}) + 2(1 - \frac{1}{\sqrt{2}})\cos(\frac{\pi k}{2}) \\ &+ 2(1 + \frac{1}{\sqrt{2}})\cos(\frac{5\pi k}{6}) + 2(-1)^k + 2\cos(\frac{2\pi k}{3}). \end{aligned}$$

3.29. (a) $N = 8$. Over one period $(0 \le n \le 7)$,

$$x[n] = 4\delta[n-1] + 4\delta[n-7] + 4j\delta[n-3] - 4j\delta[n-5].$$

(b) $N = 8$. Over one period $(0 \le n \le 7)$,

$$x[n] = \frac{1}{2j}\left[\frac{-e^{j\frac{3\pi n}{4}}\sin\{\frac{7}{2}(\frac{\pi n}{4} + \frac{\pi}{3})\}}{\sin\{\frac{1}{2}(\frac{\pi n}{4} + \frac{\pi}{3})\}} + \frac{e^{j\frac{3\pi n}{4}}\sin\{\frac{7}{2}(\frac{\pi n}{4} - \frac{\pi}{3})\}}{\sin\{\frac{1}{2}(\frac{\pi n}{4} - \frac{\pi}{3})\}}\right].$$

(c) $N = 8$. Over one period $(0 \le n \le 7)$,

$$x[n] = 1 + (-1)^n + 2\cos(\frac{\pi n}{4}) + 2\cos(\frac{3\pi n}{4}).$$

(d) $N = 8$. Over one period $(0 \le n \le 7)$,

$$x[n] = 2 + 2\cos\left(\frac{\pi n}{4}\right) + \cos\left(\frac{\pi n}{2}\right) + \frac{1}{2}\cos\left(\frac{3\pi n}{4}\right).$$

3.30. (a) The nonzero FS coefficients of $x(t)$ are $a_0 = 1$, $a_1 = a_{-1} = 1/2$.

(b) The nonzero FS coefficients of $x(t)$ are $b_1 = b_{-1}^* = e^{-j\pi/4}/2$.

(c) Using the multiplication property, we know that

$$z[n] = x[n]y[n] \overset{FS}{\longleftrightarrow} c_k = \sum_{l=-2}^{2} a_l b_{k-l}.$$

This implies that the nonzero Fourier series coefficients of $z[n]$ are $c_0 = \cos(\pi/4)/2$, $c_1 = c_{-1}^* = e^{-j\pi/4}/2$, $c_2 = c_{-2}^* = e^{-j\pi/4}/4$.

(d) We have

$$\begin{aligned} z[n] &= \sin\left(\frac{2\pi}{6}n + \frac{\pi}{4}\right) + \sin\left(\frac{2\pi}{6}n + \frac{\pi}{4}\right)\cos\left(\frac{2\pi}{6}n\right) \\ &= \sin\left(\frac{2\pi}{6}n + \frac{\pi}{4}\right) + \frac{1}{2}\left[\sin(\frac{4\pi}{6}n + \frac{\pi}{4}) + \sin(\frac{\pi}{4})\right] \end{aligned}$$

This implies that the nonzero Fourier series coefficients of $z[n]$ are $c_0 = \cos(\pi/4)/2$, $c_1 = c_{-1}^* = e^{-j\pi/4}/2$, $c_2 = c_{-2}^* = e^{-j\pi/4}/4$.

3.31. **(a)** $g[n]$ is as shown in Figure S3.31. Clearly, $g[n]$ has a fundamental period of 10.

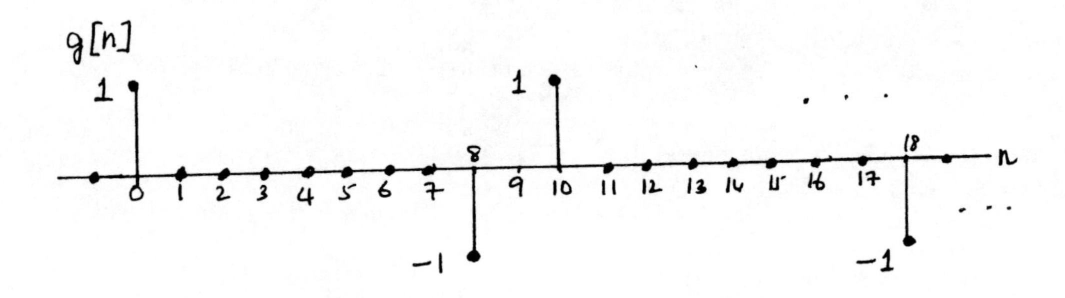

Figure S3.31

(b) The Fourier series coefficiennts of $g[n]$ are $b_k = (1/10)[1 - e^{-j(2\pi/10)8k}]$.

(c) Since $g[n] = x[n] - x[n-1]$, the FS coeffcients a_k and b_k must be related as

$$b_k = a_k - e^{-j(2\pi/10)k}a_k.$$

Therefore,

$$a_k = \frac{b_k}{1 - e^{-j(2\pi/10)k}} = \frac{(1/10)[1 - e^{-j(2\pi/10)8k}]}{1 - e^{-j(2\pi/10)k}}.$$

3.32. **(a)** The four equations are

$$a_0 + a_1 + a_2 + a_3 = 1, \qquad a_0 + ja_1 - a_2 - ja_3 = 0$$

$$a_0 - a_1 + a_2 - a_3 = 2, \qquad a_0 - ja_1 - a_2 + ja_3 = -1.$$

Solving, we get $a_0 = 1/2$, $a_1 = -\frac{1+j}{4}$, $a_2 = -1$, $a_3 = -\frac{1-j}{4}$.

(b) By direct calculation,

$$a_k = \frac{1}{4}[1 + 2e^{-jk\pi} - e^{-jk3\pi/2}].$$

This is the same as the answer we obtained in part (a) for $0 \le k \le 3$.

3.33. We will first evaluate the frequency response of the system. Consider an input $x(t)$ of the form $e^{j\omega t}$. From the discussion in Section 3.9.2 we know that the response to this input will be $y(t) = H(j\omega)e^{j\omega t}$. Therefore, substituing these in the given differential equation, we get

$$H(j\omega)j\omega e^{j\omega t} + 4e^{j\omega t} = e^{j\omega t}.$$

Therefore,

$$H(j\omega) = \frac{1}{j\omega + 4}.$$

From eq. (3.124), we know that

$$y(t) = \sum_{k=-\infty}^{\infty} a_k H(jk\omega_0)e^{jk\omega_0 t}$$

when the input is $x(t)$. $x(t)$ has the Fourier series coefficients a_k and fundamental frequency ω_0. Therefore, the Fourier series coefficients of $y(t)$ are $a_k H(jk\omega_0)$.

(a) Here, $\omega_0 = 2\pi$ and the nonzero FS coefficients of $x(t)$ are $a_1 = a_{-1} = 1/2$. Therefore, the nonzero FS coefficients of $y(t)$ are

$$b_1 = a_1 H(j2\pi) = \frac{1}{2(4 + j2\pi)}, \qquad b_{-1} = a_{-1}H(-j2\pi) = \frac{1}{2(4 - j2\pi)}.$$

(b) Here, $\omega_0 = 2\pi$ and the nonzero FS coefficients of $x(t)$ are $a_2 = a_{-2}^* = 1/2j$ and $a_3 = a_{-3}^* = e^{j\pi/4}/2$. Therefore, the nonzero FS coefficients of $y(t)$ are

$$b_2 = a_2 H(j4\pi) = \frac{1}{2j(4 + j4\pi)}, \qquad b_{-2} = a_{-2}H(-j4\pi) = -\frac{1}{2j(4 - j4\pi)},$$

$$b_3 = a_3 H(j6\pi) = \frac{e^{j\pi/4}}{2(4 + j6\pi)}, \qquad b_{-3} = a_{-3}H(-j6\pi) = -\frac{e^{-j\pi/4}}{2(4 - j6\pi)}.$$

3.34. The frequency response of the system is given by

$$H(j\omega) = \int_{-\infty}^{\infty} e^{-4|t|}e^{-j\omega t}dt = \frac{1}{4 + j\omega} + \frac{1}{4 - j\omega}.$$

(a) Here, $T = 1$ and $\omega_0 = 2\pi$ and $a_k = 1$ for all k. The FS coefficients of the output are

$$b_k = a_k H(jk\omega_0) = \frac{1}{4 + j2\pi k} + \frac{1}{4 - j2\pi k}.$$

(b) Here, $T = 2$ and $\omega_0 = \pi$ and

$$a_k = \begin{cases} 0, & k \text{ even} \\ 1, & k \text{ odd} \end{cases}.$$

Therefore, the FS coefficients of the output are

$$b_k = a_k H(jk\omega_0) = \begin{cases} 0, & k \text{ even} \\ \frac{1}{4+j\pi k} + \frac{1}{4-j\pi k}, & k \text{ odd} \end{cases}.$$

(c) Here, $T = 1$, $\omega_0 = 2\pi$ and

$$a_k = \begin{cases} 1/2, & k = 0 \\ 0, & k \text{ even}, k \neq 0 \\ \frac{\sin(\pi k/2)}{\pi k}, & k \text{ odd} \end{cases}.$$

Therefore, the FS coefficients of the output are

$$b_k = a_k H(jk\omega_0) = \begin{cases} 1/4, & k = 0 \\ 0, & k \text{ even}, k \neq 0 \\ \frac{\sin(\pi k/2)}{\pi k}\left[\frac{1}{4+j2\pi k} + \frac{1}{4-j2\pi k}\right], & k \text{ odd} \end{cases}.$$

3.35. We know that the Fourier series coefficient of $y(t)$ are $b_k = H(jk\omega_0)a_k$, where ω_0 is the fundamental frequency of $x(t)$ and a_k are the FS coefficients of $x(t)$.

If $y(t)$ is identical to $x(t)$, then $b_k = a_k$ for all k. Noting that $H(j\omega) = 0$ for $|\omega| \geq 250$, we know that $H(jk\omega_0) = 0$ for $|k| \geq 18$ (because $\omega_0 = 14$). Therefore, a_k must be zero for $|k| \geq 18$.

3.36. We will first evaluate the frequency response of the system. Consider an input $x[n]$ of the form $e^{j\omega n}$. From the discussion in Section 3.9 we know that the response to this input will be $y[n] = H(e^{j\omega})e^{j\omega n}$. Therefore, substituing these in the given difference equation, we get

$$H(e^{j\omega})e^{j\omega n} - \frac{1}{4}e^{-j\omega}e^{j\omega n}H(e^{j\omega}) = e^{j\omega n}.$$

Therefore,

$$H(j\omega) = \frac{1}{1 - \frac{1}{4}e^{-j\omega}}.$$

From eq. (3.131), we know that

$$y[n] = \sum_{k=<N>} a_k H(e^{j2\pi k/N})e^{jk(2\pi/N)n}$$

when the input is $x[n]$. $x[n]$ has the Fourier series coefficients a_k and fundamental frequency $2\pi/N$. Therefore, the Fourier series coefficients of $y[n]$ are $a_k H(e^{j2\pi k/N})$.

(a) Here, $N = 4$ and the nonzero FS coefficients of $x[n]$ are $a_3 = a_{-3}^* = 1/2j$. Therefore, the nonzero FS coefficients of $y[n]$ are

$$b_3 = a_1 H(e^{3j\pi/4}) = \frac{1}{2j(1 - (1/4)e^{-j3\pi/4})}, \qquad b_{-3} = a_{-1}H(e^{-3j\pi/4}) = \frac{-1}{2j(1 - (1/4)e^{j3\pi/4})}.$$

(b) Here, $N = 8$ and the nonzero FS coefficients of $x[n]$ are $a_1 = a_{-1} = 1/2$ and $a_2 = a_{-2} = 1$. Therefore, the nonzero FS coefficients of $y(t)$ are

$$b_1 = a_1 H(e^{j\pi/4}) = \frac{1}{2(1 - (1/4)e^{-j\pi/4})}, \qquad b_{-1} = a_{-1}H(e^{-j\pi/4}) = \frac{1}{2(1 - (1/4)e^{j\pi/4})},$$

$$b_2 = a_2 H(e^{j\pi/2}) = \frac{1}{(1 - (1/4)e^{-j\pi/2})}, \qquad b_{-2} = a_{-2}H(e^{-j\pi/2}) = \frac{1}{(1 - (1/4)e^{j\pi/2})}.$$

3.37. The frequency response of the system may be easily shown to be

$$H(e^{j\omega}) = \frac{1}{1 - \frac{1}{2}e^{-j\omega}} - \frac{1}{1 - 2e^{-j\omega}}.$$

X

(a) The Fourier series coefficients of $x[n]$ are

$$a_k = \frac{1}{4}, \qquad \text{for all } k.$$

Also, $N = 4$. Therefore, the Fourier series coefficients of $y[n]$ are

$$b_k = a_k H(e^{j2k\pi/N}) = \frac{1}{4}\left[\frac{1}{1 - \frac{1}{2}e^{-j\pi k/2}} - \frac{1}{1 - 2e^{-j\pi k/2}}\right].$$

(b) In this case, the Fourier series coefficients of $x[n]$ are

$$a_k = \frac{1}{6}[1 + 2\cos(k\pi/3)], \qquad \text{for all} k.$$

Also, $N = 6$. Therefore, the Fourier series coefficients of $y[n]$ are

$$b_k = a_k H(e^{j2k\pi/N}) = \frac{1}{6}[1 + 2\cos(k\pi/3)]\left[\frac{1}{1 - \frac{1}{2}e^{-j\pi k/3}} - \frac{1}{1 - 2e^{-j\pi k/3}}\right].$$

3.38. The frequency response of the system may be evaluated as

$$H(e^{j\omega}) = -e^{2j\omega} - e^{j\omega} + 1 + e^{-j\omega} + e^{-2j\omega}.$$

For $x[n]$, $N = 4$ and $\omega_0 = \pi/2$. The FS coefficients of the input $x[n]$ are

$$a_k = \frac{1}{4}, \qquad \text{for all } n.$$

Therefore, the FS coefficients of the output are

$$b_k = a_k H(e^{jk\omega_0}) = \frac{1}{4}[1 - e^{jk\pi/2} + e^{-jk\pi/2}].$$

3.39. Let the FS coefficients of the input be a_k. The FS coeffients of the output are of the form

$$b_k = a_k H(e^{jk\omega_0}),$$

where $\omega_0 = 2\pi/3$. Note that in the range $0 \le k \le 2$, $H(e^{jk\omega_0}) = 0$ for $k = 1, 2$. Therefore, only b_0 has a nonzero value among b_k in the range $0 \le k \le 2$.

3.40. Let the Fourier series coefficients of $x(t)$ be a_k.

(a) $x(t-t_0)$ is also periodic with period T. The Fourier series coefficients b_k of $x(t-t_0)$ are

$$\begin{aligned} b_k &= \frac{1}{T}\int_T x(t-t_0)e^{-jk(2\pi/T)t}dt \\ &= \frac{e^{-jk(2\pi/T)t_0}}{T}\int_T x(\tau)e^{-jk(2\pi/T)\tau}d\tau \\ &= e^{-jk(2\pi/T)t_0}a_k \end{aligned}$$

Similarly, the Fourier series coefficients of $x(t+t_0)$ are

$$c_k = e^{jk(2\pi/T)t_0}a_k.$$

Finally, the Fourier series coefficients of $x(t-t_0)+x(t+t_0)$ are

$$d_k = b_k + c_k = e^{-jk(2\pi/T)t_0}a_k + e^{jk(2\pi/T)t_0}a_k = 2\cos(k2\pi t_0/T)a_k.$$

(b) Note that $\mathcal{E}v\{x(t)\} = [x(t)+x(-t)]/2$. The FS coefficients of $x(-t)$ are

$$\begin{aligned} b_k &= \frac{1}{T}\int_T x(-t)e^{-jk(2\pi/T)t}dt \\ &= \frac{1}{T}\int_T x(\tau)e^{jk(2\pi/T)\tau}d\tau \\ &= a_{-k} \end{aligned}$$

Therefore, the FS coefficients of $\mathcal{E}v\{x(t)\}$ are

$$c_k = \frac{a_k+b_k}{2} = \frac{a_k+a_{-k}}{2}.$$

(c) Note that $\mathcal{R}e\{x(t)\} = [x(t)+x^*(t)]/2$. The FS coefficients of $x^*(t)$ are

$$b_k = \frac{1}{T}\int_T x^*(t)e^{-jk(2\pi/T)t}dt.$$

Conjugating both sides, we get

$$b_k^* = \frac{1}{T}\int_T x(t)e^{jk(2\pi/T)t}dt = a_{-k}.$$

Therefore, the FS coefficients of $\mathcal{R}e\{x(t)\}$ are

$$c_k = \frac{a_k+b_k}{2} = \frac{a_k+a_{-k}^*}{2}.$$

(d) The Fourier series synthesis equation gives

$$x(t) = \sum_{k=-\infty}^{\infty} a_k e^{j(2\pi/T)kt}.$$

Differentiating both sides wrt t twice, we get

$$\frac{d^2x(t)}{dt^2} = \sum_{k=-\infty}^{\infty} -k^2\frac{4\pi^2}{T^2}a_k e^{j(2\pi/T)kt}.$$

By inspection, we know that the Fourier series coefficients of $d^2x(t)/dt^2$ are $-k\frac{4\pi^2}{T^2}a_k$.

(e) The period of $x(3t)$ is a third of the period of $x(t)$. Therefore, the signal $x(3t-1)$ is periodic with period $T/3$. The Fourier series coefficents of $x(3t)$ are still a_k. Using the analysis of part (a), we know that the Fourier series coefficients of $x(3t-1)$ is $e^{-jk(6\pi/T)}a_k$.

3.41. Since $a_k = a_{-k}$, we require that $x(t) = x(-t)$. Also, note that since $a_k = a_{k+2}$, we require that

$$x(t) = x(t)e^{-j(4\pi/3)t}.$$

This in turn implies that $x(t)$ may have nonzero values only for $t = 0, \pm 1.5, \pm 3, \pm 4.5, \cdots$. Since $\int_{-0.5}^{0.5} x(t) = 1$, we may conclude that $x(t) = \delta(t)$ for $-0.5 \le t \le 0.5$. Also, since $\int_{0.5}^{1.5} x(t)dt = 2$, we may conclude that $x(t) = 2\delta(t-3/2)$ in the range $0.5 \le t \le 3/2$. Therefore, $x(t)$ may be written as

$$x(t) = \sum_{k=-\infty}^{\infty} \delta(t-k3) + 2\sum_{k=-\infty}^{\infty} \delta(t-3k-3/2).$$

3.42. **(a)** From Problem 3.40 (and Table 3.1), we know that FS coefficients of $x^*(t)$ are a^*_{-k}. Now, we know that is $x(t)$ is real, then $x(t) = x^*(t)$. Therefore, $a_k = a^*_{-k}$. Note that this implies $a_0 = a_0^*$. Therefore, a_0 must be real.

(b) From Problem 3.40 (and Table 3.1), we know that FS coefficients of $x(-t)$ are a_{-k}. If $x(t)$ is even, then $x(t) = x(-t)$. This implies that

$$a_k = a_{-k}. \quad \text{(S3.42–1)}$$

This implies that the FS coefficients are even. From the previous part, we know that if $x(t)$ is real, then

$$a_k = a^*_{-k}. \quad \text{(S3.42–2)}$$

Using eqs. (S3.42-1) and (S3.42-2), we know that $a_k = a_k^*$. Therefore, a_k is real for all k. Hence, we may conclude that a_k is real and even.

(c) From Problem 3.40 (and Table 3.1), we know that FS coefficients of $x(-t)$ are a_{-k}. If $x(t)$ is odd, then $x(t) = -x(-t)$. This implies that

$$a_k = -a_{-k}. \quad \text{(S3.42–3)}$$

This implies that the FS coefficients are odd. From the previous part, we know that if $x(t)$ is real, then

$$a_k = a_{-k}^*. \tag{S3.42–4}$$

Using eqs. (S3.42-3) and (S3.42-4), we know that $a_k = -a_k^*$. Therefore, a_k is imaginary for all k. Hence, we may conclude that a_k is real and even. Noting that eq. (S3.42-3) requires that $a_0 = -a_0$, we may also conclude that $a_0 = 0$.

(d) Note that $\mathcal{E}v\{x(t)\} = [x(t) + x(-t)]/2$. From the previous parts, we know that the FS coefficients of $\mathcal{E}v\{x(t)\}$ will be $[a_k + a_{-k}]/2$. Using eq. (S3.43-2), we may write the FS coefficients of $\mathcal{E}v\{x(t)\}$ as $[a_k + a_k^*]/2 = \mathcal{R}e\{a_k\}$.

(e) Note that $\mathcal{O}d\{x(t)\} = [x(t) - x(-t)]/2$. From the previous parts, we know that the FS coefficients of $\mathcal{O}d\{x(t)\}$ will be $[a_k - a_{-k}]/2$. Using eq. (S3.43-2), we may write the FS coefficients of $\mathcal{O}d\{x(t)\}$ as $[a_k - a_k^*]/2 = j\mathcal{I}m\{a_k\}$.

3.43. **(a)** (i) We have

$$x(t) = \sum_{\text{odd } k} a_k e^{jk\frac{2\pi}{T}t}.$$

Therefore,

$$x(t + T/2) = \sum_{\text{odd } k} a_k e^{jk\frac{2\pi}{T}t} e^{jk\pi}.$$

Since $e^{jk\pi} = -1$ for k odd,

$$x(t + T/2) = -x(t).$$

(ii) The Fourier series coefficients of $x(t)$ are

$$\begin{aligned} a_k &= \frac{1}{T}\int_0^{T/2} x(t)e^{-jk\omega_0 t}dt + \frac{1}{T}\int_{T/2}^{T} x(t)e^{-jk\omega_0 t}dt \\ &= \frac{1}{T}\int_0^{T/2} [x(t) + x(t+T/2)e^{-jk\pi}]e^{-jk\omega_0 t}dt \end{aligned}$$

Note that the right-hand side of the above equation evaluates to zero for even values of k if $x(t) = -x(t + T/2)$.

(b) The function is as shown in Figure S3.43.

Note that $T = 2$ and $\omega_0 = \pi$. Therefore,

$$a_k = \begin{cases} 0 & k \text{ even} \\ \frac{1}{jk\pi} + \frac{2}{k^2\pi^2} & k \text{ odd} \end{cases}.$$

(c) No. For an even harmonic signal we may follow the reasoning of part (a-i) to show that $x(t) = x(t + T/2)$. In this case, the fundamental period is $T/2$.

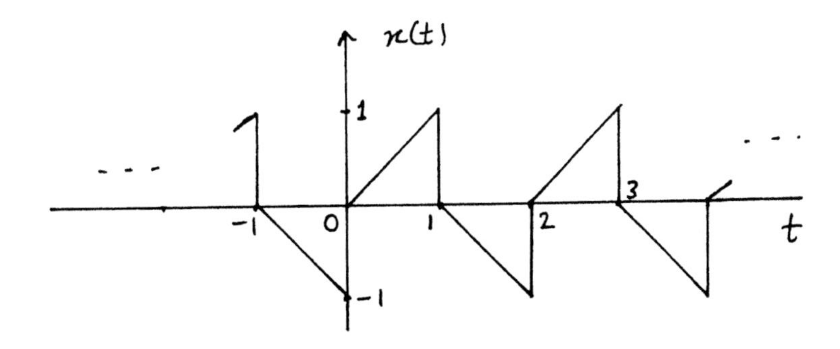

Figure S3.43

(d) (1) If a_1 or a_{-1} is nonzero, then

$$x(t) = a_{\pm 1} e^{\pm j 2\pi t/T} + \cdots.$$

and

$$x(t + t_0) = a_{\pm 1} e^{\pm j \frac{2\pi}{T}(t+t_0)} + \cdots$$

The smallest value of $|t_0|$ (other than $|t_0| = 0$ for which $e^{\pm j \frac{2\pi}{T} t_0} = 1$ is the fundamental period. Only then is

$$x(t + t_0) = a_{\pm 1} e^{\pm j 2\pi t/T} + \cdots = x(t).$$

Therefore, t_0 has to be the fundamental period.

(2) The period of $x(t)$ is the least common multiple of the periods of $e^{jk(2\pi/T)t}$ and $e^{jl(2\pi/T)t}$. The period of $e^{jk(2\pi/T)t}$ is T/k and the period of $e^{jl(2\pi/T)t}$ and T/l. Since k and l have no common factors, the least common multiple of T/k and T/l is T.

3.44. The only unknown FS coefficients are a_1, a_{-1}, a_2, and a_{-2}. Since $x(t)$ is real, $a_1 = a_{-1}^*$ and $a_2 = a_{-2}^*$. Since a_1 is real, $a_1 = a_{-1}$. Now, $x(t)$ is of the form

$$x(t) = A_1 \cos(\omega_0 t) + A_2 \cos(2\omega_0 t + \theta),$$

where $\omega_0 = 2\pi/6$. From this we get

$$x(t - 3) = A_1 \cos(\omega_0 t - 3\omega_0) + A_2 \cos(2\omega_0 t + \theta - 6\omega_0).$$

Now if we need $x(t) = -x(t-3)$, then $3\omega_0$ and $6\omega_0$ should both be odd multiples of π. Clearly, this is impossible. Therefore, $a_2 = a_{-2} = 0$ and

$$x(t) = A_1 \cos(\omega_0 t).$$

Now, using Parseval's relation on Clue 5, we get

$$\sum_{k=-\infty}^{\infty} |a_k|^2 = |a_1|^2 + |a_{-1}|^2 = \frac{1}{2}.$$

Therefore, $|a_1| = 1/2$. Since a_1 is positive, we have $a_1 = a_{-1} = 1/2$. Therefore, $x(t) = \cos(\pi t/3)$.

3.45. By inspection, we may conclude that the FS coefficients of $x(t)$ are

$$\gamma_k = \begin{cases} a_0, & k = 0 \\ B_k + jC_k, & k > 0 \\ B_k - jC_k, & k < 0 \end{cases}.$$

(a) We know from Problem 3.42 that if $x(t)$ is real, the FS coefficients of $\mathcal{E}v\{x(t)\}$ are $\mathcal{R}e\{\gamma_k\}$. Therefore,

$$\alpha_0 = a_0, \qquad \alpha_k = B_{|k|}$$

We know from Problem 3.42 that if $x(t)$ is real, the FS coefficients of $\mathcal{O}d\{x(t)\}$ are $j\mathcal{I}m\{\gamma_k\}$. Therefore,

$$\beta_0 = 0, \qquad \beta_k = \begin{cases} jC_k, & k > 0 \\ -jC_k, & k < 0 \end{cases}.$$

(b) $\alpha_k = \alpha_{-k}$ and $\beta_k = -\beta_{-k}$

(c) The signal is

$$y(t) = 1 + \mathcal{E}v\{x(t)\} + \frac{1}{2}\mathcal{E}v\{z(t)\} - \mathcal{O}d\{z(t)\}.$$

This is as shown in Figure S3.45.

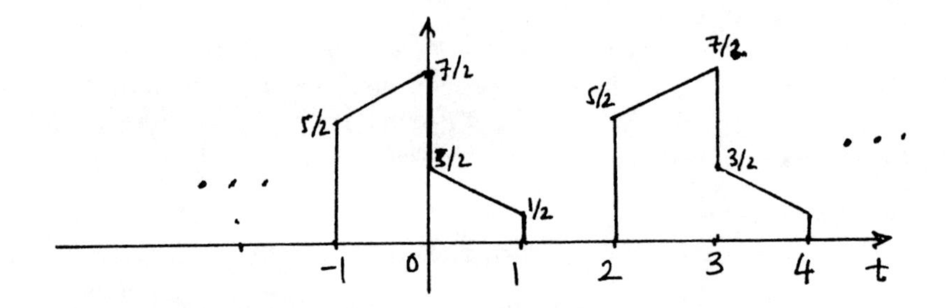

Figure S3.45

3.46. (a) The Fourier series coefficients of $z(t)$ are

$$\begin{aligned} c_k &= \frac{1}{T}\int_T \sum_n \sum_l a_n b_l e^{j(n+l)\omega_0 t} e^{-jk\omega_0 t} dt \\ &= \frac{1}{T}\sum_n \sum_l a_n b_l \delta(k - (n+l)) \\ &= \sum_n a_n b_{k-n} \end{aligned}$$

(b) (i) Here, $T_0 = 3$ and $\omega_0 = 2\pi/3$. Therefore,

$$c_k = [\frac{1}{2}\delta(k-30) + \frac{1}{2}\delta(k+30)] * \frac{2\sin(k2\pi/3)}{3k2\pi/3}.$$

Simplifying,

$$c_k = \frac{\sin\{(k-30)2\pi/3\}}{3(k-30)2\pi/3} + \frac{\sin\{(k+30)2\pi/3\}}{3(k+30)2\pi/3}$$

and $c_{\pm 30} = 1/3$.

(ii) We may express $x_2(t)$ as

$$x_2(t) = \text{sum of two shifted square waves } \times \cos(20\pi t).$$

Here, $T_0 = 3$, $\omega_0 = 2\pi/3$. Therefore,

$$\begin{aligned} c_k &= \frac{1}{3}e^{-j(k-30)(2\pi/3)}\frac{\sin\{(k-30)2\pi/3\}}{(k-30)2\pi/3} + \frac{1}{3}e^{-j(k+30)(2\pi/3)}\frac{\sin\{(k+30)2\pi/3\}}{(k+30)2\pi/3} \\ &+ \frac{1}{3}e^{-j(k-30)(\pi/3)}\frac{\sin\{(k-30)\pi/3\}}{(k-30)2\pi/3} + \frac{1}{3}e^{-j(k+30)(\pi/3)}\frac{\sin\{(k+30)\pi/3\}}{(k+30)2\pi/3} \end{aligned}$$

(iii) Here, $T_0 = 4$, $\omega_0 = \pi/2$. Therefore,

$$c_k = \left[\frac{1}{2}\delta(k-40) + \frac{1}{2}\delta(k+40)\right] * \frac{j[k\omega_0 + e^{-1}\{\sin k\omega_0 - \cos k\omega_0\}]}{2[1+(k\omega_0)^2]}.$$

Simplifying,

$$\begin{aligned} c_k &= \frac{j[(k-40)\omega_0 + e^{-1}\{\sin(k-40)\omega_0 - \cos(k-40)\omega_0\}]}{4[1+\{(k-40)\omega_0\}^2]} \\ &+ \frac{j[(k+40)\omega_0 + e^{-1}\{\sin(k+40)\omega_0 - \cos(k+40)\omega_0\}]}{4[1+\{(k+40)\omega_0\}^2]}. \end{aligned}$$

(c) From Problem 3.42, we know that $b_k = a^*_{-k}$. From part (a), we know that the FS coefficients of $z(t) = x(t)y(t) = x(t)x^*(t) = |x(t)|^2$ will be

$$c_k = \sum_{n=-\infty}^{\infty} a_n b_{n-k} = \sum_{n=-\infty}^{\infty} a_n a_{n+k}.$$

From the Fourier series analysis equation, we have

$$c_k = \frac{1}{T_0}\int_0^{T_0} |x(t)|^2 e^{-j(2\pi/T_0)kt} dt = \sum_{n=-\infty}^{\infty} a_n a^*_{n+k}.$$

Putting $k = 0$ in this equation, we get

$$\frac{1}{T_0}\int_0^{T_0} |x(t)|^2 dt = \sum_{n=-\infty}^{\infty} |a_n|^2.$$

3.47. Considering $x(t)$ to be periodic with period 1, the nonzero FS coefficients of $x(t)$ are $a_1 = a_{-1} = 1/2$. If we now consider $x(t)$ to be periodic with period 3, then the the nonzero FS coefficients of $x(t)$ are $b_3 = b_{-3} = 1/2$.

3.48. **(a)** The FS coefficients of $x[n - n_0]$ are

$$\begin{aligned}\hat{a}_k &= \frac{1}{N}\sum_{n=0}^{N-1} x[n - n_0]e^{-j2\pi nk/N} \\ &= \frac{1}{N}e^{-j\frac{2\pi n_0 k}{N}}\sum_{n=0}^{N-1} x[n]e^{-j2\pi nk/N} \\ &= e^{-j2\pi k n_0/N}a_k\end{aligned}$$

(b) Using the results of part (a), the FS coefficients of $x[n] - x[n-1]$ are given by

$$\hat{a}_k = a_k - e^{-j2\pi k/n}a_k = [1 - e^{-j2\pi k/n}]a_k.$$

(c) Using the results of part (a), the FS coefficients of $x[n] - x[n - N/2]$ are given by

$$\hat{a}_k = a_k[1 - e^{-jk\pi}] = \begin{cases} 0, & k \text{ even} \\ 2a_k, & k \text{ odd} \end{cases}.$$

(d) Note that $x[n]+x[n+N/2]$ has a period of $N/2$. The FS coefficients of $x[n]+x[n-N/2]$ are given by

$$\hat{a}_k = \frac{2}{N}\sum_{n=0}^{\frac{N}{2}-1}\left[x[n] + x[n + \frac{N}{2}]\right]e^{-j4\pi nk/N} = 2a_{2k}$$

for $0 \le k \le (N/2 - 1)$.

(e) The FS coefficients of $x^*[-n]$ are

$$\hat{a}_k = \frac{1}{N}\sum_{n=0}^{N-1} x^*[-n]e^{-j2\pi nk/N} = a_k^*.$$

(f) With N even the FS coefficients of $(-1)^n x[n]$ are

$$\hat{a}_k = \frac{1}{N}\sum_{n=0}^{N-1} x[n]e^{-j(2\pi n/N)(k-\frac{N}{2})} = a_{k-N/2}$$

(g) With N odd, the period of $(-1)^n x[n]$ is $2N$. Therefore, the FS coefficients are

$$\hat{a}_k = \frac{1}{2N}\left[\sum_{n=0}^{N-1} x[n]e^{-j\frac{2\pi n}{N}(\frac{k-N}{2})} + \sum_{n=0}^{N-1} x[n]e^{-j\frac{2\pi n}{N}(\frac{k-N}{2})}e^{-j\pi(k-N)}\right].$$

Note that for k odd $\frac{k-N}{2}$ is an integer and $k - N$ is an even integer. Also, for k even, $k - N$ is an odd integer and $e^{-j\pi(k-N)} = -1$. Therefore,

$$\hat{a}_k = \begin{cases} a_{\frac{k-N}{2}}, & k \text{ odd} \\ 0, & k \text{ even} \end{cases}.$$

(h) Here,

$$y[n] = \frac{1}{2}[x[n] + (-1)^n x[n]].$$

For N even,

$$\hat{a}_k = \frac{1}{2}[a_k + a_{k-\frac{N}{2}}].$$

For N odd,

$$\hat{a}_(k) = \begin{cases} \frac{1}{2}[a_k + a_{\frac{k-N}{2}}], & k \text{ even} \\ \frac{1}{2}a_k, & k \text{ odd} \end{cases}.$$

3.49. **(a)** The FS coefficients are given by

$$\begin{aligned} a_k &= \frac{1}{N}\sum_{n=0}^{N-1} x[n]e^{-j\frac{2\pi nk}{N}} \\ &= \frac{1}{N}\sum_{n=0}^{(N/2)-1} x[n]e^{-j\frac{2\pi nk}{N}} + \frac{1}{N}\sum_{n=N/2}^{N-1} x[n]e^{-j\frac{2\pi nk}{N}} \\ &= \frac{1}{N}\sum_{n=0}^{(N/2)-1} x[n]e^{-j\frac{2\pi nk}{N}} + \frac{e^{-j\pi k}}{N}\sum_{n=0}^{(N/2)-1} x[n+N/2]e^{-j\frac{2\pi nk}{N}} = 0 \\ &= \frac{1}{N}\sum_{n=0}^{(N/2)-1} x[n]e^{-j\frac{2\pi nk}{N}} - \frac{e^{-j\pi k}}{N}\sum_{n=0}^{(N/2)-1} x[n]e^{-j\frac{2\pi nk}{N}} \\ &= 0, \quad \text{for } k \text{ even.} \end{aligned}$$

(b) By adopting an approach similar to part (a), we may show that

$$\begin{aligned} a_k &= \frac{1}{N}\left[\sum_{n=0}^{\frac{N}{4}-1}\{1 - e^{-jk\pi/2} + e^{-j\pi k} - e^{-j\frac{3\pi k}{2}}\}x[n]e^{-j\frac{2\pi nk}{N}}\right] \\ &= 0, \quad \text{for } k = 4r, r \in \mathcal{I} \end{aligned}$$

(c) If N/M is an integer, we may generalize the approach of part (a) to show that

$$a_k = \frac{1}{N}\left[\sum_{k=0}^{B-1}\{1 - e^{-j2\pi r} + e^{-j4\pi r} - \cdots + e^{-j2\pi(M-1)r}\}x[n]e^{-j\frac{2\pi nk}{N}}\right].$$

where $B = N/M$ and $r = k/m$. From the above equation, it is clear that

$$a_k = 0, \quad \text{if } k = rM, r \in \mathcal{I}.$$

3.50. From Table 3.2, we know that if

$$x[n] \stackrel{FS}{\longleftrightarrow} a_k,$$

then

$$(-1)^n x[n] = e^{j(2\pi/N)(N/2)n} x[n] \stackrel{FS}{\longleftrightarrow} a_{k-N/2}.$$

In this case, $N = 8$. Therefore,

$$(-1)^n x[n] \stackrel{FS}{\longleftrightarrow} a_{k-4}.$$

Since it is given that $a_k = -a_{k-4}$, we have

$$x[n] = -(-1)^n x[n].$$

This implies that $x[0] = x[\pm 2] = x[\pm 4] = \cdots = 0$.

We are also given that $x[1] = x[5] = \cdots = 1$ and $x[3] = x[7] = -1$. Therefore, one period of $x[n]$ is as shown in Figure S3.50.

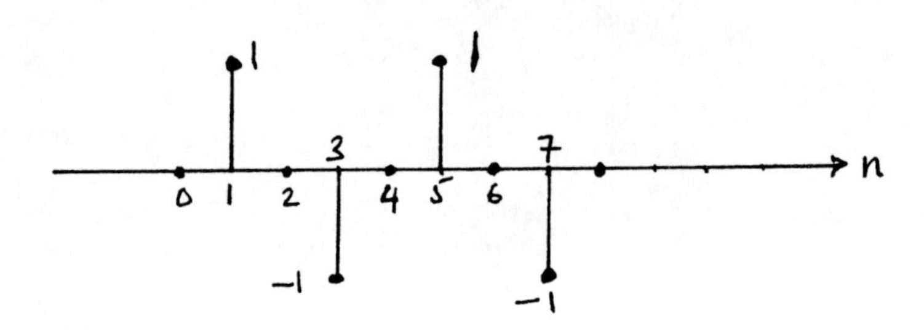

Figure S3.50

3.51. We have

$$e^{j4(2\pi/8)n} x[n] = e^{j\pi n} x[n] = (-1)^n x[n] \stackrel{FS}{\longleftrightarrow} a_{k-4}$$

and therefore,

$$(-1)^{n+1} x[n] \stackrel{FS}{\longleftrightarrow} -a_{k-4}.$$

If $a_k = -a_{k-4}$, then $x[0] = x[\pm 2] = x[\pm 4] = \cdots = 0$. Now, note that in the signal $p[n] = x[n-1]$, $p[\pm 1] = p[\pm 3] = \cdots = 0$. Now let us plot the signal $z[n] = (1 + (-1)^n)/2$. This is as shown in Figure S3.51.

Clearly, the signal $y[n] = z[n]p[n] = p[n]$ because $p[n]$ is zero whenever $z[n]$ is zero. Therefore, $y[n] = x[n-1]$. The FS coefficients of $y[n]$ are $a_k e^{-j(2\pi/8)}$.

3.52. **(a)** If $x[n]$ is real, $x[n] = x^*[n]$. Therefore,

$$a_{-k} = \sum_n x[n] e^{j2\pi nk/N} = a_k^*.$$

From this result, we get $b_{-k} = b_k$ and $c_{-k} = -c_k$.

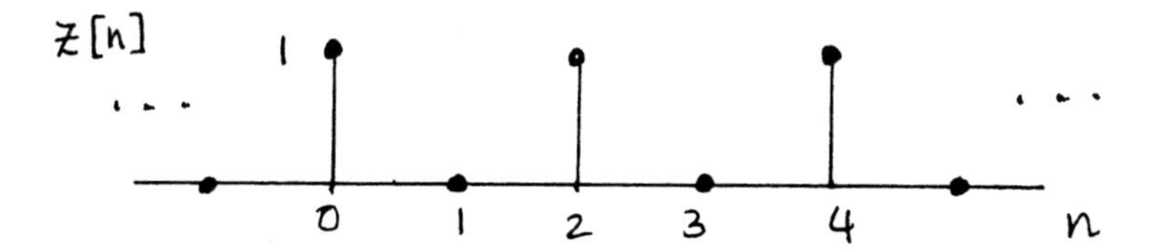

Figure S3.51

(b) If N is even, then

$$a_{N/2} = \frac{1}{N}\sum_n x[n]e^{-j\pi n} = \frac{1}{N}\sum_n (-1)^n x[n] = \text{ real.}$$

(c) If N is odd, then

$$\begin{aligned}
x[n] &= \sum_{k=-(N-1)/2}^{(N-1)/2} a_k e^{j(2\pi/N)kn} \\
&= \sum_{k=0}^{(N-1)/2} a_k e^{j(2\pi/N)kn} + \sum_{k=1}^{(N-1)/2} a_k^* e^{-j(2\pi/N)kn} \qquad \text{(From (a))} \\
&= a_0 + \sum_{k=1}^{(N-1)/2} (b_k + jc_k)e^{j(2\pi/N)kn} \sum_{k=1}^{(N-1)/2} (b_k - jc_k)e^{-j(2\pi/N)kn} \\
&= a_0 + 2\sum_{k=1}^{(N-1)/2} b_k \cos(2\pi kn/N) - c_k \sin(2\pi kn/N).
\end{aligned}$$

If N is even, then

$$\begin{aligned}
x[n] &= \sum_{k=0}^{N-1} a_k e^{j(2\pi/N)kn} \\
&= a_0 + (-1)^n a_{N/2} + 2\sum_{k=1}^{(N-2)/2} a_k e^{j(2\pi/N)kn} + a_{N-k}e^{j(2\pi/N)(N-k)n} \\
&= a_0 + (-1)^n a_{N/2} + 2\sum_{k=1}^{(N-2)/2} a_k e^{j(2\pi/N)kn} - a_k^* e^{-j(2\pi/N)kn} \qquad \text{(From (a))} \\
&= a_0 + (-1)^n a_{N/2} + 2\sum_{k=1}^{(N-1)/2} b_k \cos(2\pi kn/N) - c_k \sin(2\pi kn/N).
\end{aligned}$$

(d) If $a_k = A_k e^{j\theta_k}$, then $b_k = A\cos(\theta_k)$ and $c_k = A\sin(\theta_k)$. Substituting in the result of the previous part, we get for N odd:

$$\begin{aligned} x[n] &= a_0 + 2\sum_{k=1}^{(N-1)/2} A\cos(\theta_k)\cos(2\pi kn/N) - c_k \sin(\theta_k)\sin(2\pi kn/N) \\ &= a_0 + 2\sum_{k=1}^{(N-1)/2} A_k \cos\{\frac{2\pi nk}{N} + \theta_k\}. \end{aligned}$$

Similarly, for N even,

$$\begin{aligned} x[n] &= a_0 + (-1)^n a_{N/2} + 2\sum_{k=1}^{(N-1)/2} A\cos(\theta k)\cos(2\pi kn/N) - c_k \sin(\theta k)\sin(2\pi kn/N) \\ &= a_0 + (-1)^n a_{N/2} + 2\sum_{1}^{(N-2)/2} A_k \cos\{\frac{2\pi nk}{N} + \theta_k\}. \end{aligned}$$

(e) The signal is:

$$y[n] = d.c\{x[n]\} - d.c.\{z[n]\} + \mathcal{E}v\{z\} + \mathcal{O}d\{x\} - 2\mathcal{O}d\{z\}.$$

This is as shown Figure S3.52.

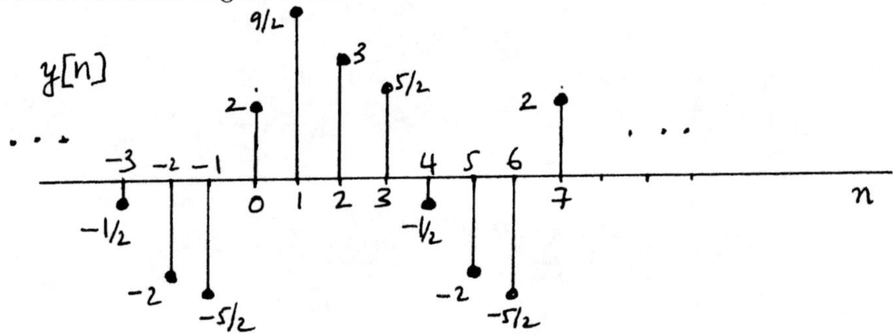

Figure S3.52

3.53. We have

$$a_k = \frac{1}{N}\sum_{<N>} x[n] e^{-j(2\pi/N)kn}.$$

Note that

$$a_0 = \frac{1}{N}\sum_{<N>} x[n]$$

which is real if $x[n]$ is real.

(a) If N is even, then

$$a_{N/2} = \frac{1}{N}\sum_{<N>} x[n]e^{-j\pi n} = \frac{1}{N}\sum_{<N>} x[n](-1)^n.$$

Clearly, $a_{N/2}$ is also real if $x[n]$ is real.

(b) If N is odd, only a_0 is guaranteed to be real.

3.54. (a) Let $k = pN$, $p \in \mathcal{I}$. Then,

$$a[pN] = \sum_{n=0}^{N-1} e^{j(2\pi/N)pNn} = \sum_{n=0}^{N-1} e^{j2\pi pn} = \sum_{n=0}^{N-1} 1 = N.$$

(b) Using the finite sum formula, we have

$$a[k] = \frac{1 - e^{j2\pi k}}{1 - e^{j(2\pi/N)k}} = 0, \qquad \text{if } k \neq pN, p \in \mathcal{I}.$$

(c) Let

$$a[k] = \sum_{n=q}^{q+N-1} e^{j(2\pi/N)kn},$$

where q is some arbitrary integer. By putting $k = pN$, we may again easily show that

$$a[pN] = \sum_{n=q}^{q+N-1} e^{j(2\pi/N)pNn} = \sum_{n=q}^{q+N-1} e^{j2\pi pn} = \sum_{n=q}^{q+N-1} 1 = N.$$

Now,

$$a[k] = e^{j(2\pi/N)kq} \sum_{n=0}^{N-1} e^{j(2\pi/N)kn}.$$

Using part (b), we may argue that $a[k] = 0$ for $k \neq pN$, $p \in \mathcal{I}$.

3.55. (a) Note that

$$x_m[n+mN] = \begin{cases} x[\frac{n}{m} + N], & n = 0, \pm m, \cdots \\ 0, & \text{otherwise} \end{cases} = \begin{cases} x[\frac{n}{m}], & n = 0, \pm m, \cdots \\ 0, & \text{otherwise} \end{cases} = x_m[n].$$

Therefore, $x_{(m)}[n]$ is periodic with period mN.

(b) The time-scaling operation discussed in this problem is a linear operation. Therefore, if $x[n] = v[n] + w[n]$, then, $x_m[n] = v_m[n] + w_m[n]$.

(c) Let us consider

$$y[n] = \frac{1}{m}\sum_{l=0}^{m-1} e^{j(2\pi/mN)(k_0+lN)n} = \frac{1}{m} e^{j(2\pi/mN)k_0 n} \sum_{l=0}^{m-1} e^{j(2\pi/m)ln}.$$

This may be written as [From Problem 3.54]

$$y[n] = \begin{cases} e^{j(2\pi/mN)k_0 n}, & n = 0, \pm N, \pm 2N, \cdots \\ 0, & \text{otherwise.} \end{cases} \tag{S3.55-1}$$

Now, also note that by applying time-scaling on $x[n]$, we get

$$x_{(m)}[n] = \begin{cases} e^{j(2\pi/mN)k_0 n}, & n = 0, \pm N, \pm 2N, \cdots \\ 0, & \text{otherwise.} \end{cases} \tag{S3.55-2}$$

Comparing eqs. (S3.55-1) and (S3.55-2), we see that $y[n] = x_{(m)}[n]$.

(d) We have

$$b_k = \frac{1}{mN} \sum_{n=0}^{mN-1} x_{(m)}[n] e^{-j(2\pi/mN)kn}.$$

We know that only every mth value in the above summation is nonzero. Therefore,

$$\begin{aligned} b_k &= \frac{1}{mN} \sum_{n=0}^{N-1} x_{(m)}[nm] e^{-j(2\pi/mN)kmn} \\ &= \frac{1}{mN} \sum_{n=0}^{N-1} x_{(m)}[nm] e^{-j(2\pi/N)kn} \end{aligned}$$

Note that $x_{(m)}[nM] = x[n]$. Therefore,

$$b_k = \frac{1}{mN} \sum_{n=0}^{N-1} x[n] e^{-j(2\pi/N)kn} = \frac{a_k}{m}.$$

3.56. **(a)** We have

$$x[n] \overset{FS}{\longleftrightarrow} a_k \quad \text{and} \quad x^*[n] \overset{FS}{\longleftrightarrow} a^*_{-k}.$$

Using the multiplication property,

$$x[n]x^*[n] = |x[n]|^2 \overset{FS}{\longleftrightarrow} \sum_{l=<N>} a_l a^*_{l+k}.$$

(b) From above, it is clear that the answer is yes.

3.57. **(a)** We have

$$x[n]y[n] = \sum_{k=0}^{N-1} \sum_{l=0}^{N-1} a_k b_l e^{j(2\pi/N)(k+l)n}.$$

Putting $l' = k + l$, we get

$$x[n]y[n] = \sum_{k=0}^{(N-1)} \sum_{l'=k}^{(k+N-1)} a_k b_{l'-k} e^{j(2\pi/N)l'n}.$$

But since both $b_{l'-k}$ and $e^{j(2\pi/N)l'n}$ are periodic with period N, we may rewrite this as

$$x[n]y[n] = \sum_{k=0}^{N-1}\sum_{l'=0}^{N-1} a_k b_{l'-k} e^{j(2\pi/N)l'n} = \sum_{l=0}^{N-1}\left[\sum_{k=0}^{N-1} a_k b_{l-k}\right] e^{j(2\pi/N)ln}.$$

Therefore,

$$c_k = \sum_{k=0}^{N-1} a_k b_{l-k}.$$

By interchanging a_k and b_k, we may show that

$$c_k = \sum_{k=0}^{N-1} b_k a_{l-k}.$$

(b) Note that since both a_k and b_k are peroidic with period N, we may rewrite the above summation as

$$c_k = \sum_{<N>} a_k b_{l-k} = \sum_{<N>} b_k a_{l-k}.$$

(c) (i) Here,

$$c_k = \sum_{l=0}^{N-1}\frac{1}{2}[\delta[l-3]+\delta[l-N+3]]a_{k-l}.$$

Therefore,

$$c_k = \frac{1}{2}a_{k-3} + \frac{1}{2}a_{k+3-N}.$$

(ii) Period=N. Also,

$$b_k = \frac{1}{N}, \qquad \text{for all } k.$$

Therefore,

$$c_k = \frac{1}{N}\sum_{l=0}^{N-1} a_l.$$

(iii) Here,

$$b_k = \frac{1}{N}[1 + e^{-j2\pi k/3} + e^{-j4\pi k/3}].$$

Therefore,

$$c_k = \frac{1}{N}\sum_{l=0}^{N-1}[1 + e^{-j2\pi l/3} + e^{-j4\pi l/3}]a_{k-l}.$$

(d) Period=12. Also,

$$x[n] \stackrel{FS}{\longleftrightarrow} a_2 = a_{10} = 1/2, \text{ All other } a_k = 0, \qquad 0 \le k \le 11$$

and

$$y[n] \stackrel{FS}{\longleftrightarrow} b_k = (\frac{1}{12})\frac{\sin 7\pi k/12}{\sin \pi k/12}, \qquad 0 \le k \le 11.$$

Therefore one period of c_k is,

$$c_k = \frac{1}{24}\left[\frac{\sin\{7\pi(k-2)/12\}}{\sin\{\pi(k-2)/12\}} + \frac{\sin\{7\pi(k-10)/12\}}{\sin\{\pi(k-10)/12\}}\right], \quad 0 \le k \le 11$$

(e) Using the FS analysis equation, we have

$$N\sum_{l=<N>} a_l b_{k-l} = \sum_{<N>} x[n]y[n]e^{-j(2\pi/N)kn}.$$

Putting $k = 0$ in this, we get

$$N\sum_{l=<N>} a_l b_{-l} = \sum_{<N>} x[n]y[n].$$

Now let $y[n] = x^*[n]$. Then $b_l = a^*_{-l}$. Therefore,

$$N\sum_{l=<N>} a_l a_l^* = \sum_{<N>} x[n]x^*[n].$$

Therefore,

$$N\sum_{l=<N>} |a_l|^2 = \sum_{<N>} |x[n]|^2.$$

3.58. **(a)** We have

$$z[n+N] = \sum_{<L>} x[r]y[n+N-r].$$

Since $y[n]$ is periodic with period N, $y[n+N-r] = y[n-r]$. Therefore,

$$z[n+N] = \sum_{<L>} x[r]y[n-r] = z[n].$$

Therefore, $z[n]$ is also periodic with period N.

(b) The FS coefficients of $z[n]$ are

$$\begin{aligned} c_l &= \frac{1}{N}\sum_{n=<N>}\sum_{k=<N>} a_k b_{n-k} e^{-j2\pi nl/N} \\ &= \frac{1}{N}\sum_{k=<N>} a_k e^{-j2\pi kl/N} \sum_{n=<N>} b_{n-k} e^{-j2\pi(n-k)l/N} \\ &= \frac{1}{N} N a_l N b_l \\ &= N a_l b_l. \end{aligned}$$

(c) Here, $n = 8$. The nonzero FS coefficients in the range $0 \le k \le 7$ for $x[n]$ are $a_3 = a_5^* = 1/2j$. Note that for $y[n]$, we need only evaluate b_3 and b_5. We have

$$b_3 = b_5^* = \frac{1}{4(1 - e^{-j3\pi/4})}.$$

Therefore, the only nonzero FS coefficients in the range $0 \le k \le 7$ for the periodic convolution of these signals are $c_3 = 8a_3b_3$ and $c_5 = 8a_5b_5$.

(d) Here,

$$x[n] \overset{FS}{\longleftrightarrow} a_k = \frac{1}{16j}\left[\frac{1 - e^{j(3\pi/7 - \pi k/4)4}}{1 - e^{-j(3\pi/7 - \pi k/4)}} - \frac{1 - e^{j(3\pi/7 + \pi k/4)4}}{1 - e^{-j(3\pi/7 + \pi k/4)}}\right]$$

and

$$y[n] \overset{FS}{\longleftrightarrow} b_k = \frac{1}{8}\left[\frac{1 - (1/2)^8}{1 - (1/2)e^{-jk\pi/4}}\right].$$

Therefore,

$$z[n] = x[n]y[n] \overset{FS}{\longleftrightarrow} 8a_kb_k.$$

3.59. **(a)** Note that the signal $x(t)$ is periodic with period NT. The FS coefficients of $x(t)$ are

$$a_k = \frac{1}{NT}\int_0^{NT}\left[\sum_{p=-\infty}^{\infty} x[p]\delta(t - pT)\right] e^{-j(2\pi/NT)kt}dt.$$

Note that the limits of the summation may be changed in accordance with the limits of the integration so that we get

$$a_k = \frac{1}{NT}\int_0^{NT}\left[\sum_{p=0}^{N-1} x[p]\delta(t - pT)\right] e^{-j(2\pi/NT)kt}dt.$$

Interchanging the summation and the integration and simplifying

$$\begin{aligned}
a_k &= (1/NT)\sum_{p=0}^{N-1} x[p]\int_0^{NT} \delta(t - pT)e^{-j(2\pi/NT)kt}dt \\
&= (1/NT)\sum_{p=0}^{N-1} x[p]e^{-j(2\pi/N)pk} \\
&= (1/T)\left[(1/N)\sum_{p=0}^{N-1} x[p]e^{-j(2\pi/N)pk}\right].
\end{aligned}$$

Note that the term within brackets on the RHS of the above equation constitutes the FS coefficients of the signal $x[n]$. Since, this is periodic with period N, a_k must also be periodic with period N.

(b) If the FS coefficients of $x(t)$ are periodic with period N, then

$$a_k = a_{k-N}.$$

This implies that

$$x(t) = x(t)e^{j(2\pi/T)Nt}.$$

This is possible only if $x(t)$ is zero for all t other than when $(2\pi/T)Nt = 2\pi k$, where $k \in \mathcal{I}$. Therefore, $x(t)$ is of the form

$$x(t) = \sum_{k=-\infty}^{\infty} g[k]\delta(t - kT/N).$$

(c) A simple example would be $x(t) = \sum_{k=-\infty}^{\infty} \delta(t - kT)$.

3.60. **(a)** The system is not LTI. $(1/2)^n$ is an eigen function of LTI systems. Therefore, the output should have been of the form $K(1/2)^n$, where K is a complex constant.

(b) It is possible to find an LTI system with this input-output relationship. The frequency response of this system would be $H(e^{j\omega}) = (1-(1/2)e^{-j\omega})/(1-(1/4)e^{-j\omega})$. The system is unique.

(c) It is possible to find an LTI system with this input-output relationship. The frequency response of this system would be $H(e^{j\omega}) = (1-(1/2)e^{-j\omega})/(1-(1/4)e^{j\omega})$. The system is unique.

(d) It is possible to find an LTI system with this input-output relationship. The system is not unique because we only require that $H(e^{j/8}) = 2$.

(e) It is possible to find an LTI system with this input-output relationship. The frequency response of this system would be $H(e^{j\omega}) = 2$. The system is unique.

(f) It is possible to find an LTI system with this input-output relationship. The system is not unique because we only require that $H(e^{j\pi/2}) = 2(1 - e^{j\pi/2})$.

(g) It is possible to find an LTI system with this input-output relationship. The system is not unique because we only require that $H(e^{j\pi/3}) = 1 - j\sqrt{3}$.

(h) Note that $x[n]$ and $y_1[n]$ are periodic with the same fundamental frequency. Therefore, it is possible to find an LTI system with this input-output relationship without violating the Eigen function property. The system is not unique because $H(e^{j\omega})$ needs to be have specific values only for $H(e^{j(2\pi/12)k})$. The rest of $H(e^{j\omega})$ may be chosen arbitrarily.

(i) Note that $x[n]$ and $y_1[n]$ are not periodic with the same fundamental frequency. Furthermore, note that $y_2[n]$ has 2/3 the period of $x[n]$. Therefore, $y[n]$ will be made up of complex exponentials which are not present in $x[n]$. This violates the eigen function property of LTI systems. Therefore, the system cannot be LTI.

3.61. **(a)** For this system,

$$x(t) \rightarrow \boxed{\delta(t)} \rightarrow x(t).$$

Therefore, all functions are eigenfunctions with an eigenvalue of one.

(b) The following is an eigen function with an eigen value of 1:

$$x(t) = \sum_k \delta(t - kT).$$

The following is an eigen function with an eigen value of 1/2:

$$x(t) = \sum_k (\frac{1}{2})^k \delta(t - kT).$$

The following is an eigen function with an eigen value of 2:

$$x(t) = \sum_k (2)^k \delta(t - kT).$$

(c) If $h(t)$ is real and even then $H(\omega)$ is real and even.

$$e^{j\omega t} \to \boxed{H(j\omega)} \to H(j\omega)e^{j\omega t}$$

and

$$e^{-j\omega t} \to \boxed{H(j\omega)} \to H(-j\omega)e^{-j\omega t} = H(j\omega)e^{-j\omega t}.$$

From these two statements, we may argue that

$$\cos(\omega t) = \frac{1}{2}[e^{j\omega t} + e^{-j\omega t}] \to \boxed{H(j\omega)} \to H(j\omega)\cos(\omega t).$$

Therefore, $\cos(\omega t)$ is an eigenfunction. We may similarly show hat $\sin(\omega t)$ is an eigenfunction.

(d) We have

$$\phi(t) \to \boxed{u(t)} \to \lambda\phi(t).$$

Therefore,

$$\lambda\phi(t) = \int_{-\infty}^{t} \phi(\tau)d\tau.$$

Differentiating both sides wrt t, we get

$$\lambda\phi'(t) = \phi(t).$$

Let $\phi(0) = \phi_0$. Then

$$\phi(t) = \phi_0 e^{t/\lambda}.$$

3.62. **(a)** The fundamental period of the input is $T = 2\pi$. The fundamental period of the input is $T = \pi$. The signals are as shown in Figure S3.62.

(b) The Fourier series coeffients of the output are

$$b_k = \frac{2(-1)^k}{\pi(1 - 4k^2)}.$$

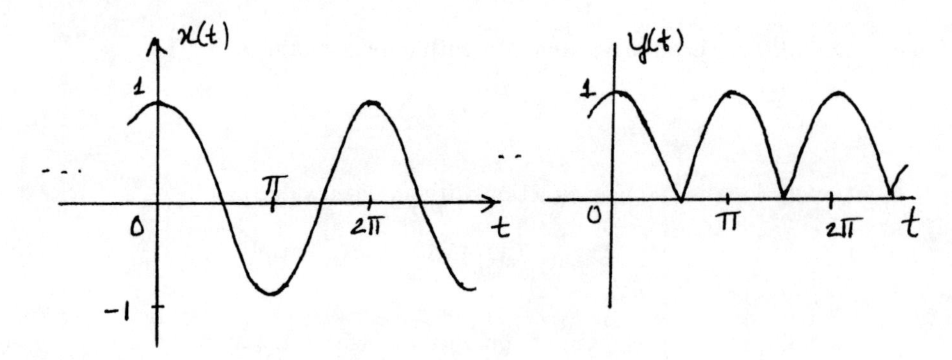

Figure S3.62

(c) The dc component of the input is 0. The dc component of the output is $2/\pi$.

3.63. The average energy per period is

$$\frac{1}{T}\int_T |x(t)|^2 dt = \sum_k |\alpha_k|^2 = \sum_k \alpha^{2|k|} = \frac{1+\alpha^2}{1-\alpha^2}.$$

We want N such that

$$\sum_{-N+1}^{N-1} |\alpha_k|^2 = 0.9\frac{1+\alpha^2}{1-\alpha^2}.$$

This implies that

$$\frac{1-2\alpha^{2N}+2\alpha^2}{1-\alpha^2} = \frac{1+\alpha^2}{1-\alpha^2}.$$

Solving,

$$N = \frac{\log[1.45\alpha^2+0.95]}{2\log\alpha},$$

and

$$\frac{\pi N}{4} < W < \frac{(N-1)\pi}{4}.$$

3.64. **(a)** Due to linearity, we have

$$y(t) = \sum_k c_k \lambda_k \phi_k(t).$$

(b) Let

$$x_1(t) \longrightarrow y_1(t) \qquad \text{and} \qquad x_2(t) \longrightarrow y_2(t).$$

Also, let

$$x_3(t) = ax_1(t) + bx_2(t) \longrightarrow y_3(t).$$

Then,

$$\begin{aligned} y_3(t) &= t^2[ax_1''(t) + bx_2''(t)] + t[ax_1'(t) + bx_2'(t)] \\ &= ay_1(t) + by_2(t) \end{aligned}$$

Therefore, the system is linear.

Now consider

$$x_4(t) = x(t - t_0) \to y_4(t).$$

We have

$$y_4(t) = t^2 \frac{d^2 x(t - t_0)}{dt^2} + t \frac{dx(t - t_0)}{dt} \neq y(t - t_0)$$

Therefore, the system is not time invariant.

(c) For inputs of the form $\phi_k(t) = t^k$, the output is

$$y(t) = k^2 t^k = k^2 \phi_k(t).$$

Therefore, $\phi_k(t)$ are eigenfunctions with eigenvalue $\lambda_k = k^2$.

(d) The output is

$$y(t) = 10^3 t^{-10} + 3t + 8t^4.$$

3.65. **(a)** Pairs (a) and (b) are orthogonal. Pairs (c) and (d) are not orthogonal.

(b) Orthogonal, but not orthonormal. $A_m = 1/\omega_0$.

(c) Orthonormal.

(d) We have

$$\int_{t_0}^{t_0+T} e^{jm\omega_0\tau} e^{-jn\omega_0\tau} d\tau = e^{j(m-n)\omega_0 t_0} \frac{[e^{j(m-n)2\pi} - 1]}{(m-n)\omega_0}$$

This evaluates to 0 when $m \neq n$ and to jT when $m = n$. Therefore, the functions are orthogonal but not orthonormal.

(e) We have

$$\begin{aligned} \int_{-T}^{T} x_e(t) x_o(t) dt &= \frac{1}{4} \int_{-T}^{T} [x(t) + x(-t)][x(t) - x(-t)] dt \\ &= \frac{1}{4} \int_{-T}^{T} x^2(t) dt - \frac{1}{4} \int_{-T}^{T} x^2(-t) dt \\ &= 0. \end{aligned}$$

(f) Consider

$$\int_a^b \frac{1}{\sqrt{A_k}} \phi_k(t) \frac{1}{\sqrt{A_l}} \phi_l^*(t) dt = \frac{1}{\sqrt{A_k A_l}} \int_a^b \int_a^b \phi_k(t) \phi_l^*(t) dt.$$

This valuates to zero for $k \neq l$. For $k = l$, it evaluates to $A_k/A_k = 1$. Therefore, the functions are orthonormal.

(g) We have

$$
\begin{aligned}
\int_a^b |x(t)|^2 dt &= \int_a^b x(t)x^*(t)dt \\
&= \int_a^b \sum_i a_i\phi_i(t) \sum_j a_j\phi_j^*(t)dt \\
&= \sum_i\sum_j a_i a_j^* \int_a^b \phi_i(t)\phi_j^*(t)dt \\
&= \sum_i |a_i|^2.
\end{aligned}
$$

(h) We have

$$
\begin{aligned}
y(T) &= \int_{-\infty}^{\infty} h_i(T-\tau)\phi_j(\tau)d\tau \\
&= \int_{-\infty}^{\infty} \phi_i(\tau)\phi_j(\tau)d\tau \\
&= \delta_{ij} = 1 \text{ for } i=j \text{ and } 0 \text{ for } i \neq j.
\end{aligned}
$$

3.66. **(a)** We have

$$
E = \int_a^b \left[x(t) - \sum_{k=-N}^{N} a_k\phi_k(t)\right]\left[x^*(t) - \sum_{k=-N}^{N} a_k^*\phi_k^*(t)\right] dt
$$

Now, let $a_i = b_i + jc_i$. Then

$$
\frac{\partial E}{\partial b_i} = 0 = -\int_a^b \phi_i^*(t)x(t)dt + 2b_i - \int_a^b \phi_i(t)x^t(t)dt
$$

and

$$
\frac{\partial E}{\partial c_i} = 0 = j\int_a^b \phi_i(t)x^*(t)dt + 2c_i - j\int_a^b \phi_i^*(t)x(t)dt.
$$

Mutliplying the last equation by j and adding to the one before, we get

$$
2b_i + 2jc_i = 2\int_a^b x(t)\phi^*(t)dt.
$$

This implies that

$$
a_i = \int_a^b x(t)\phi^*(t)dt.
$$

(b) In this case, a_i would be

$$
a_i = \frac{1}{A_i}\int_a^b x(t)\phi_i^*(t)dt.
$$

(c) Choosing

$$a_k = \frac{1}{T_0}\int_b^{b+T_0} x(t)e^{-jk\omega_0 t}dt,$$

we have

$$E = \int_{T_0}\left|x(t) - \sum_{k=-N}^{N} a_k e^{j\omega_0 kt}\right|^2 dt.$$

Putting $\frac{\partial E}{\partial a_k} = 0$, we get

$$a_k = \frac{1}{T_0}\int_{T_0} x(t)e^{-jk\omega_0 t}dt.$$

(d) $a_0 = 2/\pi$, $a_1 = a_3 = 0$, $a_2 = 2(1 - 2\sqrt{2})/\pi$, $a_4 = (1/\pi)[2 - 4\cos(\pi/8) + 4\cos(3\pi/8)]$.

(e) We have

$$\begin{aligned}\int_0^1 \sum_i (a_i\phi_i(t))^*[x(t) - \sum_i a_i\phi_i(t)]dt &= \sum_i a_i^* \int_0^1 x(t)\phi_i^*(t)dt \\ &\quad -\sum_i\sum_j a_i^* a_j \int_0^1 \phi_i^*(t)\phi_j(t)dt \\ &= \sum_i a_i^* a_i - \sum_i a_i^* a_i = 0\end{aligned}$$

(f) Not orthogonal. Example: $\int_0^1 \phi_0(t)\phi_1(t) = \int_0^1 tdt = 1 \neq 0$.

(g) Here,

$$a_0 = \int_0^1 e^t \phi_0^*(t)dt = e - 1.$$

(h) Here, $\hat{x}(t) = a_0 + a_1 t$. Therefore,

$$E = \int_0^1 (e^t - a_0 - a_1 t)(e^t - a_0 - a_1 t)dt.$$

Setting $\partial E/\partial a_0 = 0 = \partial E/\partial a_1$, we get $a_0 = 2(2e - 5)$ and $a_1 = 6(3 - e)$.

3.67. **(a)** From eq. (P3.67-1) and (P3.67-4), we get

$$\sum_{n=-\infty}^{\infty} j2\pi n b_n(x)e^{j2\pi nt} = \frac{1}{2}k^2 \sum_{n=-\infty}^{\infty} \frac{\partial^2 b_n(x)}{\partial x^2}e^{j2\pi nt}.$$

Equating coefficients of $e^{j2\pi nt}$ on both sides, we get

$$\frac{\partial^2 b_n(x)}{\partial x^2} = \frac{j4\pi n}{k^2}b_n(x).$$

(b) Since $s^2 = 4\pi jn/k^2$,

$$s = \pm\frac{2\sqrt{\pi n}e^{j\pi/4}}{k}.$$

For $n > 0$,

$$s = \frac{\sqrt{2\pi n}(1+j)}{k}$$

is a stable solution. For $n < 0$,

$$s = -\frac{\sqrt{2\pi|n|}(1-j)}{k}$$

is a stable solution. Also, $b_n(0) = a_n$ and

$$b_n(x) = \begin{cases} a_n e^{-\sqrt{2\pi n}(1+j)x/k}, & n > 0 \\ a_n e^{-\sqrt{2\pi|n|}(1-j)x/k}, & n < 0 \end{cases}.$$

(c) $b_0 = 2$. $b_1 = (1/2j)e^{-(1+j)\pi}$, $b_{-1} = -(1/2j)e^{-(1-j)\pi}$.

$$T(k\sqrt{\pi/2}, t) = 2 + e^{-\pi}\sin(2\pi t - \pi).$$

Phase reversed.

3.68. **(a)** $x(\theta) = r(\theta)\cos(\theta) = \frac{1}{2}r(\theta)e^{j\theta} + \frac{1}{2}r(\theta)e^{-j\theta}$. If

$$x(\theta) = \sum_{k=-\infty}^{\infty} b_k e^{jk\theta},$$

then $b_k = (1/2)a_{k+1} + (1/2)a_{k-1}$.

(b) $x(\theta) \stackrel{FS}{\longleftrightarrow} b_k$. Then $x(\theta) = r(\theta + \pi/4)$. The sketch is as shown in Figure S3.68.

(c) $b_0 = a_0$. Rest of b_k is all zero. Therefore, the sketch will be a circle of radius a_0 as shown in Figure S3.68.

(d) (i) $r(\theta) = r(-\theta)$. Even. Sketch as shown in Figure S3.68.

(ii) $r(\theta + k\pi) = r(\theta)$. Sketch as shown in Figure S3.68.

(iii) $r(\theta + k\pi/2) = r(\theta)$. Sketch as shown in Figure S3.68.

3.69. **(a)** $\sum_{n=-N}^{N} \phi_k[n]\phi_k^*[m] = \sum_{n=-N}^{N} \delta[n-k]\delta[n-m]$. This is 1 for $k = m$ and 0 for $k \neq m$. Therefore, orthogonal.

(b) We have

$$\sum_{n=r}^{r+N-1} \phi_k[n]\phi_m^*[n] = e^{j(2\pi/N)r(k-m)}\left[\frac{1 - e^{j2\pi(k-m)}}{1 - e^{j(2\pi/N)(k-m)}}\right] = \begin{cases} 0, & k \neq m \\ N, & k = m \end{cases}.$$

Therefore, orthogonal.

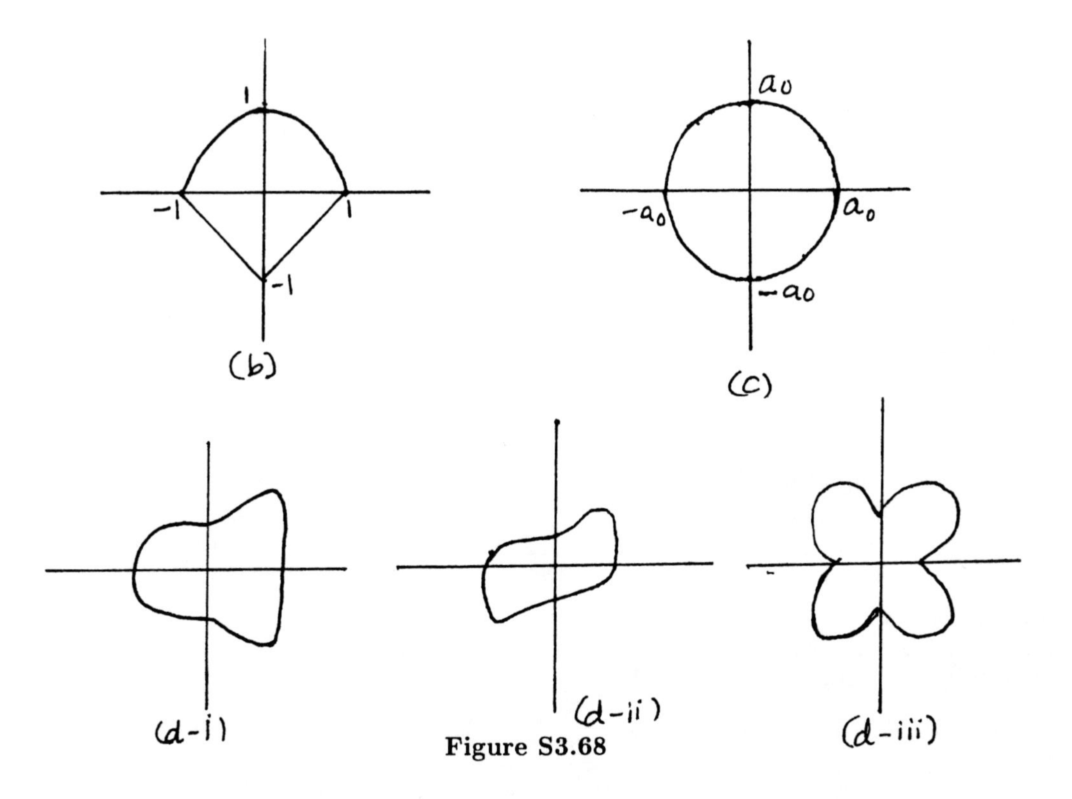

Figure S3.68

(c) We have

$$\begin{aligned}
\sum_{n=N_1}^{N_2} |x[n]|^2 &= \sum_{n=N_1}^{N_2} \sum_{i=1}^{M} a_i \phi_i[n] \sum_{k=1}^{M} a_k^* \phi_k^*[n] \\
&= \sum_{k=1}^{M} \sum_{i=1}^{M} a_i a_k^* \sum_{n=N_1}^{N_2} \phi_k^*[n] \phi_i[n] \\
&= \sum_{k=1}^{M} \sum_{i=1}^{M} a_i a_k^* A_i \delta[i-k] = \sum_{i=1}^{M} |a_i|^2 A_i
\end{aligned}$$

(d) Let $a_i = b_i + jc_i$. Then

$$\begin{aligned}
E &= \sum_{n=N_1}^{N_2} |x[n]|^2 + \sum_{i=1}^{M} (b_i^2 + c_i^2) A_i - \sum_{n=N_1}^{N_2} x[n] \sum_{i=1}^{M} (b_i - jc_i) \phi_i^*[n] \\
&\quad - \sum_{n=N_1}^{N_2} x^*[n] \sum_{i=1}^{M} (b_i + jc_i) \phi_i[n]
\end{aligned}$$

Set $\partial E/\partial b_i = 0$. Then

$$b_i = [2A_i]^{-1}\left[\sum_{n=N_1}^{N_2}\{x[n]\phi_i^*[n] + x^*[n]\phi_i[n]\}\right] = \frac{1}{A_i}\mathcal{R}e\left\{\sum_{n=N_1}^{N_2} x[n]\phi_i^*[n]\right\}.$$

Similarly,

$$c_i = \frac{1}{A_i}\mathcal{I}m\left\{\sum_{n=N_1}^{N_2} x[n]\phi_i^*[n]\right\}.$$

Therefore,

$$a_i = b_i + jc_i = \frac{1}{A_i}\sum_{n=N_1}^{N_2} x[n]\phi_i^*[n].$$

(e) $\phi_i[n] = \delta[n-i]$. Then,

$$a_i = \sum_{n=N_1}^{N_2} x[n]\delta[n-i] = x[i].$$

3.70. **(a)** We get

$$a_{mn} = \frac{1}{T_1T_2}\int_0^{T_1}\int_0^{T_2} x(t_1,t_2)e^{-jm\omega_1 t_1}e^{-jn\omega_2 t_2}dt_1dt_2.$$

(b) (i) $T_1 = 1$, $T_2 = \pi$. $a_{11} = 1/2$, $a_{-1,-1} = 1/2$. Rest of the coefficients are all zero.

(ii) Here,

$$a_{mn} = \begin{cases} 1/(\pi^2 mn), & m,n \text{ odd} \\ 0, & \text{otherwise} \end{cases}.$$

3.71. **(a)** The differential equation $f_s(t)$ and $f(t)$ is

$$\frac{B}{K}\frac{df_s(t)}{dt} + f_s(t) = f(t).$$

The frequency response of this system may be easily shown to be

$$H(j\omega) = \frac{1}{1+(B/K)j\omega}.$$

Note that for $\omega = 0$, $H(j\omega) = 1$ and for $\omega \to \infty$, $H(j\omega) = 0$. Therefore, the system approximates a lowpass filter.

(b) The differential equation $f_d(t)$ and $f(t)$ is

$$\frac{df_d(t)}{dt} + \frac{K}{B}f_d(t) = \frac{df(t)}{dt}.$$

The frequency response of this system may be easily shown to be

$$H(j\omega) = \frac{j\omega}{j\omega + (K/B)}.$$

Note that for $\omega = 0$, $H(j\omega) = 0$ and for $\omega \to \infty$, $H(j\omega) = 1$. Therefore, the system approoximates a highpass filter.

Chapter 4 Answers

4.1. (a) Let $x(t) = e^{-2(t-1)}u(t-1)$. Then the Fourier transform $X(j\omega)$ of $x(t)$ is:

$$
\begin{aligned}
X(j\omega) &= \int_{-\infty}^{\infty} e^{-2(t-1)}u(t-1)e^{-j\omega t}dt \\
&= \int_{1}^{\infty} e^{-2(t-1)}e^{-j\omega t}dt \\
&= e^{-j\omega}/(2+j\omega)
\end{aligned}
$$

$|X(j\omega)|$ is as shown in Figure S4.1.

(b) Let $x(t) = e^{-2|t-1|}$. Then the Fourier transform $X(j\omega)$ of $x(t)$ is:

$$
\begin{aligned}
X(j\omega) &= \int_{-\infty}^{\infty} e^{-2|t-1|}e^{-j\omega t}dt \\
&= \int_{1}^{\infty} e^{-2(t-1)}e^{-j\omega t}dt + \int_{-\infty}^{1} e^{2(t-1)}e^{-j\omega t}dt \\
&= e^{-j\omega}/(2+j\omega) + e^{-j\omega}/(2-j\omega) \\
&= 4e^{-j\omega}/(4+\omega^2)
\end{aligned}
$$

$|X(j\omega)|$ is as shown in Figure S4.1.

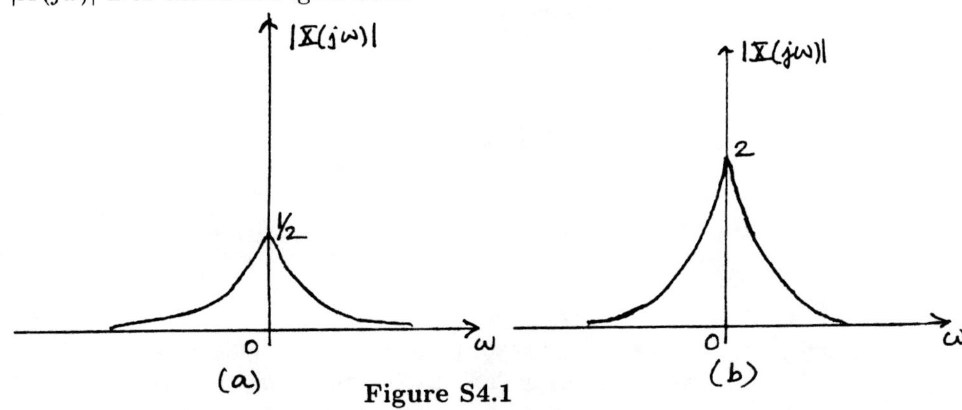

Figure S4.1

4.2. (a) Let $x_1(t) = \delta(t+1) + \delta(t-1)$. Then the Fourier transform $X_1(j\omega)$ of $x(t)$ is:

$$
\begin{aligned}
X_1(j\omega) &= \int_{-\infty}^{\infty} [\delta(t+1) + \delta(t-1)]e^{-j\omega t}dt \\
&= e^{j\omega} + e^{-j\omega} = 2\cos\omega
\end{aligned}
$$

$|X_1(j\omega)|$ is as sketched in Figure S4.2.

(b) The signal $x_2(t) = u(-2-t) + u(t-2)$ is as shown in the figure below. Clearly,

$$
\frac{d}{dt}\{u(-2-t) + u(t-2)\} = \delta(t-2) - \delta(t+2)
$$

Therefore,

$$\begin{aligned} X_2(j\omega) &= \int_{-\infty}^{\infty} [\delta(t-2) - \delta(t+2)]e^{-j\omega t}dt \\ &= e^{-2j\omega} - e^{2j\omega} = -2j\sin(2\omega) \end{aligned}$$

$|X_1(j\omega)|$ is as sketched in Figure S4.2.

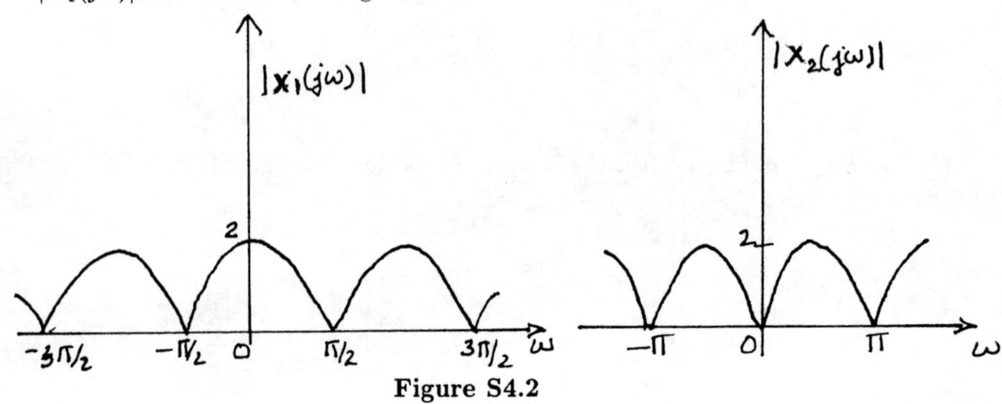

Figure S4.2

4.3. **(a)** The signal $x_1(t) = \sin(2\pi t + \pi/4)$ is periodic with a fundamental period of $T = 1$. This translates to a fundamental frequency of $\omega_0 = 2\pi$. The nonzero Fourier series coefficients of this signal may be found by writing it in the form

$$\begin{aligned} x_1(t) &= \frac{1}{2j}\left(e^{j(2\pi t+\pi/4)} - e^{-j(2\pi t+\pi/4)}\right) \\ &= \frac{1}{2j}e^{j\pi/4}e^{j2\pi t} - \frac{1}{2j}e^{-j\pi/4}e^{-j2\pi t} \end{aligned}$$

Therefore, the nonzero Fourier series coefficients of $x_1(t)$ are

$$a_1 = \frac{1}{2j}e^{j\pi/4}e^{j2\pi t}, \qquad a_{-1} = -\frac{1}{2j}e^{-j\pi/4}e^{-j2\pi t}$$

From Section 4.2, we know that for periodic signals, the Fourier transform consists of a train of impulses occurring at $k\omega_0$. Furthermore, the area under each impulse is 2π times the Fourier series coefficient a_k. Therefore, for $x_1(t)$, the corresponding Fourier transform $X_1(j\omega)$ is given by

$$\begin{aligned} X_1(j\omega) &= 2\pi a_1\delta(\omega - \omega_0) + 2\pi a_{-1}\delta(\omega + \omega_0) \\ &= (\pi/j)e^{j\pi/4}\delta(\omega - 2\pi) - (\pi/j)e^{-j\pi/4}\delta(\omega + 2\pi) \end{aligned}$$

(b) The signal $x_2(t) = 1 + \cos(6\pi t + \pi/8)$ is periodic with a fundamental period of $T = 1/3$. This translates to a fundamental frequency of $\omega_0 = 6\pi$. The nonzero Fourier series coefficients of this signal may be found by writing it in the form

$$\begin{aligned} x_2(t) &= 1 + \frac{1}{2}\left(e^{j(6\pi t+\pi/8)} - e^{-j(6\pi t+\pi/8)}\right) \\ &= 1 + \frac{1}{2}e^{j\pi/8}e^{j6\pi t} + \frac{1}{2}e^{-j\pi/8}e^{-j6\pi t} \end{aligned}$$

Therefore, the nonzero Fourier series coefficients of $x_2(t)$ are

$$a_0 = 1, \qquad a_1 = \frac{1}{2}e^{j\pi/8}e^{j6\pi t}, \qquad a_{-1} = \frac{1}{2}e^{-j\pi/8}e^{-j6\pi t}$$

From Section 4.2, we know that for periodic signals, the Fourier transform consists of a train of impulses occurring at $k\omega_0$. Furthermore, the area under each impulse is 2π times the Fourier series coefficient a_k. Therefore, for $x_2(t)$, the corresponding Fourier transform $X_2(j\omega)$ is given by

$$\begin{aligned} X_2(j\omega) &= 2\pi a_0\delta(\omega) + 2\pi a_1\delta(\omega - \omega_0) + 2\pi a_{-1}\delta(\omega + \omega_0) \\ &= 2\pi\delta(\omega) + \pi e^{j\pi/8}\delta(\omega - 6\pi) + \pi e^{-j\pi/8}\delta(\omega + 6\pi) \end{aligned}$$

4.4. (a) The inverse Fourier transform is

$$\begin{aligned} x_1(t) &= (1/2\pi)\int_{-\infty}^{\infty}[2\pi\delta(\omega) + \pi\delta(\omega - 4\pi) + \pi\delta(\omega + 4\pi)]e^{j\omega t}d\omega \\ &= (1/2\pi)[2\pi e^{j0t} + \pi e^{j4\pi t} + \pi e^{-j4\pi t}] \\ &= 1 + (1/2)e^{j4\pi t} + (1/2)e^{-j4\pi t} = 1 + \cos(4\pi t) \end{aligned}$$

(b) The inverse Fourier transform is

$$\begin{aligned} x_2(t) &= (1/2\pi)\int_{-\infty}^{\infty}X_2(j\omega)e^{j\omega t}d\omega \\ &= (1/2\pi)\int_{0}^{2}2e^{j\omega t}d\omega + (1/2\pi)\int_{-2}^{0}(-2)e^{j\omega t}d\omega \\ &= (e^{j2t} - 1)/(\pi jt) - (1 - e^{-j2t})/(\pi jt) \\ &= -(4j\sin^2 t)/(\pi t) \end{aligned}$$

4.5. From the given information,

$$\begin{aligned} x(t) &= (1/2\pi)\int_{-\infty}^{\infty}X(j\omega)e^{j\omega t}d\omega \\ &= (1/2\pi)\int_{-\infty}^{\infty}|X(j\omega)|e^{j\sphericalangle\{X(j\omega)\}}e^{j\omega t}d\omega \\ &= (1/2\pi)\int_{-3}^{3}2e^{-\frac{3}{2}\omega + \pi}e^{j\omega t}d\omega \\ &= \frac{-2}{\pi(t - 3/2)}\sin[3(t - 3/2)] \end{aligned}$$

The signal $x(t)$ is zero when $3(t - 3/2)$ is a nonzero integer multiple of π. This gives

$$t = \frac{k\pi}{2} + \frac{3}{2}, \qquad \text{for } k \in \mathcal{I}, \text{ and } k \neq 0.$$

4.6. Throughout this problem, we assume that

$$x(t) \stackrel{FT}{\longleftrightarrow} X_1(j\omega).$$

(a) Using the time reversal property (Sec. 4.3.5), we have

$$x(-t) \stackrel{FT}{\longleftrightarrow} X(-j\omega)$$

Using the time shifting property (Sec. 4.3.2) on this, we have

$$x(-t+1) \stackrel{FT}{\longleftrightarrow} e^{-j\omega t}X(-j\omega) \quad \text{and} \quad x(-t-1) \stackrel{FT}{\longleftrightarrow} e^{j\omega t}X(-j\omega)$$

Therefore,

$$\begin{aligned} x_1(t) = x(-t+1) + x(-t-1) &\stackrel{FT}{\longleftrightarrow} e^{-j\omega t}X(-j\omega) + e^{j\omega t}X(-j\omega) \\ &\stackrel{FT}{\longleftrightarrow} 2X(-j\omega)\cos\omega \end{aligned}$$

(b) Using the time scaling property (Sec. 4.3.5), we have

$$x(3t) \stackrel{FT}{\longleftrightarrow} \frac{1}{3}X\left(j\frac{\omega}{3}\right)$$

Using the time shifting property on this, we have

$$x_2(t) = x(3(t-2)) \stackrel{FT}{\longleftrightarrow} e^{-2j\omega}\frac{1}{3}X\left(j\frac{\omega}{3}\right)$$

(c) Using the differentiation in time property (Sec. 4.3.4), we have

$$\frac{dx(t)}{dt} \stackrel{FT}{\longleftrightarrow} j\omega X(j\omega)$$

Applying this property again, we have

$$\frac{d^2x(t)}{dt^2} \stackrel{FT}{\longleftrightarrow} -\omega^2 X(j\omega).$$

Using the time shifting property, we have

$$x_3(t) = \frac{d^2x(t-1)}{dt^2} \stackrel{FT}{\longleftrightarrow} -\omega^2 X(j\omega)e^{-j\omega t}.$$

4.7. **(a)** Since $X_1(j\omega)$ is not conjugate symmetric, the corresponding signal $x_1(t)$ is **not real.** Since $X_1(j\omega)$ is neither even nor odd, the corresponding signal $x_1(t)$ is **neither even nor odd.**

(b) The Fourier transform of a real and odd signal is purely imaginary and odd. Therefore, we may conclude that the Fourier transform of a purely imaginary and odd signal is real and odd. Since $X_2(j\omega)$ is real and odd, we may therefore conclude that the corresponding signal $x_2(t)$ is **purely imaginary and odd.**

(c) Consider a signal $y_3(t)$ whose magnitude of the Fourier transform is $|Y_3(j\omega)| = A(\omega)$, and whose phase of the Fourier transform is $\sphericalangle\{Y_3(j\omega)\} = 2\omega$. Since $|Y_3(j\omega)| = |Y_3(-j\omega)|$ and $\sphericalangle\{Y_3(j\omega)\} = -\sphericalangle\{Y_3(-j\omega)\}$, we may conclude that the signal $y_3(t)$ is real (See Table 4.1, Property 4.3.3).

Now, consider the signal $x_3(t)$ with Fourier transform $X_3(j\omega) = Y_3(j\omega)e^{j\pi/2} = jY_3(j\omega)$. Using the result from the previous paragraph and the linearity property of the Fourier transform, we may conclude that $x_3(t)$ has to **imaginary**. Since the Fourier transform $X_3(j\omega)$ is neither purely imaginary nor purely real, the signal $x_3(t)$ is **neither even nor odd**.

(d) Since $X_4(j\omega)$ is both real and even, the corresponding signal $x_4(t)$ is **real and even**.

4.8. **(a)** The signal $x(t)$ is as shown in the Figure S4.8.

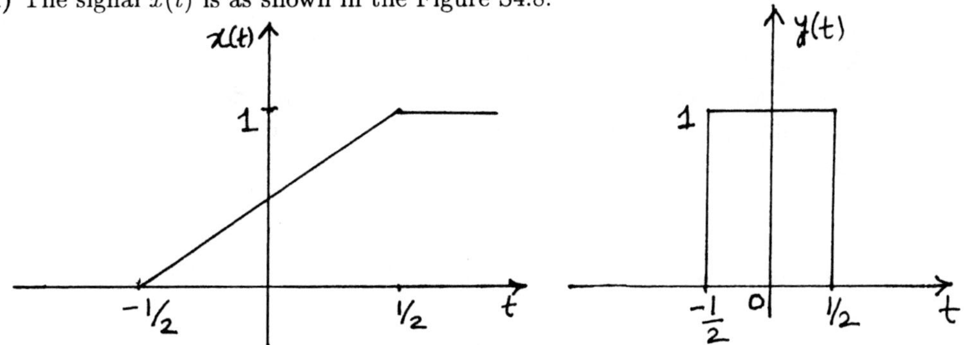

Figure S4.8

We may express this signal as

$$x(t) = \int_{-\infty}^{t} y(t)dt,$$

where $y(t)$ is the rectangular pulse shown in Figure S4.8. Using the integration property of the Fourier transform, we have

$$x(t) \stackrel{FT}{\longleftrightarrow} X(j\omega) = \frac{1}{j\omega}Y(j\omega) + \pi Y(j0)\delta(\omega)$$

We know from Table 4.2 that

$$Y(j\omega) = \frac{2\sin(\omega/2)}{\omega}$$

Therefore,

$$X(j\omega) = \frac{2\sin(\omega/2)}{j\omega^2} + \pi\delta(\omega)$$

(b) If $g(t) = x(t) - 1/2$, then the Fourier transform $G(j\omega)$ of $g(t)$ is given by

$$G(j\omega) = X(j\omega) - (1/2)2\pi\delta(\omega) = \frac{2\sin(\omega/2)}{j\omega^2}.$$

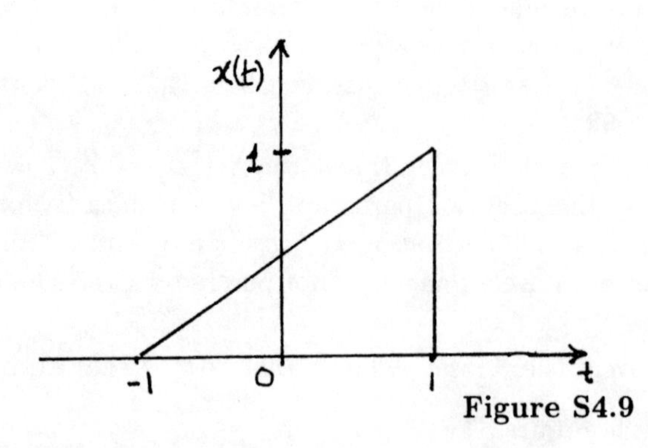

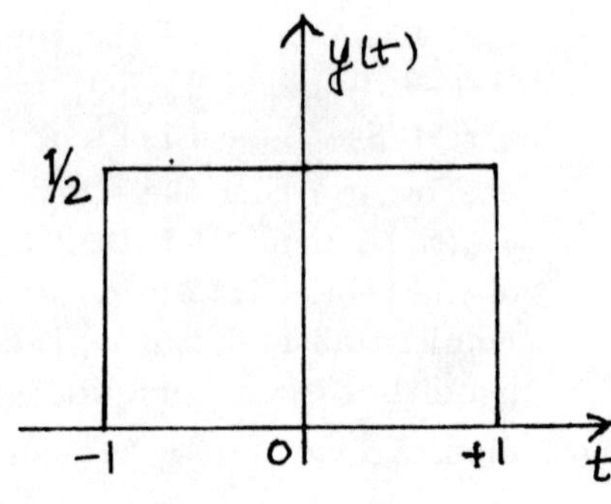

Figure S4.9

4.9. **(a)** The signal $x(t)$ is plotted in Figure S4.9.

We see that this signal is very similar to the one considered in the previous problem. In fact we may again express the signal $x(t)$ in terms of the rectangular pulse $y(t)$ shown above as follows

$$x(t) = \int_{-\infty}^{t} y(t)dt - u(t - \frac{1}{2}).$$

Using the result obtained in part (a) of the previous problem, the Fourier transform $X(j\omega)$ of $x(t)$ is

$$\begin{aligned} X(j\omega) &= \frac{2\sin(\omega/2)}{j\omega^2} + \pi\delta(\omega) - \mathcal{FT}\{u(t - \frac{1}{2})\} \\ &= \frac{\sin\omega}{j\omega^2} - \frac{e^{-j\omega}}{j\omega} \end{aligned}$$

(b) The even part of $x(t)$ is given by

$$\mathcal{E}v\{x(t)\} = \frac{x(t) + x(-t)}{2}.$$

This is as shown in the Figure S4.9.

Therefore,

$$\mathcal{FT}\{\mathcal{E}v\{x(t)\}\} = \frac{\sin\omega}{\omega}.$$

Now the real part of the answer to part (a) is

$$\mathcal{R}e\left\{-\frac{e^{-j\omega}}{j\omega}\right\} = (1/\omega)\mathcal{R}e\{j(\cos\omega - j\sin\omega)\} = \frac{\sin\omega}{\omega}$$

(c) The Fourier transform of the odd part of $x(t)$ is same as j times imaginary part of the answer to part (a). We have

$$\mathcal{I}m\left\{\frac{\sin\omega}{j\omega^2} - \frac{e^{-j\omega}}{j\omega}\right\} = -\frac{\sin\omega}{\omega^2} + \frac{\cos\omega}{\omega}$$

Therefore, the desired result is

$$\mathcal{FT}\{\text{Odd part of} x(t)\} = \frac{\sin\omega}{j\omega^2} - \frac{\cos\omega}{j\omega}$$

4.10. **(a)** We know from Table 4.2 that

$$\frac{\sin t}{\pi t} \stackrel{FT}{\longleftrightarrow} \text{Rectangular function } Y(j\omega) \text{ [See Figure S4.10]}$$

Therefore

$$\left(\frac{\sin t}{\pi t}\right)^2 \stackrel{FT}{\longleftrightarrow} (1/2\pi)\,[\text{Rectangular function } Y(j\omega) * \text{Rectangular function } Y(j\omega)]$$

This is a triangular function $Y_1(j\omega)$ as shown in the Figure S4.10.

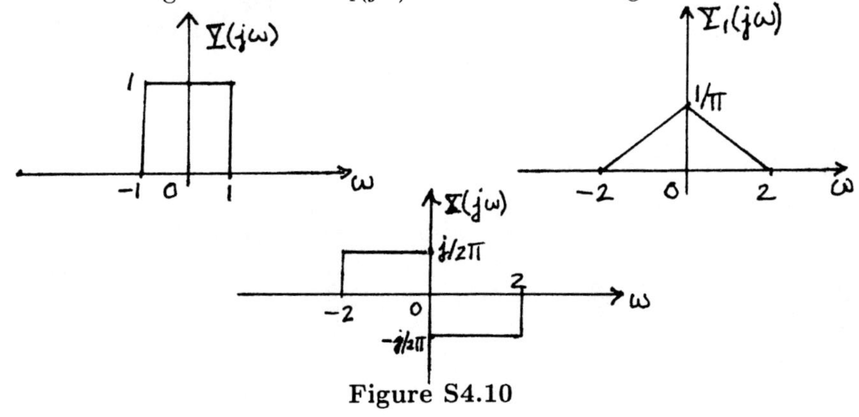

Figure S4.10

Using Table 4.1, we may write

$$t\left(\frac{\sin t}{\pi t}\right)^2 \stackrel{FT}{\longleftrightarrow} X(j\omega) = j\frac{d}{d\omega}Y_1(j\omega)$$

This is as shown in the figure above. $X(j\omega)$ may be expressed mathematically as

$$X(j\omega) = \begin{cases} j/2\pi, & -2 \le \omega < 0 \\ -j/2\pi, & 0 \le \omega 2 \\ 0. & \text{otherwise} \end{cases}$$

(b) Using Parseval's relation,

$$\int_{-\infty}^{\infty} t^2 \left(\frac{\sin t}{\pi t}\right)^4 dt = \frac{1}{2\pi}\int_{-\infty}^{\infty} |X(j\omega)|^2 d\omega = \frac{1}{2\pi^3}$$

4.11. We know that

$$x(3t) \xleftrightarrow{FT} \frac{1}{3}X(j\frac{\omega}{3}), \qquad h(3t) \xleftrightarrow{FT} \frac{1}{3}H(j\frac{\omega}{3})$$

Therefore,

$$G(j\omega) = \mathcal{FT}\{x(3t) * h(3t)\} = \frac{1}{9}X(j\frac{\omega}{3})H(j\frac{\omega}{3})$$

Now note that

$$Y(j\omega) = \mathcal{FT}\{x(t) * h(t)\} = X(j\omega)H(j\omega)$$

From this, we may write

$$Y(j\frac{\omega}{3}) = X\left(j\frac{\omega}{3}\right) H\left(j\frac{\omega}{3}\right)$$

Using this in eq. (**), we have

$$G(j\omega) = \frac{1}{9}Y(j\frac{\omega}{3})$$

and

$$g(t) = \frac{1}{3}y(3t).$$

Therefore, $A = \frac{1}{3}$ and $B = 3$.

4.12. **(a)** From Example 4.2 we know that

$$e^{-|t|} \xleftrightarrow{FT} \frac{2}{1+\omega^2}.$$

Using the differentiation in frequency property, we have

$$te^{-|t|} \xleftrightarrow{FT} j\frac{d}{d\omega}\left\{\frac{2}{1+\omega^2}\right\} = -\frac{4j\omega}{(1+\omega^2)^2}.$$

(b) The duality property states that if

$$g(t) \xleftrightarrow{FT} G(j\omega)$$

then

$$G(t) \xleftrightarrow{FT} 2\pi g(j\omega).$$

Now, since

$$te^{-|t|} \xleftrightarrow{FT} -\frac{4j\omega}{(1+\omega^2)^2}$$

we may use duality to write

$$-\frac{4jt}{(1+t^2)^2} \xleftrightarrow{FT} 2\pi\omega e^{-|\omega|}$$

Multiplying both sides by j, we obtain

$$\frac{4t}{(1+t^2)^2} \xleftrightarrow{FT} j2\pi\omega e^{-|\omega|}.$$

4.13. (a) Taking the inverse Fourier transform of $X(j\omega)$, we obtain

$$x(t) = \frac{1}{2\pi} + \frac{1}{2\pi}e^{j\pi t} + \frac{1}{2\pi}e^{j5t}$$

The signal $x(t)$ is therefore a constant summed with two complex exponentials whose fundamental frequencies are $2\pi/5$ rad/sec and 2 rad/sec. These two complex exponentials are not harmonically related. That is, the fundamental frequencies of these complex complex exponentials can never be integral multiples of a common fundamental frequency. Therefore, the signal is **not periodic**.

(b) Consider the signal $y(t) = x(t) * h(t)$. From the convolution property, we know that $Y(j\omega) = X(j\omega)H(j\omega)$. Also, from $h(t)$, we know that

$$H(j\omega) = e^{-j\omega}\frac{2\sin\omega}{\omega}.$$

The function $H(j\omega)$ is zero when $\omega = k\pi$, where k is a nonzero integer. Therefore,

$$Y(j\omega) = X(j\omega)H(j\omega) = \delta(\omega) + \delta(\omega - 5)$$

This gives

$$y(t) = \frac{1}{2\pi} + \frac{1}{2\pi}e^{j5t}$$

Therefore, $y(t)$ is a complex exponential summed with a constant. We know that a complex exponential is periodic. Adding a constant to a complex exponential does not affect its periodicity. Therefore, $y(t)$ will be a signal with a fundamental frequency of $2\pi/5$.

(c) From the results of parts (a) and (b), we see that the answer is yes.

4.14. Taking the Fourier transform of both sides of the equation

$$\mathcal{F}^{-1}\{(1 + j\omega)X(j\omega)\} = A2^{-2t}u(t),$$

we obtain

$$X(j\omega) = \frac{A}{(1 + j\omega)(2 + j\omega)} = A\left\{\frac{1}{1 + j\omega} - \frac{1}{2 + j\omega}\right\}.$$

Taking the inverse Fourier transform of the above equation

$$x(t) = Ae^{-t}u(t) - Ae^{-2t}u(t)$$

Using Parseval's relation, we have

$$\int_{-\infty}^{\infty}|X(j\omega)|^2 d\omega = 2\pi\int_{-\infty}^{\infty}|x(t)|^2 dt$$

Using the fact that $\int_{-\infty}^{\infty}|X(j\omega)|^2 d\omega = 2\pi$, we have

$$\int_{-\infty}^{\infty}|x(t)|^2 dt = 1$$

Substituting the previously obtained expression for $x(t)$ in the above equation, we have

$$\int_{-\infty}^{\infty} [A^2 e^{-2t} + A^2 e^{-4t} - 2A^2 e^{-3t}] u(t) dt = 1$$
$$\int_{0}^{\infty} [A^2 e^{-2t} + A^2 e^{-4t} - 2A^2 e^{-3t}] dt = 1$$
$$A^2/12 = 1$$
$$\Rightarrow A = \sqrt{12}$$

We choose A to be $\sqrt{12}$ instead of $-\sqrt{12}$ because we know that $x(t)$ is non negative.

4.15. Since $x(t)$ is real,

$$\mathcal{E}v\{x(t)\} = \frac{x(t) + x(-t)}{2} \overset{FT}{\longleftrightarrow} \mathcal{R}e\{X(j\omega)\}.$$

We are given that

$$\mathcal{IFT}\{\mathcal{R}e\{X(j\omega)\}\} = |t|e^{-|t|}.$$

Therefore,

$$\mathcal{E}v\{x(t)\} = \frac{x(t) + x(-t)}{2} = |t|e^{-|t|}.$$

We also know that $x(t) = 0$ for $t \leq 0$. This implies that $x(-t)$ is zero for $t > 0$. We may conclude that

$$x(t) = 2|t|e^{-|t|} \qquad \text{for } t \geq 0$$

Therefore,

$$x(t) = 2te^{-t}u(t)$$

4.16. **(a)** We may write

$$\begin{aligned} x(t) &= \sum_{k=-\infty}^{\infty} \frac{\sin(k\pi/4)}{k\pi/4} \delta(t - k\pi/4) \\ &= \frac{\sin t}{\pi t} \sum_{k=-\infty}^{\infty} \pi\delta(t - k\pi/4) \end{aligned}$$

Therefore, $g(t) = \sum_{k=-\infty}^{\infty} \pi\delta(t - k\pi/4)$.

(b) Since $g(t)$ is an impulse train, its Fourier transform $G(j\omega)$ is also an impulse train. From Table 4.2,

$$\begin{aligned} G(j\omega) &= \pi \frac{2\pi}{\pi/4} \sum_{k=-\infty}^{\infty} \delta\left(\omega - \frac{2\pi k}{\pi/4}\right) \\ &= 8\pi \sum_{k=-\infty}^{\infty} \delta(\omega - 8k) \end{aligned}$$

We see that $G(j\omega)$ is periodic with a period of 8. Using the multiplication property, we know that

$$X(j\omega) = \frac{1}{2\pi}\left[\mathcal{FT}\left\{\frac{\sin t}{\pi t}\right\} * G(j\omega)\right]$$

If we denote $\mathcal{FT}\left\{\frac{\sin t}{\pi t}\right\}$ by $A(j\omega)$, then

$$\begin{aligned} X(j\omega) &= (1/2\pi)[A(j\omega) * 8\pi \sum_{k=-\infty}^{\infty} \delta(\omega - 8k) \\ &= 4 \sum_{k=-\infty}^{\infty} A(j\omega - 8k) \end{aligned}$$

$X(j\omega)$ may thus be viewed as a replication of $4A(j\omega)$ every 8 rad/sec. This is obviously periodic.

Using Table 4.2, we obtain

$$A(j\omega) = \begin{cases} 1, & |\omega| \leq 1 \\ 0, & \text{otherwise} \end{cases}$$

Therefore, we may specify $X(j\omega)$ over one period as

$$X(j\omega) = \begin{cases} 4, & |\omega| \leq 1 \\ 0, & 1 < |\omega \leq 4 \end{cases}$$

4.17. **(a)** From Table 4.1, we know that a real and odd signal signal $x(t)$ has a purely imaginary and odd Fourier transform $X(j\omega)$. Let us now consider the purely imaginary and odd signal $jx(t)$. Using linearity, we obtain the Fourier transform of this signal to be $jX(j\omega)$. The function $jX(j\omega)$ will clearly be real and odd. Therefore, the given statement is **false**.

(b) An odd Fourier transform corresponds to an odd signal, while an even Fourier transform corresponds to an even signal. The convolution of an even Fourier transform with an odd Fourier may be viewed in the time domain as a multiplication of an even and odd signal. Such a multiplication will always result in an odd time signal. The Fourier transform of this odd signal will always be odd. Therefore, the given statement is **true**.

4.18. Using Table 4.2, we see that the rectangular pulse $x_1(t)$ shown in Figure S4.18 has a Fourier transform $X_1(j\omega) = \sin(3\omega)/\omega$. Using the convolution property of the Fourier transform, we may write

$$x_2(t) = x_1(t) * x_1(t) \stackrel{FT}{\longleftrightarrow} X_2(j\omega) = X_1(j\omega)X_1(j\omega) = \left(\frac{\sin(3\omega)}{\omega}\right)^2$$

The signal $x_2(t)$ is shown in Figure S4.18. Using the shifting property, we also note that

$$\frac{1}{2}x_2(t+1) \stackrel{FT}{\longleftrightarrow} \frac{1}{2}e^{j\omega}\left(\frac{\sin(3\omega)}{\omega}\right)^2$$

and

$$\frac{1}{2}x_2(t-1) \overset{FT}{\longleftrightarrow} \frac{1}{2}e^{-j\omega}\left(\frac{\sin(3\omega)}{\omega}\right)^2.$$

Adding the two above equations, we obtain

$$h(t) = \frac{1}{2}x_2(t+1) + \frac{1}{2}x_2(t-1) \overset{FT}{\longleftrightarrow} \cos(\omega)\left(\frac{\sin(3\omega)}{\omega}\right)^2.$$

The signal $h(t)$ is as shown in Figure S4.18. We note that $h(t)$ has the given Fourier transform $H(j\omega)$.

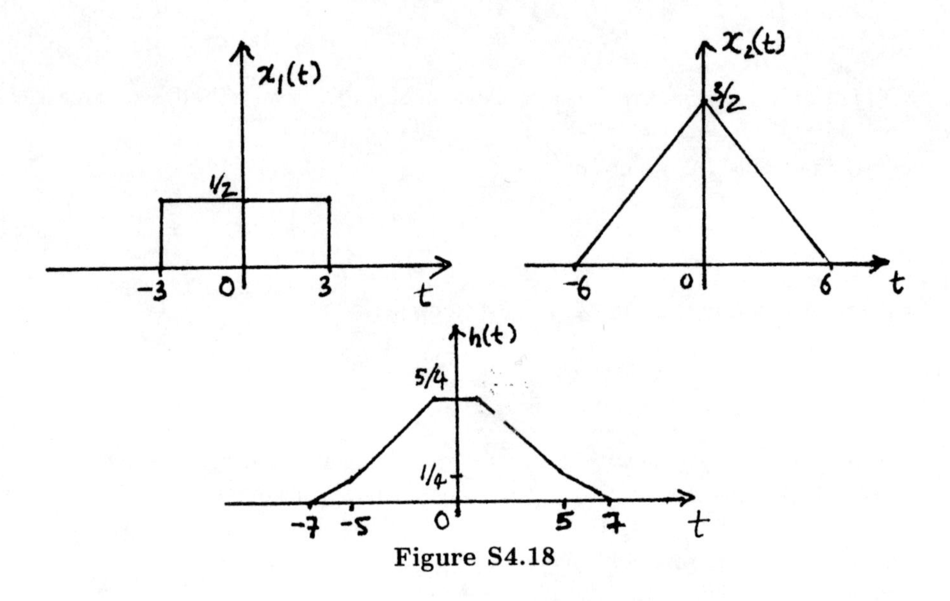

Figure S4.18

Mathematically $h(t)$ may be expressed as

$$h(t) = \begin{cases} \frac{5}{4}, & |t| < 1 \\ -\frac{|t|}{4} + \frac{3}{2}, & 1 \le |t| \le 5 \\ -\frac{|t|}{8} + \frac{7}{8}, & 5 < |t| \le 7 \\ 0, & \text{otherwise} \end{cases}$$

4.19. We know that

$$H(j\omega) = \frac{Y(j\omega)}{X(j\omega)}.$$

Since it is given that $y(t) = e^{-3t}u(t) - e^{-4t}u(t)$, we can compute $Y(j\omega)$ to be

$$Y(j\omega) = \frac{1}{3+j\omega} - \frac{1}{4+j\omega} = \frac{1}{(3+j\omega)(4+j\omega)}.$$

Since, $H(j\omega) = 1/(3 + j\omega)$, we have

$$X(j\omega) = \frac{Y(j\omega)}{H(j\omega)} = 1/(4 + j\omega)$$

Taking the inverse Fourier transform of $X(j\omega)$, we have

$$x(t) = e^{-4t}u(t).$$

4.20. From the answer to Problem 3.20, we know that the frequency response of the circuit is

$$H(j\omega) = \frac{1}{-\omega^2 + j\omega + 1}.$$

Breaking this up into partial fractions, we may write

$$H(j\omega) = -\frac{1}{j\sqrt{3}}\left[\frac{-1}{\frac{1}{2} - \frac{\sqrt{3}}{2}j + j\omega} + \frac{-1}{\frac{1}{2} + \frac{\sqrt{3}}{2}j + j\omega}\right]$$

Using the Fourier transform pairs provided in Table 4.2, we obtain the Fourier transform of $H(j\omega)$ to be

$$h(t) = -\frac{1}{j\sqrt{3}}\left[-e^{(-\frac{1}{2}+\frac{\sqrt{3}}{2}j)t} + +e^{(-\frac{1}{2}-\frac{\sqrt{3}}{2}j)t}\right]u(t).$$

Simplifying,

$$h(t) = \frac{2}{\sqrt{3}}e^{-\frac{1}{2}t}\sin(\frac{\sqrt{3}}{2}t)u(t).$$

4.21. **(a)** The given signal is

$$e^{-\alpha t}\cos(\omega_0 t)u(t) = \frac{1}{2}e^{-\alpha t}e^{j\omega_0 t}u(t) + \frac{1}{2}e^{-\alpha t}e^{-j\omega_0 t}u(t).$$

Therefore,

$$X(j\omega) = \frac{1}{2(\alpha - j\omega_0 + j\omega)} - \frac{1}{2(\alpha - j\omega_0 + j\omega)}.$$

(b) The given signal is

$$x(t) = e^{-3t}\sin(2t)u(t) + e^{3t}\sin(2t)u(-t).$$

We have

$$x_1(t) = e^{-3t}\sin(2t)u(t) \stackrel{FT}{\longleftrightarrow} X_1(j\omega) = \frac{1/2j}{3 - j2 + j\omega} - \frac{1/2j}{3 + j2 + j\omega}.$$

Also,

$$x_2(t) = e^{3t}\sin(2t)u(-t) = -x_1(-t) \stackrel{FT}{\longleftrightarrow} X_2(j\omega) = -X_1(-j\omega) = \frac{1/2j}{3 - j2 - j\omega} - \frac{1/2j}{3 + j2 - j\omega}.$$

Therefore,

$$X(j\omega) = X_1(j\omega + X_2(j\omega) = \frac{3j}{9 + (\omega + 2)^2} - \frac{3j}{9 + (\omega - 2)^2}.$$

(c) Using the Fourier transform analysis equation (4.9) we have

$$X(j\omega) = \frac{2\sin\omega}{\omega} + \frac{\sin\omega}{\pi-\omega} - \frac{\sin\omega}{\pi+\omega}.$$

(d) Using the Fourier transform analysis equation (4.9) we have

$$X(j\omega) = \frac{1}{1-\alpha e^{-j\omega T}}.$$

(e) We have

$$x(t) = (1/2j)te^{-2t}e^{j4t}u(t) - (1/2j)te^{-2t}e^{-j4t}u(t).$$

Therefore,

$$X(j\omega) = \frac{1/2j}{(2-j4+j\omega)^2} - \frac{1/2j}{(2+j4-j\omega)^2}.$$

(f) We have

$$x_1(t) = \frac{\sin \pi t}{\pi t} \overset{FT}{\longleftrightarrow} X_1(j\omega) = \begin{cases} 1, & |\omega| < \pi \\ 0, & \text{otherwise} \end{cases}.$$

Also

$$x_2(t) = \frac{\sin 2\pi(t-1)}{\pi(t-1)} \overset{FT}{\longleftrightarrow} X_2(j\omega) = \begin{cases} e^{-2\omega}, & |\omega| < 2\pi \\ 0, & \text{otherwise} \end{cases}.$$

$$x(t) = x_1(t)x_2(t) \overset{FT}{\longleftrightarrow} X(j\omega) = \frac{1}{2\pi}\{X_1(j\omega) * X_2(j\omega)\}.$$

Therefore,

$$X(j\omega) = \begin{cases} e^{-j\omega}, & |\omega| < \pi \\ (1/2\pi)(3\pi+\omega)e^{-j\omega}, & -3\pi < \omega < -\pi \\ (1/2\pi)(3\pi-\omega)e^{-j\omega}, & \pi < \omega < 3\pi \\ 0, & \text{otherwise} \end{cases}.$$

(g) Using the Fourier transform analysis eq. (4.9) we obtain

$$X(j\omega) = \frac{2j}{\omega}\left[\cos 2\omega - \frac{\sin\omega}{\omega}\right].$$

(h) If

$$x_1(t) = \sum_{k=-\infty}^{\infty} \delta(t-2k),$$

then

$$x(t) = 2x_1(t) + x_1(t-1).$$

Therefore,

$$X(j\omega) = X_1(j\omega)[2+e^{-\omega}] = \pi \sum_{k=-\infty}^{\infty} \delta(\omega - k\pi)[2+(-1)^k].$$

(i) Using the Fourier transform analysis eq. (4.9) we obtain

$$X(j\omega) = \frac{1}{j\omega} + \frac{2e^{-j\omega}}{-\omega^2} - \frac{2e^{-j\omega} - 2}{j\omega^2}.$$

(j) $x(t)$ is periodic with period 2. Therefore,

$$X(j\omega) = \pi \sum_{k=-\infty}^{\infty} \tilde{X}(jk\pi)\delta(\omega - k\pi),$$

where $\tilde{X}(j\omega)$ is the Fourier transform of one period of $x(t)$. That is,

$$\tilde{X}(j\omega) = \frac{1}{1-e^{-2}} \left[\frac{1 - e^{-2(1+j\omega)}}{1+j\omega} - \frac{e^{-2}[1 - e^{-2(1+j\omega)}]}{1-j\omega} \right].$$

4.22. (a) $x(t) = \begin{cases} e^{j2\pi t}, & |t| < 3 \\ 0, & \text{otherwise} \end{cases}$

(b) $x(t) = \frac{1}{2}e^{-j\pi/3}\delta(t-4) + \frac{1}{2}e^{j\pi/3}\delta(t+4)$.

(c) The Fourier transform synthesis eq. (4.8) may be written as

$$x(t) = \frac{1}{2\pi}\int_{-\infty}^{\infty} |X(j\omega)| e^{j\sphericalangle X(j\omega)} e^{j\omega t} d\omega.$$

From the given figure we have

$$x(t) = \frac{1}{\pi}\left[\frac{\sin(t-3)}{t-3} + \frac{\cos(t-3) - 1}{(t-3)^2}\right].$$

(d) $x(t) = \frac{2j}{\pi}\sin t + \frac{3}{\pi}\cos(2\pi t)$

(e) Using the Fourier transform synthesis equation (4.8),

$$x(t) = \frac{\cos 3t}{j\pi t} + \frac{\sin t - \sin 2t}{j\pi t^2}.$$

4.23. For the given signal $x_0(t)$, we use the Fourier transform analysis eq. (4.8) to evaluate the corresponding Fourier transform

$$X_0(j\omega) = \frac{1 - e^{-(1+j\omega)}}{1+j\omega}.$$

(i) We know that

$$x_1(t) = x_0(t) + x_0(-t).$$

Using the linearity and time reversal properties of the Fourier transform we have

$$X_1(j\omega) = X_0(j\omega) + X_0(-j\omega) = \frac{2 - 2e^{-1}\cos\omega - 2\omega e^{-1}\sin\omega}{1+\omega^2}.$$

(ii) We know that

$$x_2(t) = x_0(t) - x_0(-t).$$

Using the linearity and time reversal properties of the Fourier transform we have

$$X_2(j\omega) = X_0(j\omega) - X_0(-j\omega) = j\left[\frac{-2\omega + 2e^{-1}\sin\omega + 2\omega e^{-1}\cos\omega}{1+\omega^2}\right].$$

(iii) We know that

$$x_3(t) = x_0(t) + x_0(t+1).$$

Using the linearity and time shifting properties of the Fourier transform we have

$$X_3(j\omega) = X_0(j\omega) + e^{j\omega}X_0(-j\omega) = \frac{1 + e^{j\omega} - e^{-1}(1+e^{-j\omega})}{1+j\omega}.$$

(iv) We know that

$$x_4(t) = tx_0(t).$$

Using the differentiation in frequency property

$$X_4(j\omega) = j\frac{d}{d\omega}X_0(j\omega).$$

Therefore,

$$X_4(j\omega) = \frac{1 - 2e^{-1}e^{-j\omega} - j\omega e^{-1}e^{-j\omega}}{(1+j\omega)^2}.$$

4.24. **(a)** (i) For $\mathcal{R}e\{X(j\omega)\}$ to be 0, the signal $x(t)$ must be real and odd. Therefore, signals in figures (a) and (c) have this property.

(ii) For $\mathcal{I}m\{X(j\omega)\}$ to be 0, the signal $x(t)$ must be real and even. Therefore, signals in figures (e) and (f) have this property.

(iii) For there to exist a real α such that $e^{j\alpha\omega}X(j\omega)$ is real, we require that $x(t+\alpha)$ be a real and even signal. Therefore, signals in figures (a), (b), (e), and (f) have this property.

(iv) For this condition to be true, $x(0) = 0$. Therefore, signals in figures (a), (b), (c), (d), and (f) have this property.

(v) For this condition to be true the derivative of $x(t)$ has to be zero at $t = 0$. Therefore, signals in figures (b), (c), (e), and (f) have this property.

(vi) For this to be true, the signal $x(t)$ has to be periodic. Only the signal in figure (a) has this property.

(b) For a signal to satisfy only properties (i), (iv), and (v), it must be real and odd, and

$$x(t) = 0, \qquad x'(0) = 0.$$

The signal shown below is an example of that.

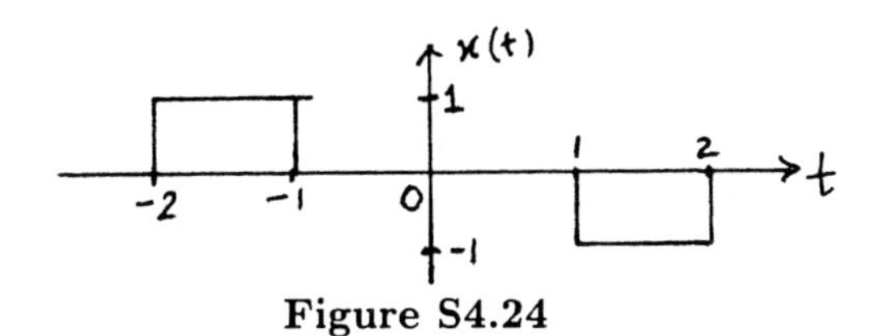

Figure S4.24

4.25. **(a)** Note that $y(t) = x(t+1)$ is a real and even signal. Therefore, $Y(j\omega)$ is also real and even. This implies that $\sphericalangle Y(j\omega) = 0$. Also, since $Y(j\omega) = e^{j\omega}X(j\omega)$, we know that $\sphericalangle X(j\omega) = -\omega$.

(b) We have

$$X(j0) = \int_{-\infty}^{\infty} x(t)dt = 7.$$

(c) We have

$$\int_{-\infty}^{\infty} X(j\omega)d\omega = 2\pi x(0) = 4\pi.$$

(d) Let $Y(j\omega) = \frac{2\sin\omega}{\omega}e^{2j\omega}$. The corresponding signal $y(t)$ is

$$y(t) = \begin{cases} 1, & -3 < t < -1 \\ 0, & \text{otherwise} \end{cases}.$$

Then the given integral is

$$\int_{-\infty}^{\infty} X(j\omega)Y(j\omega)d\omega = 2\pi\{x(t) * y(t)\}_{t=0} = 7\pi.$$

(e) We have

$$\int_{-\infty}^{\infty} |X(j\omega)|^2 d\omega = 2\pi \int_{-\infty}^{\infty} |x(t)|^2 dt = 26\pi.$$

(f) The inverse Fourier transform of $\mathcal{R}e\{X(j\omega)\}$ is the $\mathcal{E}v\{x(t)\}$ which is $[x(t)+x(-t)]/2$. This is as shown in the figure below.

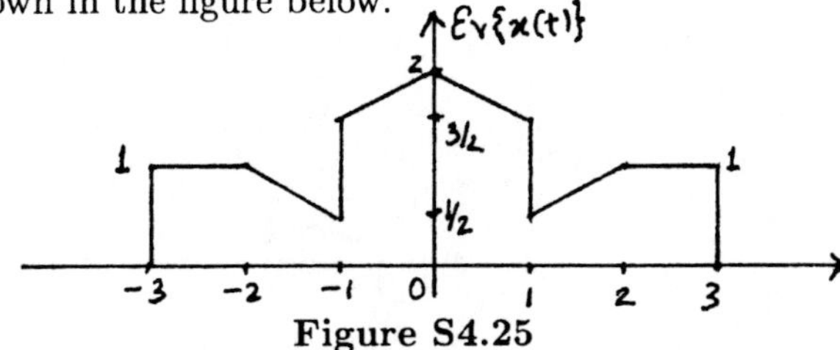

Figure S4.25

4.26. **(a)** (i) We have

$$\begin{aligned} Y(j\omega) &= X(j\omega)H(j\omega) = \left[\frac{1}{(2+j\omega)^2}\right]\left[\frac{1}{4+j\omega}\right] \\ &= \frac{(1/4)}{4+j\omega} - \frac{(1/4)}{2+j\omega} + \frac{(1/2)}{(2+j\omega)^2} \end{aligned}$$

Taking the inverse Fourier transform we obtain

$$y(t) = \frac{1}{4}e^{-4t}u(t) - \frac{1}{4}e^{-2t}u(t) + \frac{1}{2}te^{-2t}u(t).$$

(ii) We have

$$\begin{aligned} Y(j\omega) &= X(j\omega)H(j\omega) = \left[\frac{1}{(2+j\omega)^2}\right]\left[\frac{1}{(4+j\omega)^2}\right] \\ &= \frac{(1/4)}{2+j\omega} + \frac{(1/4)}{(2+j\omega)^2} - \frac{(1/4)}{4+j\omega} + \frac{(1/4)}{(4+j\omega)^2} \end{aligned}$$

Taking the inverse Fourier transform we obtain

$$y(t) = \frac{1}{4}e^{-2t}u(t) + \frac{1}{4}te^{-2t}u(t) - \frac{1}{4}e^{-4t}u(t) + \frac{1}{4}te^{-4t}u(t).$$

(iii) We have

$$\begin{aligned} Y(j\omega) &= X(j\omega)H(j\omega) \\ &= \left[\frac{1}{1+j\omega}\right]\left[\frac{1}{1-j\omega}\right] \\ &= \frac{1/2}{1+j\omega} + \frac{1/2}{1-j\omega} \end{aligned}$$

Taking the inverse Fourier transform, we obtain

$$y(t) = \frac{1}{2}e^{-|t|}.$$

(b) By direct convolution of $x(t)$ with $h(t)$ we obtain

$$y(t) = \begin{cases} 0, & t < 1 \\ 1 - e^{-(t-1)}, & 1 < t \leq 5 \\ e^{-(t-5)} - e^{-(t-1)}, & t > 5 \end{cases}.$$

Taking the Fourier transform of $y(t)$,

$$\begin{aligned} Y(j\omega) &= \frac{2e^{-j3\omega}\sin(2\omega)}{\omega(1+j\omega)} \\ &= \left[\frac{e^{-j2\omega}}{1+j\omega}\right]\frac{e^{-j\omega}2\sin(2\omega)}{\omega} \\ &= X(j\omega)H(j\omega) \end{aligned}$$

4.27. **(a)** The Fourier transform $X(j\omega)$ is

$$\begin{aligned} X(j\omega) &= \int_{-\infty}^{\infty} x(t)e^{-j\omega t}dt = \int_1^2 e^{-j\omega t}dt - \int_2^3 e^{-j\omega t}dt \\ &= 2\frac{\sin(\omega/2)}{\omega}\{1 - e^{-j\omega}\}e^{-j3\omega/2} \end{aligned}$$

(b) The Fourier series coefficients a_k are

$$\begin{aligned} a_k &= \frac{1}{T}\int_{<T>} \tilde{x}(t)e^{-j\frac{2\pi}{T}kt} \\ &= \frac{1}{2}\{\int_1^2 e^{-j\frac{2\pi}{T}kt}dt - \int_2^3 e^{-j\frac{2\pi}{T}kt}dt\} \\ &= \frac{\sin(k\pi/2)}{k\pi}\{1 - e^{-jk\pi}\}e^{-j3k\pi/2} \end{aligned}$$

Comparing the answers to parts (a) and (b), it is clear that

$$a_k = \frac{1}{T}X(j\frac{2\pi k}{T}),$$

where $T = 2$.

4.28. (a) From Table 4.2 we know that

$$p(t) = \sum_{n=-\infty}^{\infty} a_n e^{jn\omega_0 t} \stackrel{FT}{\longleftrightarrow} P(j\omega) = 2\pi \sum_{k=-\infty}^{\infty} a_k \delta(\omega - k\omega_0).$$

From this,

$$Y(j\omega) = \frac{1}{2\pi}\{X(j\omega) * H(j\omega)\} = \sum_{k=-\infty}^{\infty} a_k X(j(\omega - k\omega_0)).$$

(b) The spectra are sketched in Figure S4.28.

4.29. (i) We have

$$X_a(j\omega) = |X(j\omega)|e^{j\sphericalangle X(j\omega) - ja\omega} = X(j\omega)e^{-ja\omega}.$$

From the time shifting property we know that

$$x_a(t) = x(t-a).$$

(ii) We have

$$X_b(j\omega) = |X(j\omega)|e^{j\sphericalangle X(j\omega) + jb\omega} = X(j\omega)e^{jb\omega}.$$

From the time shifting property we know that

$$x_b(t) = x(t+b).$$

(iii) We have

$$X_c(j\omega) = |X(j\omega)|e^{-j\sphericalangle X(j\omega)} = X^*(j\omega).$$

From the conjugation and time reversal properties we know that

$$x_c(t) = x^*(-t).$$

Since $x(t)$ is real, $x_c(t) = x(-t)$.

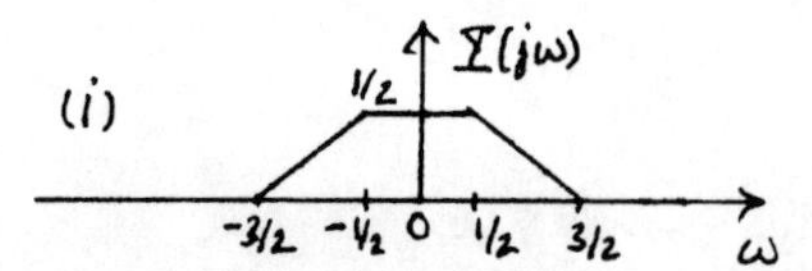

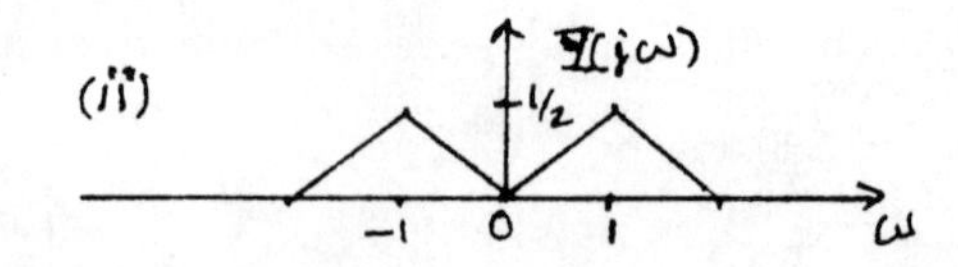

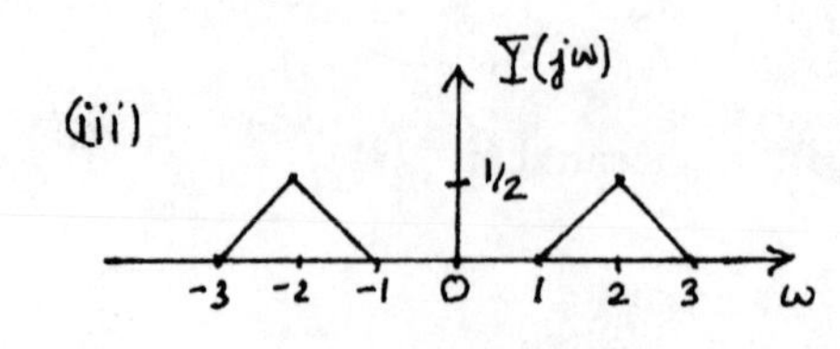

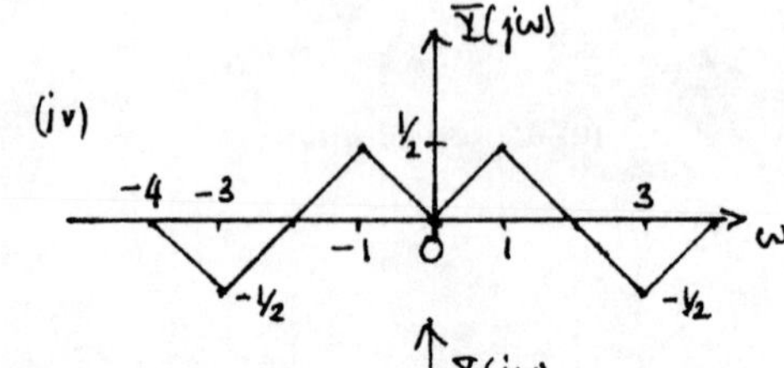

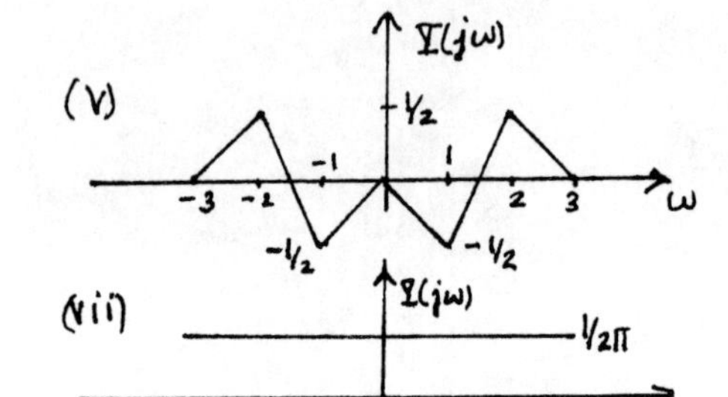

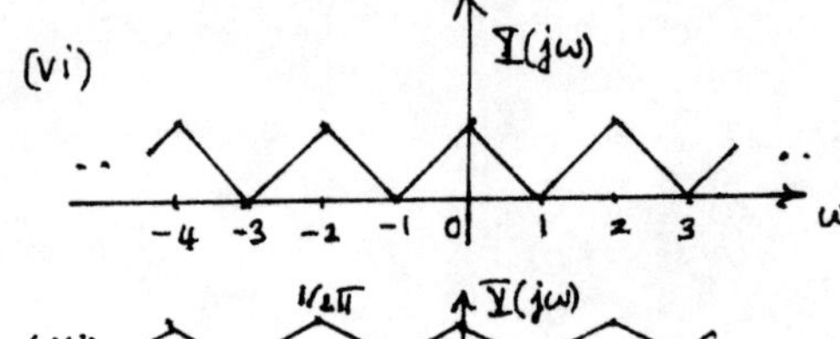

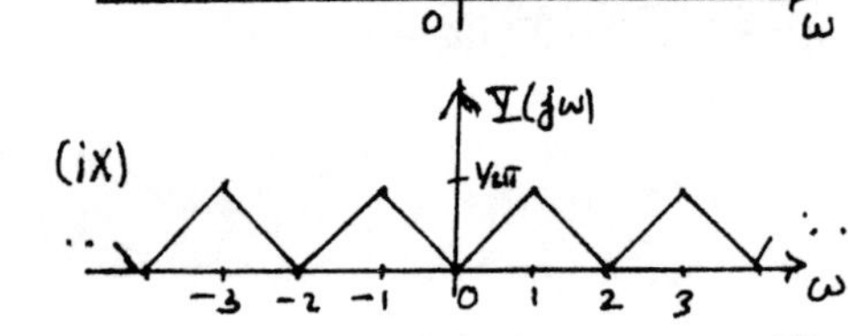

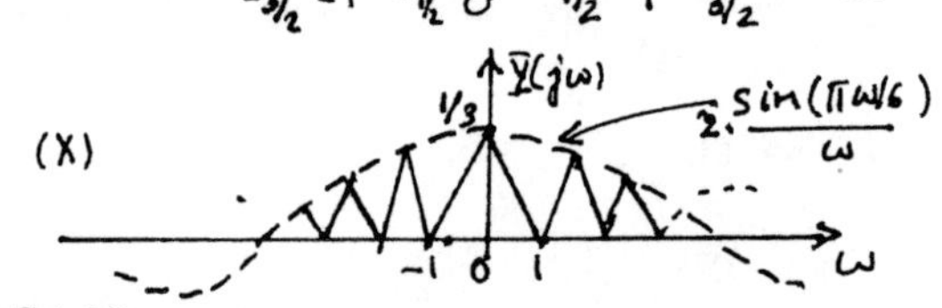

Figure S4.28

(iv) We have

$$X_d(j\omega) = |X(j\omega)|e^{-j\sphericalangle X(j\omega)+jd\omega} = X^*(j\omega)e^{jb\omega}.$$

From the conjugation, time reversal, and time shifting properties, we know that

$$x_d(t) = x^*(-t-d).$$

Since $x(t)$ is real, $x_d(t) = x(-t-d)$.

4.30. **(a)** We know that

$$w(t) = \cos t \overset{FT}{\longleftrightarrow} W(j\omega) = \pi[\delta(\omega-1) + \delta(\omega+1)$$

and

$$g(t) = x(t)\cos t \overset{FT}{\longleftrightarrow} G(j\omega) = \frac{1}{2\pi}\{X(j\omega) * W(j\omega)\}.$$

Therefore,

$$G(j\omega) = \frac{1}{2}X(j(\omega-1)) + \frac{1}{2}X(j(\omega+1)).$$

Since $G(j\omega)$ is as shown in Figure S4.30, it is clear from the above equation that $X(j\omega)$ is as shown in the Figure S4.30.

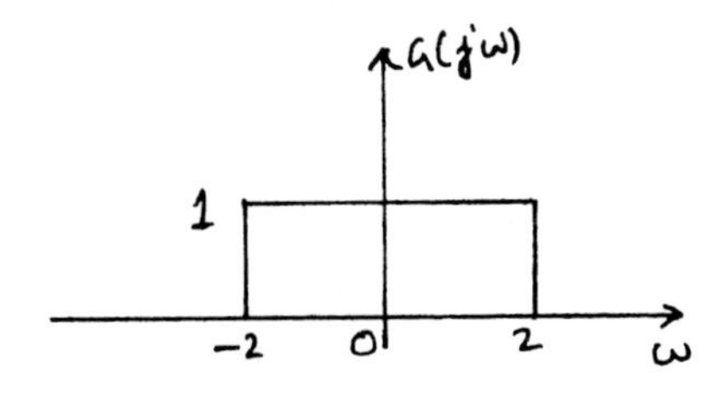

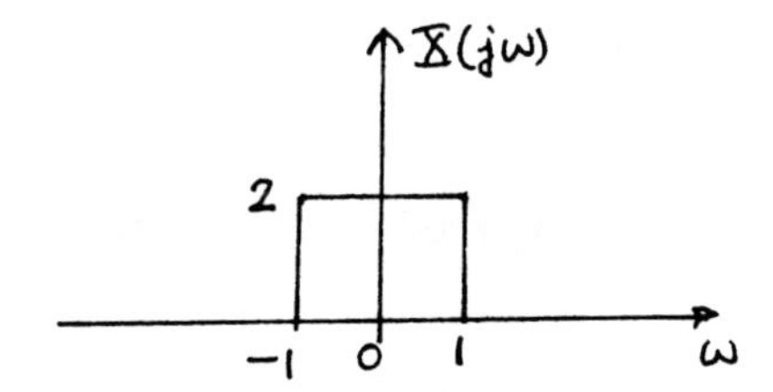

Figure S4.30

Therefore,

$$x(t) = \frac{2\sin t}{\pi t}.$$

(b) $X_1(j\omega)$ is as shown in Figure S4.30.

4.31. **(a)** We have

$$x(t) = \cos t \stackrel{FT}{\longleftrightarrow} X(j\omega) = \pi[\delta(\omega+1) + \delta(\omega-1)].$$

(i) We have

$$h_1(t) = u(t) \stackrel{FT}{\longleftrightarrow} H_1(j\omega) = \frac{1}{j\omega} + \pi\delta(\omega).$$

Therefore,

$$Y(j\omega) = X(j\omega)H_1(j\omega) = \frac{\pi}{j}[\delta(\omega+1) - \delta(\omega-1)].$$

Taking the inverse Fourier transform, we obtain

$$y(t) = \sin(t).$$

(ii) We have

$$h_2(t) = -2\delta(t) + 5e^{-2t}u(t) \stackrel{FT}{\longleftrightarrow} H_2(j\omega) = -2 + \frac{5}{2+j\omega}.$$

Therefore,

$$Y(j\omega) = X(j\omega)H_1(j\omega) = \frac{\pi}{j}[\delta(\omega+1) - \delta(\omega-1)].$$

Taking the inverse Fourier transform, we obtain

$$y(t) = \sin(t).$$

(iii) We have

$$h_3(t) = 2te^{-t}u(t) \stackrel{FT}{\longleftrightarrow} H_2(j\omega) = \frac{2}{(1+j\omega)^2}.$$

Therefore,

$$Y(j\omega) = X(j\omega)H_1(j\omega) = \frac{\pi}{j}[\delta(\omega+1) - \delta(\omega-1)].$$

Taking the inverse Fourier transform, we obtain

$$y(t) = \sin(t).$$

(b) An LTI system with impulse response

$$h_4(t) = \frac{1}{2}\left[h_1(t) + h_2(t)\right]$$

will have the same response to $x(t) = \cos(t)$. We can find other such impulse responses by suitably scaling and linearly combining $h_1(t)$, $h_2(t)$, and $h_3(t)$.

4.32. Note that $h(t) = h_1(t-1)$, where

$$h_1(t) = \frac{\sin 4t}{\pi t}.$$

The Fourier transform $H_1(j\omega)$ of $h_1(t)$ is as shown in Figure S4.32.

From the above figure it is clear that $h_1(t)$ is the impulse response of an ideal lowpass filter whose passband is in the range $|\omega| < 4$. Therefore, $h(t)$ is the impulse response of an ideal lowpass filter shifted by one to the right. Using the shift property,

$$H(j\omega) = \begin{cases} e^{-j\omega}, & |\omega| < 4 \\ 0, & \text{otherwise} \end{cases}.$$

(a) We have

$$X_1(j\omega) = \pi e^{j\frac{\pi}{12}}\delta(\omega-6) + \pi e^{j\frac{\pi}{12}}\delta(\omega+6).$$

It is clear that

$$Y_1(j\omega) = X_1(j\omega)H(j\omega) = 0 \Rightarrow y_1(t) = 0.$$

This result is equivalent to saying that $X_1(j\omega)$ is zero in the passband of $H(j\omega)$.

(b) We have

$$X_2(j\omega) = \frac{\pi}{j}\left[\sum_{k=0}^{\infty}(\frac{1}{2})^k\{\delta(\omega-3k) - \delta(\omega+3k)\}\right].$$

Therefore,

$$Y_2(j\omega) = X_2(j\omega)H(j\omega) = \frac{\pi}{j}\left[(1/2)\{\delta(\omega-3) - \delta(\omega+3)\}e^{-j\omega}\right].$$

This implies that

$$y_2(t) = \frac{1}{2}\sin(3t-1).$$

We may have obtained the same result by noting that only the sinusoid with frequency 3 in $X_2(j\omega)$ lies in the passband of $H(j\omega)$.

(c) We have

$$X_3(j\omega) = \begin{cases} e^{j\omega}, & |\omega| < 4 \\ 0, & \text{otherwise} \end{cases}.$$

$$Y_3(j\omega) = X_3(j\omega)H(j\omega) = X_3(j\omega)e^{-j\omega}.$$

This implies that

$$y_3(t) = x_3(t-1) = \frac{\sin(4t)}{\pi t}.$$

We may have obtained the same result by noting that $X_3(j\omega)$ lies entirely in the passband of $H(j\omega)$.

(d) $X_4(j\omega)$ is as shown in Figure S4.32.

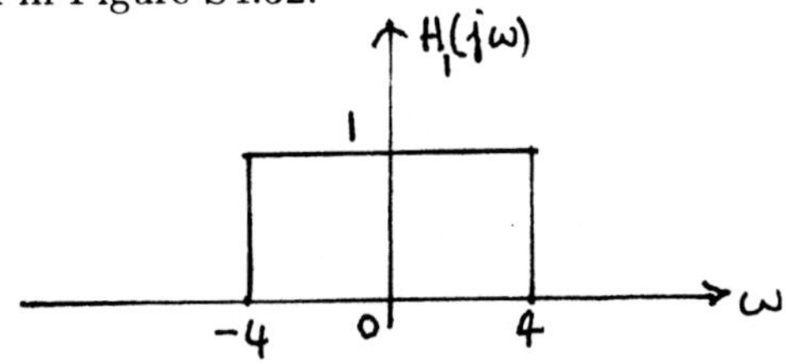

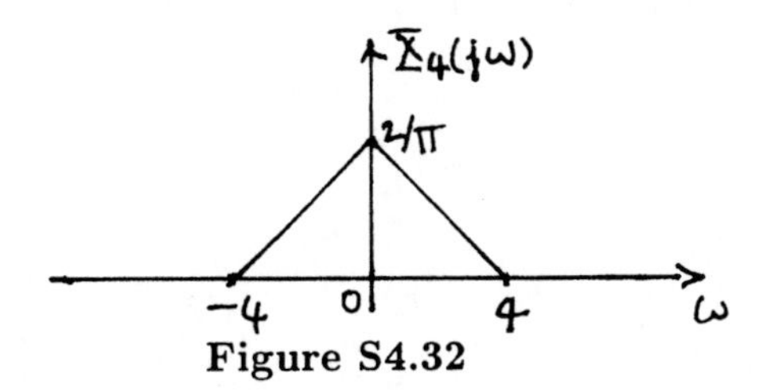

Figure S4.32

Therefore,

$$Y_4(j\omega) = X_4(j\omega)H(j\omega) = X_4(j\omega)e^{-j\omega}.$$

This implies that

$$y_4(t) = x_4(t-1) = \left(\frac{\sin(2(t-1))}{\pi(t-1)}\right)^2.$$

We may have obtained the same result by noting that $X_4(j\omega)$ lies entirely in the passband of $H(j\omega)$.

4.33. (a) Taking the Fourier transform of both sides of the given differential equation, we obtain

$$H(j\omega) = \frac{Y(j\omega)}{X(j\omega)} = \frac{2}{-\omega^2 + 2j\omega + 8}.$$

Using partial fraction expansion, we obtain

$$H(j\omega) = \frac{1}{j\omega + 2} - \frac{1}{j\omega + 4}.$$

Taking the inverse Fourier transform,

$$h(t) = e^{-2t}u(t) - e^{-4t}u(t).$$

(b) For the given signal $x(t)$, we have

$$X(j\omega) = \frac{1}{(2+j\omega)^2}.$$

Therefore,

$$Y(j\omega) = X(j\omega)H(j\omega) = \frac{2}{(-\omega^2+2j\omega+8)}\frac{1}{(2+j\omega)^2}.$$

Using partial fraction expansion, we obtain

$$Y(j\omega) = \frac{1/4}{j\omega+2} - \frac{1/2}{(j\omega+2)^2} + \frac{1}{(j\omega+2)^3} - \frac{1/4}{j\omega+4}.$$

Taking the inverse Fourier transform,

$$y(t) = \frac{1}{4}e^{-2t}u(t) - \frac{1}{2}te^{-2t}u(t) + t^2e^{-2t}u(t) - \frac{1}{4}e^{-4t}u(t).$$

(c) Taking the Fourier transform of both sides of the given differential equation, we obtain

$$H(j\omega) = \frac{Y(j\omega)}{X(j\omega)} = \frac{2(-\omega^2-1)}{-\omega^2+\sqrt{2}j\omega+1}.$$

Using partial fraction expansion, we obtain

$$H(j\omega) = 2 + \frac{-\sqrt{2}-2\sqrt{2}j}{j\omega - \frac{-\sqrt{2}+j\sqrt{2}}{2}} + \frac{-\sqrt{2}+2\sqrt{2}j}{j\omega - \frac{-\sqrt{2}-j\sqrt{2}}{2}}.$$

Taking the inverse Fourier transform,

$$h(t) = 2\delta(t) - \sqrt{2}(1+2j)e^{-(1+j)t/\sqrt{2}}u(t) - \sqrt{2}(1-2j)e^{-(1-j)t/\sqrt{2}}u(t).$$

4.34. **(a)** We have

$$\frac{Y(j\omega}{X(j\omega)} = \frac{j\omega+4}{6-\omega^2+5j\omega}.$$

Cross-multiplying and taking the inverse Fourier transform, we obtain

$$\frac{d^2y(t)}{dt^2} + 5\frac{dy(t)}{dt} + 6y(t) = \frac{dx(t)}{dt} + 4x(t).$$

(b) We have

$$H(j\omega) = \frac{2}{2+j\omega} - \frac{1}{3+j\omega}.$$

Taking the inverse Fourier transform we obtain,

$$h(t) = 2e^{-2t}u(t) - e^{-3t}u(t).$$

(c) We have

$$X(j\omega) = \frac{1}{4+j\omega} - \frac{1}{(4+j\omega)^2}.$$

Therefore,

$$Y(j\omega) = X(j\omega)H(j\omega) = \frac{1}{(4+j\omega)(2+j\omega)}.$$

Finding the partial fraction expansion of $Y(j\omega)$ and taking the inverse Fourier transform,

$$y(t) = \frac{1}{2}e^{-2t}u(t) - \frac{1}{2}e^{-4t}u(t).$$

4.35. (a) From the given information,

$$|H(j\omega)| = \frac{\sqrt{a^2+\omega^2}}{\sqrt{a^2+\omega^2}} = 1.$$

Also,

$$\sphericalangle H(j\omega) = -\tan^{-1}\frac{\omega}{a} - \tan^{-1}\frac{\omega}{a} = -2\tan^{-1}\frac{\omega}{a}.$$

Also,

$$H(j\omega) = -1 + \frac{2a}{a+j\omega} \quad \Rightarrow \quad h(t) = -\delta(t) + 2ae^{-at}u(t).$$

(b) If $a = 1$, we have

$$|H(j\omega)| = 1, \qquad \sphericalangle H(j\omega) = -2\tan^{-1}\omega.$$

Therefore,

$$y(t) = \cos(\frac{t}{\sqrt{3}} - \frac{\pi}{3}) - \cos(t - \frac{\pi}{2}) + \cos(\sqrt{3}t - \frac{2\pi}{3}).$$

4.36. (a) The frequency response is

$$H(j\omega) = \frac{Y(j\omega)}{X(j\omega)} = \frac{3(3+j\omega)}{(4+j\omega)(2+j\omega)}.$$

(b) Finding the partial fraction expansion of answer in part (a) and taking its inverse Fourier transform, we obtain

$$h(t) = \frac{3}{2}\left[e^{-4t} + e^{-2t}\right]u(t).$$

(c) We have

$$\frac{Y(j\omega)}{X(j\omega)} = \frac{(9+3j\omega)}{8+6j\omega-\omega^2}.$$

Cross-multiplying and taking the inverse Fourier transform, we obtain

$$\frac{d^2y(t)}{dt^2} + 6\frac{dy(t)}{dt} + 8y(t) = 3\frac{dx(t)}{dt} + 9x(t).$$

4.37. **(a)** Note that

$$x(t) = x_1(t) * x_1(t),$$

where

$$x_1(t) = \begin{cases} 1, & |\omega| < \frac{1}{2} \\ 0, & \text{otherwise} \end{cases}$$

Also, the Fourier transform $X_1(j\omega)$ of $x_1(t)$ is

$$X_1(j\omega) = 2\frac{\sin(\omega/2)}{\omega}.$$

Using the convolution property we have

$$X(j\omega) = X_1(j\omega)X_1(j\omega) = \left[2\frac{\sin(\omega/2)}{\omega}\right]^2.$$

(b) The signal $\tilde{x}(t)$ is as shown in Figure S4.37

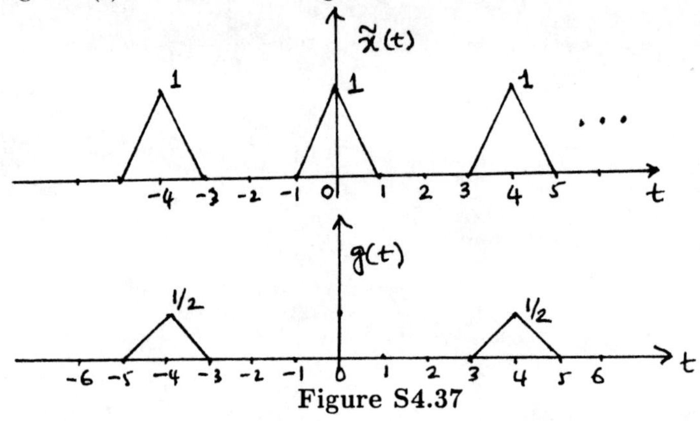

Figure S4.37

(c) One possible choice of $g(t)$ is as shown in Figure S4.37.

(d) Note that

$$\tilde{X}(j\omega) = X(j\omega)\frac{\pi}{2}\sum_{k=-\infty}^{\infty}\delta(j(\omega - k\frac{\pi}{2})) = G(j\omega)\frac{\pi}{2}\sum_{k=-\infty}^{\infty}\delta(j(\omega - k\frac{\pi}{2})).$$

This may also be written as

$$\tilde{X}(j\omega) = \frac{\pi}{2}\sum_{k=-\infty}^{\infty}X(j\pi k/2)\delta(j(\omega - k\frac{\pi}{2})) = \frac{\pi}{2}\sum_{k=-\infty}^{\infty}G(j\pi k/2)\delta(j(\omega - k\frac{\pi}{2})).$$

Clearly, this is possible only if

$$G(j\pi k/2) = X(j\pi k/2).$$

4.38. (a) Applying a frequency shift to the analysis equation, we have

$$X(j(\omega-\omega_0)) = \int_{-\infty}^{\infty} x(t)e^{-j(\omega-\omega_0)t}dt = \int_{-\infty}^{\infty} x(t)e^{j\omega_0 t}e^{-j\omega t}dt = \mathcal{FT}\{x(t)e^{j\omega_0 t}\}.$$

(b) We have

$$w(t) = e^{j\omega_0 t} \stackrel{FT}{\longleftrightarrow} W(j\omega) = 2\pi\delta(\omega-\omega_0).$$

Also,

$$\begin{aligned} x(t)w(t) \stackrel{FT}{\longleftrightarrow} & \quad \frac{1}{2\pi}[X(j\omega) * W(j\omega)] \\ = & \quad X(j\omega) * \delta(\omega-\omega_0) \\ = & \quad X(j(\omega-\omega_0)) \end{aligned}$$

4.39. (a) From the Fourier transform analysis equation, we have

$$G(j\omega) = \int_{-\infty}^{\infty} g(t)e^{-j\omega t}dt = \int_{-\infty}^{\infty} X(jt)e^{-j\omega t}dt. \qquad \text{(S4.39–1)}$$

Also from the Fourier transform synthesis equation, we have

$$x(t) = \frac{1}{2\pi}\int_{-\infty}^{\infty} X(j\omega)e^{j\omega t}d\omega.$$

Switching the variables t and ω, we have

$$x(\omega) = \frac{1}{2\pi}\int_{-\infty}^{\infty} X(jt)e^{j\omega t}dt.$$

We may also write this equation as

$$2\pi x(-\omega) = \int_{-\infty}^{\infty} X(jt)e^{-j\omega t}dt.$$

Substituting this equation in eq. (S4.39-1), we obtain

$$G(j\omega) = 2\pi x(-\omega).$$

(b) If in part (a) we have $x(t) = \delta(t+B)$, then we would have $g(t) = X(jt) = e^{jBt}$ and $G(j\omega) = 2\pi x(-\omega) = 2\pi\delta(-\omega+B) = 2\pi\delta(\omega-B)$.

4.40. When $n=1$, $x_1(t) = e^{-at}u(t)$ and $X_1(j\omega) = 1/(a+j\omega)$.
When $n=2$, $x_2(t) = te^{-at}u(t)$ and $X_2(j\omega) = 1/(a+j\omega)^2$.

Now, let us assume that the given statement is true when $n=m$, that is,

$$x_m(t) = \frac{t^{m-1}}{(m-1)!}e^{-at}u(t) \stackrel{FT}{\longleftrightarrow} X_m(j\omega) = \frac{1}{(1+j\omega)^m}.$$

For $n = m+1$ we may use the differentiation in frequency property to write,

$$x_{m+1}(t) = \frac{t}{m} x_m(t) \stackrel{FT}{\longleftrightarrow} X_{m+1}(j\omega) = \frac{1}{m} j \frac{dX_m(j\omega)}{d\omega} = \frac{1}{(1+j\omega)^{m+1}}.$$

This shows that if we assume that the given statement is true for $n = m$, then it is true for $n = m+1$. Since we also shown that the given statement is true for $n = 2$, we may argue that it is true for $n = 2+1 = 3$, $n = 3+1 = 4$, and so on. Therefore, the given statement is true for any n.

4.41. **(a)** We have

$$\begin{aligned} g(t) &= \frac{1}{2\pi}\int_{-\infty}^{\infty} \frac{1}{2\pi}\left[X(j\omega) * Y(j\omega)\right] e^{j\omega t} d\omega \\ &= \frac{1}{2\pi}\int_{-\infty}^{\infty} \frac{1}{2\pi}\left[\int_{-\infty}^{\infty} X(j\theta)Y(j(\omega-\theta))d\theta\right] e^{j\omega t} d\omega \\ &= \frac{1}{2\pi}\int_{-\infty}^{\infty} X(j\theta)\left[\frac{1}{2\pi}\int_{-\infty}^{\infty} Y(j(\omega-\theta))e^{j\omega t} d\omega\right] d\theta \end{aligned}$$

(b) Using the frequency shift property of the Fourier transform we have

$$\frac{1}{2\pi}\int_{-\infty}^{\infty} Y(j(\omega-\theta))e^{j\omega t} d\omega = e^{j\theta t} y(t).$$

(c) Combining the results of parts (a) and (b),

$$\begin{aligned} g(t) &= \frac{1}{2\pi}\int_{-\infty}^{\infty} X(j\theta)e^{j\theta t} y(t) d\theta \\ &= y(t)\frac{1}{2\pi}\int_{-\infty}^{\infty} X(j\theta)e^{j\theta t} d\theta \\ &= y(t)x(t). \end{aligned}$$

4.42. $x(t)$ is a periodic signal with Fourier series coefficients a_k. The fundamental frequency of $x(t)$ is $\omega_f = 100$ rad/sec. From Section 4.2 we know that the Fourier transform $X(j\omega)$ of $x(t)$ is

$$X(j\omega) = \sum_{k=-\infty}^{\infty} 2\pi a_k \delta(\omega - 100k).$$

(a) Since

$$y_1(t) = x(t)\cos(\omega_0 t) \stackrel{FT}{\longleftrightarrow} Y_1(j\omega) = \frac{1}{2}\{X(j(\omega-\omega_0)) + X(j(\omega+\omega_0))\}$$

we have

$$\begin{aligned} Y_1(j\omega) &= \pi \sum_{k=-\infty}^{\infty} \left[a_k\delta(\omega - 100k - \omega_0) + a_k\delta(\omega - 100k + \omega_0)\right] \\ &= \pi \sum_{k=-\infty}^{\infty} \left[a_{-k}\delta(\omega + 100k - \omega_0) + a_k\delta(\omega - 100k + \omega_0)\right] \quad \text{(S4.42-1)} \end{aligned}$$

If $\omega_0 = 500$, then the term in the above summation with $k = 5$ becomes

$$\pi a_{-5}\delta(\omega) + \pi a_5 \delta(\omega).$$

Since $x(t)$ is real, $a_k = a^*_{-k}$. Therefore, the above expression becomes $2\pi \mathcal{R}e\{a_5\}\delta(\omega)$, which is an impulse at $\omega = 0$. Note that the inverse Fourier transform of $2\pi\mathcal{R}e\{a_5\}\delta(\omega)$ is $g_1(t) = \mathcal{R}e\{a_5\}$. Therefore, we now need to find a $H(j\omega)$ such that

$$Y_1(j\omega)H(j\omega) = G_1(j\omega) = 2\pi\mathcal{R}e\{a_5\}\delta(\omega).$$

We may easily obtain such a $H(j\omega)$ by noting that the other terms (other than that for $k = 5$) in the summation of eq. (S4.42-1) result in impulses at $\omega = 100m$, $m \neq 0$. Therefore, we my choose any $H(j\omega)$ which is zero for $\omega = 100m$, where $m = \pm 1, \pm 2, \cdots$.

Similarly since

$$y_2(t) = x(t)\sin(\omega_0 t) \stackrel{FT}{\longleftrightarrow} Y_2(j\omega) = \frac{1}{2j}\{X(j(\omega - \omega_0)) - X(j(\omega + \omega_0))\},$$

we have

$$\begin{aligned} Y_2(j\omega) &= \frac{\pi}{j}\sum_{k=-\infty}^{\infty}[a_k\delta(\omega - 100k - \omega_0) - a_k\delta(\omega - 100k + \omega_0)] \\ &= \frac{\pi}{j}\sum_{k=-\infty}^{\infty}[a_{-k}\delta(\omega + 100k - \omega_0) - a_k\delta(\omega - 100k + \omega_0)] \quad (S4.42-2) \end{aligned}$$

If $\omega_0 = 500$, then the term in the above summation with $k = 5$ becomes

$$\frac{\pi}{j}a_{-5}\delta(\omega) - \frac{\pi}{j}a_5\delta(\omega).$$

Since $x(t)$ is real, $a_k = a^*_{-k}$. Therefore, the above expression becomes $2\pi\mathcal{I}m\{a_5\}\delta(\omega)$, which is an impulse at $\omega = 0$. Note that the inverse Fourier transform of $2\pi\mathcal{I}m\{a_5\}\delta(\omega)$ is $g_2(t) = \mathcal{I}m\{a_5\}$. Therefore, we now need to find a $H(j\omega)$ such that

$$Y_2(j\omega)H(j\omega) = G_2(j\omega) = 2\pi\mathcal{R}e\{a_5\}\delta(\omega).$$

We may easily obtain such a $H(j\omega)$ by noting that the other terms (other than that for $k = 5$) in the summation of eq. (S4.42-2) result in impulses at $\omega = 100m$, $m \neq 0$. Therefore, we my choose any $H(j\omega)$ which is zero for $\omega = 100m$, where $m = \pm 1, \pm 2, \cdots$.

(b) An example of a valid $H(j\omega)$ would be the frequency response of an ideal lowpass filter with passband gain of unity and cutoff frequency of 50 rad/sec. In this case,

$$h(t) = \frac{\sin(50t)}{\pi t}.$$

4.43. Since

$$y_1(t) = \cos^2 t = \frac{1 + \cos(2t)}{2},$$

we obtain

$$Y_1(j\omega) = \pi\delta(\omega) + \frac{\pi}{2}\delta(\omega - 2) + \frac{\pi}{2}\delta(\omega + 2).$$

Therefore,

$$y_2(t) = x(t)y_1(t) = x(t)\cos^2(t) \stackrel{FT}{\longleftrightarrow} Y_2(j\omega) = \frac{1}{2\pi}\{X(j\omega) * Y_1(j\omega)\}.$$

This gives

$$Y_2(j\omega) = \frac{1}{2}X(j\omega) + \frac{1}{4}X(j(\omega - 2)) + \frac{1}{4}X(j(\omega + 2)).$$

$X(j\omega)$ and $Y_2(j\omega)$ are as shown in Figure S4.43.

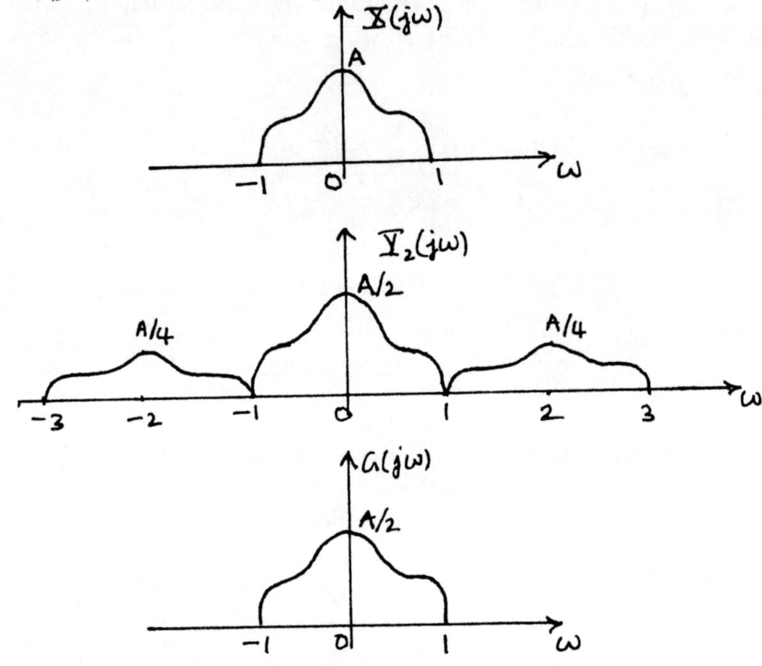

Figure S4.43

Now,

$$y_3(t) = \frac{\sin t}{\pi t} \stackrel{FT}{\longleftrightarrow} Y_3(j\omega) = \begin{cases} 1, & |\omega| < 1 \\ 0, & \text{otherwise} \end{cases}.$$

Also,

$$g(t) = y_2(t) * y_3(t) \stackrel{FT}{\longleftrightarrow} G(j\omega) = Y_2(j\omega)Y_3(j\omega).$$

From Figure S4.43 it is clear that

$$G(j\omega) = \frac{1}{2}X(j\omega).$$

Therefore, an LTI system with impulse response $h(t) = \frac{1}{2}\delta(t)$ may be used to obtain $g(t)$ from $x(t)$.

4.44. (a) Taking the Fourier transform of both sides of the given differential equation, we have

$$Y(j\omega)[10 + j\omega] = X(j\omega)[Z(j\omega) - 1].$$

Since, $Z(j\omega) = \frac{1}{1+j\omega} + 3$, we obtain from the above equation

$$H(j\omega) = \frac{Y(j\omega)}{X(j\omega)} = \frac{3 + 2j\omega}{(1 + j\omega)(10 + j\omega)}.$$

(b) Finding the partial fraction expansion of $H(j\omega)$ and then taking its inverse Fourier transform we obtain

$$h(t) = \frac{1}{9}e^{-t}u(t) + \frac{17}{9}e^{-10t}u(t).$$

4.45. We have

$$y(t) = x(t) * h(t) \quad \Rightarrow \quad Y(j\omega) = X(j\omega)H(j\omega).$$

From Parseval's relation the total energy in $y(t)$ is

$$\begin{aligned}
E &= \int_{-\infty}^{\infty} |y(t)|^2 dt = \frac{1}{2\pi}\int_{-\infty}^{\infty} |Y(j\omega)|^2 d\omega \\
&= \frac{1}{2\pi}\int_{-\infty}^{\infty} |X(j\omega)|^2 |H(j\omega)|^2 d\omega \\
&= \frac{1}{2\pi}\int_{-\omega_0-\Delta/2}^{-\omega_0+\Delta/2} |X(j\omega)|^2 d\omega + \frac{1}{2\pi}\int_{\omega_0-\Delta/2}^{\omega_0+\Delta/2} |X(j\omega)|^2 d\omega \\
&\approx \frac{1}{2\pi}|X(-j\omega_0)|^2\Delta + \frac{1}{2\pi}|X(j\omega_0)|^2\Delta
\end{aligned}$$

For real $x(t)$, $|X(-j\omega_0)|^2 = |X(j\omega_0)|^2$. Therefore,

$$E = \frac{1}{\pi}|X(j\omega_0)|^2\Delta.$$

4.46. Let $g_1(t)$ be the response of $H_1(j\omega)$ to $x(t)\cos\omega_c t$. Let $g_2(t)$ be the response of $H_2(j\omega)$ to $x(t)\sin\omega_c t$. Then, with reference to Figure 4.30,

$$y(t) = x(t)e^{j\omega_c t} = x(t)\cos\omega_c t + jx(t)\sin\omega_c t,$$

and

$$w(t) = g_1(t) + jg_2(t).$$

Also,

$$f(t) = e^{-j\omega_c t}w(t) = [\cos\omega_c t - j\sin\omega_c t][g_1(t) + jg_2(t)].$$

Therefore,

$$\mathcal{R}e\{f(t)\} = g_1(t)\cos\omega_c t + g_2(t)\sin\omega_c t.$$

This is exactly what Figure P4.46 implements.

4.47. (a) We have

$$h_e(t) = \frac{h(t) + h(-t)}{2}.$$

Since $h(t)$ is causal, the non-zero portions of $h(t)$ and $h(-t)$ overlap *only* at $t = 0$. Therefore,

$$h(t) = \begin{cases} 0, & t < 0 \\ h_e(t), & t = 0 \\ 2h_e(t), & t > 0 \end{cases}. \qquad \text{(S4.47-1)}$$

Also, from Table 4.1 we have

$$h_e(t) \stackrel{FT}{\longleftrightarrow} \mathcal{R}e\{H(j\omega)\}.$$

Given $\mathcal{R}e\{H(j\omega)$, we can obtain $h_e(t)$. From $h_e(t)$, we can recover $h(t)$ (and consequently $H(j\omega)$) by using eq. (S4.47-1). Therefore, $H(j\omega)$ is completely specified by $\mathcal{R}e\{H(j\omega)$.

(b) If

$$\mathcal{R}e\{H(j\omega)\} = \cos t = \frac{1}{2}e^{j\omega t} + \frac{1}{2}e^{-j\omega t}$$

then,

$$h_e(t) = \frac{1}{2}\delta(t+1) + \frac{1}{2}\delta(t-1).$$

Therefore from eq. (S4.47-1),

$$h(t) = \delta(t-1).$$

(c) We have

$$h_o(t) = \frac{h(t) + h(-t)}{2}.$$

Since $h(t)$ is causal, the non-zero portions of $h(t)$ and $h(-t)$ overlap *only* at $t = 0$ and $h_o(t)$ will be zero at $t = 0$. Therefore,

$$h(t) = \begin{cases} 0, & t < 0 \\ \text{unknown}, & t = 0 \\ 2h_o(t), & t > 0 \end{cases}. \qquad \text{(S4.47-2)}$$

Also, from Table 4.1 we have

$$h_o(t) \stackrel{FT}{\longleftrightarrow} \mathcal{I}m\{H(j\omega)\}.$$

Given $\mathcal{I}m\{H(j\omega)$, we can obtain $h_o(t)$. From $h_o(t)$, we can recover $h(t)$ except for $t = 0$ by using eq. (S4.47-1). If there are no singularities in $h(t)$ at $t = 0$, then $H(j\omega)$ can be recovered from $h(t)$ even if $h(0)$ is unknown. Therefore $H(j\omega)$ is completely specified by $\mathcal{I}m\{H(j\omega)$ in this case.

4.48. **(a)** Using the multiplication property we have

$$h(t) = h(t)u(t) \overset{FT}{\longleftrightarrow} H(j\omega) = \frac{1}{2\pi}\left\{H(j\omega) * \left[\frac{1}{j\omega} + \pi\delta(\omega)\right]\right\}.$$

The right-hand side may be written as

$$H(j\omega) = \frac{1}{2}H(j\omega) + \frac{1}{2\pi j}\left[H(j\omega) * \frac{1}{\omega}\right].$$

That is,

$$H(j\omega) = \frac{1}{\pi j}\int_{-\infty}^{\infty}\frac{H(j\eta)}{\omega - \eta}d\eta.$$

Breaking up $H(j\omega)$ into real and imaginary parts,

$$H_R(j\omega) + jH_I(j\omega) = \frac{1}{\pi j}\int_{-\infty}^{\infty}\frac{H_R(j\eta) + jH_I(j\eta)}{\omega - \eta}d\eta = \frac{1}{\pi}\int_{-\infty}^{\infty}\frac{H_I(j\eta) - jH_R(j\eta)}{\omega - \eta}d\eta.$$

Comparing real and imaginary parts on both sides, we obtain

$$H_R(j\omega) = \frac{1}{\pi}\int_{-\infty}^{\infty}\frac{H_I(j\eta)}{\omega - \eta}d\eta \quad \text{and} \quad H_I(j\omega) = -\frac{1}{\pi}\int_{-\infty}^{\infty}\frac{H_R(j\eta)}{\omega - \eta}d\eta.$$

(b) From eq. (P4.48-3), we may write

$$y(t) = x(t) * \frac{1}{\pi t} \quad \Rightarrow \quad Y(j\omega) = X(j\omega)\mathcal{FT}\{1/(\pi t)\} \tag{S4.48–1}$$

Also, from Table 4.2

$$u(t) \overset{FT}{\longleftrightarrow} \frac{1}{j\omega} + \pi\delta(\omega).$$

Therefore,

$$2u(t) - 1 \overset{FT}{\longleftrightarrow} 2\frac{1}{j\omega}.$$

Using the duality property, we have

$$\frac{2}{jt} \overset{FT}{\longleftrightarrow} 2\pi[2u(-\omega) - 1]$$

or

$$\frac{1}{\pi t} \overset{FT}{\longleftrightarrow} j[2u(-\omega) - 1].$$

Therefore, from eq.(S4.48-1), we have

$$Y(j\omega) = X(j\omega)H(j\omega)$$

where

$$H(j\omega) = j[2u(-\omega) - 1] = \begin{cases} -j, & \omega > 0 \\ j, & \omega > 0 \end{cases}.$$

(c) Let $y(t)$ be the Hilbert transform of $x(t) = \cos(3t)$. Then,

$$Y(j\omega) = X(j\omega)H(j\omega) = \pi[\delta(\omega-3)+\delta(\omega+3)]H(j\omega) = -j\pi\delta(\omega-3)+j\pi\delta(\omega+3).$$

Therefore,

$$y(t) = \sin(3t).$$

4.49. (a) (i) Since $H(j\omega)$ is real and even, $h(t)$ is also real and even.

(ii)

$$|h(t)| = \left|\frac{1}{2\pi}\int_{-\infty}^{\infty} H(j\omega)e^{j\omega t}d\omega\right| \le \frac{1}{2\pi}\int_{-\infty}^{\infty}|H(j\omega)|e^{j\omega t}d\omega.$$

Since $H(j\omega)$ is real and positive,

$$|h(t)| \le \frac{1}{2\pi}\int_{-\infty}^{\infty} H(j\omega)e^{j\omega t}d\omega = h(0).$$

Therefore,

$$\max[|h(t)|] = h[0].$$

(b) The bandwidth of this system is $2W$.

(c) We have

$$B_w H(j0) = \text{Area under } H(j\omega).$$

Therefore,

$$B_w = \frac{1}{H(j0)}\int_{-\infty}^{\infty} H(j\omega)d\omega.$$

(d) We have

$$t_r = \frac{s(\infty)}{h(0)} = \frac{\int_{-\infty}^{\infty} h(t)dt}{\frac{1}{2\pi}\int_{-\infty}^{\infty} H(j\omega)d\omega} = \frac{H(j0)}{\frac{1}{2\pi}\int_{-\infty}^{\infty} H(j\omega)d\omega} = \frac{2\pi}{B_w}.$$

(e) Therefore,

$$B_w t_r = B_w\frac{2\pi}{B_w} = 2\pi.$$

4.50. (a) We know from problems 1.45 and 2.67 that

$$\phi_{xy}(t) = \phi_{yx}(-t).$$

Therefore,

$$\Phi_{xy}(j\omega) = \Phi_{yx}(-j\omega).$$

Since $\phi_{yx}(t)$ is real,

$$\Phi_{xy}(j\omega) = \Phi_{yx}^{*}(j\omega).$$

(b) We may write

$$\phi_{xy}(t) = \int_{-\infty}^{\infty} x(t+\tau)y(\tau)d\tau = x(t) * y(-t).$$

Therefore,

$$\Phi_{xy}(j\omega) = X(j\omega)Y(-j\omega).$$

Since $y(t)$ is real, we may write this as

$$\Phi_{xy}(j\omega) = X(j\omega)Y^*(j\omega).$$

(c) Using the results of part (b) with $y(t) = x(t)$,

$$\Phi_{xx}(j\omega) = X(j\omega)X^*(j\omega) = |X(j\omega)|^2 \geq 0.$$

(d) From part (b) we have

$$\begin{aligned}
\Phi_{xy}(j\omega) &= X(j\omega)Y^*(j\omega) \\
&= X(j\omega)[H(j\omega)X(j\omega)]^* \\
&= \Phi_{xx}(j\omega)H^*(j\omega)
\end{aligned}$$

Also,

$$\begin{aligned}
\Phi_{yy}(j\omega) &= Y(j\omega)Y^*(j\omega) \\
&= [H(j\omega)X(j\omega)][H(j\omega)X(j\omega)]^* \\
&= \Phi_{xx}(j\omega)|H(j\omega)|^2
\end{aligned}$$

(e) From the given information, we have

$$X(j\omega) = \frac{e^{-j\omega}-1}{\omega^2} - j\frac{e^{-j\omega}}{\omega}$$

and

$$H(j\omega) = \frac{1}{a+j\omega}.$$

Therefore,

$$\Phi_{xx}(j\omega) = |X(j\omega)|^2 = \frac{2-2\cos\omega}{\omega^4} - \frac{2\sin\omega}{\omega^2} + \frac{1}{\omega^2},$$

$$\Phi_{xy}(j\omega) = \Phi_{xx}(j\omega)H^*(j\omega) = \left[\frac{2-2\cos\omega}{\omega^4} - \frac{2\sin\omega}{\omega^2} + \frac{1}{\omega^2}\right]\left[\frac{1}{a-j\omega}\right],$$

and

$$\Phi_{yy}(j\omega) = \Phi_{xx}(j\omega)|H(j\omega)|^2 = \left[\frac{2-2\cos\omega}{\omega^4} - \frac{2\sin\omega}{\omega^2} + \frac{1}{\omega^2}\right]\left[\frac{1}{a^2+\omega^2}\right].$$

(f) We require that

$$|H(j\omega)|^2 = \frac{\omega^2 + 100}{\omega^2 + 25}.$$

The possible causal and stable choices for $H(j\omega)$ are

$$H_1(j\omega) = \frac{10 + j\omega}{5 + j\omega} \quad \text{and} \quad H_2(j\omega) = \frac{10 - j\omega}{5 + j\omega}.$$

The corresponding impulse responses are

$$h_1(t) = \delta(t) + 5e^{-5t}u(t) \quad \text{and} \quad h_2(t) = -\delta(t) + 15e^{-5t}u(t).$$

Only the system with impulse response $h_1(t)$ has a causal and stable inverse.

4.51. **(a)** $H(j\omega) = 1/G(j\omega)$.

(b) (i) If we denote the output by $y(t)$, then we have

$$Y(j0) = \frac{1}{2}.$$

Since $H(j0) = 0$, it is impossible for us to have $Y(j0) = X(j0)H(j0)$. Therefore, we cannot find an $x(t)$ which produces an output which looks like Figure P4.50.

(ii) This system is not invertible because $1/H(j\omega)$ is not defined for all ω.

(c) We have

$$H(j\omega) = \sum_{k=0}^{\infty} e^{-kT} e^{-j\omega kT} = \frac{1}{1 - e^{-(1+j\omega)T}}.$$

We now need to find a $G(j\omega)$ such that

$$H(j\omega)G(j\omega) = 1.$$

Thus $G(j\omega)$ is the inverse system of $H(j\omega)$, and is given by

$$G(j\omega) = 1 - e^{-(1+j\omega)T}.$$

(d) Since $H(j\omega) = 2 + j\omega$,

$$G(j\omega) = \frac{Y(j\omega)}{X(j\omega)} = \frac{1}{2 + j\omega}.$$

Cross-multiplying and taking the inverse Fourier transform, we obtain

$$\frac{dy(t)}{dt} + 2y(t) = x(t).$$

(e) We have

$$H(j\omega) = \frac{-\omega^2 + 3j\omega + 2}{-\omega^2 + 6j\omega + 9}.$$

Therefore, the frequency response of the inverse is

$$G(j\omega) = \frac{1}{H(j\omega)} = \frac{-\omega^2 + 6j\omega + 9}{-\omega^2 + 3j\omega + 2}.$$

The differential equation describing the inverse system is

$$\frac{d^2y(t)}{dt} + 3\frac{dy(t)}{dt} + 2y(t) = \frac{d^2x(t)}{dt} + 6\frac{dx(t)}{dt} + 9x(t).$$

Using partial fraction expansion followed by application of the inverse Fourier transform, we find the impulse responses to be

$$h(t) = \delta(t) - 3e^{-3t}u(t) + 2te^{-3t}u(t)$$

and

$$g(t) = \delta(t) - e^{-2t}u(t) + 4e^{-t}u(t).$$

4.52. **(a)** Since the step response is $s(t) = (1 - e^{-t/2})u(t)$, the impulse response has to be

$$h(t) = \frac{1}{2}e^{-t/2}u(t).$$

The frequency response of the system is

$$H(j\omega) = \frac{1/2}{\frac{1}{2} + j\omega}.$$

We now desire to build an inverse for the above system. Therefore, the frequency response of the inverse system has to be

$$G(j\omega) = \frac{1}{H(j\omega)} = 2\left[\frac{1}{2} + j\omega\right].$$

Taking the inverse Fourier transform we obtain

$$g(t) = \delta(t) + 2u_1(t).$$

(b) When $\sin(\omega t)$ passes through the inverse system, the output will be

$$y(t) = \sin(\omega t) + 2\omega\cos(\omega t).$$

We see that the output is directly proportional to ω. Therefore, as ω increases, the contribution to the output due to the noise also increases.

(c) In this case we require that $|H(j\omega)| \leq \frac{1}{4}$ when $\omega = 6$. Since

$$|H(j\omega)|^2 = \frac{1}{a^2 + \omega^2},$$

we require that

$$\frac{1}{a^2 + 36} \leq \frac{1}{16}.$$

Therefore, $a \leq \frac{6}{\sqrt{15}}$.

4.53. **(a)** From the given definition we obtain

$$\begin{aligned} X(j\omega_1, j\omega_2) &= \int_{-\infty}^{\infty}\int_{-\infty}^{\infty} x(t_1,t_2)e^{-j(\omega_1 t_1+\omega_2 t_2)}dt_1 dt_2 \\ &= \int_{-\infty}^{\infty}\left[\int_{-\infty}^{\infty} x(t_1,t_2)e^{-j\omega_1 t_1}dt_1\right] e^{-j\omega_2 t_2}dt_2 \\ &= \int_{-\infty}^{\infty} X(\omega_1,t_2)e^{-j\omega_2 t_2}dt_2 \end{aligned}$$

(b) From the result of part (a) we may write

$$x(t_1,t_2) = \mathcal{FT}_{\omega_1}^{-1}\{\mathcal{FT}_{\omega_2}^{-1}\{X(j\omega_1,j\omega_2)\}\} = \frac{1}{4\pi^2}\int_{-\infty}^{\infty}\int_{-\infty}^{\infty} X(j\omega_1,j\omega_2)e^{j(\omega_1 t_1+\omega_2 t_2)}d\omega_1 d\omega_2.$$

(c) (i) $X(j\omega_1,\omega_2) = \frac{e^{-(1+j\omega_1)}}{(1+j\omega_1)}\frac{e^{2(2-j\omega_2)}}{(2-j\omega_2)}$

(ii) $X(j\omega_1,\omega_2) = \frac{[1-e^{-(1+j\omega_1)}][1-e^{-(1-j\omega_2)}]}{(1+j\omega_1)(1-j\omega_2)} + \frac{[1-e^{-(1+j\omega_1)}][1-e^{-(1+j\omega_2)}]}{(1+j\omega_1)(1+j\omega_2)}$

(iii) $X(j\omega_1,\omega_2) = \frac{2-e^{-(1+j\omega_1)}-e^{-(1+j\omega_2)}-[1-e^{-(1+j\omega_1)}][1-e^{-(1+j\omega_2)}]}{(1+j\omega_1)(1+j\omega_2)} +$
$\frac{1-e^{-(1+j\omega_1)}}{(1+j\omega_1)(1-j\omega_2)} + \frac{1-e^{-(1+j\omega_2)}}{(1-j\omega_1)(1+j\omega_2)}$

(iv) $X(\omega_1,\omega_2) = -\frac{1}{j\omega_2}\left[\frac{e^{-j\omega_2}(1-e^{j(\omega_1+\omega_2)})+e^{j\omega_2}(1-e^{-j(\omega_1+\omega_2)})}{-j(\omega_1+\omega_2)}\right.$
$\left.+\frac{e^{j\omega_2}(1-e^{j(\omega_1-\omega_2)})+e^{-j\omega_2}(e^{-j(\omega_1-\omega_2)}-1)}{-j(\omega_1-\omega_2)}\right]$

(v) As shown in the Figure S4.53, this signal has six different regions in the (t_1, t_2) plane.

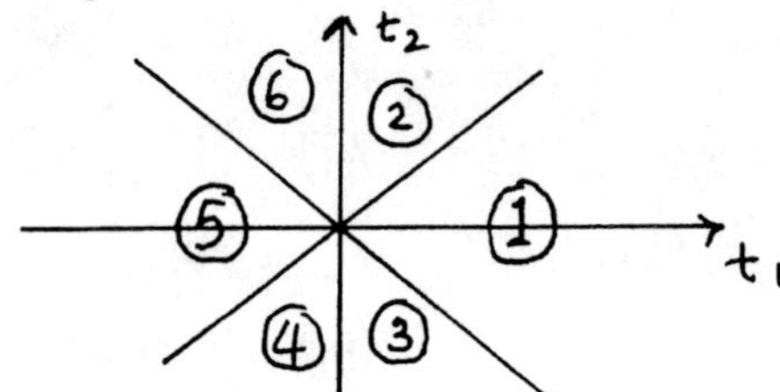

Figure S4.53

The signal $x(t_1,t_2)$ is given by

$$x(t_1,t_2) = \begin{cases} e^{-2t_1}, & \text{in region 1} \\ e^{-2t_2}, & \text{in region 2} \\ e^{2t_2}, & \text{in region 3} \\ e^{2t_2}, & \text{in region 4} \\ e^{2t_1}, & \text{in region 5} \\ e^{-2t_2}, & \text{in region 6} \end{cases}.$$

Therefore,

$$\begin{aligned}
X(\omega_1, \omega_2) &= \frac{1}{(2 + j\omega_1 + j\omega_2)(2 + j\omega_1 - j\omega_2)} + \frac{1}{(2 + j\omega_2)(2 + j\omega_1 + j\omega_2)} \\
&+ \frac{1}{(2 - j\omega_2)(2 + j\omega_1 - j\omega_2)} + \frac{1}{(2 - j\omega_2)(2 - j\omega_1 - j\omega_2)} \\
&- \frac{2}{(2 - j\omega_1 - j\omega_2)(2 - j\omega_1 + j\omega_2)} + \frac{1}{(j\omega_2)(2 - j\omega_1 - j\omega_2)}
\end{aligned}$$

(d) $x(t_1, t_2) = e^{-4(t_1 + 2t_2)} u(t_1 + 2t_2)$

(e) (i) $e^{-j\omega_1 T_1} e^{-j\omega_2 T_2} X(j\omega_1, j\omega_2)$

(ii) $\frac{1}{|ab|} X(j\frac{\omega_1}{a}, j\frac{\omega_2}{b})$

(iii) $X(j\omega_1, j\omega_2) H(j\omega_1, j\omega_2)$

Chapter 5 Answers

5.1. **(a)** Let $x[n] = (1/2)^{n-1}u[n-1]$. Using the Fourier transform analysis equation (5.9), the Fourier transform $X(e^{j\omega})$ of this signal is

$$\begin{aligned} X(e^{j\omega}) &= \sum_{n=-\infty}^{\infty} x[n]e^{-j\omega n} \\ &= \sum_{n=1}^{\infty}(1/2)^{n-1}e^{-j\omega n} \\ &= \sum_{n=0}^{\infty}(1/2)^{n}e^{-j\omega(n+1)} \\ &= e^{-j\omega}\frac{1}{(1-(1/2)e^{-j\omega})} \end{aligned}$$

(b) Let $x[n] = (1/2)^{|n-1|}$. Using the Fourier transform analysis equation (5.9), the Fourier transform $X(e^{j\omega})$ of this signal is

$$\begin{aligned} X(e^{j\omega}) &= \sum_{n=-\infty}^{\infty} x[n]e^{-j\omega n} \\ &= \sum_{n=-\infty}^{0} (1/2)^{-(n-1)}e^{-j\omega n} + \sum_{n=1}^{\infty}(1/2)^{n-1}e^{-j\omega n} \end{aligned}$$

The second summation in the right-hand side of the above equation is exactly the same as the result of part (a). Now,

$$\sum_{n=-\infty}^{0} (1/2)^{-(n-1)}e^{-j\omega n} = \sum_{n=0}^{\infty}(1/2)^{(n+1)}e^{j\omega n} = \left(\frac{1}{2}\right)\frac{1}{1-(1/2)e^{j\omega}}.$$

Therefore,

$$X(e^{j\omega}) = \left(\frac{1}{2}\right)\frac{1}{1-(1/2)e^{j\omega}} + e^{-j\omega}\frac{1}{(1-(1/2)e^{-j\omega})} = \frac{0.75e^{-j\omega}}{1.25-\cos\omega}.$$

5.2. **(a)** Let $x[n] = \delta[n-1] + \delta[n+1]$. Using the Fourier transform analysis equation (5.9), the Fourier transform $X(e^{j\omega})$ of this signal is

$$\begin{aligned} X(e^{j\omega}) &= \sum_{n=-\infty}^{\infty} x[n]e^{-j\omega n} \\ &= e^{-j\omega} + e^{j\omega} = 2\cos\omega \end{aligned}$$

(b) Let $x[n] = \delta[n+2] - \delta[n-2]$. Using the Fourier transform analysis equation (5.9), the Fourier transform $X(e^{j\omega})$ of this signal is

$$\begin{aligned} X(e^{j\omega}) &= \sum_{n=-\infty}^{\infty} x[n]e^{-j\omega n} \\ &= e^{2j\omega} - e^{-2j\omega} = 2j\sin(2\omega) \end{aligned}$$

5.3. We note from Section 5.2 that a periodic signal $x[n]$ with Fourier series representation

$$x[n] = \sum_{k=<N>} a_k e^{jk(2\pi/N)n}$$

has a Fourier transform

$$X(e^{j\omega}) = \sum_{k=-\infty}^{\infty} 2\pi a_k \delta\left(\omega - \frac{2\pi k}{N}\right).$$

(a) Consider the signal $x_1[n] = \sin(\frac{\pi}{3}n + \frac{\pi}{4})$. We note that the fundamental period of the signal $x_1[n]$ is $N = 6$. The signal may be written as

$$x_1[n] = (1/2j)e^{j(\frac{\pi}{3}n+\frac{\pi}{4})} - (1/2j)e^{-j(\frac{\pi}{3}n+\frac{\pi}{4})} = (1/2j)e^{j\frac{\pi}{4}}e^{j\frac{2\pi}{6}n} - (1/2j)e^{-j\frac{\pi}{4}}e^{-j\frac{2\pi}{6}n}.$$

From this, we obtain the non-zero Fourier series coefficients a_k of $x_1[n]$ in the range $-2 \le k \le 3$ as

$$a_1 = (1/2j)e^{j\frac{\pi}{4}}, \qquad a_{-1} = -(1/2j)e^{-j\frac{\pi}{4}}.$$

Therefore, in the range $-\pi \le \omega \le \pi$, we obtain

$$\begin{aligned} X(e^{j\omega}) &= 2\pi a_1 \delta(\omega - \frac{2\pi}{6}) + 2\pi a_{-1}\delta(\omega + \frac{2\pi}{6}) \\ &= (\pi/j)\{e^{j\pi/4}\delta(\omega - 2\pi/6) - e^{-j\pi/4}\delta(\omega + 2\pi/6)\} \end{aligned}$$

(b) Consider the signal $x_2[n] = 2 + \cos(\frac{\pi}{6}n + \frac{\pi}{8})$. We note that the fundamental period of the signal $x_1[n]$ is $N = 12$. The signal may be written as

$$x_1[n] = 2 + (1/2)e^{j(\frac{\pi}{6}n+\frac{\pi}{8})} + (1/2)e^{-j(\frac{\pi}{6}n+\frac{\pi}{8})} = 2 + (1/2)e^{j\frac{\pi}{8}}e^{j\frac{2\pi}{12}n} + (1/2)e^{-j\frac{\pi}{8}}e^{-j\frac{2\pi}{12}n}.$$

From this, we obtain the non-zero Fourier series coefficients a_k of $x_2[n]$ in the range $-5 \le k \le 6$ as

$$a_0 = 2, \qquad a_1 = (1/2)e^{j\frac{\pi}{8}}, \qquad a_{-1} = (1/2)e^{-j\frac{\pi}{8}}.$$

Therefore, in the range $-\pi \le \omega \le \pi$, we obtain

$$\begin{aligned} X(e^{j\omega}) &= 2\pi a_0 \delta(\omega) + 2\pi a_1 \delta(\omega - \frac{2\pi}{12}) + 2\pi a_{-1}\delta(\omega + \frac{2\pi}{12}) \\ &= 4\pi\delta(\omega) + \pi\{e^{j\pi/8}\delta(\omega - \pi/6) + e^{-j\pi/8}\delta(\omega + \pi/6)\} \end{aligned}$$

5.4. (a) Using the Fourier transform synthesis equation (5.8),

$$\begin{aligned}
x_1[n] &= (1/2\pi)\int_{-\pi}^{\pi} X_1(e^{j\omega})e^{j\omega n}d\omega \\
&= (1/2\pi)\int_{-\pi}^{\pi} [2\pi\delta(\omega) + \pi\delta(\omega - \pi/2) + \pi\delta(\omega + \pi/2)]e^{j\omega n}d\omega \\
&= e^{j0} + (1/2)e^{j(\pi/2)n} + (1/2)e^{-j(\pi/2)n} \\
&= 1 + \cos(\pi n/2)
\end{aligned}$$

(b) Using the Fourier transform synthesis equation (5.8),

$$\begin{aligned}
x_2[n] &= (1/2\pi)\int_{-\pi}^{\pi} X_2(e^{j\omega})e^{j\omega n}d\omega \\
&= -(1/2\pi)\int_{-\pi}^{0} 2je^{j\omega n}d\omega + (1/2\pi)\int_{0}^{\pi} 2je^{j\omega n}d\omega \\
&= (j/\pi)\left[-\frac{1 - e^{-jn\pi}}{jn} + \frac{e^{jn\pi} - 1}{jn}\right] \\
&= -(4/(n\pi))\sin^2(n\pi/2)
\end{aligned}$$

5.5. From the given information,

$$\begin{aligned}
x[n] &= (1/2\pi)\int_{-\pi}^{\pi} X(e^{j\omega})e^{j\omega n}d\omega \\
&= (1/2\pi)\int_{-\pi}^{\pi} |X(e^{j\omega})|e^{j\sphericalangle\{X(e^{j\omega})\}}e^{j\omega n}d\omega \\
&= (1/2\pi)\int_{-\pi/4}^{\pi/4} e^{-\frac{3}{2}\omega}e^{j\omega n}d\omega \\
&= \frac{\sin(\frac{\pi}{4}(n - 3/2))}{\pi(n - 3/2)}
\end{aligned}$$

The signal $x[n]$ is zero when $\frac{\pi}{4}(n - 3/2)$ is a nonzero integer multiple of π or when $|n| \to \infty$. The value of $\frac{\pi}{4}(n - 3/2)$ can never be such that it is a nonzero integer multiple of π. Therefore, $x[n] = 0$ only for $n = \pm\infty$.

5.6. Throughout this problem, we assume that

$$x[n] \stackrel{FT}{\longleftrightarrow} X_1(e^{j\omega}).$$

(a) Using the time reversal property (Sec. 5.3.6), we have

$$x[-n] \stackrel{FT}{\longleftrightarrow} X(e^{-j\omega})$$

Using the time shifting property (Sec. 5.3.3) on this, we have

$$x[-n+1] \stackrel{FT}{\longleftrightarrow} e^{-j\omega n}X(e^{-j\omega}) \quad \text{and} \quad x[-n-1] \stackrel{FT}{\longleftrightarrow} e^{j\omega n}X(e^{-j\omega})$$

Therefore,

$$\begin{aligned} x_1[n] = x[-n+1] + x[-n-1] &\stackrel{FT}{\longleftrightarrow} e^{-j\omega n}X(e^{-j\omega}) + e^{j\omega n}X(e^{-j\omega}) \\ &\stackrel{FT}{\longleftrightarrow} 2X(e^{-j\omega})\cos\omega \end{aligned}$$

(b) Using the time reversal property (Sec. 5.3.6), we have

$$x[-n] \stackrel{FT}{\longleftrightarrow} X(e^{-j\omega})$$

Using the conjugation property on this, we have

$$x^*[-n] \stackrel{FT}{\longleftrightarrow} X^*(e^{j\omega})$$

Therefore,

$$\begin{aligned} x_2[n] = (1/2)(x^*[-n] + x[n]) &\stackrel{FT}{\longleftrightarrow} (1/2)(X(e^{j\omega}) + X^*(e^{j\omega}) \\ &\stackrel{FT}{\longleftrightarrow} \mathcal{R}e\{X(e^{j\omega})\} \end{aligned}$$

(c) Using the differentiation in frequency property (Sec. 5.3.8), we have

$$nx[n] \stackrel{FT}{\longleftrightarrow} j\frac{dX(e^{j\omega})}{d\omega}$$

Using the same property a second time,

$$n^2x[n] \stackrel{FT}{\longleftrightarrow} -\frac{d^2X(e^{j\omega})}{d\omega^2}$$

Therefore,

$$x_3[n] = n^2x[n] - 2nx[n] + 1 \stackrel{FT}{\longleftrightarrow} -\frac{d^2X(e^{j\omega})}{d\omega^2} - 2j\frac{dX(e^{j\omega})}{d\omega} + X(e^{j\omega})$$

5.7. **(a)** Consider the signal $y_1[n]$ with Fourier transform

$$Y_1(e^{j\omega}) = \sum_{k=1}^{10} \sin(k\omega).$$

We see that $Y_1(e^{j\omega})$ is real and odd. From Table 5.1, we know that the Fourier transform of a real and odd signal is purely imaginary and odd. Therefore, we may say that the Fourier transform of a purely imaginary and odd signal is real and odd. Using this observation, we conclude that $y_1[n]$ is purely imaginary and odd.

Note now that

$$X_1(e^{j\omega}) = e^{-j\omega}Y_1(e^{j\omega}).$$

Therefore, $x_1[n] = y_1[n-1]$. Therefore, $x_1[n]$ is also **purely imaginary**. But $x_1[n]$ is **neither even nor odd**.

(b) We note that $X_2(e^{j\omega})$ is purely imaginary and odd. Therefore, $x_2[n]$ has to be **real and odd**.

(c) Consider a signal $y_3[n]$ whose magnitude of the Fourier transform is $|Y_3(e^{j\omega})| = A(\omega)$, and whose phase of the Fourier transform is $\sphericalangle\{Y_3(e^{j\omega})\} = -(3/2)\omega$. Since $|Y_3(e^{j\omega})| = |Y_3(e^{-j\omega})|$ and $\sphericalangle\{Y_3(e^{j\omega})\} = -\sphericalangle\{Y_3(e^{-j\omega})\}$, we may conclude that the signal $y_3[n]$ is real (See Table 5.1, Property 5.3.4).

Now, consider the signal $x_3[n]$ with Fourier transform $X_3(e^{j\omega}) = Y_3(e^{j\omega})e^{j\pi} = -Y_3(j\omega)$. Using the result from the previous paragraph and the linearity property of the Fourier transform, we may conclude that $x_3[n]$ has to **real**. Since the Fourier transform $X_3(e^{j\omega})$ is neither purely imaginary nor purely real, the signal $x_3[n]$ is **neither even nor odd**.

5.8. Consider the signal

$$x_1[n] = \begin{cases} 1, & |n| \leq 1 \\ 0, & |n| > 1 \end{cases}$$

From Table 5.2, we know that

$$x_1[n] \overset{FT}{\longleftrightarrow} X_1(e^{j\omega}) = \frac{\sin(3\omega/2)}{\sin(\omega/2)}$$

Using the accumulation property (Table 5.1, Property 5.3.5), we have

$$\sum_{k=-\infty}^{n} x_1[k] \overset{FT}{\longleftrightarrow} \frac{1}{1-e^{-j\omega}} X_1(e^{j\omega}) + \pi X_1(e^{j0}) \sum_{k=-\infty}^{\infty} \delta(\omega - 2\pi k).$$

Therefore, in the range $-\pi < \omega \leq \pi$,

$$\sum_{k=-\infty}^{n} x_1[k] \overset{FT}{\longleftrightarrow} \frac{1}{1-e^{-j\omega}} X_1(e^{j\omega}) + 3\pi\delta(\omega).$$

Also, in the range $-\pi < \omega \leq \pi$,

$$1 \overset{FT}{\longleftrightarrow} 2\pi\delta(\omega)$$

Therefore, in the range $-\pi < \omega \leq \pi$,

$$x[n] = 1 + \sum_{k=-\infty}^{n} x_1[k] \overset{FT}{\longleftrightarrow} \frac{1}{1-e^{-j\omega}} X_1(e^{j\omega}) + 5\pi\delta(\omega).$$

The signal $x[n]$ has the desired Fourier transform. We may express $x[n]$ mathematically as

$$x[n] = 1 + \sum_{k=-\infty}^{n} x_1[k] = \begin{cases} 1, & n \leq -2 \\ n+3, & -1 \leq n \leq 1 \\ 4, & n \geq 2 \end{cases}$$

5.9. From Property 5.3.4 in Table 5.1, we know that for a real signal $x[n]$,

$$\mathcal{O}d\{x[n]\} \stackrel{FT}{\longleftrightarrow} j\mathcal{I}m\{X(e^{j\omega})\}$$

From the given information,

$$\begin{aligned} j\mathcal{I}m\{X(e^{j\omega})\} &= j\sin\omega - j\sin 2\omega \\ &= (1/2)(e^{j\omega} - e^{-j\omega} - e^{2j\omega} + e^{-2j\omega}) \end{aligned}$$

Therefore,

$$\mathcal{O}d\{x[n]\} = \mathcal{IFT}\{j\mathcal{I}m\{X(e^{j\omega})\}\} = (1/2)(\delta[n+1] - \delta[n-1] - \delta[n+2] + \delta[n-2])$$

We also know that

$$\mathcal{O}d\{x[n]\} = \frac{x[n] - x[-n]}{2}$$

and that $x[n] = 0$ for $n > 0$. Therefore,

$$x[n] = 2\mathcal{O}d\{x[n]\} = \delta[n+1] - \delta[n+2], \qquad \text{for } n < 0.$$

Now we only have to find $x[0]$. Using Parseval's relation, we have

$$\frac{1}{2\pi}\int_{-\infty}^{\infty} |X(e^{j\omega})|^2 d\omega = \sum_{n=-\infty}^{\infty} |x[n]|^2.$$

From the given information, we can write

$$3 = (x[0])^2 + \sum_{n=-\infty}^{-1} |x[n]|^2 = (x[0])^2 + 2$$

This gives $x[0] = \pm 1$. But since we are given that $x[0] > 0$, we conclude that $x[0] = 1$. Therefore,

$$x[n] = \delta[n] + \delta[n+1] - \delta[n+2].$$

5.10. From Table 5.2, we know that

$$\left(\frac{1}{2}\right)^n u[n] \stackrel{FT}{\longleftrightarrow} \frac{1}{1 - \frac{1}{2}e^{-j\omega}}$$

Using Property 5.3.8 in Table 5.1,

$$x[n] = n\left(\frac{1}{2}\right)^n u[n] \stackrel{FT}{\longleftrightarrow} X(e^{j\omega}) = j\frac{d}{d\omega}\left\{\frac{1}{1 - \frac{1}{2}e^{-j\omega}}\right\} = \frac{\frac{1}{2}e^{-j\omega}}{(1 - \frac{1}{2}e^{-j\omega})^2}$$

Therefore,

$$\sum_{n=0}^{\infty} n\left(\frac{1}{2}\right)^n = \sum_{n=-\infty}^{\infty} x[n] = X(e^{j0}) = 2$$

5.11. We know from the time expansion property (Table 5.1, Property 5.3.7) that

$$g[n] = x_{(2)}[n] \overset{FT}{\longleftrightarrow} G(e^{j\omega}) = X(e^{j2\omega}).$$

Therefore, $G(e^{j\omega})$ is obtained by compressing $X(e^{j\omega})$ by a factor of 2. Since we know that $X(e^{j\omega})$ is periodic with a period of 2π, we may conclude that $G(e^{j\omega})$ has a period which is $(1/2)2\pi = \pi$. Therefore,

$$G(e^{j\omega}) = G(e^{j(\omega-\pi)}) \quad \text{and } \alpha = \pi.$$

5.12. Consider the signal

$$x_1[n] = \left(\frac{\sin\frac{\pi}{4}n}{\pi n}\right).$$

From Table 5.2, we obtain the Fourier transform of $x_1[n]$ to be

$$X_1(e^{j\omega}) = \begin{cases} 1, & 0 \le |\omega| \le \frac{\pi}{4} \\ 0, & \frac{\pi}{4} < |\omega| < \pi \end{cases}.$$

The plot of $X_1(e^{j\omega})$ is as shown in the Figure S5.12. Now consider the signal $x_2[n] = (x_1[n])^2$. Using the multiplication property (Table 5.1, Property 5.5), we obtain the Fourier tranform of $x_2[n]$ to be

$$X_2(e^{j\omega}) = (1/2\pi)[X_1(e^{j\omega}) * X_1(e^{j\omega})].$$

This is plotted in the Figure S5.12.

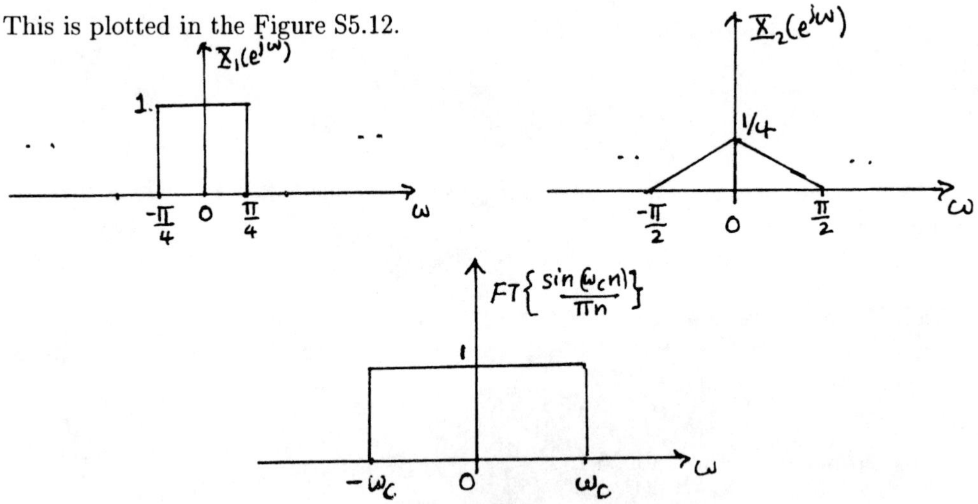

Figure S5.12

From Figure S5.12 it is clear that $X_2(e^{j\omega})$ is zero for $|\omega| > \pi/2$. By using the convolution property (Table 5.1, Property 5.4), we note that

$$Y(e^{j\omega}) = X_2(e^{j\omega})\mathcal{FT}\left\{\frac{\sin(\omega_c n)}{\pi n}\right\}.$$

The plot of $\mathcal{FT}\left\{\frac{\sin(\omega_c n)}{\pi n}\right\}$ is shown in Figure S5.12. It is clear that if $Y(e^{j\omega}) = X_2(e^{j\omega})$, then $(\pi/2) \leq \omega_c \leq \pi$.

5.13. When two LTI systems are connected in parallel, the impulse response of the overall system is the sum of the impulse responses of the individual systems. Therefore,

$$h[n] = h_1[n] + h_2[n].$$

Using the linearity property (Table 5.1, Property 5.3.2),

$$H(e^{j\omega}) = H_1(e^{j\omega}) + H_2(e^{j\omega})$$

Given that $h_1[n] = (1/2)^n u[n]$, we obtain

$$H_1(e^{j\omega}) = \frac{1}{1 - \frac{1}{2}e^{-j\omega}}$$

Therefore,

$$H_2(e^{j\omega}) = \frac{-12 + 5^{-j\omega}}{12 - 7e^{-j\omega} + e^{-2j\omega}} - \frac{1}{1 - \frac{1}{2}e^{-j\omega}} = \frac{-2}{1 - \frac{1}{4}e^{-j\omega}}.$$

Taking the inverse Fourier transform,

$$h_2[n] = -2\left(\frac{1}{4}\right)^n u[n].$$

5.14. From the given information, we have the Fourier transform $G(e^{j\omega})$ of $g[n]$ to be

$$G(e^{j\omega}) = g[0] + g[1]e^{-j\omega}.$$

Also, when the input to the system is $x[n] = (1/4)^n u[n]$, the output is $g[n]$. Therefore,

$$H(e^{j\omega}) = \frac{G(e^{j\omega})}{X(e^{j\omega})}.$$

From Table 5.2, we obtain

$$X(e^{j\omega}) = \frac{1}{1 - \frac{1}{4}e^{-j\omega}}.$$

Therefore,

$$H(e^{j\omega}) = \{g[0] + g[1]e^{-j\omega}\}\{1 - \frac{1}{4}e^{-j\omega}\} = g[0] + \{g[1] - \frac{1}{4}g[0]\}e^{-j\omega} - g[1]e^{-2j\omega}.$$

Clearly, $h[n]$ is a three point sequence.

We have

$$H(e^{j\omega}) = h[0] + h[1]e^{-j\omega} + h[2]e^{-2j\omega}$$

and

$$\begin{aligned} H(e^{j(\omega-\pi)}) &= h[0] + h[1]e^{-j(\omega-\pi)} + h[2]e^{-2j(\omega-\pi)} \\ &= h[0] - h[1]e^{-j\omega} + h[2]e^{-2j\omega} \end{aligned}$$

We see that $H(e^{j\omega}) = H(e^{j(\omega-\pi)})$ only if $h[1] = 0$.

We also have

$$\begin{aligned} H(e^{j\pi/2}) &= h[0] + h[1]e^{-j\pi/2} + h[2]e^{-2j\pi/2} \\ &= h[0] - h[2] \end{aligned}$$

Since we are also given that $H(e^{j\pi/2}) = 1$, we have

$$h[0] - h[2] = 1. \qquad \text{(S5.14–1)}$$

Now note that

$$\begin{aligned} g[n] &= h[n] * \{(1/4)^n u[n]\} \\ &= \sum_{k=0}^{2} h[k](1/4)^{n-k} u[n-k] \end{aligned}$$

Evaluating this equation at $n = 2$, we have

$$g[2] = 0 = \frac{1}{16}h[0] + \frac{1}{4}h[1] + h[2]$$

Since $h[1] = 0$,

$$\frac{1}{16}h[0] + h[2] = 0. \qquad \text{(S5.14–2)}$$

Solving equations (S5.14-1) and (S5.14-2), we obtain

$$h[0] = \frac{16}{17}, \quad \text{and} \quad h[2] = -\frac{1}{17}.$$

Therefore,

$$h[n] = \frac{16}{17}\delta[n] - \frac{1}{17}\delta[n-2].$$

5.15. Consider $x[n] = \sin(\omega_c n)/(\pi n)$. The Fourier transform $X(e^{j\omega})$ of $x[n]$ is as shown in Figure S5.15. We note that the given signal $y[n] = x[n]x[n]$. Therefore, the Fourier transform $Y(e^{j\omega})$ of $y[n]$ is

$$Y(e^{j\omega}) = \frac{1}{2\pi}\int_{<2\pi>} X(e^{j\theta})X(e^{j(\omega-\theta)})d\theta.$$

Employing the approach used in Example 5.15, we can convert the above periodic convolution into an aperiodic signal by defining

$$\hat{X}(e^{j\omega}) = \begin{cases} X(e^{j\omega}), & -\pi < \omega \leq \pi \\ 0, & \text{otherwise} \end{cases}$$

Then we may write

$$Y(e^{j\omega}) = \frac{1}{2\pi}\int_{-\infty}^{\infty} \hat{X}(e^{j\theta})X(e^{j(\omega-\theta)})d\theta.$$

This is the aperiodic convolution of the rectangular pulse $\hat{X}(e^{j\omega})$ shown in Figure S5.15 with the periodic square wave $X(e^{j\omega})$. The result of this convolution is as shown in the Figure S5.15.

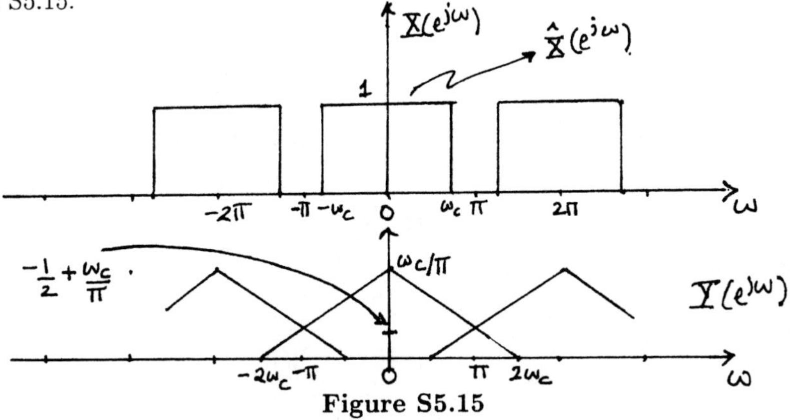

Figure S5.15

From the figure, it is clear that we require $-1+(2\omega_c/\pi)$ to be $1/2$. Therefore, $\omega_c = 3\pi/4$.

5.16. We may write

$$X(e^{j\omega}) = \frac{1}{2\pi}\left\{\frac{1}{1-\frac{1}{4}e^{-j\omega}} * \left[2\pi\sum_{k=0}^{3}\delta(\omega - \frac{\pi k}{2})\right]\right\}$$

where $*$ denotes aperiodic convolution. We may also rewrite this as a periodic convolution

$$X(e^{j\omega}) = \frac{1}{2\pi}\int_0^{2\pi} G(e^{j\theta})Q(e^{j(\omega-\theta)})d\theta$$

where

$$G(e^{j\omega}) = \frac{1}{1-\frac{1}{4}e^{-j\omega}}$$

and

$$Q(e^{j\omega}) = 2\pi\sum_{k=0}^{3}\delta(\omega - \frac{\pi k}{2}) \qquad \text{for } 0 \le \omega < 2\pi.$$

(a) Taking the inverse Fourier transform of $G(e^{j\omega})$ (see Table 5.2), we get $g[n] = (1/4)^n u[n]$. Therefore, $\alpha = \frac{1}{4}$.

(b) Taking the inverse Fourier transform of $Q(e^{j\omega})$ (see Table 5.2), we get

$$q[n] = 1 + \frac{1}{2}e^{j(\pi/2)n} + \frac{1}{4}e^{j\pi n} + \frac{1}{8}e^{j(3\pi/2)n}.$$

This signal is periodic with a fundamental period of $N = 4$.

(c) We can easily show that $X(e^{j\omega})$ is not conjugate symmetric. Therefore, $x[n]$ is not real.

5.17. Using the duality property, we have

$$(-1)^n \stackrel{FS}{\longleftrightarrow} a_k \Rightarrow a_n \stackrel{FS}{\longleftrightarrow} \frac{1}{N}(-1)^{-k} = \frac{1}{2}(-1)^k.$$

5.18. Knowing that

$$\left(\frac{1}{2}\right)^{|n|} \stackrel{FT}{\longleftrightarrow} \frac{1-\frac{1}{4}}{1-\cos\omega+\frac{1}{4}} = \frac{3}{5-4\cos\omega},$$

we may use the Fourier transform analysis equation to write

$$\frac{3}{5-4\cos\omega} = \sum_{n=-\infty}^{\infty}\left(\frac{1}{2}\right)^{|n|} e^{-j\omega n}$$

Putting $\omega = -2\pi t$ in this equation, and replacing the variable n by the variable k

$$\frac{1}{5-4\cos(2\pi t)} = \sum_{k=-\infty}^{\infty}\frac{1}{3}\left(\frac{1}{2}\right)^{|k|} e^{j2\pi kt}.$$

By comparing this with the continuous-time Fourier series synthesis equation, it is immediately apparent that $a_k = \frac{1}{3}\left(\frac{1}{2}\right)^{|k|}$ are the Fourier series coefficients of the signal $1/(5-4\cos(2\pi t))$.

5.19. (a) Taking the Fourier transform of both sides of the difference equation, we have

$$Y(e^{j\omega})\left[1-\frac{1}{6}e^{-j\omega}-\frac{1}{6}e^{-2j\omega}\right] = X(e^{j\omega}).$$

Therefore,

$$H(ej\omega) = \frac{Y(e^{j\omega})}{X(e^{j\omega})} = \frac{1}{1-\frac{1}{6}e^{-j\omega}-\frac{1}{6}e^{-2j\omega}} = \frac{1}{(1-\frac{1}{2}e^{-j\omega})(1+\frac{1}{3}e^{-j\omega})}.$$

(b) Using Partial fraction expansion,

$$H(ej\omega) = \frac{3/5}{1-\frac{1}{2}e^{-j\omega}} + \frac{2/5}{1+\frac{1}{3}e^{-j\omega}}.$$

Using Table 5.2, and taking the inverse Fourier trasform, we obtain

$$h[n] = \frac{3}{5}\left(\frac{1}{2}\right)^n u[n] + \frac{2}{5}\left(-\frac{1}{3}\right)^n u[n].$$

5.20. (a) Since the LTI system is causal and stable, a single input-output pair is sufficient to determine the frequency response of the system. In this case, the input is $x[n] = (4/5)^n u[n]$ and the output is $y[n] = n(4/5)^n u[n]$. The frequency response is given by

$$H(e^{j\omega}) = \frac{Y(e^{j\omega})}{X(e^{j\omega})}$$

where $X(e^{j\omega})$ and $Y(e^{j\omega})$ are the Fourier transforms of $x[n]$ and $y[n]$ respectively. Using Table 5.2, we have

$$x[n] = \left(\frac{4}{5}\right)^n u[n] \stackrel{FT}{\longleftrightarrow} X(e^{j\omega}) = \frac{1}{1 - \frac{4}{5}e^{-j\omega}}.$$

Using the differentiation in frequency property (Table 5.1, Property 5.3.8), we have

$$y[n] = n\left(\frac{4}{5}\right)^n u[n] \stackrel{FT}{\longleftrightarrow} Y(e^{j\omega}) = j\frac{dX(e^{j\omega})}{d\omega} = \frac{(4/5)e^{-j\omega}}{(1 - \frac{4}{5}e^{-j\omega})^2}.$$

Therefore,

$$H(e^{j\omega}) = \frac{(4/5)e^{-j\omega}}{1 - \frac{4}{5}e^{-j\omega}}.$$

(b) Since $H(ej\omega) = Y(e^{j\omega})/X(e^{j\omega})$, we may write

$$Y(e^{j\omega})\left[1 - \frac{4}{5}e^{-j\omega}\right] = X(e^{j\omega})\left[(4/5)e^{-j\omega}\right].$$

Taking the inverse Fourier tranform of both sides

$$y[n] - \frac{4}{5}y[n-1] = \frac{4}{5}x[n].$$

5.21. (a) The given signal is

$$x[n] = u[n-2] - u[n-6] = \delta[n-2] + \delta[n-3] + \delta[n-4] + \delta[n-5].$$

Using the Fourier transform analysis eq. (5.9), we obtain

$$X(e^{j\omega}) = e^{-2j\omega} + e^{-3j\omega} + e^{-4j\omega} + e^{-5j\omega}.$$

(b) Using the Fourier transform analysis eq. (5.9), we obtain

$$\begin{aligned} X(e^{j\omega}) &= \sum_{n=-\infty}^{-1} (\frac{1}{2})^{-n} e^{-j\omega n} \\ &= \sum_{n=1}^{\infty} (\frac{1}{2}e^{j\omega})^n \\ &= \frac{e^{j\omega}}{2} \frac{1}{(1 - \frac{1}{2}e^{j\omega})} \end{aligned}$$

(c) Using the Fourier transform analysis eq. (5.9), we obtain

$$\begin{aligned} X(e^{j\omega}) &= \sum_{n=-\infty}^{-2} (\frac{1}{3})^{-n} e^{-j\omega n} \\ &= \sum_{n=2}^{\infty} (\frac{1}{3} e^{j\omega})^{n} \\ &= \frac{e^{2j\omega}}{9} \frac{1}{(1-\frac{1}{3}e^{j\omega})} \end{aligned}$$

(d) Using the Fourier transform analysis eq. (5.9), we obtain

$$\begin{aligned} X(e^{j\omega}) &= \sum_{n=-\infty}^{0} 2^{n} \sin(\pi n/4) e^{-j\omega n} \\ &= -\sum_{n=0}^{\infty} 2^{-n} \sin(\pi n/4) e^{j\omega n} \\ &= -\frac{1}{2j} \sum_{n=0}^{\infty} [(1/2)^{n} e^{j\pi n/4} e^{j\omega n} - (1/2)^{n} e^{-j\pi n/4} e^{j\omega n}] \\ &= -\frac{1}{2j} \left[\frac{1}{1-(1/2)e^{j\pi/4}e^{j\omega}} - \frac{1}{1-(1/2)e^{-j\pi/4}e^{j\omega}} \right] \end{aligned}$$

(e) Using the Fourier transform analysis eq. (5.9), we obtain

$$\begin{aligned} X(e^{j\omega}) &= \sum_{n=-\infty}^{\infty} (1/2)^{|n|} \cos[\pi(n-1)/8] e^{-j\omega n} \\ &= \frac{1}{2} \left[\frac{e^{-j\pi/8}}{1-(1/2)e^{j\pi/8}e^{-j\omega}} + \frac{e^{j\pi/8}}{1-(1/2)e^{-j\pi/8}e^{-j\omega}} \right] \\ &+ \frac{1}{4} \left[\frac{e^{j\pi/4}e^{j\omega}}{1-(1/2)e^{j\pi/8}e^{j\omega}} + \frac{e^{-j\pi/4}e^{j\omega}}{1-(1/2)e^{-j\pi/8}e^{j\omega}} \right] \end{aligned}$$

(f) The given signal is

$$x[n] = -3\delta[n+3] - 2\delta[n+2] - \delta[n+1] + \delta[n-1] + 2\delta[n-2] + 3\delta[n-3].$$

Using the Fourier transform analysis eq. (5.9), we obtain

$$X(e^{j\omega}) = -3e^{3j\omega} - 2e^{2j\omega} - e^{j\omega} + e^{-j\omega} + 2e^{-2j\omega} + 3e^{-3j\omega}.$$

(g) The given signal is

$$x[n] = \sin(\pi n/2) + \cos(n) = \frac{1}{2j}[e^{j\pi n/2} - e^{-j\pi n/2}] + \frac{1}{2}[e^{jn} + e^{-jn}].$$

Therefore,

$$X(e^{j\omega}) = \frac{\pi}{j}[\delta(\omega - \pi/2) - \delta(\omega + \pi/2)] + \pi[\delta(\omega-1) + \delta(\omega+1)], \qquad \text{in } 0 \le |\omega| < \pi.$$

(h) The given signal is

$$\begin{aligned} x[n] &= \sin(5\pi n/3) + \cos(7\pi n/3) \\ &= -\sin(\pi n/3) + \cos(\pi n/3) \\ &= -\frac{1}{2j}[e^{j\pi n/3} - e^{-j\pi n/3}] + \frac{1}{2}[e^{j\pi n/3} + e^{-j\pi n/3}]. \end{aligned}$$

Therefore,

$$X(e^{j\omega}) = -\frac{\pi}{j}[\delta(\omega - \pi/3) - \delta(\omega + \pi/3)] + \pi[\delta(\omega - \pi/3) + \delta(\omega + \pi/3)], \qquad \text{in } 0 \le |\omega| < \pi.$$

(i) $x[n]$ is periodic with period 6. The Fourier series coefficients of $x[n]$ are given by

$$\begin{aligned} a_k &= \frac{1}{6}\sum_{n=0}^{5} x[n]e^{-j(2\pi/6)kn} \\ &= \frac{1}{6}\sum_{n=0}^{4} e^{-j(2\pi/6)kn} \\ &= \frac{1}{6}\left[\frac{1 - e^{-j5\pi k/3}}{1 - e^{-j(2\pi/6)k}}\right] \end{aligned}$$

Therefore, from the results of Section 5.2

$$X(e^{j\omega}) = \sum_{l=-\infty}^{\infty} 2\pi\left(\frac{1}{6}\right)\left[\frac{1 - e^{-j5\pi k/3}}{1 - e^{-j(2\pi/6)k}}\right]\delta(\omega - \frac{2\pi}{6} - 2\pi l).$$

(j) Using the Fourier transform analysis eq. (5.9) we obtain

$$\left(\frac{1}{3}\right)^{|n|} \stackrel{FT}{\longleftrightarrow} \frac{4}{5 - 3\cos\omega}.$$

Using the differentiation in frequency property of the Fourier transform,

$$n\left(\frac{1}{3}\right)^{|n|} \stackrel{FT}{\longleftrightarrow} -j\frac{12\sin\omega}{(5 - 3\cos\omega)^2}.$$

Therefore,

$$x[n] = n\left(\frac{1}{3}\right)^{|n|} - \left(\frac{1}{3}\right)^{|n|} \stackrel{FT}{\longleftrightarrow} \frac{4}{5 - 3\cos\omega} - j\frac{12\sin\omega}{(5 - 3\cos\omega)^2}.$$

(k) We have

$$x_1[n] = \frac{\sin(\pi n/5)}{\pi n} \stackrel{FT}{\longleftrightarrow} X_1(e^{j\omega}) = \begin{cases} 1, & |\omega| < \frac{\pi}{5} \\ 0, & \frac{\pi}{5} \le |\omega| < \pi \end{cases}.$$

Also,

$$x_2[n] = \cos(7\pi n/2) = \cos(\pi n/2) \overset{FT}{\longleftrightarrow} X_2(e^{j\omega}) = \pi\{\delta(\omega - \pi/2) + \delta(\omega + \pi/2)\},$$

in the range $0 \leq |\omega| < \pi$. Therefore, if $x[n] = x_1[n]x_2[n]$, then

$$X(e^{j\omega}) = \text{Periodic convolution of } X_1(e^{j\omega}) \text{ and } X_2(e^{j\omega}).$$

Using the mechanics of periodic convolution demosntrated in Example 5.15, we obtain in the range $0 \leq |\omega| < \pi$,

$$X(e^{j\omega}) = \begin{cases} 1, & \frac{3\pi}{10} < |\omega| < \frac{7\pi}{10} \\ 0, & \text{otherwise} \end{cases}.$$

5.22. **(a)** Using the Fourier transform synthesis eq. (5.8), we obtain

$$\begin{aligned} x[n] &= \frac{1}{2\pi}\int_{-3\pi/4}^{-\pi/4} e^{j\omega n} d\omega + \frac{1}{2\pi}\int_{\pi/4}^{3\pi/4} e^{j\omega n} d\omega \\ &= \frac{1}{\pi n}[\sin(3\pi n/4) - \sin(\pi n/4)] \end{aligned}$$

(b) Comparing the given Fourier transform with the analysis eq. (5.8), we obtain

$$x[n] = \delta[n] + 3\delta[n-1] + 2\delta[n-2] - 4\delta[n-3] + \delta[n-10].$$

(c) Using the Fourier transform synthesis eq. (5.8), we obtain

$$\begin{aligned} x[n] &= \frac{1}{2\pi}\int_{-\pi}^{\pi} e^{-j\omega/2} e^{j\omega n} d\omega \\ &= \frac{(-1)^{n+1}}{\pi(n - \frac{1}{2})} \end{aligned}$$

(d) The given Fourier transform is

$$\begin{aligned} X(e^{j\omega}) &= \cos^2\omega + \sin^2(3\omega) \\ &= \frac{1+\cos(2\omega)}{2} + \frac{1-\cos(3\omega)}{2} \\ &= 1 + \frac{1}{4}e^{2j\omega} + \frac{1}{4}e^{-2j\omega} + -\frac{1}{4}e^{3j\omega} - \frac{1}{4}e^{-3j\omega} \end{aligned}$$

Comparing the given Fourier transform with the analysis eq. (5.8), we obtain

$$x[n] = \delta[n] + \frac{1}{4}\delta[n-2] + \frac{1}{4}\delta[n+2] - \frac{1}{4}\delta[n-3] - \frac{1}{4}\delta[n+3].$$

(e) This is the Fourier transform of a periodic signal with fundamental frequency $\pi/2$. Therefore, its fundamental period is 4. Also, the Fourier series coefficients of this signal are $a_k = (-1)^k$. Therefore, the signal is given by

$$x[n] = \sum_{k=0}^{3}(-1)^k e^{jk(\pi/2)n} = 1 - e^{j\pi n/2} + e^{j\pi n} - e^{j3\pi n/2}.$$

(f) The given Fourier transform may be written as

$$\begin{aligned} X(e^{j\omega}) &= e^{-j\omega}\sum_{n=0}^{\infty}(1/5)^n e^{-j\omega n} - (1/5)\sum_{n=0}^{\infty}(1/5)^n e^{-j\omega n} \\ &= 5\sum_{n=1}^{\infty}(1/5)^n e^{-j\omega n} - (1/5)\sum_{n=0}^{\infty}(1/5)^n e^{-j\omega n} \end{aligned}$$

Comparing each of the two terms in the right-hand side of the above equation with the Fourier transform analysis eq. (5.9) we obtain

$$x[n] = \left(\frac{1}{5}\right)^{n-1} u[n-1] - \left(\frac{1}{5}\right)^{n+1} u[n].$$

(g) The given Fourier transform may be written as

$$X(e^{j\omega}) = \frac{2/9}{1 - \frac{1}{2}e^{-j\omega}} + \frac{7/9}{1 + \frac{1}{4}e^{-j\omega}}.$$

Therefore,

$$x[n] = \frac{2}{9}\left(\frac{1}{2}\right)^n u[n] + \frac{7}{9}\left(-\frac{1}{4}\right)^n u[n].$$

(h) The given Fourier transform may be written as

$$X(e^{j\omega}) = 1 + \frac{1}{3}e^{-j\omega} + \frac{1}{3^2}e^{-j2\omega} + \frac{1}{3^3}e^{-j3\omega} + \frac{1}{3^4}e^{-j4\omega} + \frac{1}{3^5}e^{-j5\omega}.$$

Comparing the given Fourier transform with the analysis eq. (5.8), we obtain

$$x[n] = \delta[n] + \frac{1}{3}\delta[n-1] + \frac{1}{9}\delta[n-2] + \frac{1}{27}\delta[n-3] + \frac{1}{81}\delta[n-4] + \frac{1}{243}\delta[n-5].$$

5.23. (a) We have from eq. (5.9)

$$X(e^{j0}) = \sum_{n=-\infty}^{\infty} x[n] = 6.$$

(b) Note that $y[n] = x[n+2]$ is an even signal. Therefore, $Y(e^{j\omega})$ is real and even. This implies that $\sphericalangle Y(e^{j\omega}) = 0$. Furthermore, from the time shifting property of the Fourier transform we have $Y(e^{j\omega}) = e^{j2\omega}X(e^{j\omega})$. Therefore, $\sphericalangle X(e^{j\omega}) = e^{-j2\omega}$.

(c) We have from eq. (5.8)

$$2\pi x[0] = \int_{-\pi}^{\pi} X(e^{j\omega})d\omega.$$

Therefore,

$$\int_{-\pi}^{\pi} X(e^{j\omega})d\omega = 4\pi.$$

(d) We have from eq. (5.9)

$$X(e^{j\pi}) = \sum_{n=-\infty}^{\infty} x[n](-1)^n = 2.$$

(e) From Table 5.1, we have

$$\mathcal{E}v\{x[n]\} \stackrel{FT}{\longleftrightarrow} \mathcal{R}e\{X(e^{j\omega})\}.$$

Therefore, the desired signal is $\mathcal{E}v\{x[n]\} = (x[n]+x[-n])/2$. This is as shown in Figure S5.23.

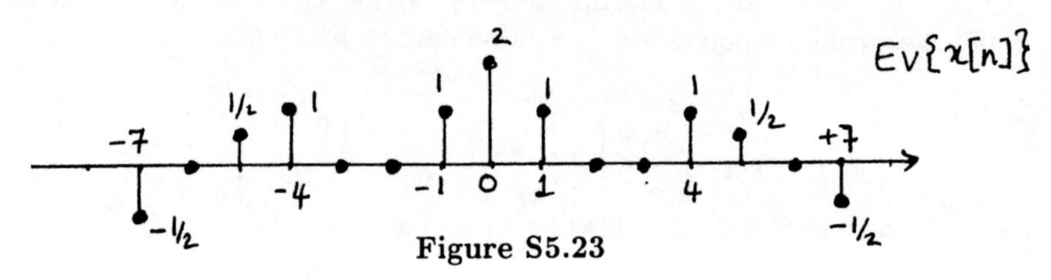

Figure S5.23

(f) (i) From Parseval's theorem we have

$$\int_{-\infty}^{\infty} |X(e^{j\omega})|^2 = 2\pi \sum_{n=-\infty}^{\infty} |x[n]|^2 = 28\pi.$$

(ii) Using the differentiation in frequency property of the Fourier transform we obtain

$$nx[n] \stackrel{FT}{\longleftrightarrow} j\frac{dX(e^{j\omega})}{d\omega}.$$

Again using Parseval's theorem, we obtain

$$\int_{-\infty}^{\infty} |\frac{dX(e^{j\omega})}{d\omega}|^2 = 2\pi \sum_{n=-\infty}^{\infty} |n|^2|x[n]|^2 = 316\pi.$$

5.24. (1) For $\mathcal{R}e\{X(e^{j\omega})\}$ to be zero, the signal must be real and odd. Only signals (b) and (i) are real and odd.

(2) For $\mathcal{I}m\{X(e^{j\omega})\}$ to be zero, the signal must be real and even. Only signals (d) and (h) are real and even.

(3) Assume $Y(e^{j\omega}) = e^{j\alpha\omega}X(e^{j\omega})$. Using the time shifting property of the Fourier transform we have $y[n] = x[n+\alpha]$. If $Y(e^{j\omega})$ is real, then $y[n]$ is real and even (assuming that $x[n]$ is real). Therefore, $x[n]$ has to be symmetric about α. This is true only for signals (a), (b), (d), (e), (f), and (h).

(4) Since $\int_{-\pi}^{\pi} X(e^{j\omega})d\omega = 2\pi x[0]$, the given condition is satisfied only if $x[0] = 0$. This is true for signals (b), (e), (f), (h), and (i).

(5) $X(e^{j\omega})$ is always periodic with period 2π. Therefore, all signals satisfy this condition.

(6) Since $X(e^{j0}) = \sum_{n=-\infty}^{\infty} x[n]$, the given condition is satisfied only if the samples of the signal add up to zero. This is true for signals (b), (g), and (i).

5.25. If the inverse Fourier transform of $X(e^{j\omega})$ is $x[n]$, then

$$x_e[n] = \mathcal{E}v\{x[n]\} = \frac{x[n] + x[-n]}{2} \stackrel{FT}{\longleftrightarrow} A(\omega)$$

and

$$x_o[n] = \mathcal{O}d\{x[n]\} = \frac{x[n] - x[-n]}{2} \stackrel{FT}{\longleftrightarrow} jB(\omega)$$

Therefore, the inverse Fourier transform of $B(\omega)$ is $-jx_o[n]$. Also, the inverse Fourier transform of $A(\omega)e^{j\omega}$ is $x_e[n+1]$. Therefore, the time function corresponding to the inverse Fourier transform of $B(\omega) + A(\omega)e^{j\omega}$ will be $x_e[n+1] - jx_o[n]$. This is as shown in the Figure S5.25.

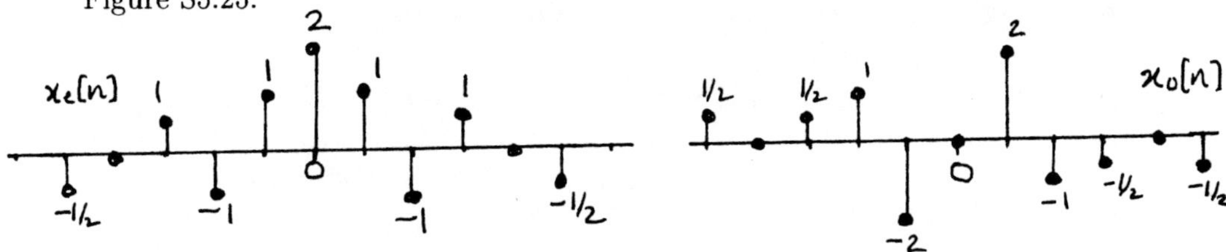

Figure S5.25

5.26. (a) We may express $X_2(e^{j\omega})$ as

$$X_2(e^{j\omega}) = \mathcal{R}e\{X_1(e^{j\omega})\} + \mathcal{R}e\{X_1(e^{j(\omega-2\pi/3)})\} + \mathcal{R}e\{X_1(e^{j(\omega+2\pi/3)})\}.$$

Therefore,

$$x_2[n] = \mathcal{E}v\{x_1[n]\}\left[1 + e^{j2\pi/3} + e^{-j2\pi/3}\right].$$

(b) We may express $X_3(e^{j\omega})$ as

$$X_3(e^{j\omega}) = \mathcal{I}m\{X_1(e^{j(\omega-\pi)})\} + \mathcal{I}m\{X_1(e^{j(\omega+\pi)})\}.$$

Therefore,

$$x_3[n] = \mathcal{O}d\{x_1[n]\}\left[e^{j\pi n} + e^{-j\pi n}\right] = 2(-1)^n \mathcal{O}d\{x_1[n]\}.$$

(c) We may express α as

$$\alpha = \frac{j\left.\frac{dX_1(e^{j\omega})}{d\omega}\right|_{\omega=0}}{\left.X_1(e^{j\omega})\right|_{\omega=0}} = \frac{j(-6j/\pi)}{1} = \frac{6}{\pi}.$$

(d) Using the fact that $H(e^{j\omega})$ is the frequency response of an ideal lowpass filter with cutoff frequency $\pi/6$, we may draw $X_4(e^{j\omega})$ as shown in Figure S5.26.

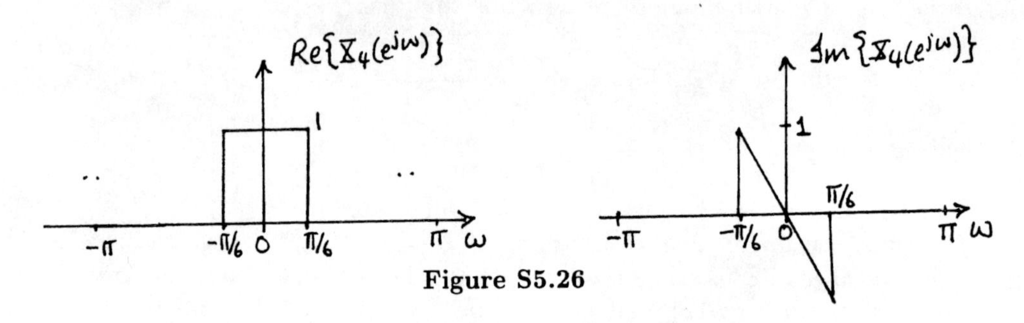

Figure S5.26

5.27. (a) $W(e^{j\omega})$ will be the periodic convolution of $X(e^{j\omega})$ with $P(e^{j\omega})$. The Fourier transforms are sketched in Figure S5.27.

(b) The Fourier transform of $Y(e^{j\omega})$ of $y[n]$ is $Y(e^{j\omega}) = P(e^{j\omega})H(e^{j\omega})$. The LTI system with unit sample response $h[n]$ is an ideal lowpass filter with cutoff frequency $\pi/2$. Therefore, $Y(e^{j\omega})$ for each choice of $p[n]$ are as shown in Figure S5.27. Therefore, $y[n]$ in each case is:

(i) $y[n] = 0$

(ii) $y[n] = \frac{\sin(\pi n/2)}{2\pi n} - \frac{1-\cos(\pi n/2)}{\pi^2 n^2}$

(iii) $y[n] = \frac{\sin(\pi n/2)}{\pi^2 n^2} - \frac{\cos(\pi n/2)}{2\pi n}$

(iv) $y[n] = 2\left[\frac{\sin(\pi n/4)}{\pi n}\right]^2$

(v) $y[n] = \frac{1}{4}\left[\frac{\sin(\pi n/2)}{\pi n}\right]$

5.28. Let

$$\frac{1}{2\pi}\int_{-\pi}^{\pi} X(e^{j\theta})G(e^{j(\omega-\theta)})d\theta = 1 + e^{-j\omega} = Y(e^{j\omega}).$$

Taking the inverse Fourier transform of the above equation, we obtain

$$g[n]x[n] = \delta[n] + \delta[n-1] = y[n].$$

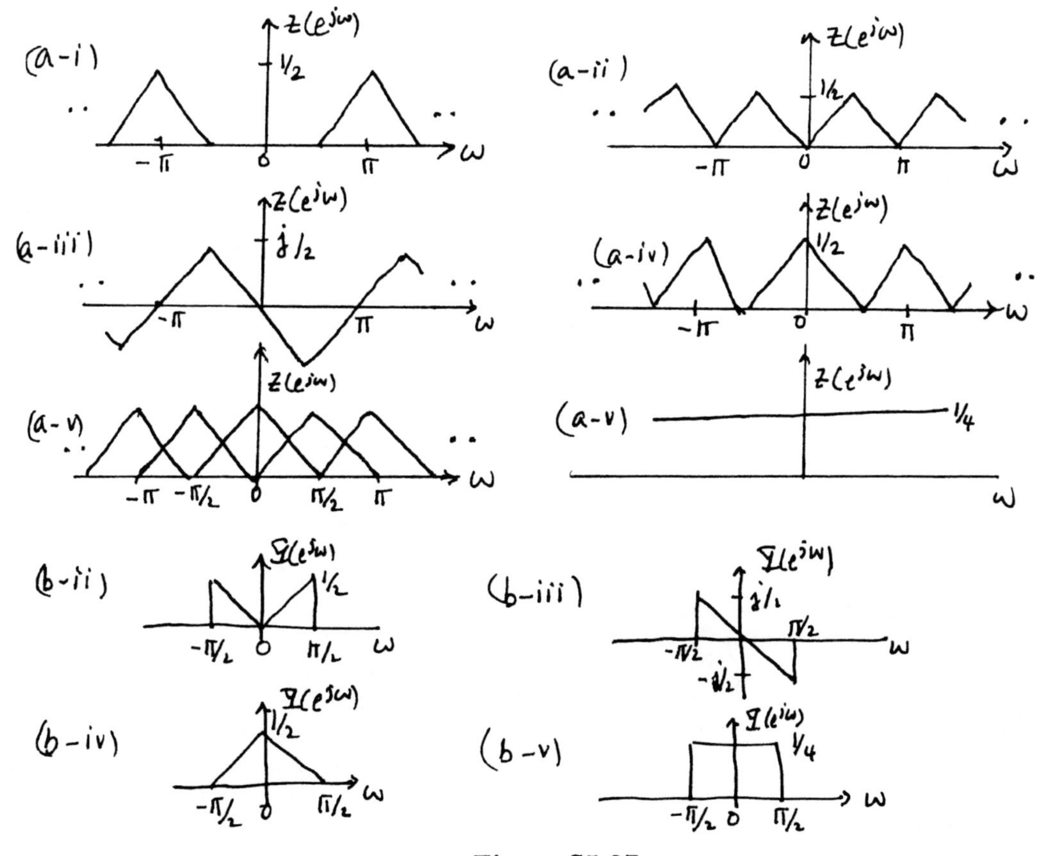

Figure S5.27

(a) If $x[n] = (-1)^n$,

$$g[n] = \delta[n] - \delta[n-1].$$

(b) If $x[n] = (1/2)^n u[n]$, $g[n]$ has to be chosen such that

$$g[n] = \begin{cases} 1, & n = 0 \\ 2, & n = 1 \\ 0, & n > 1 \\ \text{any value,} & \text{otherwise} \end{cases}$$

Therefore, there are many possible choices for $g[n]$.

5.29. **(a)** Let the output of the system be $y[n]$. We know that

$$Y(e^{j\omega}) = X(e^{j\omega})H(e^{j\omega}).$$

In this part of the problem

$$H(e^{j\omega}) = \frac{1}{1 - \frac{1}{2}e^{-j\omega}}.$$

(i) We have

$$X(e^{j\omega}) = \frac{1}{1 - \frac{3}{4}e^{-j\omega}}.$$

Therefore,

$$\begin{aligned} Y(e^{j\omega}) &= \left[\frac{1}{1 - \frac{3}{4}e^{-j\omega}}\right]\left[\frac{1}{1 - \frac{1}{2}e^{-j\omega}}\right] \\ &= \frac{-2}{1 - \frac{1}{2}e^{-j\omega}} + \frac{3}{1 - \frac{3}{4}e^{-j\omega}} \end{aligned}$$

Taking the inverse Fourier transform, we obtain

$$y[n] = 3\left(\frac{3}{4}\right)^n u[n] - 2\left(\frac{1}{2}\right)^n u[n].$$

(ii) We have

$$X(e^{j\omega}) = \frac{1}{(1 - \frac{1}{4}e^{-j\omega})^2}.$$

Therefore,

$$\begin{aligned} Y(e^{j\omega}) &= \left[\frac{1}{(1 - \frac{1}{4}e^{-j\omega})^2}\right]\left[\frac{1}{1 - \frac{1}{2}e^{-j\omega}}\right] \\ &= \frac{4}{1 - \frac{1}{2}e^{-j\omega}} - \frac{2}{1 - \frac{1}{4}e^{-j\omega}} - \frac{3}{(1 - \frac{1}{4}e^{-j\omega})^2} \end{aligned}$$

Taking the inverse Fourier transform, we obtain

$$y[n] = 4\left(\frac{1}{2}\right)^n u[n] - 2\left(\frac{1}{4}\right)^n u[n] - 3(n+1)\left(\frac{1}{4}\right)^n u[n].$$

(iii) We have

$$X(e^{j\omega}) = 2\pi \sum_{k=-\infty}^{\infty} \delta(\omega - (2k+1)\pi).$$

Therefore,

$$\begin{aligned} Y(e^{j\omega}) &= \left[2\pi \sum_{k=-\infty}^{\infty} \delta(\omega - (2k+1)\pi)\right]\left[\frac{1}{1 - \frac{1}{2}e^{-j\omega}}\right] \\ &= \frac{4\pi}{3} \sum_{k=-\infty}^{\infty} \delta(\omega - (2k+1)\pi) \end{aligned}$$

Taking the inverse Fourier transform, we obtain

$$x[n] = \frac{2}{3}(-1)^n.$$

(b) Given

$$h[n] = \frac{1}{2}\left(\frac{1}{2}e^{j\pi/2}\right)^n u[n] + \frac{1}{2}\left(\frac{1}{2}e^{-j\pi/2}\right)^n u[n],$$

we obtain

$$H(e^{j\omega}) = \frac{1/2}{1 - \frac{1}{2}e^{j\pi/2}e^{-j\omega}} + \frac{1/2}{1 - \frac{1}{2}e^{-j\pi/2}e^{-j\omega}}.$$

(i) We have

$$X(e^{j\omega}) = \frac{1}{1 - \frac{1}{2}e^{-j\omega}}.$$

Therefore,

$$\begin{aligned} Y(e^{j\omega}) &= \left[\frac{1/2}{1 - \frac{1}{2}e^{j\pi/2}e^{-j\omega}} + \frac{1/2}{1 - \frac{1}{2}e^{-j\pi/2}e^{-j\omega}}\right]\left[\frac{1}{1 - \frac{1}{2}e^{-j\omega}}\right] \\ &= \frac{A}{1 - (1/2)e^{j\pi/2}e^{-j\omega}} + \frac{B}{1 - (1/2)e^{-j\omega}} + \frac{C}{1 - (1/2)e^{-j\pi/2}e^{-j\omega}}, \end{aligned}$$

where $A = -j/[2(1-j)]$, $B = 1/2$, and $C = 1/[2(1+j)]$. Therefore,

$$y[n] = \frac{-j}{2(1-j)}\left(\frac{j}{2}\right)^n u[n] + \frac{1}{2(1+j)}\left(-\frac{j}{2}\right)^n u[n] + \frac{1}{2}\left(\frac{1}{2}\right)^n u[n].$$

(ii) In this case,

$$y[n] = \frac{\cos(\pi n/2)}{3}\left[4 - (\tfrac{1}{2})^n\right] u[n].$$

(c) Here,

$$\begin{aligned} Y(e^{j\omega}) &= X(e^{j\omega})H(e^{j\omega}) = -3e^{-2j\omega} - e^{j\omega} + 1 - 2e^{-j2\omega} \\ &\quad +6e^{-j\omega} + 2e^{-j2\omega} - 2e^{-j3\omega} + 4e^{-j5\omega} \\ &\quad +3e^{j5\omega} + e^{j4\omega} - e^{j3\omega} + 2e^{j\omega} \end{aligned}$$

Therefore,

$$\begin{aligned} y[n] &= 3\delta[n+5] + \delta[n+4] - \delta[n+3] - 3\delta[n+2] \\ &\quad +\delta[n+1] + \delta[n] + 6\delta[n-1] - 2\delta[n-3] + 4\delta[n-5]. \end{aligned}$$

5.30. **(a)** The frequency response of the system is as shown in Figure S5.30.

(b) The Fourier transform $X(e^{j\omega})$ of $x[n]$ is as shown in Figure S5.30.

(i) The frequency response $H(e^{j\omega})$ is as shown in Figure S5.30. Therefore, $y[n] = \sin(\pi n/8)$.

(ii) The frequency response $H(e^{j\omega})$ is as shown in Figure S5.30. Therefore, $y[n] = 2\sin(\pi n/8) - 2\cos(\pi n/4)$.

(iii) The frequency response $H(e^{j\omega})$ is as shown in Figure S5.30. Therefore, $y[n] = \frac{1}{6}\sin(\pi n/8) - \frac{1}{4}\cos(\pi n/4)$.

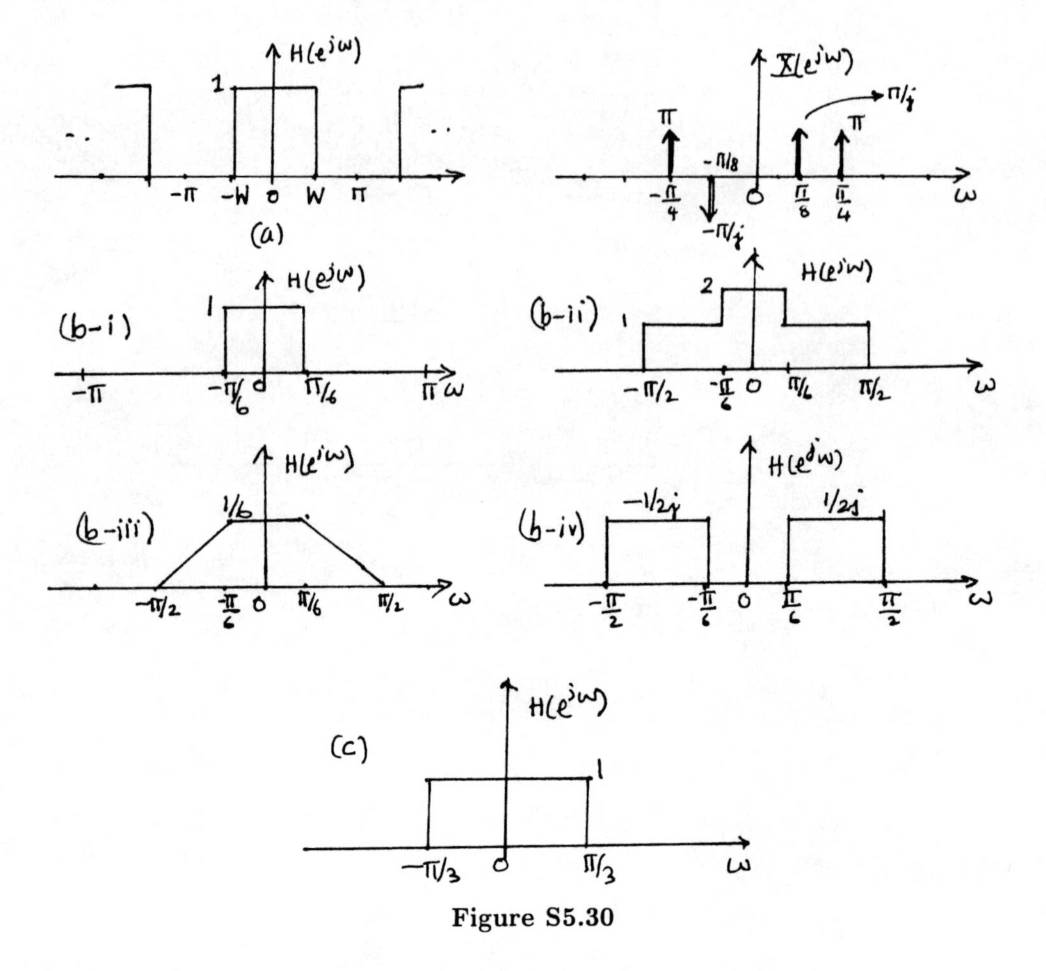

Figure S5.30

(iv) The frequency response $H(e^{j\omega})$ is as shown in Figure S5.30. Therefore, $y[n] = -\sin(\pi n/4)$.

(c) The frequency response $H(e^{j\omega})$ is as shown in Figure S5.30.

(i) The signal $x[n]$ is periodic with period 8. The Fourier series coefficients of the signal are

$$a_k = \frac{1}{8}\sum_{n=0}^{7} x[n]e^{-j(2\pi/8)kn}.$$

The Fourier transform of this signal is

$$X(e^{j\omega}) = \sum_{k=-\infty}^{\infty} 2\pi a_k \delta(\omega - 2\pi k/8).$$

The Fourier transform $Y(e^{j\omega})$ of the output is $Y(e^{j\omega}) = X(e^{j\omega})H(e^{j\omega})$. Therefore,

$$Y(e^{j\omega}) = 2\pi[a_0\delta(\omega) + a_1\delta(\omega - \pi/4) + a_{-1}\delta(\omega + \pi/4)]$$

in the range $0 \le |\omega| \le \pi$. Therefore,

$$y[n] = a_0 + a_1 e^{j\pi n/4} + a_{-1} e^{-j\pi/4} = \frac{5}{8} + [(1/4) + (1/2)(1/\sqrt{2})]\cos(\pi n/4).$$

(ii) The signal $x[n]$ is periodic with period 8. The Fourier series coefficients of the signal are

$$a_k = \frac{1}{8}\sum_{n=0}^{7} x[n] e^{-j(2\pi/8)kn}.$$

The Fourier transform of this signal is

$$X(e^{j\omega}) = \sum_{k=-\infty}^{\infty} 2\pi a_k \delta(\omega - 2\pi k/8).$$

The Fourier transform $Y(e^{j\omega})$ of the output is $Y(e^{j\omega}) = X(e^{j\omega})H(e^{j\omega})$. Therefore,

$$Y(e^{j\omega}) = 2\pi[a_0\delta(\omega) + a_1\delta(\omega - \pi/4) + a_{-1}\delta(\omega + \pi/4)]$$

in the range $0 \le |\omega| \le \pi$. Therefore,

$$y[n] = a_0 + a_1 e^{j\pi n/4} + a_{-1} e^{-j\pi/4} = \frac{1}{8} + \frac{1}{4}\cos(\pi n/4).$$

(iii) Again in this case, the Fourier transform $X(e^{j\omega})$ of the signal $x[n]$ is of the form shown in part (i). Therefore,

$$y[n] = a_0 + a_1 e^{j\pi n/4} + a_{-1} e^{-j\pi/4} = \frac{1}{8} + [(1/4) - (1/2)(1/\sqrt{2})]\cos(\pi n/4).$$

(iv) In this case, the output is

$$y[n] = h[n] * x[n] = \frac{\sin[\pi/3(n-1)]}{\pi(n-1)} + \frac{\sin[\pi/3(n+1)]}{\pi(n+1)}.$$

5.31. **(a)** From the given information, it is clear that when the input to the system is a complex exponential of frequency ω_0, the output is a complex exponential of the same frequency but scaled by the $|\omega_0|$. Therefore, the frequency response of the system is

$$H(e^{j\omega}) = |\omega|, \qquad \text{for } 0 \le |\omega| \le \pi.$$

(b) Taking the inverse Fourier transform of the frequency response, we obtain

$$\begin{aligned} h[n] &= \frac{1}{2\pi}\int_{-\pi}^{\pi} H(e^{j\omega})e^{j\omega n}d\omega \\ &= \frac{1}{2\pi}\int_{-\pi}^{0} -\omega e^{j\omega n}d\omega + \frac{1}{2\pi}\int_{0}^{\pi} \omega e^{j\omega n}d\omega \\ &= \frac{1}{\pi}\int_{0}^{\pi} \omega\cos(\omega n)d\omega \\ &= \frac{1}{\pi}\left[\frac{\cos(n\pi) - 1}{n^2}\right] \end{aligned}$$

5.32. From the synthesis equation (5.8) we have

$$\left[\frac{1}{2\pi}\int_{-\pi}^{\pi} H_1(e^{j\omega})d\omega\right]\left[\frac{1}{2\pi}\int_{-\pi}^{\pi} H_2(e^{j\omega})d\omega\right] = h_1[0]h_2[0].$$

Also, since

$$h_1[n] * h_2[n] \stackrel{FT}{\longleftrightarrow} H_1(e^{j\omega})H_2(e^{j\omega}),$$

we have

$$\frac{1}{2\pi}\int_{-\pi}^{\pi} H_1(e^{j\omega})H_2(e^{j\omega})d\omega = [h_1[n] * h_2[n]]_{n=0}.$$

Therefore, the question here amounts to asking whether it is true that

$$h_1[0]h_2[0] = [h_1[n] * h_2[n]]_{n=0}.$$

Since $h_1[n]$ and $h_2[n]$ are causal, this is indeed true.

5.33. **(a)** Taking the Fourier transform of the given difference equation we have

$$H(e^{j\omega}) = \frac{Y(e^{j\omega})}{X(e^{j\omega})} = \frac{1}{1+\frac{1}{2}e^{-j\omega}}.$$

(b) The Fourier transform of the output will be $Y(e^{j\omega}) = X(e^{j\omega})H(e^{j\omega})$.

(i) In this case

$$X(e^{j\omega}) = \frac{1}{1-\frac{1}{2}e^{-j\omega}}.$$

Therefore,

$$\begin{aligned} Y(e^{j\omega}) &= \left[\frac{1}{1-\frac{1}{2}e^{-j\omega}}\right]\left[\frac{1}{1+\frac{1}{2}e^{-j\omega}}\right] \\ &= \frac{1/2}{1-\frac{1}{2}e^{-j\omega}} + \frac{1/2}{1+\frac{1}{2}e^{-j\omega}} \end{aligned}$$

Taking the inverse Fourier transform, we obtain

$$y[n] = \frac{1}{2}\left(\frac{1}{2}\right)^n u[n] + \frac{1}{2}\left(-\frac{1}{2}\right)^n u[n].$$

(ii) In this case

$$X(e^{j\omega}) = \frac{1}{1+\frac{1}{2}e^{-j\omega}}.$$

Therefore,

$$Y(e^{j\omega}) = \left[\frac{1}{1-\frac{1}{2}e^{-j\omega}}\right]^2.$$

Taking the inverse Fourier transform, we obtain

$$y[n] = (n+1)\left(-\frac{1}{2}\right)^n u[n].$$

(iii) In this case

$$X(e^{j\omega}) = 1 + \frac{1}{2}e^{-j\omega}.$$

Therefore,

$$Y(e^{j\omega}) = 1.$$

Taking the inverse Fourier transform, we obtain

$$y[n] = \delta[n].$$

(iv) In this case

$$X(e^{j\omega}) = 1 - \frac{1}{2}e^{-j\omega}.$$

Therefore,

$$\begin{aligned} Y(e^{j\omega}) &= \left[1 - \frac{1}{2}e^{-j\omega}\right]\left[\frac{1}{1+\frac{1}{2}e^{-j\omega}}\right] \\ &= -1 + \frac{2}{1+\frac{1}{2}e^{-j\omega}} \end{aligned}$$

Taking the inverse Fourier transform, we obtain

$$y[n] = -\delta[n] + 2\left(-\frac{1}{2}\right)^n u[n].$$

(c) (i) We have

$$\begin{aligned} Y(e^{j\omega}) &= \left[\frac{1-\frac{1}{4}e^{-j\omega}}{1+\frac{1}{2}e^{-j\omega}}\right]\left[\frac{1}{1+\frac{1}{2}e^{-j\omega}}\right] \\ &= \frac{1}{(1+\frac{1}{2}e^{-j\omega})^2} - \frac{\frac{1}{4}e^{-j\omega}}{(1+\frac{1}{2}e^{-j\omega})^2} \end{aligned}$$

Taking the inverse Fourier transform, we obtain

$$y[n] = (n+1)\left(-\frac{1}{2}\right)^n u[n] - \frac{1}{4}n\left(-\frac{1}{2}\right)^{n-1} u[n-1].$$

(ii) We have

$$\begin{aligned} Y(e^{j\omega}) &= \left[\frac{1+\frac{1}{2}e^{-j\omega}}{1-\frac{1}{4}e^{-j\omega}}\right]\left[\frac{1}{1+\frac{1}{2}e^{-j\omega}}\right] \\ &= \frac{1}{1-\frac{1}{4}e^{-j\omega}} \end{aligned}$$

Taking the inverse Fourier transform, we obtain

$$y[n] = \left(\frac{1}{4}\right)^n u[n].$$

(iii) We have

$$\begin{aligned} Y(e^{j\omega}) &= \left[\frac{1}{(1+\frac{1}{2}e^{-j\omega})(1-\frac{1}{4}e^{-j\omega})}\right]\left[\frac{1}{1+\frac{1}{2}e^{-j\omega}}\right] \\ &= \frac{2/3}{(1+\frac{1}{2}e^{-j\omega})^2} + \frac{2/9}{1+\frac{1}{2}e^{-j\omega}} + \frac{1/9}{1-\frac{1}{4}e^{-j\omega}} \end{aligned}$$

Taking the inverse Fourier transform, we obtain

$$y[n] = \frac{2}{3}(n+1)\left(-\frac{1}{2}\right)^n u[n] + \frac{2}{9}\left(-\frac{1}{2}\right)^n u[n] + \frac{1}{9}\left(\frac{1}{4}\right)^n u[n].$$

(iv) We have

$$\begin{aligned} Y(e^{j\omega}) &= \left[1+2e^{-3j\omega}\right]\left[\frac{1}{1+\frac{1}{2}e^{-j\omega}}\right] \\ &= \frac{1}{1+\frac{1}{2}e^{-j\omega}} + \frac{2e^{-3j\omega}}{1+\frac{1}{2}e^{-j\omega}} \end{aligned}$$

Taking the inverse Fourier transform, we obtain

$$y[n] = \left(-\frac{1}{2}\right)^n u[n] + 2\left(-\frac{1}{2}\right)^{n-3} u[n-3].$$

5.34. **(a)** Since the two systems are cascaded, the frequency response of the overall system is

$$\begin{aligned} H(e^{j\omega}) &= H_1(e^{j\omega})H_2(e^{j\omega}) \\ &= \frac{2-e^{-j\omega}}{1+\frac{1}{8}e^{-j3\omega}} \end{aligned}$$

Therefore, the Fourier transforms of the input and output of the overall system are related by

$$\frac{Y(e^{j\omega})}{X(e^{j\omega})} = \frac{2-e^{-j\omega}}{1+\frac{1}{8}e^{-j3\omega}}.$$

Cross-multiplying and taking the inverse Fourier transform, we get

$$y[n] + \frac{1}{8}y[n-3] = 2x[n] - x[n-1].$$

(b) We may rewrite the overall frequency response as

$$H(e^{j\omega}) = \frac{4/3}{1+\frac{1}{2}e^{-j\omega}} + \frac{(1+j\sqrt{3})/3}{1-\frac{1}{2}e^{j120}e^{-j\omega}} + \frac{(1-j\sqrt{3})/3}{1-\frac{1}{2}e^{-j120}e^{-j\omega}}.$$

Taking the inverse Fourier transform we get

$$h[n] = \frac{4}{3}\left(-\frac{1}{2}\right)^n u[n] + \frac{1+j\sqrt{3}}{3}\left(\frac{1}{2}e^{j120}\right)^n u[n] + \frac{1-j\sqrt{3}}{3}\left(\frac{1}{2}e^{-j120}\right)^n u[n].$$

5.35. (a) Taking the Fourier transform of both sides of the given difference equation we obtain

$$H(e^{j\omega}) = \frac{Y(e^{j\omega})}{X(e^{j\omega})} = \frac{b + e^{-j\omega}}{1 - ae^{-j\omega}}.$$

In order for $|H(e^{j\omega})|$ to be one, we must ensure that

$$\begin{aligned} |b + e^{-j\omega}| &= |1 - ae^{-j\omega}| \\ 1 + b^2 + 2b\cos\omega &= 1 + a^2 - 2a\cos\omega \end{aligned}$$

This is possible only if $b = -a$.

(b) The plot is as shown Figure S5.35.

(c) The plot is as shown Figure S5.35.

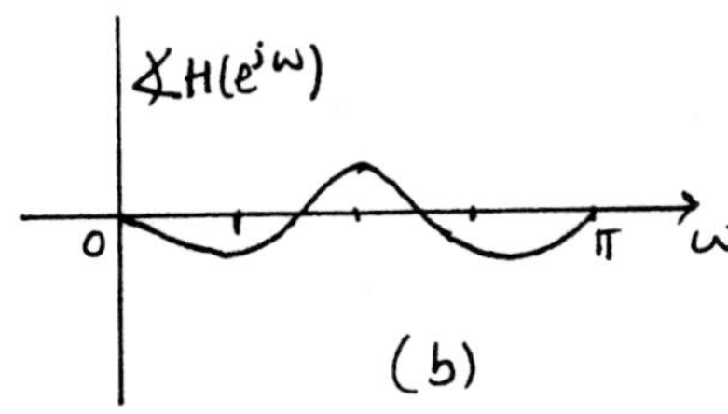

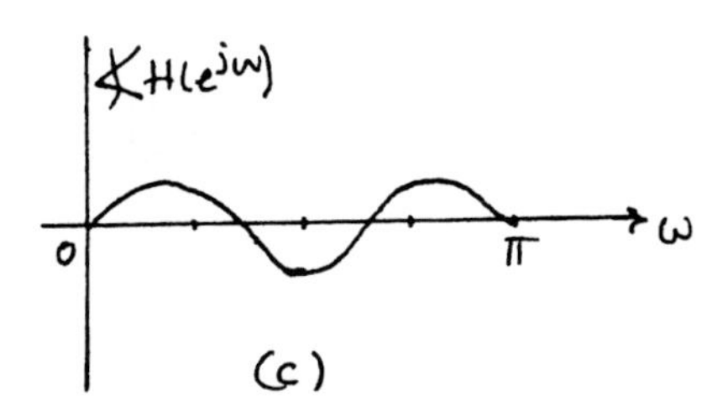

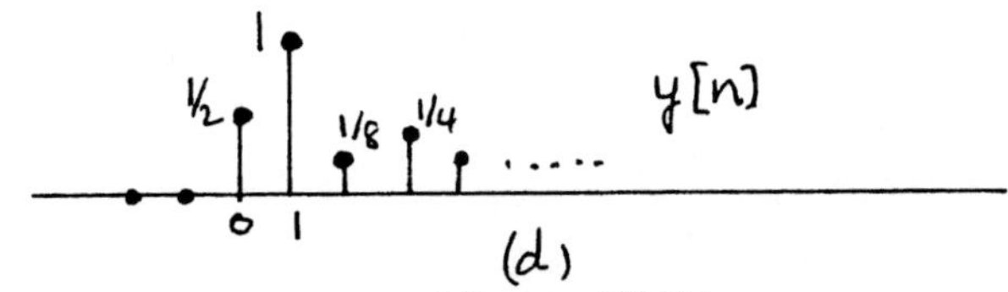

Figure S5.35

(d) When $a = -\frac{1}{2}$,

$$H(e^{j\omega}) = \frac{\frac{1}{2} + e^{-j\omega}}{1 + \frac{1}{2}e^{-j\omega}}.$$

Also,

$$X(e^{j\omega}) = \frac{1}{1 - \frac{1}{2}e^{-j\omega}}.$$

Therefore,

$$\begin{aligned} Y(e^{j\omega}) &= \frac{\frac{1}{2} + e^{-j\omega}}{(1 + \frac{1}{2}e^{-j\omega})(1 - \frac{1}{2}e^{-j\omega})} \\ &= \frac{5/4}{1 - \frac{1}{2}e^{-j\omega}} - \frac{3/4}{1 + \frac{1}{2}e^{-j\omega}} \end{aligned}$$

Taking the inverse Fourier transform we obtain

$$y[n] = \frac{5}{4}\left(\frac{1}{2}\right)^n u[n] - \frac{3}{4}\left(-\frac{1}{2}\right)^n u[n].$$

This is as sketched in Figure S5.35.

5.36. **(a)** The frequency responses are related by the following expression:

$$G(e^{j\omega}) = \frac{1}{H(e^{j\omega})}.$$

(b) (i) Here, $H(e^{j\omega}) = 1 - \frac{1}{4}e^{-j\omega}$. Therefore, $G(e^{j\omega}) = 1/(1 - \frac{1}{4}e^{-j\omega})$ and $g[n] = (\frac{1}{4})^n u[n]$. Since

$$G(e^{j\omega}) = \frac{Y(e^{j\omega})}{X(e^{j\omega})} = \frac{1}{1 - \frac{1}{4}e^{-j\omega}},$$

the difference equation relating the input $x[n]$ and output $y[n]$ is

$$y[n] - \frac{1}{4}y[n-1] = x[n].$$

(ii) Here, $H(e^{j\omega}) = 1/(1 + \frac{1}{2}e^{-j\omega})$. Therefore, $G(e^{j\omega}) = 1 + \frac{1}{2}e^{-j\omega}$ and $g[n] = \delta[n] + \frac{1}{2}\delta[n-1]$. Since

$$G(e^{j\omega}) = \frac{Y(e^{j\omega})}{X(e^{j\omega})} = 1 + \frac{1}{2}e^{-j\omega},$$

the difference equation relating the input $x[n]$ and output $y[n]$ is

$$y[n] = x[n] + \frac{1}{2}x[n-1].$$

(iii) Here, $H(e^{j\omega}) = (1 - \frac{1}{4}e^{-j\omega})/(1 + \frac{1}{2}e^{-j\omega})$. Therefore, $G(e^{j\omega}) = (1 + \frac{1}{2}e^{-j\omega})/(1 - \frac{1}{4}e^{-j\omega})$ and $g[n] = (\frac{1}{4})^n u[n] + \frac{1}{2}(\frac{1}{4})^{n-1}u[n-1]$. Since

$$G(e^{j\omega}) = \frac{Y(e^{j\omega})}{X(e^{j\omega})} = \frac{1 + \frac{1}{2}e^{-j\omega}}{1 - \frac{1}{4}e^{-j\omega}},$$

the difference equation relating the input $x[n]$ and output $y[n]$ is

$$y[n] - \frac{1}{4}y[n-1] = x[n] + \frac{1}{2}x[n-1].$$

(iv) Here, $H(e^{j\omega}) = (1 - \frac{1}{4}e^{-j\omega} - \frac{1}{8}e^{-2j\omega})/(1 + \frac{5}{4}e^{-j\omega} - \frac{1}{8}e^{-2j\omega})$. Therefore, $G(e^{j\omega}) = (1 + \frac{5}{4}e^{-j\omega} - \frac{1}{8}e^{-2j\omega})/(1 - \frac{1}{4}e^{-j\omega} - \frac{1}{8}e^{-2j\omega})$. Therefore,

$$G(e^{j\omega}) = 1 + \frac{2}{1 - (1/2)e^{-j\omega}} - \frac{2}{1 + (1/4)e^{-j\omega}}$$

and

$$g[n] = \delta[n] + 2\left(\frac{1}{2}\right)^n u[n] - 2\left(-\frac{1}{4}\right)^n u[n].$$

Since

$$G(e^{j\omega}) = \frac{Y(e^{j\omega})}{X(e^{j\omega})} = \frac{(1 + \frac{5}{4}e^{-j\omega} - \frac{1}{8}e^{-2j\omega})}{(1 - \frac{1}{4}e^{-j\omega} - \frac{1}{8}e^{-2j\omega})},$$

the difference equation relating the input $x[n]$ and output $y[n]$ is

$$y[n] - \frac{1}{4}y[n-1] - \frac{1}{8}y[n-1] = x[n] + \frac{5}{4}x[n-1] - \frac{1}{8}x[n-2].$$

(v) Here, $H(e^{j\omega}) = (1-\frac{1}{2}e^{-j\omega})/(1+\frac{5}{4}e^{-j\omega}-\frac{1}{8}e^{-2j\omega})$. Therefore, $G(e^{j\omega}) = (1+\frac{5}{4}e^{-j\omega}-\frac{1}{8}e^{-2j\omega})/(1-\frac{1}{2}e^{-j\omega})$ Since

$$G(e^{j\omega}) = \frac{Y(e^{j\omega})}{X(e^{j\omega})} = \frac{(1+\frac{5}{4}e^{-j\omega}-\frac{1}{8}e^{-2j\omega})}{(1-\frac{1}{2}e^{-j\omega})},$$

the difference equation relating the input $x[n]$ and output $y[n]$ is

$$y[n] - \frac{1}{2}y[n-1] = x[n] + \frac{5}{4}x[n-1] - \frac{1}{8}x[n-2].$$

(vi) Here, $H(e^{j\omega}) = 1/(1+\frac{5}{4}e^{-j\omega}-\frac{1}{8}e^{-2j\omega})$. Therefore, $G(e^{j\omega}) = (1+\frac{5}{4}e^{-j\omega}-\frac{1}{8}e^{-2j\omega})$ Since

$$G(e^{j\omega}) = \frac{Y(e^{j\omega})}{X(e^{j\omega})} = (1+\frac{5}{4}e^{-j\omega}-\frac{1}{8}e^{-2j\omega})$$

we have

$$g[n] = \delta[n] + \frac{5}{4}\delta[n-1] - \frac{1}{8}\delta[n-2]$$

and the difference equation relating the input $x[n]$ and output $y[n]$ is

$$y[n] = x[n] + \frac{5}{4}x[n-1] - \frac{1}{8}x[n-2].$$

(c) The frequency response of the given system is

$$H(e^{j\omega}) = \frac{e^{-j\omega}-\frac{1}{2}e^{-2j\omega}}{1+e^{-j\omega}+\frac{1}{4}e^{-2j\omega}}.$$

The frequency response of the inverse system is

$$G(e^{j\omega}) = \frac{1}{H(e^{j\omega})} = \frac{e^{j\omega}+1+\frac{1}{4}e^{-j\omega}}{1-\frac{1}{2}e^{-j\omega}}.$$

Therefore,

$$g[n] = \left(\frac{1}{2}\right)^{n+1}u[n+1] + \left(\frac{1}{2}\right)^{n}u[n] + \frac{1}{4}\left(\frac{1}{2}\right)^{n-1}u[n-1].$$

Clearly, $g[n]$ is not a causal impulse response.

If we delay this impulse response by 1 sample, then it becomes causal. Furthermore, the output of the inverse system will then be $x[n-1]$. The impulse response of this causal system is

$$g_1[n] = g[n-1] = \left(\frac{1}{2}\right)^{n}u[n] + \left(\frac{1}{2}\right)^{n-1}u[n-1] + \frac{1}{4}\left(\frac{1}{2}\right)^{n-2}u[n-2].$$

5.37. Given that

$$x[n] \stackrel{FT}{\longleftrightarrow} X(e^{j\omega}).$$

(i) Since

$$X(e^{j\omega}) = \sum_{n=-\infty}^{\infty} x[n]e^{-j\omega n},$$

we may write

$$X^*(e^{-j\omega}) = \sum_{n=-\infty}^{\infty} x^*[n]e^{-j\omega n}.$$

Comparing with the analysis eq. (5.9), we conclude that

$$x^*[n] \overset{FT}{\longleftrightarrow} X^*(e^{-j\omega}).$$

Therefore,

$$\mathcal{R}e\{x[n]\} = \frac{x[n] + x^*[n]}{2} \overset{FT}{\longleftrightarrow} \frac{X(e^{j\omega}) + X^*(e^{-j\omega})}{2}.$$

(ii) Since

$$X(e^{j\omega}) = \sum_{n=-\infty}^{\infty} x[n]e^{-j\omega n},$$

we may write

$$X(e^{-j\omega}) = \sum_{n=-\infty}^{\infty} x[-n]e^{-j\omega n}.$$

Therefore,

$$x[-n] \overset{FT}{\longleftrightarrow} X(e^{-j\omega}).$$

From the previous part we know that

$$x^*[n] \overset{FT}{\longleftrightarrow} X^*(e^{-j\omega}).$$

Therefore, putting these two statements together we get

$$x^*[-n] \overset{FT}{\longleftrightarrow} X^*(e^{j\omega}).$$

(iii) From our previous results we know that

$$\mathcal{E}v\{x[n]\} = \frac{x[n] + x[-n]}{2} \overset{FT}{\longleftrightarrow} \frac{X(e^{j\omega}) + X(e^{-j\omega})}{2}.$$

5.38. From the synthesis equation (5.8) we obtain

$$\begin{aligned} x[n] &= \frac{1}{2\pi}\int_{-\pi}^{\pi} X(e^{j\omega})e^{j\omega n}d\omega \\ &= \frac{1}{2\pi}\int_{0}^{\pi} X(e^{j\omega})e^{j\omega n}d\omega + \frac{1}{2\pi}\int_{0}^{\pi} X(e^{-j\omega})e^{-j\omega n}d\omega \end{aligned}$$

Since $x[n]$ is real, $X(e^{-j\omega}) = X^*(e^{j\omega})$. Therefore,

$$\begin{aligned} x[n] &= \frac{1}{2\pi}\int_0^{\pi} \mathcal{R}e\{X(e^{j\omega})\}\{e^{\omega n} + e^{-j\omega n}\}d\omega + \frac{j}{2\pi}\int_0^{\pi} \mathcal{I}m\{X(e^{j\omega})\}\{e^{\omega n} - e^{-j\omega n}\}d\omega \\ &= \frac{1}{\pi}\int_0^{\pi} \mathcal{R}e\{X(e^{j\omega})\}2\cos(\omega n)d\omega - \frac{}{\pi}\int_0^{\pi} \mathcal{I}m\{X(e^{j\omega})\}\sin(\omega n)d\omega \end{aligned}$$

Therefore,

$$B(\omega) = \frac{1}{\pi}\mathcal{R}e\{X(e^{j\omega})\}\cos(\omega n), \qquad \text{and} \qquad -\frac{1}{\pi}\mathcal{I}m\{X(e^{j\omega})\}\sin(\omega n).$$

5.39. Let $y[n] = x[n] * h[n]$. Then

$$\begin{aligned} Y(e^{j\omega}) &= \sum_{n=-\infty}^{\infty} \{x[n] * h[n]\}e^{-j\omega n} \\ &= \sum_{n=-\infty}^{\infty}\sum_{k=-\infty}^{\infty} x[k]h[n-k]e^{-j\omega n} \\ &= \sum_{k=-\infty}^{\infty} x[k] \sum_{n=-\infty}^{\infty} h[n-k]e^{-j\omega n} \\ &= \sum_{k=-\infty}^{\infty} x[k]e^{-j\omega k}H(e^{j\omega}) \\ &= H(e^{j\omega})\sum_{k=-\infty}^{\infty} x[k]e^{-j\omega k} \\ &= H(e^{j\omega})X(e^{j\omega}) \end{aligned}$$

5.40. Let $y[n] = x[n] * h[n]$. Then using the convolution sum

$$y[0] = \sum_{k=-\infty}^{\infty} x[k]h[-k] \qquad \text{(S5.40–1)}$$

Using the convolution property of the Fourier transform,

$$y[0] = \frac{1}{2\pi}\int_{-\pi}^{\pi} X(e^{j\omega})H(e^{j\omega})d\omega \qquad \text{(S5.40–2)}$$

Now let $h[n] = x^*[-n]$. Then $H(e^{j\omega}) = X^*(e^{j\omega})$. Substituting in the right-hand sides of equations (S5.40-1) and (S5.40-2) and equating them,

$$\sum_{k=-\infty}^{\infty} x[k]x^*[k] = \frac{1}{2\pi}\int_{-\infty}^{\infty} X(e^{j\omega})X^*(e^{j\omega})d\omega.$$

Therefore,

$$\sum_{n=-\infty}^{\infty} |x[n]|^2 = \frac{1}{2\pi}\int_{-\pi}^{\pi} |X(e^{j\omega})|^2 d\omega.$$

Now let $h[n] = z^*[-n]$. Then $H(e^{j\omega}) = Z^*(e^{j\omega})$. Substituting in the right-hand sides of equations (S5.40-1) and (S5.40-2) and equating them,

$$\sum_{k=-\infty}^{\infty} x[k]z^*[k] = \frac{1}{2\pi}\int_{-\pi}^{\pi} X(e^{j\omega})Z^*(e^{j\omega})d\omega.$$

5.41. **(a)** The Fourier transform $X(e^{j\omega})$ of the signal $x[n]$ is

$$X(e^{j\omega}) = \sum_{n=-\infty}^{\infty} x[n]e^{-j\omega n} = \sum_{n_0}^{n_0+N-1} x[n]e^{-j\omega n}.$$

Therefore,

$$X(e^{j2\pi k/N}) = \sum_{n=n_0}^{n_0+N-1} x[n]e^{-j(2\pi/N)kn}. \qquad \text{(S5.41-1)}$$

Now, we may write the expression for the FS coefficients of $\tilde{x}[n]$ as

$$a_k = \frac{1}{N}\sum_{<N>} \tilde{x}[n]e^{-j(2\pi/N)kn} = \frac{1}{N}\sum_{n=n_0}^{n_0+N-1} x[n]e^{-j(2\pi/N)kn}.$$

(Because $x[n] = \tilde{x}[n]$ in the range $n_0 \le n \le n_0+N-1$). Comparing the above equation with eq. (S5.41-1), we get

$$a_k = \frac{1}{N}X(e^{j2\pi k/N}).$$

(b) (i) From the given information,

$$\begin{aligned}
X(e^{j\omega}) &= 1+e^{-j\omega}+e^{-2j\omega}+e^{-3j\omega} \\
&= e^{-j(3/2)\omega}\{e^{j(3/2)\omega}+e^{-j(3/2)\omega}\}+e^{-j(3/2)\omega}\{e^{j(1/2)\omega}+e^{-j(1/2)\omega}\} \\
&= 2e^{-j(3/2)\omega}\{\cos(3\omega/2)+\cos(\omega/2)\}
\end{aligned}$$

(ii) From part (a),

$$a_k = \frac{1}{N}X(e^{j2\pi k/N}) = \frac{1}{N}2e^{-j(3/2)2\pi k/N}\{\cos(6\pi k/(2N)) + \cos(\pi k/N)\}.$$

5.42. **(a)** $P(e^{j\omega}) = 2\pi\delta(\omega-\omega_0)$ for $|\omega| < \pi$. This is as shown in Figure S5.42.

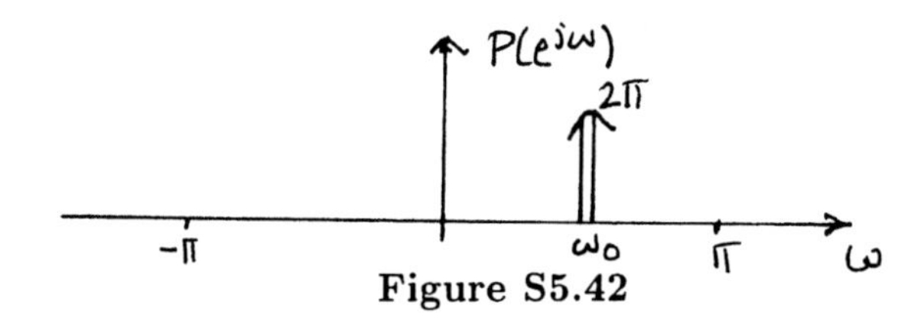

Figure S5.42

(b) From the multiplication property of the Foutier transform we have

$$\begin{aligned} G(e^{j\omega}) &= \frac{1}{2\pi}\int_{-\pi}^{\pi} X(e^{j\theta})P(e^{j(\omega-\theta)})d\theta \\ &= \frac{1}{2\pi}\int_{-\pi}^{\pi} X(e^{j\theta})2\pi\delta(\omega-\theta-\omega_0)d\theta \\ &= X(e^{j(\omega-\omega_0)}) \end{aligned}$$

5.43. **(a)** Using the frequency shift and linearity properties,

$$V(e^{j\omega}) = \frac{X(e^{j(\omega-\pi)}) + X(e^{j\omega})}{2}.$$

(b) Let $y[n] = v[2n]$. Then

$$Y(e^{j\omega}) = \sum_{n=-\infty}^{\infty} v[2n]e^{-j\omega n}.$$

Since the odd-indexed samples of $v[n]$ are zero, we may put $m = 2n$ in the above equation to get

$$Y(e^{j\omega}) = \sum_{m=-\infty}^{\infty} v[m]e^{-j\omega m/2} = V(e^{j\omega/2}).$$

(Note that the substitution of n by $2m$ is valid *only* if the odd-indexed samples in the summation are zero.)

(c) $x[2n]$ is a new sequence which consists of only the even indexed samples of $x[n]$. $v[n]$ is a sequence whose even-indexed samples are equal to $x[n]$. The odd-indexed samples of $v[n]$ are zero. $v[2n]$ is a new sequence which consists of only the even indexed samples of $v[n]$. This implies that $v[2n]$ is a sequence which consists of only the even indexed samples of $x[n]$. This idea is illustrated in Figure S5.43.

From part (a),

$$G(e^{j\omega}) = \frac{X(e^{j(\omega/2-\pi)}) + X(e^{j\omega/2})}{2}.$$

5.44. **(a)** The signal $x_1[n]$ is as shown in Figure S5.44.

(i) Taking the inverse Fourier transform, the signal $x_2[n]$ is

$$x_2[n] = x_1[n+1].$$

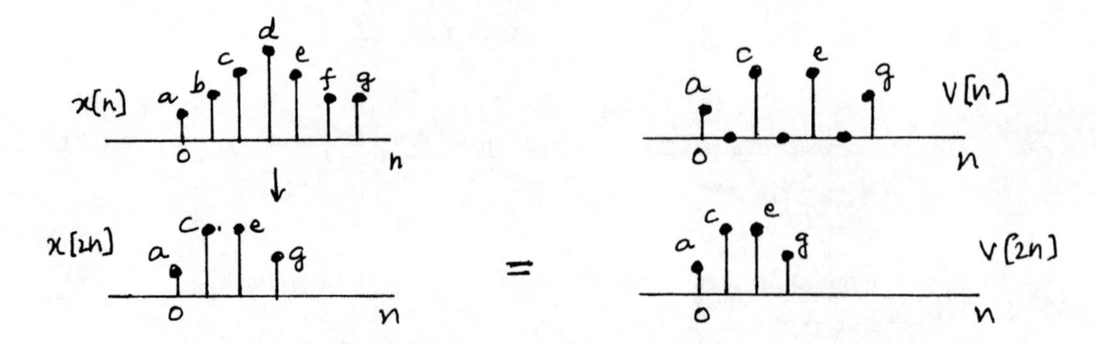

Figure S5.43

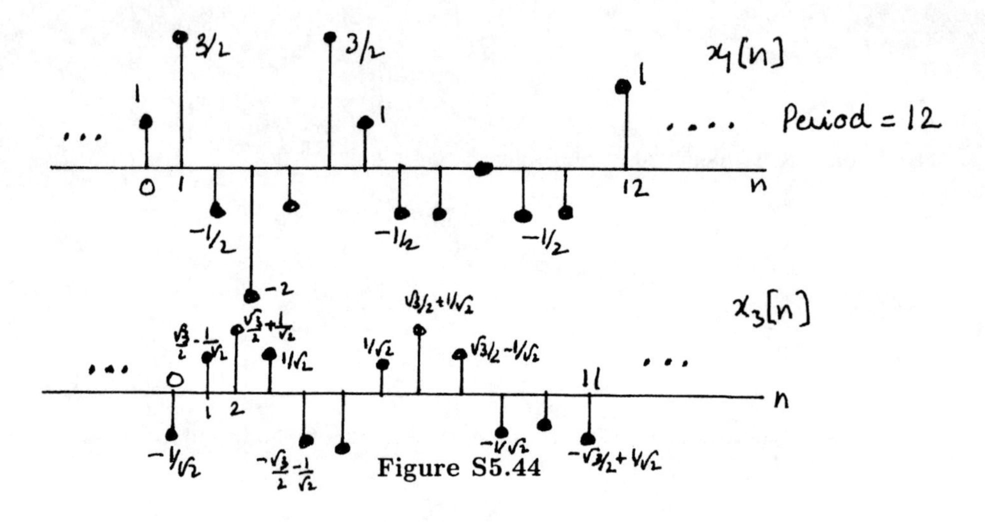

Figure S5.44

(ii) Taking the inverse Fourier transform, the signal $x_3[n]$ is

$$x_2[n] = x_1[n - 3/2] = \sin(\pi n/3) + \sin(\pi n/2)\cos(3\pi/4) - \cos(\pi n/2)\sin(3\pi/4).$$

This is as shown in Figure S5.44.

(b) From part (a),

$$x_2[n] = x_1[n+1] = w(nT + T).$$

Also,

$$x_3[n] = x_1[n - 3/2] = w(nT - 3T/2).$$

Therefore, $\alpha = -1$ and $\beta = 3/2$.

5.45. From the Fourier transform analysis equation

$$X(e^{j\omega}) = \sum_{n=-\infty}^{\infty} x[n]e^{-j\omega n}.$$

(a) Comparing the equation for $x_1(t)$ with the above equation, we obtain

$$x_1(t) = X(e^{-j(2\pi/10)t}).$$

Therefore $x_1(t)$ is as shown in Figure S5.45.

(b) Comparing the equation for $x_2(t)$ with the equation for $X(e^{j\omega})$, we obtain

$$x_2(t) = X(e^{j(2\pi/10)t}) = x_1(-t).$$

Therefore $x_2(t)$ is as shown in Figure S5.45.

(c) We know that $\mathcal{O}d\{x[n]\} = (x[n] - x[-n])/2$. Therefore,

$$\frac{X(e^{j\omega}) - X(e^{-j\omega})}{2} = \sum_{n=-\infty}^{\infty} \mathcal{O}d\{x[n]\}e^{-j\omega n}.$$

Comparing this with the given equation for $x_3(t)$, we obtain

$$x_3(t) = \frac{X(e^{-j(2\pi/8)t}) - X(e^{j(2\pi/8)t})}{2}.$$

Therefore $x_3(t)$ is as shown in Figure S5.45.

(d) We know that $\mathcal{R}e\{x[n]\} = (x[n] + x^*[n])/2$. Therefore,

$$\frac{X(e^{j\omega}) - X^*(e^{-j\omega})}{2} = \sum_{n=-\infty}^{\infty} \mathcal{R}e\{x[n]\}e^{-j\omega n}.$$

Comparing this with the given equation for $x_4(t)$, we obtain

$$x_4(t) = \frac{X(e^{-j(2\pi/6)t}) + X^*(e^{j(2\pi/6)t})}{2}.$$

Therefore $x_4(t)$ is as shown in the Figure S5.45.

5.46. **(a)** Let $x[n] = \alpha^n u[n]$. Then $X(e^{j\omega}) = \frac{1}{1-\alpha e^{-j\omega}}$. Using the differentiation in frequency property,

$$n\alpha^n u[n] \overset{FT}{\longleftrightarrow} j\frac{dX(e^{j\omega})}{d\omega} = \frac{\alpha e^{-j\omega}}{(1-\alpha e^{-j\omega})^2}.$$

Therefore,

$$(n+1)\alpha^n u[n] \overset{FT}{\longleftrightarrow} j\frac{dX(e^{j\omega})}{d\omega} + X(e^{j\omega}) = \frac{1}{(1-\alpha e^{-j\omega})^2}.$$

(b) From part (a), it is clear that the result is true for $r = 1$ and $r = 2$. Let us assume that it is also true for $k = r - 1$. We will now attempt to prove that the result is true for $k = r$. We have

$$x_{r-1}[n] = \frac{(n+r-2)!}{n!(r-2)!}\alpha^n u[n] \overset{FT}{\longleftrightarrow} X_{r-1}(e^{j\omega}) = \frac{1}{(1-\alpha e^{-j\omega})^{r-1}}.$$

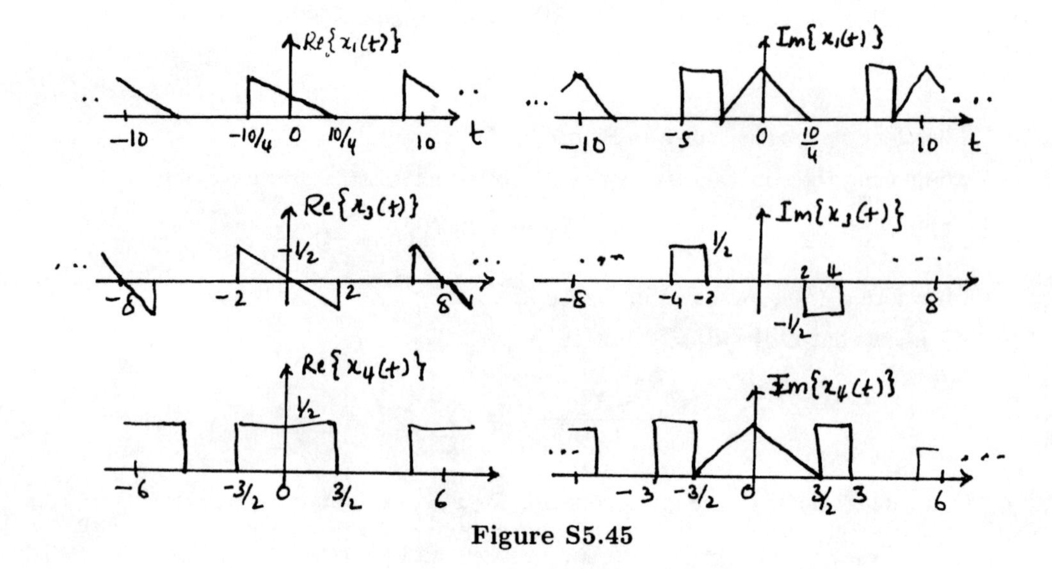

Figure S5.45

From the differentiation in frequency property,

$$nx_{r-1}[n] \stackrel{FT}{\longleftrightarrow} \frac{\alpha(r-1)e^{-j\omega}}{(1-\alpha e^{-j\omega})^{r-1}}.$$

Therefore,

$$\frac{(n+1)x_{r-1}[n+1]}{\alpha(r-1)} \stackrel{FT}{\longleftrightarrow} \frac{1}{(1-\alpha e^{-j\omega})^{r}}.$$

The left hand side of the above expression is

$$\frac{(n+1)x_{r-1}[n+1]}{\alpha(r-1)} = \frac{(n+r-1)!}{n!(r-1)!}\alpha^n u[n] = x_r[n].$$

Therefore, we have shown that the result is valid for r if it is valid for $r-1$. Since, we know that the result is valid for $r=2$, we may conclude that it is valid for $r=3$, $r=4$, and so on.

5.47. **(a)** If $X(e^{j\omega}) = X(e^{j(\omega-1)})$ then $X(e^{j\omega})$ is periodic with a period of 1. But we already know that $X(e^{j\omega})$ is periodic with a period of 2π. This is only possible if $X(e^{j\omega})$ is a constant for all ω. This implies that $x[n]$ is of the form $k\delta[n]$ where k is a constant. Therefore, the given statemet is true.

(b) If $X(e^{j\omega}) = X(e^{j(\omega-\pi)})$ then $X(e^{j\omega})$ is periodic with a period of π. We also know that $X(e^{j\omega})$ is periodic with a period of 2π. Both these conditions can be satisfied even if $X(e^{j\omega})$ has some arbitrary shape in the region $0 \leq |\omega| \leq \pi/2$. Therefore, $X(e^{j\omega})$ need not necessarily be a constant. Consequently, $x[n]$ need not be just an impulse. Therefore, the given statement is false.

(c) We know from Problem 5.43 that the inverse Fourier transform of $X(e^{j\omega/2})$ is the sequence $v[n] = (x[n] + e^{j\pi n}x[n])/2$. The even-indexed samples of $v[n]$ are identical to the even-indexed samples of $x[n]$. The odd-indexed samples of $v[n]$ are zero. If $X(e^{j\omega}) = X(e^{j\omega/2})$, then $x[n] = v[n]$. This implies that the even-indexed samples of $x[n]$ are zero. Consequently, $x[n]$ does not necessarily have to be an impulse. Therefore, the given statement is false.

(d) From Table 5.1 we know that the inverse Fourier transform of $X(e^{j2\omega})$ is the time-expanded signal

$$x_{(2)}[n] = \begin{cases} x[n/2], & n = 0, \pm 2, \pm 4, \cdots \\ 0, & \text{otherwise} \end{cases}.$$

If $X(e^{j\omega}) = X(e^{j2\omega})$, then $x[n] = x_{(2)}[n]$. This is possible only if $x[n]$ is an impulse. Therefore, the given statement is true.

5.48. (a) Taking the Fourier transform of both equations and eliminating $W(e^{j\omega})$, we obtain

$$H(e^{j\omega}) = \frac{Y(e^{j\omega})}{X(e^{j\omega})} = \frac{3 - \frac{1}{2}e^{-j\omega}}{(1 - \frac{1}{2}e^{-j\omega})(1 - \frac{1}{4}e^{-j\omega})}.$$

Taking the inverse Fourier transform of the partial fraction expansion of the above expression, we obtain

$$h[n] = 4\left(\frac{1}{2}\right)^n u[n] - \left(\frac{1}{4}\right)^n u[n].$$

(b) We know that

$$H(e^{j\omega}) = \frac{Y(e^{j\omega})}{X(e^{j\omega})} = \frac{3 - \frac{1}{2}e^{-j\omega}}{(1 - \frac{1}{2}e^{-j\omega})(1 - \frac{1}{4}e^{-j\omega})}.$$

Cross-multiplying and taking the inverse Fourier transform, we obtain

$$y[n] - \frac{3}{4}y[n-1] + \frac{1}{8}y[n-2] = 3x[n] - \frac{1}{2}x[n-1].$$

5.49. (a) (i) Consider the signal $x[n] = ax_1[n] + bx_2[n]$, where a and b are constants. Then, $X(e^{j\omega}) = aX_1(e^{j\omega}) + bX_2(e^{j\omega})$. Also let the responses of the system to $x_1[n]$ and $x_2[n]$ be $y_1[n]$ and $y_2[n]$, respectively. Substituting for $X(e^{j\omega})$ in the equation given in the problem and simplifying we obtain $Y(e^{j\omega}) = aY_1(e^{j\omega}) + bY_2(e^{j\omega})$. Therefore, the system is linear

(ii) Consider the signal $x_1[n] = x[n-1]$. Then, $X_1(e^{j\omega}) = e^{-j\omega}X(e^{j\omega})$. Let the response of the system to this signal be $y_1[n]$. From the given equation,

$$\begin{aligned} Y_1(e^{j\omega}) &= 2X_1(e^{j\omega}) + e^{-j\omega}X_1(e^{j\omega}) - \frac{dX_1(e^{j\omega})}{d\omega} \\ &= e^{-j\omega}\left[2X(e^{j\omega}) + e^{-j\omega}X(e^{j\omega}) - \frac{dX(e^{j\omega})}{d\omega}\right] + je^{-j\omega}X(e^{j\omega}) \\ &\neq e^{-j\omega}Y(e^{j\omega}) \end{aligned}$$

Therefore, the system is not time invariant.

(iii) If $x[n] = \delta[n]$, $X(e^{j\omega}) = 1$. Then,

$$Y(e^{j\omega}) = 2 + e^{-j\omega}.$$

Therefore, $y[n] = 2\delta[n] + \delta[n-1]$.

(b) We may write

$$Y(e^{j\omega}) = \frac{1}{2\pi}\int_{\omega-\pi/4}^{\omega+\pi/4} X(e^{j\theta})H(e^{j(\omega-\theta)})d\theta,$$

where $H(e^{j\omega})$ is as shown in the Figure S5.49.

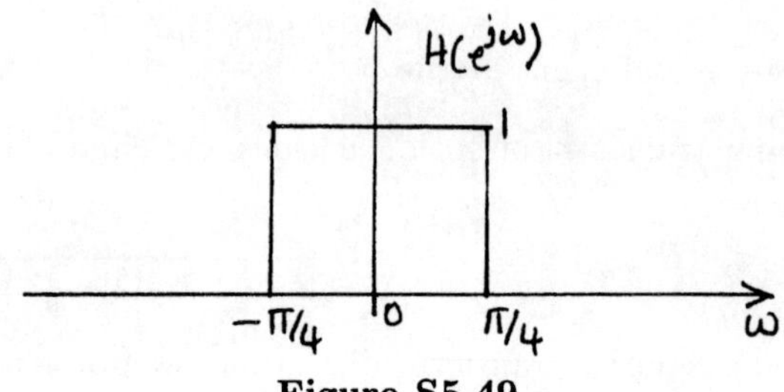

Figure S5.49

Using the multiplication property of the Fourier transform and Table 5.2, we obtain

$$y[n] = 2x[n]\frac{\sin(\pi n/4)}{n}.$$

5.50. **(a)** (i) From the given information,

$$H(e^{j\omega}) = \frac{Y(e^{j\omega})}{X(e^{j\omega})} = \frac{1-\frac{1}{2}e^{-j\omega}}{(1-\frac{1}{3}e^{-j\omega})(1-\frac{1}{4}e^{-j\omega})}.$$

Taking the inverse Fourier transform, we obtain

$$h[n] = 3\left(\frac{1}{4}\right)^n u[n] - 2\left(\frac{1}{3}\right)^n u[n].$$

(ii) From part (a), we know that

$$\frac{Y(e^{j\omega})}{X(e^{j\omega})} = \frac{1-\frac{1}{2}e^{-j\omega}}{(1-\frac{1}{3}e^{-j\omega})(1-\frac{1}{4}e^{-j\omega})}.$$

Cross-multiplying and taking the inverse Fourier transform

$$y[n] - \frac{7}{12}y[n-1] + \frac{1}{12}y[n-2] = x[n] - \frac{1}{2}x[n-1].$$

(b) From the given information,

$$H(e^{j\omega}) = \frac{Y(e^{j\omega})}{X(e^{j\omega})} = \frac{(1-\frac{1}{2}e^{-j\omega})^2}{2(1-\frac{1}{4}e^{-j\omega})^2}.$$

We now want to find $X(e^{j\omega})$ when $Y(e^{j\omega}) = (1/2)e^{-j\omega}/(1+\frac{1}{2}e^{-j\omega})$. From the above equation we obtain

$$X(e^{j\omega}) = \frac{e^{-j\omega}(1-\frac{1}{4}e^{-j\omega})^2}{(1-\frac{1}{2}e^{-j\omega})^2(1+\frac{1}{2}e^{-j\omega})}.$$

Taking the inverse Fourier transform of the partial fraction expansion of the above expression, we obtain

$$x[n] = \frac{3}{8}\left(-\frac{1}{2}\right)^{n-1}u[n-1] + \frac{3}{8}\left(\frac{1}{2}\right)^{n-1}u[n-1] + \frac{1}{8}n\left(\frac{1}{2}\right)^{n-1}u[n-1].$$

5.51. **(a)** Taking the Fourier transform of $h[n]$ we obtain

$$H(e^{j\omega}) = Y(e^{j\omega})/X(e^{j\omega}) = \frac{\frac{3}{2}-\frac{1}{2}e^{-j\omega}}{1-\frac{3}{4}e^{-j\omega}+\frac{1}{8}e^{-j2\omega}}.$$

Cross-multiplying and taking the inverse Fourier transform we obtain

$$y[n] - \frac{3}{4}y[n-1] + \frac{1}{8}y[n-2] = \frac{3}{2}x[n] - \frac{1}{2}x[n-1].$$

(b) (i) Let us name the intemediate output $w[n]$ (See Figure S5.51).

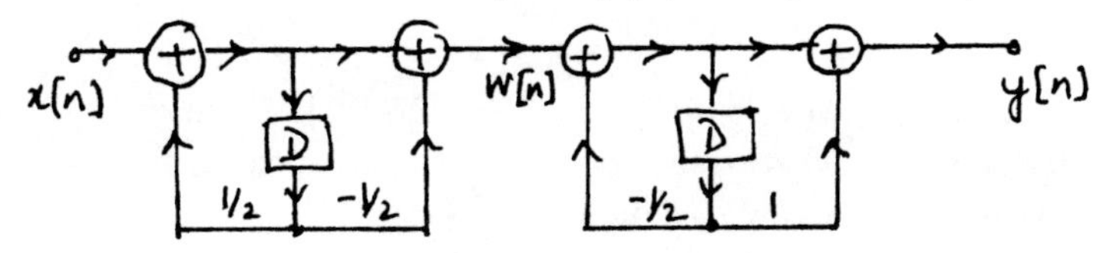

Figure S5.51

We may then write the following difference equations:

$$y[n] + \frac{1}{2}y[n-1] = \frac{1}{4}w[n] + w[n-1]$$

and

$$w[n] - \frac{1}{3}w[n-1] = x[n] - \frac{1}{2}x[n-1].$$

Taking the Fourier transform of both these equations and eliminating $W(e^{j\omega})$, we obtain

$$H(e^{j\omega}) = \frac{Y(e^{j\omega})}{X(e^{j\omega})} = \frac{\frac{1}{4}+\frac{7}{8}e^{-j\omega}-\frac{1}{2}e^{-2j\omega}}{1-\frac{1}{4}e^{-2j\omega}}.$$

Cross-multiplying and taking the inverse Fourier transform we obtain

$$y[n] - \frac{1}{4}y[n-2] = \frac{1}{4}x[n] + \frac{7}{8}x[n-1] - \frac{1}{2}x[n-2].$$

(ii) From (i)

$$H(e^{j\omega}) = \frac{Y(e^{j\omega})}{X(e^{j\omega})} = \frac{\frac{1}{4} + \frac{7}{8}e^{-j\omega} - \frac{1}{2}e^{2j\omega}}{1 - \frac{1}{4}e^{-2j\omega}}.$$

(iii) Taking the inverse Fourier transform of the partial fraction expansion of $H(e^{j\omega})$, we obtain

$$h[n] = 2\delta[n] - \frac{21}{16}\left(-\frac{1}{2}\right)^n u[n] + \frac{7}{16}\left(\frac{1}{2}\right)^n u[n].$$

5.52. **(a)** Since $h[n]$ is causal, the nonzero sample values of $h[n]$ and $h[-n]$ overlap only at $n = 0$. Therefore,

$$\mathcal{E}v\{h[n]\} = \frac{h[n] + h[-n]}{2} = \begin{cases} h[n]/2, & n > 0 \\ h[0], & n = 0 \\ h[-n]/2, & n < 0 \end{cases}.$$

In other words,

$$h[n] = \begin{cases} 2\mathcal{E}v\{h[n]\}, & n > 0 \\ \mathcal{E}v\{h[0]\}, & n = 0 \\ 0, & n < 0 \end{cases} \qquad \text{(S5.52–1)}$$

Now note that if

$$h[n] \overset{FT}{\longleftrightarrow} H(e^{j\omega})$$

then

$$\mathcal{E}v\{h[n]\} = \frac{h[n] + h[-n]}{2} \overset{FT}{\longleftrightarrow} \mathcal{R}e\{H(e^{j\omega})\}.$$

Clearly, we can recover $\mathcal{E}v\{h[n]\}$ from $\mathcal{R}e\{H(e^{j\omega})\}$. From $\mathcal{E}v\{h[n]\}$ we can use eq.(S5.52-1) to recover $h[n]$. Obviously, from $h[n]$ we can once again obtain $H(e^{j\omega})$. Therefore, the system is completely specified by $\mathcal{R}e\{H(e^{j\omega})\}$.

(b) Taking the inverse Fourier transform of $\mathcal{R}e\{H(e^{j\omega})\}$, we obtain

$$\mathcal{E}v\{h[n]\} = \delta[n] + \frac{\alpha}{2}\delta[n-2] + \frac{\alpha}{2}\delta[n+2].$$

Therefore,

$$h[n] = \delta[n] + \alpha\delta[n-2],$$

and

$$H(e^{j\omega}) = 1 + \alpha e^{-j2\omega}.$$

(c) Since $h[n]$ is causal, the nonzero sample values of $h[n]$ and $h[-n]$ overlap only at $n = 0$. Therefore,

$$\mathcal{O}d\{h[n]\} = \frac{h[n] - h[-n]}{2} = \begin{cases} h[n]/2, & n > 0 \\ 0, & n = 0 \\ -h[-n]/2, & n < 0 \end{cases}.$$

In other words,

$$h[n] = \begin{cases} 2\mathcal{O}d\{h[n]\}, & n > 0 \\ \text{some value}, & n = 0 \\ 0, & n < 0 \end{cases} \qquad \text{(S5.52–2)}$$

Now note that if

$$h[n] \overset{FT}{\longleftrightarrow} H(e^{j\omega})$$

then

$$\mathcal{O}d\{h[n]\} = \frac{h[n] - h[-n]}{2} \overset{FT}{\longleftrightarrow} j\mathcal{I}m\{H(e^{j\omega})\}.$$

Clearly, we can recover $\mathcal{O}d\{h[n]\}$ from $\mathcal{I}m\{H(e^{j\omega})\}$. From $\mathcal{O}d\{h[n]\}$ we can use eq.(S5.52-2) to recover $h[n]$ (provided $h[0]$ is given). Obviously, from $h[n]$ we can once again obtain $H(e^{j\omega})$. Therefore, the system is completely specified by $\mathcal{I}m\{H(e^{j\omega})\}$ and $h[0]$.

(d) Let $\mathcal{I}m\{H(e^{j\omega})\} = \sin\omega$. Then,

$$\mathcal{O}d\{x[n]\} = \frac{1}{2}\delta[n-1] - \frac{1}{2}\delta[n+1].$$

Therefore,

$$h[n] = h[0]\delta[n] + \delta[n-1].$$

We may choose two different values for $h[0]$ (say 1 and 2) to obtain two different systems whose frequncy responses have imaginary parts equal to $\sin\omega$.

5.53. **(a)** The analysis equation of the Fourier transform is

$$X(e^{j\omega}) = \sum_{n=-\infty}^{\infty} x[n]e^{-j\omega n}.$$

Comparing with eq. (P5.53-2), we have

$$\tilde{X}[k] = \frac{1}{N}X(e^{j(2\pi k/N)}).$$

(b) From the figures we obtain

$$X_1(e^{j\omega}) = 1 - e^{-j\omega} + 2e^{-3j\omega}$$

and

$$X_2(e^{j\omega}) = -e^{2j\omega} - e^{j\omega} - 1 + e^{-2j\omega} + e^{-3j\omega} + 2e^{-j4\omega} - e^{-j5\omega} + 2e^{-j7\omega}.$$

Now,

$$X_1(e^{j(2\pi k/4)}) = 1 - e^{-j\pi k/2} + 2e^{-3j\pi k/2}$$

and

$$X_2(e^{j(2\pi k/4)}) = 1 - e^{-j\pi k/2} + 2e^{-3j\pi k/2} = X_1(e^{j(2\pi k/4)}).$$

5.54. **(a)** From eq. (P5.54-1) it is clear that to compute $\tilde{X}[k]$ for one particular value of k, we need to perform N complex multiplications. Therefore, in order to compute $\tilde{X}[k]$ for N different values of k, we need to perform $N.N = N^2$ complex multiplications.

(b) (i) Since $f[n] = x[2n]$, we have $f[0] = x[0]$, $f[1] = x[2]$, $\cdots$, $f[(N/2)-1] = x[N-2]$. Since $x[n]$ is nonzero only in the range $0 \le n \le N-1$, $f[n]$ is nonzero only in the range $0 \le n \le (N/2)-1$.

Similarly, since $g[n] = x[2n+1]$, we have $g[0] = x[1]$, $g[1] = x[3]$, $\cdots$, $g[(N/2)-1] = x[N]$. Since $x[n]$ is nonzero only in the range $0 \le n \le N-1$, $g[n]$ is nonzero only in the range $0 \le n \le (N/2)-1$.

(ii) We may rewrite eq. (5.54-1) as

$$\tilde{X}[k] = \frac{1}{N}\sum_{n=0}^{(N/2)-1} x[2n]W_N^{2nk} + W_N^k \frac{1}{N}\sum_{n=0}^{(N/2)-1} x[2n+1]W_N^{2nk}.$$

Since $W_N^{2nk} = W_{N/2}^{nk}$, we may rewrite the above equation as

$$\begin{aligned}\tilde{X}[k] &= \tfrac{1}{N}\sum_{n=0}^{(N/2)-1} f[n]W_{N/2}^{nk} + W_N^k \tfrac{1}{N}\sum_{n=0}^{(N/2)-1} g[n]W_{N/2}^{nk} \\ &= \tfrac{1}{2}\tilde{F}[k] + \tfrac{1}{2}W_N^k\tilde{G}[k] \end{aligned} \qquad \text{(S5.54-1)}$$

(iii) We have

$$\tilde{F}[k+N/2] = \frac{2}{N}\sum_{n=0}^{(N/2)-1} f[n]W_{N/2}^{kn}W_{N/2}^{nN/2} = \tilde{F}[k].$$

Similarly,

$$\tilde{G}[k+N/2] = \tilde{G}[k].$$

(iv) Since $\tilde{F}[k]$ is a $N/2$ point DFT, we may use an approach similar to the one in part (a) to show that we need $N^2/4$ complex multplications to compute it. Similarly we may show that the computation of $\tilde{F}[k]$ requires $N^2/4$ multiplications. From eq. (S5.54-1), it is clear that we need $N^2/2 + N$ complex multiplcations to compute $\tilde{X}[k]$.

(c) By decomposing $g[n]$ and $f[n]$ into their odd and even indexed samples, we can bring down the number of computations to $N^2/4 + N/2$. Repeating this decomposition $\log_2 N$ times, we make the required computation $N \log_2 N$. We tabulate below the computations required by the direct method and the FFT method for values of N.

N	Direct method	FFT method
32	1024	160
256	65536	2048
1024	1048576	10240
4096	16777216	49152

5.55. **(a)** (i) From Table 5.2, we have

$$X(e^{j\omega}) = 2\pi \sum_{k=-\infty}^{\infty} \delta(\omega - 2\pi k).$$

(ii) When $M = 1$, $P(e^{j\omega}) = e^{j\omega} + 1 + e^{-j\omega} = 1 + 2\cos\omega$.

(iii) When $M = 10$, we may use Table 5.2 to find that

$$P(e^{j\omega}) = \frac{\sin(21\omega/2)}{\omega/2}.$$

(b) The plots are as shown in Figure S5.55.

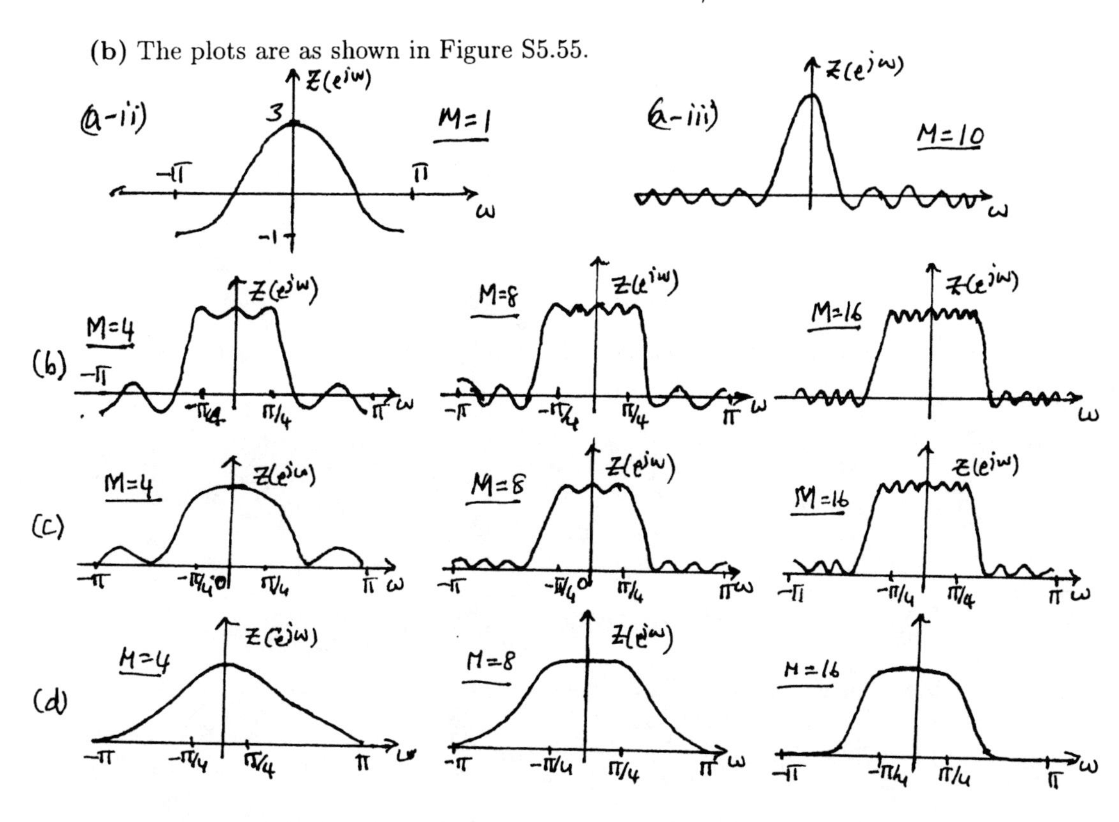

Figure S5.55

(c) We have $W(e^{j\omega}) = \frac{\sin^2[(M+1)\omega/2]}{\sin^2(\omega/2)}$. The plots are as shown in Figure S5.55.

(d) The plots are as shown Figure S5.55.

5.56. **(a)** We have

$$\begin{aligned} X(e^{j\omega_1}, e^{j\omega_2}) &= \sum_{n=-\infty}^{\infty}\sum_{m=-\infty}^{\infty} x[m,n]e^{-j(\omega_1 m+\omega_2 n)} \\ &= \sum_{n=-\infty}^{\infty}\left[\sum_{m=-\infty}^{\infty} x[m,n]e^{-j\omega_1 m}\right] e^{-j\omega_2 n} \\ &= \sum_{n=-\infty}^{\infty} X(e^{j\omega_1}, n)e^{-j\omega_2 n} \end{aligned}$$

Therefore, we may write

$$X(e^{j\omega_1}, n) = \frac{1}{2\pi}\int_{-\pi}^{\pi} X(e^{j\omega_1}, e^{j\omega_2})e^{j\omega_2 n}d\omega_2.$$

From this we obtain

$$x[m,n] = \frac{1}{4\pi^2}\int_{-\pi}^{\pi}\int_{-\pi}^{\pi} X(e^{j\omega_1}, e^{j\omega_2})e^{j\omega_1 m}e^{j\omega_2 n}d\omega_1 d\omega_2.$$

(b) We may easily show that

$$X(e^{j\omega_1}, e^{j\omega_2}) = A(e^{j\omega})B(e^{j\omega}).$$

(c) We use the result of the previous part in many of the problems of this part.

(i) $X(e^{j\omega_1}, e^{j\omega_2}) = e^{-j\omega_1}e^{4j\omega_2}$.

(ii) $X(e^{j\omega_1}, e^{j\omega_2}) = \left[\frac{e^{-j2\omega_2}}{(1-\frac{1}{2}e^{-j\omega_2})}\right]\left[\frac{1}{(1-\frac{1}{2}e^{-j\omega_1})}\right]$

(iii) $X(e^{j\omega_1}, e^{j\omega_2}) = \left[\frac{1}{(1-\frac{1}{2}e^{-j\omega_2})}\right]\left[\pi\sum_{k=-\infty}^{\infty}\delta(\omega_1 - \frac{2\pi}{3} - 2\pi k) + \pi\sum_{k=-\infty}^{\infty}\delta(\omega_1 + \frac{2\pi}{3} - 2\pi k)\right]$.

(iv) Here $x[n,m] = \{u[m+1] - u[m-2]\}\{u[n+4] - u[n-5]\}$. Therefore,

$$X(e^{j\omega_1}, e^{j\omega_2}) = \left[\frac{\sin(7\omega_2/2)}{\sin(\omega_2/2)}\right]\left[\frac{\sin(3\omega_1/2)}{\sin(\omega_1/2)}\right].$$

(v) From the definition of the 2D Fourier transform we obtain

$$X(e^{j\omega_1}, e^{j\omega_2}) = \frac{e^{j(\omega_1+3\omega_2)}}{1-e^{-j\omega_1}}\left[\frac{1-e^{-j7(\omega_1+\omega_2)}}{1-e^{-j(\omega_1+\omega_2)}} - e^{-j\omega_1}\left(\frac{1-e^{-j7(3\omega_1+\omega_2)}}{1-e^{-j(3\omega_1+\omega_2)}}\right)\right].$$

(vi) From the definition of the 2D Fourier transform we obtain

$$\begin{aligned} X(e^{j\omega_1}, e^{j\omega_2}) &= \frac{\pi}{j}\sum_{l=-\infty}^{\infty}\sum_{r=-\infty}^{\infty}\left[\delta(\omega_1 - \tfrac{2\pi}{5} + 2\pi l)\delta(\omega_2 - \tfrac{\pi}{3} + 2\pi r) - \right. \\ &\quad \left.\delta(\omega_1 + \tfrac{2\pi}{5} + 2\pi l)\delta(\omega_2 + \tfrac{\pi}{3} + 2\pi r)\right]. \end{aligned}$$

(d) (i) $X(e^{j(\omega_1-W_1)}, e^{j(\omega_2-W_2)})$

(ii) $X(e^{2\omega_1}, e^{3\omega_2})$

(iii) $\frac{1}{4\pi^2}\int_{-\pi}^{\pi}\int_{-\pi}^{\pi} X(e^{j\zeta}, e^{j\theta})H(e^{j(\omega_1-\zeta)}, e^{j(\omega_1-\theta)})d\zeta d\theta$

Chapter 6 Answers

6.1. The signal $x(t)$ may be broken up into a sum of the two complex exponentials $x_1(t) = (1/2)e^{j\omega_0 t+\phi_0}$ and $x_2(t) = (1/2)e^{-j\omega_0 t-\phi_0}$. Since complex exponentials are Eigen functions of LTI systems, we know that when $x_1(t)$ passes through the LTI system, the output is

$$\begin{aligned} y_1(t) &= x_1(t)H(j\omega_0) = x_1(t)|H(j\omega_0)|e^{j\sphericalangle H(j\omega_0)} \\ &= (1/2)|H(j\omega_0)|e^{j(\omega_0 t+\phi_0+\sphericalangle H(j\omega_0))} \end{aligned}$$

Similarly, when the input is $x_2(t)$, the output is

$$y_2(t) = (1/2)|H(-j\omega_0)|e^{-j(\omega_0 t+\phi_0-\sphericalangle H(-j\omega_0))}.$$

But since $h[n]$ is given to be real, $|H(j\omega_0)| = |H(-j\omega_0)|$ and $\sphericalangle H(j\omega_0) = -\sphericalangle H(-j\omega_0)$. Therefore,

$$y_2(t) = (1/2)|H(j\omega_0)|e^{-j(\omega_0 t+\phi_0+\sphericalangle H(j\omega_0))}.$$

Using linearity we may argue that when the input to the LTI system is $x(t) = x_1(t)+x_2(t)$, the output will be $y(t) = y_1(t) = y_2(t)$. Therefore,

$$y(t) = |H(j\omega_0)|\cos(\omega_0 t + \phi_0 + \sphericalangle H(j\omega_0)) = |H(j\omega_0)|\cos\left(\omega_0(t - \frac{-\sphericalangle H(j\omega_0)}{\omega_0}) + \phi_0\right)$$

(a) From $y(t)$, we have $A = |H(j\omega_0)|$.

(b) From $y(t)$, we have $t_0 = \frac{-\sphericalangle H(j\omega_0)}{\omega_0}$.

6.2. The signal $x[n]$ may be broken up into a sum of the two complex exponentials $x_1[n] = (1/2j)e^{j\omega_0 n+\phi_0}$ and $x_2[n] = (-1/2j)e^{-j\omega_0 n-\phi_0}$. Since complex exponentials are Eigen functions of LTI systems, we know that when $x_1[n]$ passes through the LTI system, the output is

$$\begin{aligned} y_1[n] &= x_1[n]H(e^{j\omega_0}) = x_1[n]|H(e^{j\omega_0})|e^{j\sphericalangle H(e^{j\omega_0})} \\ &= (1/2j)|H(e^{j\omega_0})|e^{j(\omega_0 n+\phi_0+\sphericalangle H(e^{j\omega_0}))} \end{aligned}$$

Similarly, when the input is $x_2[n]$, the output is

$$y_2[n] = (-1/2j)|H(e^{-j\omega_0})|e^{-j(\omega_0 n+\phi_0-\sphericalangle H(e^{-j\omega_0}))}.$$

But since $h(t)$ is given to be real, $|H(e^{j\omega_0})| = |H(e^{-j\omega_0})|$ and $\sphericalangle H(e^{j\omega_0}) = -\sphericalangle H(e^{-j\omega_0})$. Therefore,

$$y_2[n] = (-1/2j)|H(e^{j\omega_0})|e^{-j(\omega_0 t+\phi_0+\sphericalangle H(e^{j\omega_0}))}.$$

Using linearity we may argue that when the input to the LTI system is $x[n] = x_1[n]+x_2[n]$, the output will be $y[n] = y_1[n] + y_2[n]$. Therefore,

$$y[n] = |H(e^{j\omega_0})|\sin(\omega_0 n + \phi_0 + \sphericalangle H(e^{j\omega_0})) = |H(e^{j\omega_0})|\sin\left(\omega_0(n - \frac{-\sphericalangle H(e^{j\omega_0})}{\omega_0}) + \phi_0\right)$$

Now note that if we require that $y[n] = |H(e^{j\omega_0})|x[n-n_0]$, then $n_0 = -\sphericalangle H(e^{j\omega_0})/\omega_0$ has to be an integer. Therefore, $\sphericalangle H(e^{j\omega_0}) = -n_0\omega_0$. Now also, note that if we add an integer multiple of 2π to this $\sphericalangle H(e^{j\omega_0})$, it does not make any difference. Therefore, we require in general that $\sphericalangle H(e^{j\omega_0}) = -n_0(\omega_0 + 2k\pi)$.

6.3. (a) We have

$$|H(j\omega)| = \frac{|1-j\omega|}{|1+j\omega|} = \frac{\sqrt{1+\omega^2}}{\sqrt{1+\omega^2}} = 1.$$

Therefore, $A = 1$.

(b) We have

$$\sphericalangle H(j\omega) = \tan^{-1}(-\omega) - \tan^{-1}(\omega) = 2\tan^{-1}(\omega).$$

Therefore, the group delay is

$$\tau(\omega) = -\frac{d}{d\omega}\sphericalangle H(j\omega) = \frac{2}{1+\omega^2}.$$

Clearly, $\tau(\omega) > 0$ for $\omega > 0$. Therefore, statement 2 is true.

6.4. (a) The signal $\cos(\pi n/2)$ can be broken up into a sum of two complex exponentials $x_1[n] = (1/2)e^{j\pi n/2}$ and $x_2[n] = (1/2)e^{-j\pi n/2}$. From the given information, we know that when $x_1[n]$ passes through the given LTI system, it experiences a delay of 2 samples. Since the system has a real impulse response, it has an even group delay function. Therefore, the complex exponential $x_2[n]$ with frequency $-\omega_0$ also experiences a group delay of 2 samples. The output $y[n]$ of the LTI system when the input is $x[n] = x_1[n] + x_2[n]$ is therefore

$$y[n] = 2x_1[n-2] + 2x_2[n-2] = 2\cos\left(\frac{\pi}{2}(n-2)\right) = 2\cos\left(\frac{\pi}{2}n - \pi\right)$$

(b) The signal $x[n] = \sin(\frac{7\pi}{2}n + \frac{\pi}{4})$ is the same as $-\sin(\frac{\pi}{2}n - \frac{\pi}{4})$. This signal may once again be broken up into complex exponentials of frequency $\pi/2$ and $-\pi/2$. We may then use an argument similar to the one used in part (a) to argue that the output $y[n]$ is

$$\begin{aligned}
y[n] &= 2x[n-2] = 2\sin\left(\frac{7\pi}{2}(n-2) + \frac{\pi}{4}\right) \\
&= 2\sin\left(\frac{7\pi}{2}n - 7\pi + \frac{\pi}{4}\right) \\
&= 2\sin\left(\frac{7\pi}{2}n - \pi + \frac{\pi}{4}\right) \\
&= 2\sin\left(\frac{7\pi}{2}n - \frac{3\pi}{4}\right)
\end{aligned}$$

6.5. The frequency response $H(j\omega)$ is as shown in Figure S6.5.

(a) Consider the signal $h_1(t) = \sin(\omega_c t)/(\pi t)$. Its Fourier transform $H_1(j\omega)$ is as shown in Figure S6.5.

Clearly,

$$H(j\omega) = H_1(j(\omega - 2\omega_c)) + H_1(j(\omega + 2\omega_c)).$$

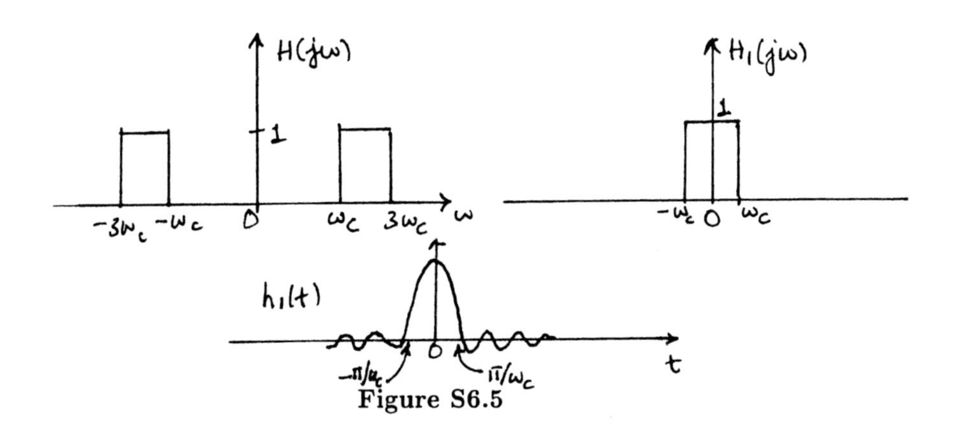

Figure S6.5

Taking the inverse Fourier transform, we have

$$\begin{aligned} h(t) &= h_1(t)e^{j2\omega_c t} + h_1(t)e^{-j2\omega_c t} \\ &= 2h_1(t)\cos(2\omega_c t) \end{aligned}$$

Therefore, $g(t) = \cos(2\omega_c t)$.

(b) The impulse response $h_1(t)$ is as shown in Figure S6.5. As ω_c increases, it is clear that the significant central lobe of $h_1(t)$ becomes more concentrated around the origin. Consequently $h(t) = 2h_1(t)\cos(2\omega_c t)$ also becomes more concentrated about the origin.

6.6. The frequency response $H(e^{j\omega})$ is as shown in Figure S6.6.

(a) Consider the signal $h_1[n] = \sin(\omega_c n)/(\pi n)$. Its Fourier transform $H_1(e^{j\omega})$ is as shown in the figure below.

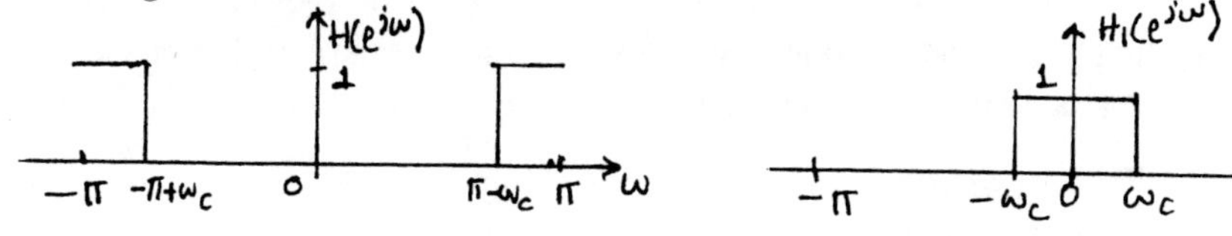

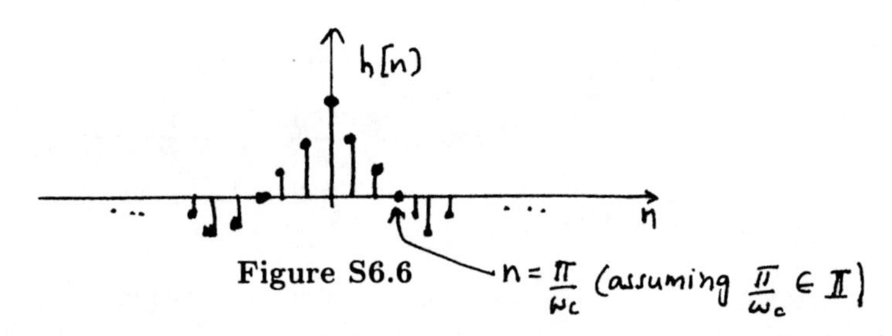

Figure S6.6

Clearly,

$$H(e^{j\omega}) = H_1(e^{j(\omega-\pi)}).$$

Taking the inverse Fourier transform, we have

$$h[n] = h_1[n]e^{j\pi n} = h_1[n](-1)^n.$$

Therefore, $g[n] = (-1)^n$.

(b) The impulse response $h_1[n]$ is as shown in Figure S6.6. As ω_c increases, it is clear that the significant central lobe of $h_1[n]$ becomes more concentrated around the origin. Consequently $h[n] = h_1[n](-1)^n$ also becomes more concentrated about the origin.

6.7. The frequency response magnitude $|H(j\omega)|$ is as shown in Figure S6.7. The frequency response of the bandpass filter $G(j\omega)$ will be given by

$$
\begin{aligned}
G(j\omega) &= \mathcal{FT}\{2h(t)\cos(4000\pi t)\} \\
&= H(j(\omega - 4000\pi)) + H(j(\omega + 4000\pi))
\end{aligned}
$$

This is as shown in Figure S6.7

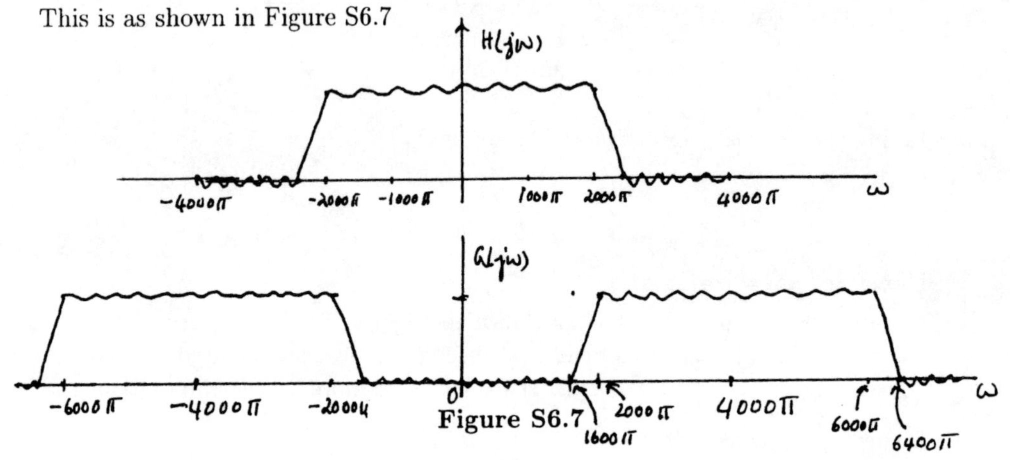

Figure S6.7

(a) From the figure, it is obvious that the passband edges are at 2000π rad/sec and 6000π rad/sec. This translates to 1000 Hz and 3000 Hz, respectively.

(b) From the figure, it is obvious that the stopband edges are at 1600π rad/sec and 6400π rad/sec. This translates to 800 Hz and 3200 Hz, respectively.

6.8. Taking the Fourier transform of both sides of the first difference equation and simplifying, we obtain the frequency response $H(e^{j\omega})$ of the first filter.

$$
H(e^{j\omega}) = \frac{Y(e^{j\omega})}{X(e^{j\omega})} = \frac{\sum_{k=0}^{M} b_k e^{-j\omega k}}{1 - \sum_{k=1}^{N} a_k e^{-j\omega k}}.
$$

Taking the Fourier transform of both sides of the second difference equation and simplifying, we obtain the frequency response $H_1(e^{j\omega})$ of the second filter.

$$
H_1(e^{j\omega}) = \frac{Y(e^{j\omega})}{X(e^{j\omega})} = \frac{\sum_{k=0}^{M} (-1)^k b_k e^{-j\omega k}}{1 - \sum_{k=1}^{N} (-1)^k a_k e^{-j\omega k}}.
$$

This may also be written as

$$H_1(e^{j\omega}) = \frac{\sum_{k=0}^{M} b_k e^{-j(\omega-\pi)k}}{1 - \sum_{k=1}^{N} a_k e^{-j(\omega-\pi)k}} = H(e^{j(\omega-\pi)}).$$

Therefore, the frequency response of the second filter is obtained by shifting the frequency response of the first filter by π. Although the location of the passband changes, the tolerances will be the same in the second filter. The first filter has its passband between $-\omega_p$ and ω_p. Therefore, the second filter will have its passband between $\pi - \omega_p$ and $\pi + \omega_p$.

6.9. Taking the Fourier transform of the given differential equation and simplifying, we obtain the frequency response of the LTI system to be

$$H(j\omega) = \frac{Y(j\omega)}{X(j\omega)} = \frac{2}{5 + j\omega}.$$

Taking the inverse Fourier transform, we obtain the impulse response to be

$$h(t) = 2e^{-5t}u(t).$$

Using the result derived in Section 6.5.1, we have the step response of the system

$$s(t) = h(t) * u(t) = \frac{2}{5}[1 - e^{-5t}]u(t).$$

The final value of the step response is

$$s(\infty) = \frac{2}{5}.$$

We also have

$$s(t_0) = \frac{2}{5}[1 - e^{-5t_0}].$$

Substituting $s(t_0) = (2/5)[1 - 1/e^2]$, in the above equation, we obtain $t_0 = \frac{2}{5}$ sec.

6.10. We use Example 6.5 to guide us through this problem.

(a) We may rewrite $H_1(j\omega)$ to be

$$H(j\omega) = \left(\frac{1}{j\omega + 40}\right)(j\omega + 0.1).$$

We may then treat each of the two factors as individual first order systems and draw their Bode magnitude plots. The final Bode magnitude plot will then be a sum of these two Bode plots. This is shown in the Figure S6.10.

Mathematically, the straight-line approximation of the Bode magnitude plot is

$$20\log_{10}|H(j\omega)| \approx \begin{cases} -20, & \omega << 0.1 \\ 20\log_{10}(\omega), & 0.1 << \omega << 40 \\ 32, & \omega >> 40 \end{cases}.$$

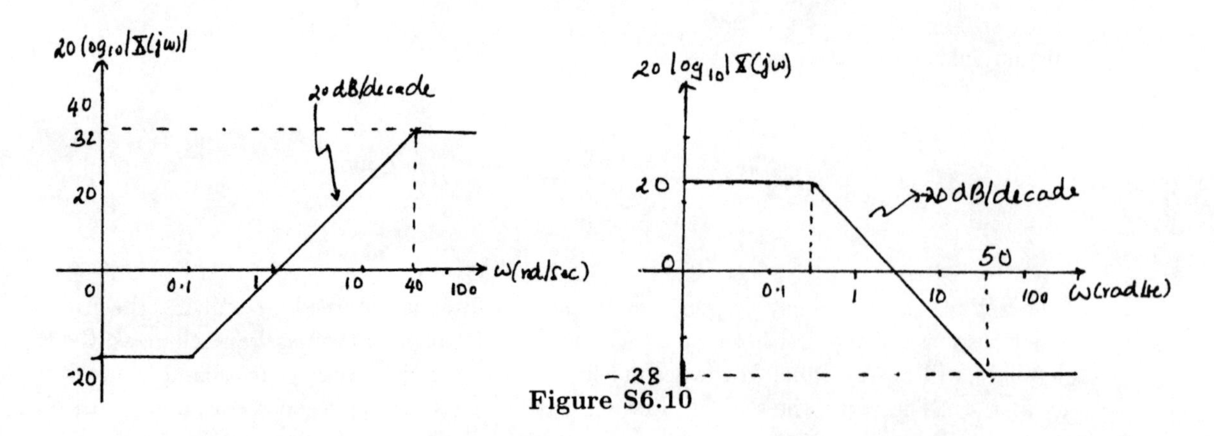

Figure S6.10

(b) Using a similar approach as in part (a), we obtain the Bode plot to be as shown in Figure S6.10.

Mathematically, the straight-line approximation of the Bode magnitude plot is

$$20\log_{10}|H(j\omega)| \approx \begin{cases} 20, & \omega << 0.2 \\ -20\log_{10}(\omega)+6, & 0.2 << \omega << 50 \\ -28, & \omega >> 50 \end{cases}.$$

6.11. **(a)** We may rewrite the given frequency response $H_1(j\omega)$ as

$$H_1(j\omega) = \frac{250}{(j\omega)^2 + 50.5j\omega + 25} = \frac{250}{(j\omega+0.5)(j\omega+50)}.$$

We may then use an approach similar to the one used in Example 6.5 and in Problem 6.10 to obtain the Bode magnitude plot (with straight line approximations) shown in Figure S6.11.

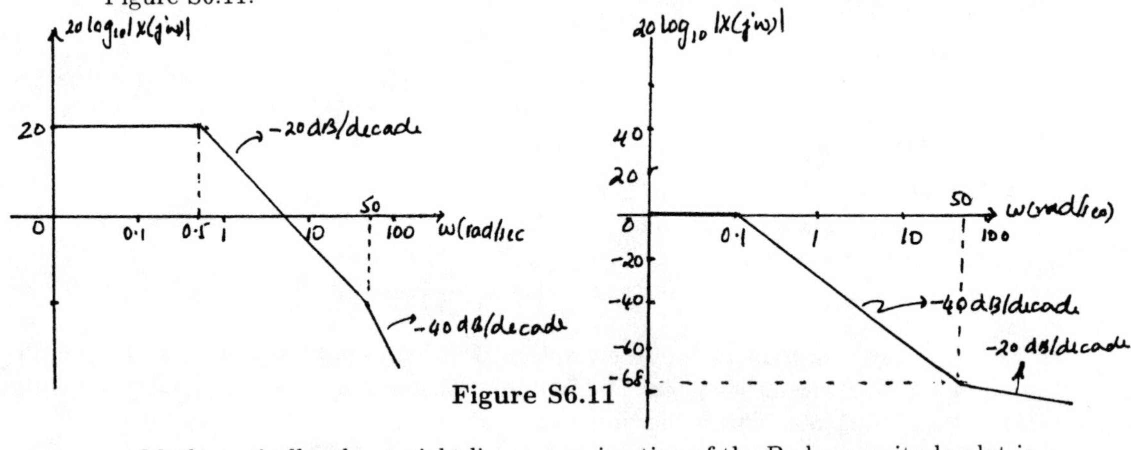

Figure S6.11

Mathematically, the straight-line approximation of the Bode magnitude plot is

$$20\log_{10}|H(j\omega)| \approx \begin{cases} 20, & \omega << 0.5 \\ -20\log_{10}(\omega)+14, & 0.5 << \omega << 50 \\ -40\log_{10}(\omega)+48, & \omega >> 50 \end{cases}.$$

(b) We may rewrite the frequency response $H_2(j\omega)$ as

$$H_2(j\omega) = (j\omega + 50)\left(\frac{0.02}{(j\omega)^2 + 0.2j\omega + 1}\right).$$

Again using an approach similar to the one used in Example 6.5, we may draw the Bode magnitude plot by treating the first and second order factors separately. This gives us a Bode magnitude plot (using straight line) approximations as shown below:

Mathematically, the straight-line approximation of the Bode magnitude plot is

$$20\log_{10}|H(j\omega)| \approx \begin{cases} 0, & \omega << 1 \\ -40\log_{10}\omega, & 1 << \omega << 50 \\ -20\log_{10}\omega - 34, & \omega >> 50 \end{cases}.$$

6.12. Using the Bode magnitude plot specified in Figure P6.12(a), we may obtain an expression for $H_1(j\omega)$. The figure shows that $H_1(j\omega)$ has the break frequencies $\omega_1 = 1$, $\omega_2 = 8$, and $\omega_3 = 40$. The frequency response rises at 20 dB/decade after ω_1. At ω_2, this rise is canceled by a -20 dB/decade contribution. Finally, at ω_3, an additional -20 dB/decade contribution results in the subsequent decay at the rate of -20 dB/decade. Therefore, we may conclude that

$$H_1(j\omega) = \frac{A(j\omega + \omega_1)}{(j\omega + \omega_2)(j\omega + \omega_3)}. \qquad \text{(S6.12–1)}$$

We now need to find A. Note that when $\omega = 0$, $20\log_{10}|H_1(j0)| = 2$. Therefore, $H_1(j0) = 0.05$. From eq. (S6.12-1), we know that

$$H_1(j0) = A/320.$$

Therefore, $A = 640$. This gives us

$$H_1(j\omega) = \frac{640(j\omega + 1)}{(j\omega + 8)(j\omega + 40)}.$$

Using a similar approach on Figure P6.12(b), we obtain

$$H(j\omega) = \frac{6.4}{(j\omega + 8)^2}.$$

Since the overall system (with frequency response $H(j\omega)$) is constructed by cascading systems with frequency responses $H_1(j\omega)$ and $H_2(j\omega)$,

$$H(j\omega) = H_1(j\omega)H_2(j\omega).$$

Using the previously obtained expressions for $H(j\omega)$ and $H_1(j\omega)$,

$$H_2(j\omega) = \frac{H(j\omega)}{H_1(j\omega)} = \frac{0.01(j\omega + 40)}{(j\omega + 1)(j\omega + 8)}.$$

6.13. Using an approach similar to the one used in the previous problem, we obtain

$$H(j\omega) = \frac{320}{(j\omega + 2)(j\omega + 80)}.$$

(a) Let us assume that we desire to construct this system by cascading two systems with frequency responses $H_1(j\omega)$ and $H_2(j\omega)$, respectively. We require that

$$H(j\omega) = H_1(j\omega)H_2(j\omega).$$

We see that $H_1(j\omega)$ and $H_2(j\omega)$ may be defined in different ways to obtain $H(j\omega)$. For instance

$$H_1(j\omega) = \frac{40}{(j\omega + 2)} \quad \text{and} \quad H_2(j\omega) = \frac{8}{(j\omega + 80)}$$

and

$$H_1(j\omega) = \frac{32}{(j\omega + 2)} \quad \text{and} \quad H_2(j\omega) = \frac{10}{(j\omega + 80)}$$

are both valid combinations.

(b) Let us assume that we desire to construct this system by connecting two systems with frequency responses $H_1(j\omega)$ and $H_2(j\omega)$ in parallel. We require that

$$H(j\omega) = H_1(j\omega) + H_2(j\omega).$$

Using partial fraction expansion on $H(j\omega)$, we obtain

$$H(j\omega) = \frac{160/39}{(j\omega + 2)} - \frac{160/39}{(j\omega + 80)}$$

From the above expression it is clear that we can define $H_1(j\omega)$ and $H_2(j\omega)$ in only one way.

6.14. Using an approach similar to the one used in Problem 6.12, we have

$$H(j\omega) = \frac{50000(j\omega + 0.2)^2}{(j\omega + 50)(j\omega + 10)}.$$

The inverse to this system has a frequency response

$$H_I(j\omega) = \frac{1}{H(j\omega)} = \frac{0.2 \times 10^{-4}(j\omega + 50)(j\omega + 10)}{(j\omega + 0.2)^2}.$$

6.15. We will use the results from Section 6.5 in this problem.

(a) We may write the frequency response of the system described by the given differential equation as

$$H_1(j\omega) = \frac{1}{(j\omega)^2 + 4j\omega + 4}.$$

This may be rewritten as

$$H_1(j\omega) = \frac{1/4}{(j\omega/2)^2 + 2j(\omega/2) + 1}.$$

From this we obtain the damping ratio to be $\zeta = 1$. Therefore, the system is **critically damped**.

(b) We may write the frequency response of the system described by the given differential equation as

$$H_2(j\omega) = \frac{7}{5(j\omega)^2 + 4j\omega + 5}.$$

This may be rewritten as

$$H_2(j\omega) = \frac{7/5}{(j\omega)^2 + 2(2/5)j(\omega) + 1}.$$

From this we obtain the damping ratio to be $\zeta = 2/5$. Therefore, the system is **underdamped**.

(c) We may write the frequency response of the system described by the given differential equation as

$$H_3(j\omega) = \frac{1}{(j\omega)^2 + 20j\omega + 1}.$$

This may be rewritten as

$$H_3(j\omega) = \frac{1}{(j\omega)^2 + 2(10)j(\omega) + 1}.$$

From this we obtain the damping ratio to be $\zeta = 10$. Therefore, the system is **overdamped**.

(d) We may write the frequency response of the system described by the given differential equation as

$$H_3(j\omega) = \frac{7 + (1/3)j\omega}{5(j\omega)^2 + 4j\omega + 5}.$$

The terms in the numerator do not affect the ringing behavior of the impulse response of this system. Therefore, we need to only consider the denominator in order to determine if the system is critically damped, underdamped, or overdamped. We see that this frequency response has the same denominator as the one obtained in part (b). Therefore, this system is still **underdamped**.

6.16. The system of interest will have a difference equation of the form

$$y[n] - ay[n-1] = bx[n].$$

Making slight modifications to the results obtained in Section 6.6.1, we determine the step response of this system to be

$$b\left(\frac{1 - a^{n+1}}{1 - a}\right)u[n].$$

The final value of the step response will be $b/(1-a)$. The step response exhibits oscillatory behavior only if $|a| < 1$. Using this fact, we may easily show that the maximum overshoot in the step response occurs when $n = 0$. Therefore, the maximum value of the step response is

$$\frac{b}{1-a}(1-a) = b.$$

Since we are given that the maximum, overshoot is 1.5 times the final value, we have

$$1.5\frac{b}{1-a} = b \quad \Rightarrow \quad a = -\frac{1}{2}.$$

Also, since we are given that the final value us 1,

$$\frac{b}{1-a} = 1 \quad \Rightarrow \quad b = \frac{3}{2}.$$

Therefore, the difference equation relating the input and output will be

$$y[n] + \frac{1}{2}y[n-1] = \frac{3}{2}x[n].$$

6.17. We will use the results derived in Section 6.6.2 to solve this problem.

(a) Comparing the given difference equation with eq. (6.56), we obtain

$$r = \frac{1}{2}, \quad \text{and} \quad \cos\theta = -1.$$

Therefore, $\theta = \pi$, and the system has an **oscillatory** step response.

(b) Comparing the given difference equation with eq. (6.56), we obtain

$$r = \frac{1}{2}, \quad \text{and} \quad \cos\theta = 1.$$

Therefore, $\theta = 0$, and the system has a **non-oscillatory** step response.

6.18. Let us first find the differential equation governing the input and output of this circuit.
Current through resistor = Current through capacitor = $C\frac{dy(t)}{dt}$.
Voltage across resistor = $RC\frac{dy(t)}{dt}$.
Total input voltage = Voltage across resistor + Voltage across capacitor
Therefore,

$$x(t) = RC\frac{dy(t)}{dt} + y(t).$$

The frequency response of this circuit is therefore

$$H(j\omega) = \frac{1}{RCj\omega + 1}.$$

Since this is a first order system, the step response has to be non oscillatory.

6.19. Let us first find the differential equation governing the input and output of this circuit.
Current through resistor and inductor = Current through capacitor = $C\frac{dy(t)}{dt}$.
Voltage across resistor = $RC\frac{dy(t)}{dt}$.
Voltage across inductor = $LC\frac{d^2y(t)}{dt^2}$.
Total input voltage = Voltage across inductor + Voltage across resistor + Voltage across capacitor
Therefore,

$$x(t) = LC\frac{d^2y(t)}{dt^2} + RC\frac{dy(t)}{dt} + y(t).$$

The frequency response of this circuit is therefore

$$H(j\omega) = \frac{1}{LC(j\omega)^2 + RCj\omega + 1}.$$

We may rewrite this to be

$$H(j\omega) = \frac{1}{(\frac{j\omega}{1/\sqrt{LC}})^2 + 2(R/2)\sqrt{C/L}\frac{j\omega}{1/\sqrt{LC}} + 1}.$$

Therefore, the damping constant $\zeta = (R/2)\sqrt{C/L}$. In order for the step response to have no oscillations, we must have $\zeta \geq 1$. Therefore, we require

$$R \geq 2\sqrt{\frac{L}{C}}.$$

6.20. Let us call the given impulse response $h[n]$. It is easily observed that the signal $h_1[n] = h[n+2]$ is real and even. Therefore, (using properties of the Fourier transform) we know that the Fourier transform $H_1(e^{j\omega})$ of $h_1[n]$ is real and even. Therefore $H_1(e^{j\omega})$ has zero phase. We also know that the Fourier transform $H(e^{j\omega}) = H_1(e^{j\omega})e^{-2j\omega}$. Since $H_1(e^{j\omega})$ is zero phase, we have

$$\sphericalangle H(e^{j\omega}) = -2\omega.$$

Therefore, the group delay is

$$\tau(\omega) = \frac{d}{d\omega}\sphericalangle H(e^{j\omega}) = 2.$$

6.21. Note that in all parts of this problem $Y(j\omega) = H(j\omega)X(j\omega) = -2j\omega X(j\omega)$. Therefore, $y(t) = -2dx(t)/dt$.

(a) Here, $x(t) = e^{jt}$. Therefore, $y(t) = -2dx(t)/dt = -2je^{jt}$. This part could also have been solved by noting that complex exponentials are Eigen functions of LTI systems. Then, when $x(t) = e^{jt}$, $y(t)$ should be $y(t) = H(j1)e^{jt} = -2je^{jt}$.

(b) Here, $x(t) = \sin(\omega_0 t)u(t)$. Then, $dx(t)/dt = \omega_0\cos(\omega_0 t)u(t) + \sin(\omega_0 t)\delta(t) = \omega_0\cos(\omega_0 t)u(t)$. Therefore, $y(t) = -2dx(t)/dt = -2\omega_0\cos(\omega_0 t)u(t)$.

(c) Here, $Y(j\omega) = X(j\omega)H(j\omega) = -2/(6+j\omega)$. Taking the inverse Fourier transform we obtain $y(t) = -2e^{-6t}u(t)$.

(d) Here, $X(j\omega) = 1/(2 + j\omega)$. From this we obtain $x(t) = e^{-2t}u(t)$. Therefore, $y(t) = -2dx(t)/dt = 4e^{-2t}u(t) - 2\delta(t)$.

6.22. Note that

$$H(j\omega) = \begin{cases} \frac{j\omega}{3\pi}, & -3\pi \leq \omega \leq 3\pi \\ 0, & \text{otherwise} \end{cases}.$$

(a) Since $x(t) = \cos(2\pi t + \theta)$, $X(j\omega) = e^{j\theta}\pi\delta(\omega - 2\pi) + e^{-j\theta}\pi\delta(\omega + 2\pi)$. This is zero outside the region $-3\pi < \omega < 3\pi$. Thus, $Y(j\omega) = H(j\omega)X(j\omega) = (j\omega/3\pi)X(j\omega)$. This implies that $y(t) = (1/3\pi)dx(t)/dt = (-2/3)\sin(2\pi t + \theta)$.

(b) Since $x(t) = \cos(4\pi t + \theta)$, $X(j\omega) = e^{j\theta}\pi\delta(\omega - 4\pi) + e^{-j\theta}\pi\delta(\omega + 4\pi)$. Therefore, the nonzero portions of $X(j\omega)$ lie outside the range $-3\pi < \omega < 3\pi$. This implies that $Y(j\omega) = X(j\omega)H(j\omega) = 0$. Therefore, $y(t) = 0$.

(c) The Fourier series coefficients of the signal $x(t)$ are given by

$$a_k = \frac{1}{T_0}\int_{<T_0>} x(t)e^{-jk\omega_0 t},$$

where $T_0 = 1$ and $\omega_0 = 2\pi/T_0 = 2\pi$. Also,

$$X(j\omega) = 2\pi \sum_{k=-\infty}^{\infty} a_k\delta(\omega - k\omega_0).$$

The only impulses of $X(j\omega)$ which lie in the region $-3\pi < \omega < 3\pi$ are at $\omega = 0,\ 2\pi$, and 2π. Defining the signal $x_{lp}(t) = a_0 + a_1e^{j2\pi t} + a_{-1}e^{-j2\pi t}$, we note that $y(t) = (1/3\pi)dx_{lp}(t)/dt$. We can also easily show that $a_0 = 1/\pi$, $a_1 = a_{-1}^* = -1/(4j)$. Putting these into the expression for $x_{lp}(t)$ we obtain $x_{lp}(t) = (1/\pi) + (1/2)\sin(2\pi t)$. Finally, $y(t) = (1/3\pi)dx_{lp}(t)/dt = (1/3)\cos(2\pi t)$.

6.23. **(a)** From the given information, we have

$$H_a(j\omega) = \begin{cases} 1, & |\omega| \leq \omega_c \\ 0, & \text{otherwise} \end{cases}.$$

Using Table 4.2, we get

$$h_a(t) = \frac{\sin(\omega_c t)}{\pi t}.$$

(b) Here,

$$H_b(j\omega) = H_a(j\omega)e^{j\omega T}.$$

Using Table 4.1, we get

$$h_b(t) = h_a(t + T).$$

Therefore,

$$h_b(t) = \frac{\sin[\omega_c(t + T)]}{\pi(t + T)}.$$

(c) Let us consider a frequency response $H_0(j\omega)$ given by

$$H_0(j\omega) = \begin{cases} 1, & |\omega| \le \omega_c/2 \\ 0, & \text{otherwise} \end{cases}.$$

Clearly,

$$H_c(j\omega) = \frac{1}{2\pi}[H_0(j\omega) * W(j\omega)],$$

where

$$W(j\omega) = j2\pi\delta(\omega - \omega_c/2) - j2\pi\delta(\omega + \omega_c/2).$$

Therefore, from Table 4.1

$$h_c(t) = h_0(t)w(t) = \left[\frac{\sin(\omega_c t/2)}{\pi t}\right][-2\sin(\omega_c t/2)].$$

6.24. If $\tau(\omega) = k_1$, where k_1 is a constant, then

$$\sphericalangle H(j\omega) = -k_1\omega + k_2 \qquad \text{(S6.24-1)}$$

where k_2 is another constant.

(a) Note that if $h(t)$ is real, the phase of the Fourier transform $\sphericalangle H(j\omega)$ has to be an odd function. Therefore, the value of k_2 in eq. (S6.24-1) will be zero.

Also, let us define $H_0(j\omega) = |H(j\omega)|$. Then,

$$h_0(t) = \frac{\sin(200\pi t)}{\pi t}.$$

(i) Here $k_1 = 5$. Hence, $\sphericalangle H(j\omega) = -5\omega$. Then,

$$H(j\omega) = |H(j\omega)|e^{j\sphericalangle H(j\omega)} = H_0(j\omega)e^{-j5\omega}.$$

Therefore,

$$h(t) = h_0(t-5) = \frac{\sin[200\pi(t-5)]}{\pi(t-5)}.$$

(ii) Here $k_1 = 5/2$. Hence, $\sphericalangle H(j\omega) = -(5/2)\omega$. Then,

$$H(j\omega) = |H(j\omega)|e^{j\sphericalangle H(j\omega)} = H_0(j\omega)e^{-j(5/2)\omega}.$$

Therefore,

$$h(t) = h_0(t-5/2) = \frac{\sin[200\pi(t-5/2)]}{\pi(t-5/2)}.$$

(iii) Here $k_1 = -5/2$. Hence, $\sphericalangle H(j\omega) = (5/2)\omega$. Then,

$$H(j\omega) = |H(j\omega)|e^{j\sphericalangle H(j\omega)} = H_0(j\omega)e^{j(5/2)\omega}.$$

Therefore,

$$h(t) = h_0(t+5/2) = \frac{\sin[200\pi(t+5/2)]}{\pi(t+5/2)}.$$

(b) If $h(t)$ is not specified to be real, then $\sphericalangle H(j\omega)$ does not have to be an odd function. Therefore, the value of k_2 in eq. (S6.24-1) does not have to be zero. Given only $|H(j\omega)|$ and $\tau(\omega)$, k_2 cannot be determined uniquely. Therefore, $h(t)$ cannot be determined uniquely.

6.25. **(a)** We may write $H_a(j\omega)$ as

$$H_a(j\omega) = \frac{(1-j\omega)}{(1+j\omega)(1-j\omega)} = \frac{1-j\omega}{2}.$$

Therefore,

$$\sphericalangle H_a(j\omega) = \tan^{-1}[-\omega].$$

and

$$\tau_a(\omega) = -\frac{d\sphericalangle H_a(j\omega)}{d\omega} = \frac{1}{1+\omega^2}.$$

Since $\tau_a(0) = 1 \neq 2 = \tau_a(1)$, $\tau_a(\omega)$ is not a constant for all ω. Therefore, the frequency response has nonlinear phase.

(b) In this case, $H_b(j\omega)$ is the frequency response of a system which is a cascade combination of two systems, each of which has a frequency response $H_a(j\omega)$. Therefore,

$$\sphericalangle H_b(j\omega) = \sphericalangle H_a(j\omega) + \sphericalangle H_a(j\omega)$$

and

$$\tau_b(\omega) = -2\frac{d\sphericalangle H_a(j\omega)}{d\omega} = \frac{2}{1+\omega^2}.$$

Since $\tau_b(0) = 2 \neq 4 = \tau_b(1)$, $\tau_b(\omega)$ is not a constant for all ω. Therefore, the frequency response has nonlinear phase.

(c) In this case, $H_c(j\omega)$ is again the frequency response of a system which is a cascade combination of two systems. The first system has a frequency response $H_a(j\omega)$, while the second system has a frequency response $H_0(j\omega) = 1/(2+j\omega)$. Therefore,

$$\sphericalangle H_b(j\omega) = \sphericalangle H_a(j\omega) + \sphericalangle H_0(j\omega)$$

and

$$\tau_c(\omega) = -\frac{d\sphericalangle H_a(j\omega)}{d\omega} - \frac{d\sphericalangle H_0(j\omega)}{d\omega} = \frac{1}{1+\omega^2} + \frac{2}{4+\omega^2}.$$

Since $\tau_c(0) = (3/2) \neq (3/5) = \tau_c(1)$, $\tau_b(\omega)$ is not a constant for all ω. Therefore, the frequency response has nonlinear phase.

6.26. **(a)** Note that $H(j\omega) = 1 - H_0(j\omega)$, where $H_0(j\omega)$ is

$$H_0(j\omega) = \begin{cases} 1, & 0 \leq |\omega| \leq \omega_c \\ 0, & \text{otherwise} \end{cases}.$$

Therefore,

$$h(t) = \delta(t) - h_0(t).$$

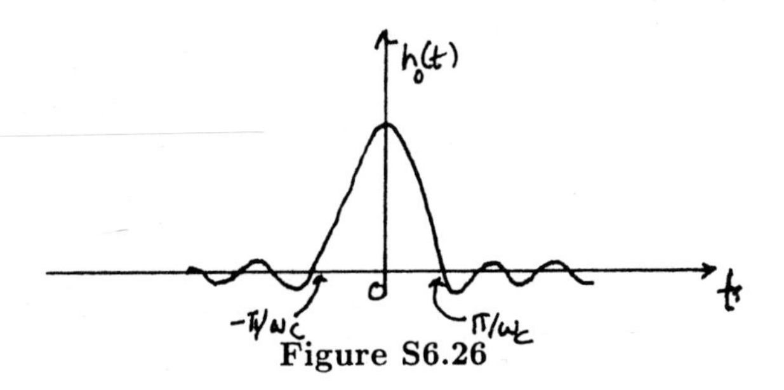

Figure S6.26

From Table 4.2, we have

$$h_0(t) = \frac{\sin(\omega_c t)}{\pi t}.$$

Therefore,

$$h(t) = \delta(t) - \frac{\sin(\omega_c t)}{\pi t}.$$

(b) A sketch of $h_0(t)$ is Figure S6.26. Clearly, as ω_c increases, $h(t)$ becomes more concentrated about the origin.

(c) Note that the step response is given by

$$s(t) = h(t) * u(t) = u(t) - u(t) * h_0(t).$$

Also, note that $h_0(t)$ is the impulse response of an ideal lowpass filter. If $s_0(t) = u(t) * h_0(t)$ denotes the step response of the lowpass filter, we know from Figure 6.14 that $s_0(0) = 0$ and $s_\infty = 1$. Therefore,

$$s(0+) = u(0+) - s_0(0+) = 1 - (1/2) = 1/2$$

and

$$s(\infty) = u(\infty) - s_0(\infty) = 0.$$

6.27. **(a)** Taking the Fourier transform of both sides of the given differential equation, we obtain

$$H(j\omega) = \frac{Y(j\omega)}{X(j\omega)} = \frac{1}{2 + j\omega}.$$

The Bode plot is as shown in Figure S6.27.

(b) From the expression for $H(j\omega)$ we obtain

$$\sphericalangle H(j\omega) = -\tan^{-1}(\omega/2).$$

Therefore,

$$\tau(\omega) = -\frac{d\sphericalangle H(j\omega)}{d\omega} = \frac{2}{4 + \omega^2}.$$

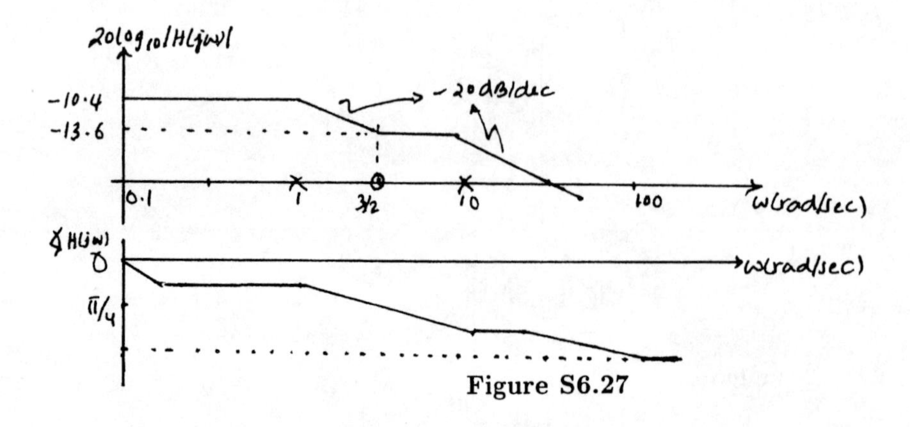

Figure S6.27

(c) Since $x(t) = e^{-t}u(t)$,

$$X(j\omega) = \frac{1}{1+j\omega}.$$

Therefore,

$$Y(j\omega) = X(j\omega)H(j\omega) = \frac{1}{(1+j\omega)(2+j\omega)}.$$

(d) Taking the inverse Fourier transform of the partial fraction expansion of $Y(j\omega)$, we obtain

$$y(t) = e^{-t}u(t) - e^{-2t}u(t).$$

(e) (i) Here,

$$Y(j\omega) = \frac{1+j\omega}{(2+j\omega)^2}.$$

Taking the inverse Fourier transform of the partial fraction expansion of $Y(j\omega)$, we obtain

$$y(t) = e^{-2t}u(t) - te^{-2t}u(t).$$

(ii) Here,

$$Y(j\omega) = \frac{1}{(1+j\omega)}.$$

Taking the inverse Fourier transform of $Y(j\omega)$, we obtain

$$y(t) = e^{-t}u(t).$$

(iii) Here,

$$Y(j\omega) = \frac{1}{(1+j\omega)(2+j\omega)^2}.$$

Taking the inverse Fourier transform of the partial fraction expansion of $Y(j\omega)$, we obtain

$$y(t) = e^{-t}u(t) + \frac{1}{2}e^{-2t}u(t) - te^{-2t}u(t).$$

6.28. **(a)** The Bode plots are as shown below

(b) We may write the frequency response of (iv) as

$$H(j\omega) = \frac{11/10}{1+j\omega} - \frac{1}{10}.$$

Therefore,

$$h(t) = \frac{11}{10}e^{-t}u(t) - \frac{1}{10}\delta(t)$$

and

$$s(t) = h(t) * u(t) = \frac{11}{10}(1-e^{-t})u(t) - \frac{1}{10}u(t).$$

Both $h(t)$ and $s(t)$ are as shown in Figure S6.28.

We may write the frequency response of (vi) as

$$H(j\omega) = \frac{9/10}{1+j\omega} + \frac{1}{10}.$$

Therefore,

$$h(t) = \frac{9}{10}e^{-t}u(t) + \frac{1}{10}\delta(t)$$

and

$$s(t) = h(t) * u(t) = \frac{9}{10}(1-e^{-t})u(t) + \frac{1}{10}u(t).$$

Both $h(t)$ and $s(t)$ are as shown in Figure S6.28.

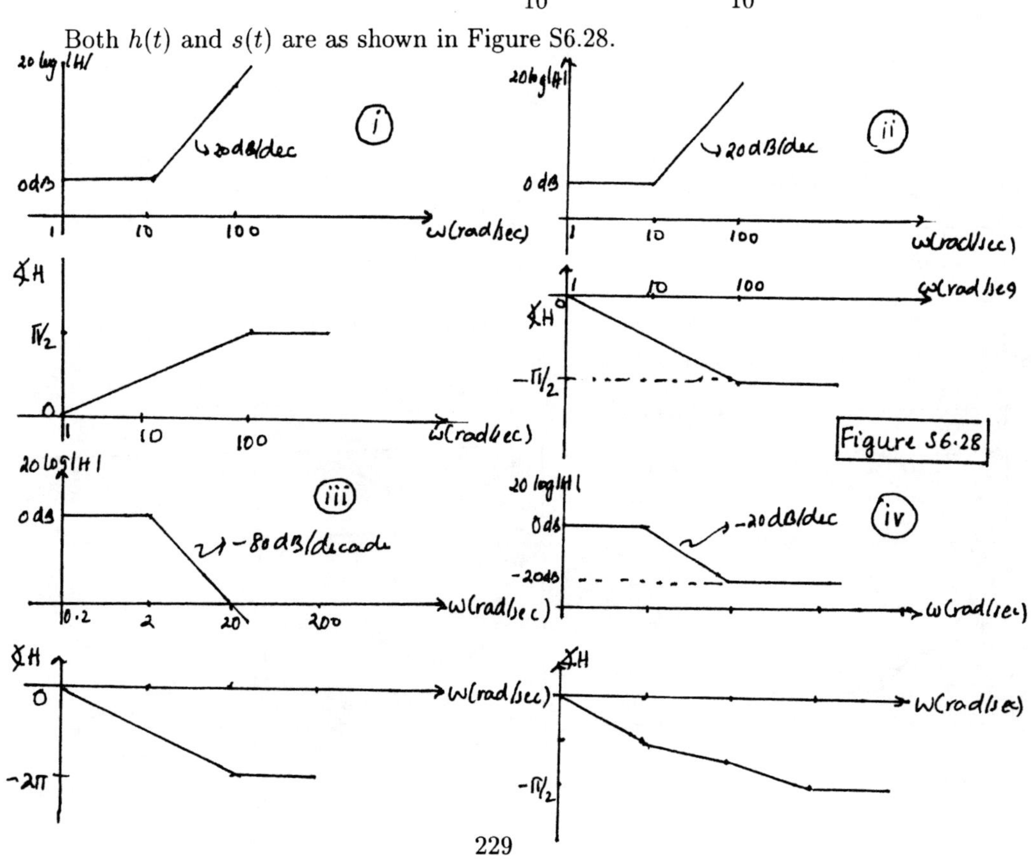

Figure S6.28

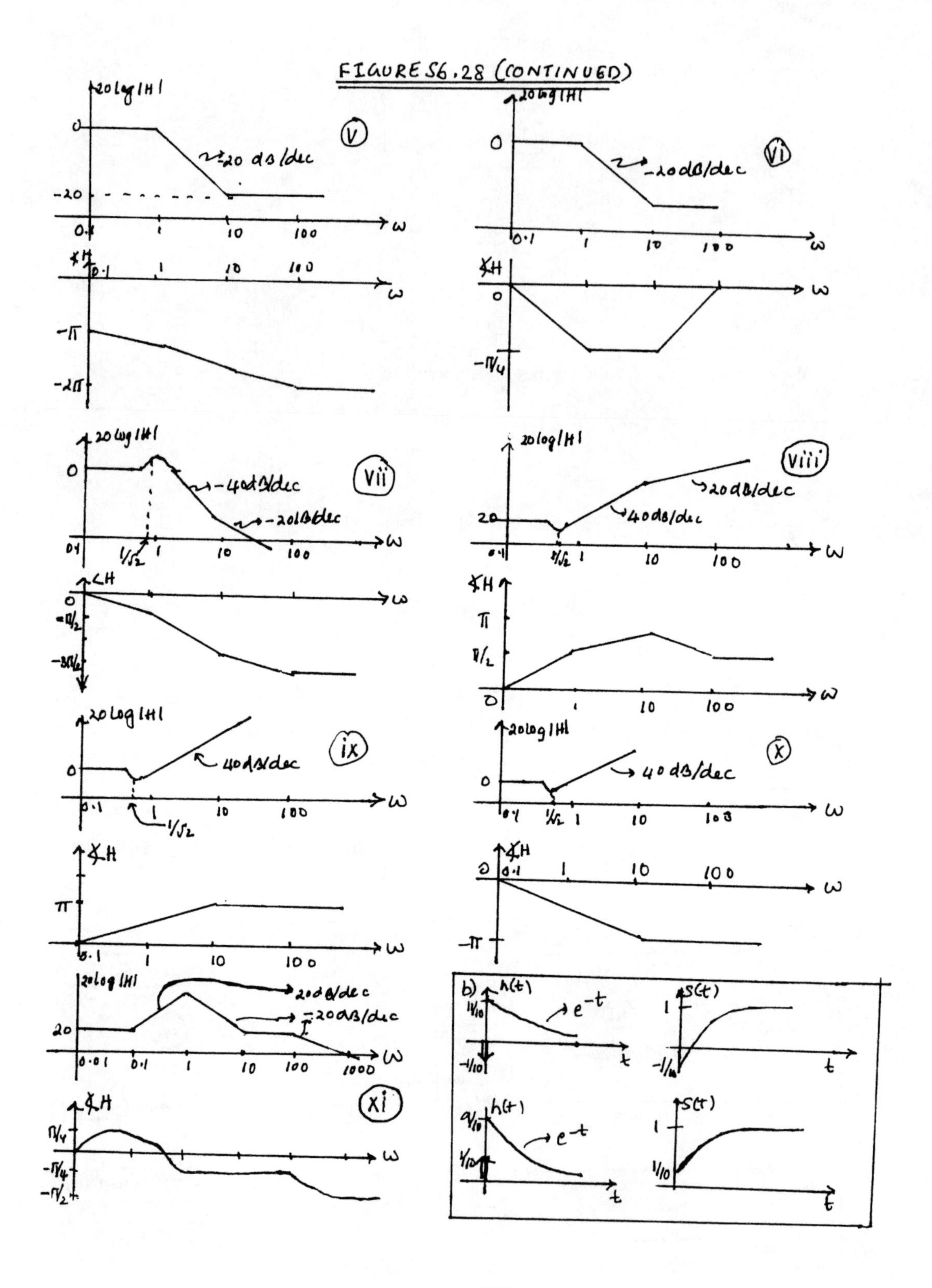
FIGURE S6.28 (CONTINUED)
20 log|H|
-20 dB/dec
-20 dB/dec
-40 dB/dec
-20 dB/dec
20 dB/dec
40 dB/dec
40 dB/dec
40 dB/dec
20 dB/dec
-20 dB/dec
b)
h(t)
s(t)
e^{-t}

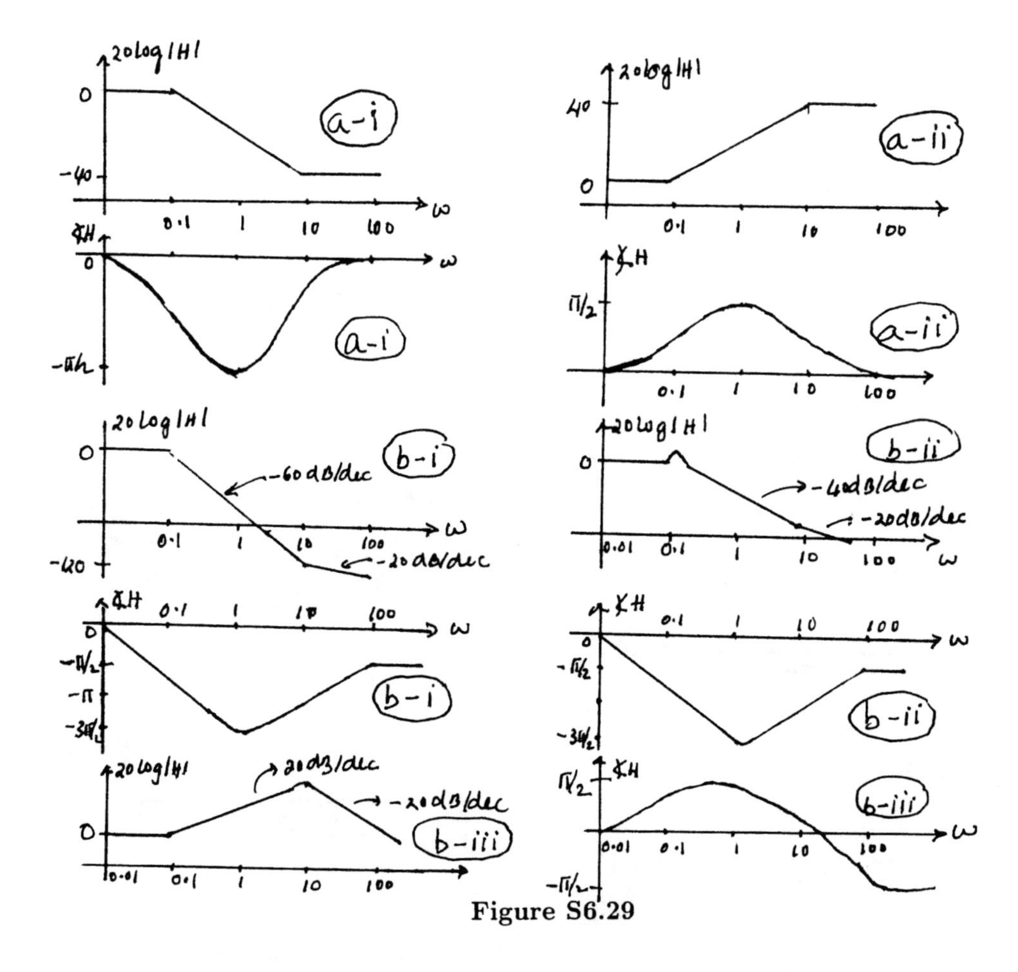

Figure S6.29

6.29. **(a)** (i) The Bode plot is as shown in Figure S6.29. Clearly, the system has phase lag. It also has no amplification at any frequencies (i.e., $|H(j\omega|$ never exceeds 0 dB).

(ii) The Bode plot is as shown in Figure S6.29. Clearly, the system has phase lead. It has amplification at approximately frequencies which exceed 0.1 rad/sec.

(b) (i) The Bode plot is as shown in Figure S6.29. Clearly, the system has phase lag. It also has no amplification at any frequencies (i.e., $|H(j\omega|$ never exceeds 0 dB).

(ii) The Bode plot is as shown in Figure S6.29. Clearly, the system has phase lag. It has some amplification at frequencies near 0.1 rad/sec.

(iii) The Bode plot is as shown in Figure S6.29. Clearly, the system has both phase lag and phase lead. It also has amplification for a band of frequencies.

6.30. We know that

$$10x(10t) \stackrel{FT}{\longleftrightarrow} X\left(j\frac{\omega}{10}\right).$$

Therefore, the Bode plot shifts by 1 decade to the left. The shape remains unaltered.

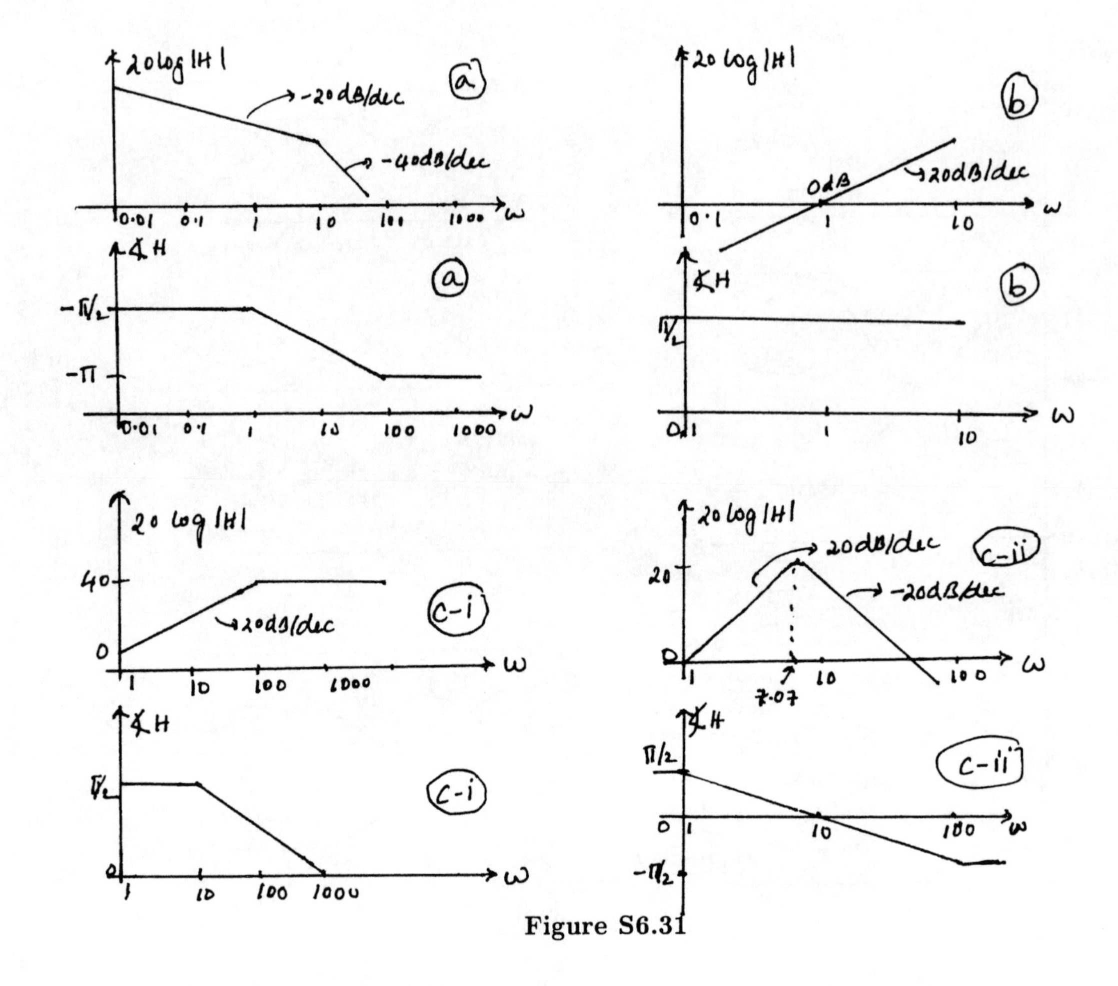

Figure S6.31

6.31. **(a)** The Bode plot is as shown in Figure S6.31.

(b) Since

$$\frac{dx(t)}{dt} \stackrel{FT}{\longleftrightarrow} j\omega X(j\omega),$$

the frequency response of a differentiator is $H(j\omega) = j\omega$. Therefore, its Bode plot is as shown in the figure below.

(c) (i) The Bode plot is as shown in Figure S6.31.

(ii) Here, $\omega_n = 10$ and $\zeta = \frac{1}{2}$. The Bode plot is as shown in Figure S6.31.

6.32. **(a)** One possible choice for the compensator frequency response is

$$H_c(j\omega) = \frac{50(\frac{j\omega}{50} + 1)}{(\frac{j\omega}{100} + 1)^2}.$$

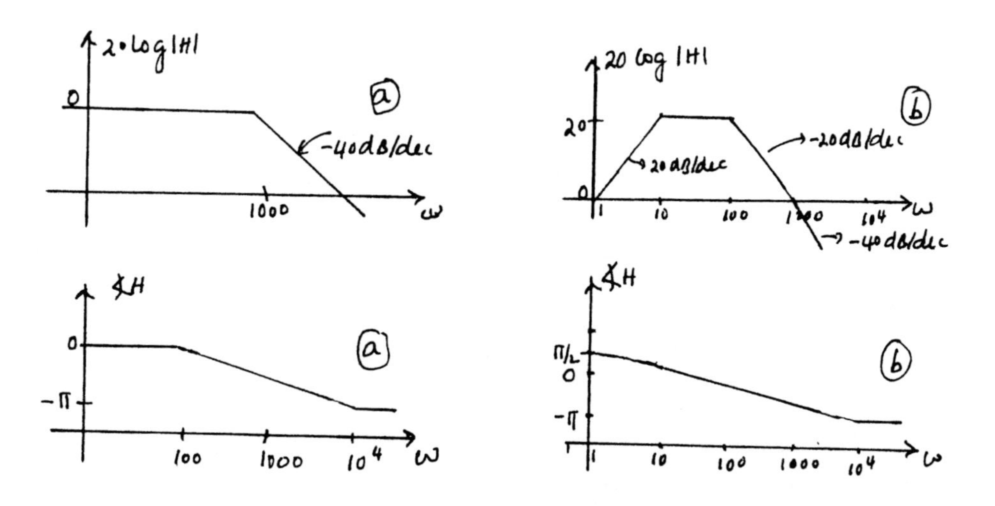

Figure S6.32

Therefore, the overall frequency response is

$$H(j\omega) = \frac{1}{(1 + \frac{j\omega}{100})^2}.$$

The Bode plot for this frequency response is as shown in Figure S6.32.

(b) One possible choice for the compensator frequency response is

$$H_c(j\omega) = \frac{50j\omega(\frac{j\omega}{50} + 1)}{(\frac{j\omega}{10} + 1)(\frac{j\omega}{100} + 1)(\frac{j\omega}{1000} + 1)}.$$

Therefore, the overall frequency response is

$$H(j\omega) = \frac{j\omega}{(\frac{j\omega}{10} + 1)(\frac{j\omega}{100} + 1)(\frac{j\omega}{1000} + 1)}.$$

The Bode plot for this frequency response is as shown in Figure S6.32.

6.33. **(a)** From Figure P6.33, we may write

$$Y(j\omega) = X(j\omega) - H(j\omega)H(j\omega) = H_{ov}(j\omega)X(j\omega).$$

Therefore,

$$H_{ov}(j\omega) = 1 - H(j\omega) \qquad \text{(S6.33-1)}$$

If $H(j\omega)$ corresponds to an ideal lowpass filter with cutoff frequency ω_{lp}, then $H_{ov}(j\omega)$ is as shown in Figure S6.33.

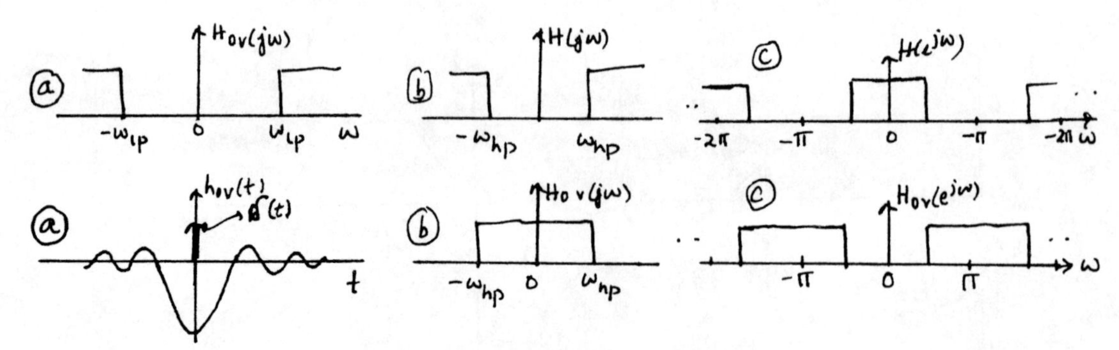

Figure S6.33

Clearly, $H_{ov}(j\omega)$ corresponds to an ideal highpass filter with cutoff frequency ω_{lp}. Also,

$$h_{ov}(t) = \delta(t) - h(t) = \delta(t) - \frac{\sin(\omega_{lp}t)}{\pi t}.$$

This is as shown in Figure S6.33.

(b) If $H(j\omega)$ corresponds to an ideal highpass filter with cutoff frequency ω_{hp}, then from eq.(S6.33-1) it is clear that $H_{ov}(j\omega)$ is as shown in Figure S6.33. Clearly, $H_{ov}(j\omega)$ corresponds to an ideal lowpass filter with cutoff frequency ω_{hp}.

(c) If we replace $H(j\omega)$ with a discrete-time lowpass filter with frequency response $H(e^{j\omega})$ as shown in Figure S6.33, then the overall frequency response still is

$$H_{ov}(e^{j\omega}) = 1 - H(e^{j\omega}).$$

Therefore, $H(e^{j\omega})$ is as shown in Figure S6.33. Clearly, it is highpass.

6.34. **(a)** From the previous problem,

$$H_{ov}(j\omega) = 1 - H(j\omega).$$

This is sketched in Figure S6.34. Clearly, it is approximately highpass.

(b) We have $H(j\omega) = H_1(j\omega)e^{j\theta(\omega)}$. Therefore, $|H(j\omega)| = |H_1(j\omega)|$. Therefore, it is still lowpass.

(c) We have

$$H_{ov}(j\omega) = 1 - H(j\omega) = 1 - H_1(j\omega)e^{j\theta(\omega)}.$$

Therefore,

$$|H_{ov}(j\omega)| = |1 - H_1(j\omega)e^{j\theta(\omega)}|.$$

We also have

$$1 - |H_1(j\omega)| \leq |1 - H_1(j\omega)e^{j\theta(\omega)}| \leq 1 + |H_1(j\omega)|.$$

Therefore, $H_{ov}(j\omega)$ is between the two curves sketched in Figure S6.34.

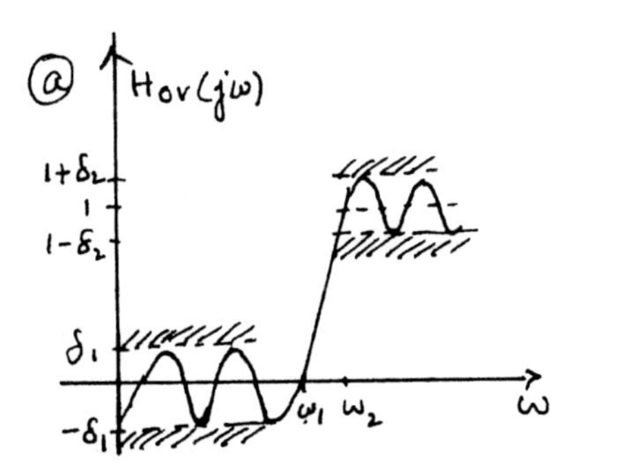

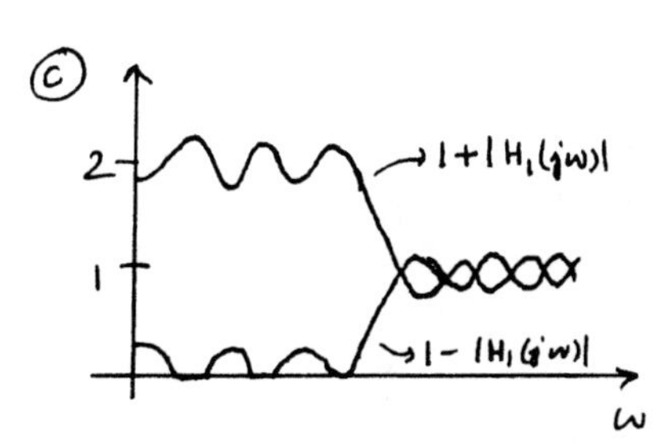

Figure S6.34

(d) From the tolerances derived in the previous part, it is clear that $H_{ov}(j\omega)$ is not necessarily highpass.

6.35. Since $x[n] = \cos(\omega_0 n + \theta)$, we have

$$X(e^{j\omega}) = \pi \sum_{l=-\infty}^{\infty} [e^{j\theta}\delta(\omega - \omega_0 - 2\pi l) + e^{-j\theta}\delta(\omega + \omega_0 - 2\pi l)].$$

Let ω_0' be the principal value of ω_0 in $[-\pi, \pi]$. Then

$$Y(e^{j\omega}) = X(e^{j\omega})H(e^{j\omega}) = \pi \sum_{l=-\infty}^{\infty} [e^{j\theta} j\omega_0'\delta(\omega - \omega_0 - 2\pi l) - e^{-j\theta} j\omega_0'\delta(\omega + \omega_0 - 2\pi l)].$$

It follows that

$$y[n] = -\omega_0' \sin(\omega_0 n + \theta).$$

If $-\pi \le \omega_0 \le \pi$, then

$$y[n] = -\omega_0 \sin(\omega_0 n + \theta).$$

6.36. Let $H_1(e^{j\omega}) = |H(e^{j\omega})|$. Then from Table 5.2 we know that

$$h_1[n] = \frac{\sin(\pi n/2)}{\pi n}.$$

If $\tau(\omega) = -\frac{d}{d\omega}\sphericalangle H(e^{j\omega}) = k$ (where k is a constant), then $\sphericalangle H(e^{j\omega}) = -k\omega + k_1$, where k_1 is a constant. If $h[n]$ is real, then $\sphericalangle H(e^{j\omega})$ is an odd function, and therefore we may conclude that $k_1 = 0$. Therefore,

$$H(e^{j\omega}) = |H(e^{j\omega})| e^{j\sphericalangle H(e^{j\omega})} = H_1(e^{j\omega}) e^{-jk\omega}.$$

Taking the inverse Fourier transform we obtain

$$h[n] = h_1[n-k] = \frac{\sin[\pi(n-k)/2]}{\pi(n-k)}.$$

(a) If $\tau(\omega) = 5$, then from the above result,

$$h[n] = \frac{\sin[\pi(n-5)/2]}{\pi(n-5)}.$$

(b) If $\tau(\omega) = 5/2$, then from the result derived at the beginning of this problem

$$h[n] = \frac{\sin[\pi(n-5/2)/2]}{\pi(n-5/2)}.$$

(c) If $\tau(\omega) = -5/2$, then from the result derived at the beginning of this problem

$$h[n] = \frac{\sin[\pi(n+5/2)/2]}{\pi(n+5/2)}.$$

The results of all the parts of this problem are sketched in Figure S6.36.

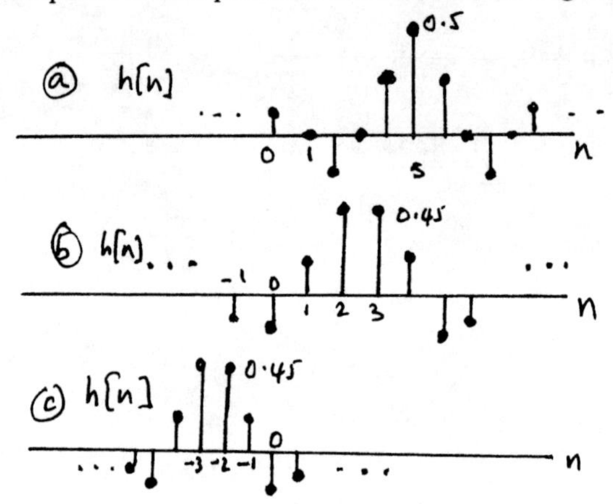

Figure S6.36

6.37. **(a)** We have

$$|H(e^{j\omega})| = \frac{|1 - \frac{1}{2}e^{j\omega}|}{|1 - \frac{1}{2}e^{-j\omega}|} = 1.$$

(b) We have

$$\begin{aligned}
\sphericalangle H(e^{j\omega}) &= \sphericalangle[e^{-j\omega}] + \sphericalangle\left[1 - \frac{1}{2}e^{j\omega}\right] - \sphericalangle\left[1 - \frac{1}{2}e^{-j\omega}\right] \\
&= \sphericalangle[e^{-j\omega}] + \sphericalangle\left[1 - \frac{1}{2}\cos(\omega) - \frac{j}{2}\sin(\omega)\right] - \sphericalangle\left[1 - \frac{1}{2}\cos(\omega) + \frac{j}{2}\sin(\omega)\right] \\
&= -\omega - 2\tan^{-1}\left[\frac{\frac{1}{2}\sin(\omega)}{1 - \frac{1}{2}\cos(\omega)}\right]
\end{aligned}$$

(c) Using the result of the previous part, we can show with some algebraic manipulation that

$$\tau(\omega) = -\frac{d\sphericalangle H(e^{j\omega})}{d\omega} = \frac{\frac{3}{4}}{\frac{5}{4} - \cos\omega}.$$

This is as sketched below

(d) Let $x[n] = \cos(\pi n/3)$. We may write this as $x[n] = e^{j\pi n/3}/2 + e^{-j\pi n/3}/2$. From the result of part (c), we know that the delay suffered by a complex exponential of frequency $\pi/3$ is

$$\frac{\frac{3}{4}}{\frac{5}{4} - \cos(\pi/3)} = 1.$$

Similarly, we know that the delay suffered by a complex exponential of frequency $-\pi/3$ is also 1. Therefore, the output of the system is $y[n] = e^{j\pi(n-1)/3}/2 + e^{-j\pi(n-1)/3}/2 = \cos(\pi(n-1)/3)$.

6.38. We may express $H(e^{j\omega})$ as

$$H(e^{j\omega}) = \frac{1}{2\pi}\left[H_1(e^{j\omega}) * \{2\pi\delta(\omega - \pi/2) + 2\pi\delta(\omega + \pi/2)\}\right],$$

and

$$H_1(e^{j\omega}) = \begin{cases} 1, & |\omega| < \omega_c \\ 0, & \omega_c < |\omega| < \pi \end{cases}.$$

Using the properties of the Fourier transform, we obtain

$$h[n] = h_1[n]\left[2\cos(\pi n/2)\right],$$

where

$$h_1[n] = \frac{\sin(\omega_c n)}{\pi n}.$$

(a) When $\omega_c = \pi/5$, $h[n] = 2\frac{\sin(\pi n/5)}{\pi n}\cos(\pi n/2)$. This is as shown in Figure S6.38.

(b) When $\omega_c = \pi/4$, $h[n] = 2\frac{\sin(\pi n/4)}{\pi n}\cos(\pi n/2)$. This is as shown in Figure S6.38.

(c) When $\omega_c = \pi/3$, $h[n] = 2\frac{\sin(\pi n/3)}{\pi n}\cos(\pi n/2)$. This is as shown in Figure S6.38.
As ω_c increases, $h[n]$ becomes more concentrated about the origin.

6.39. The plots are as shown in Figure S6.39.

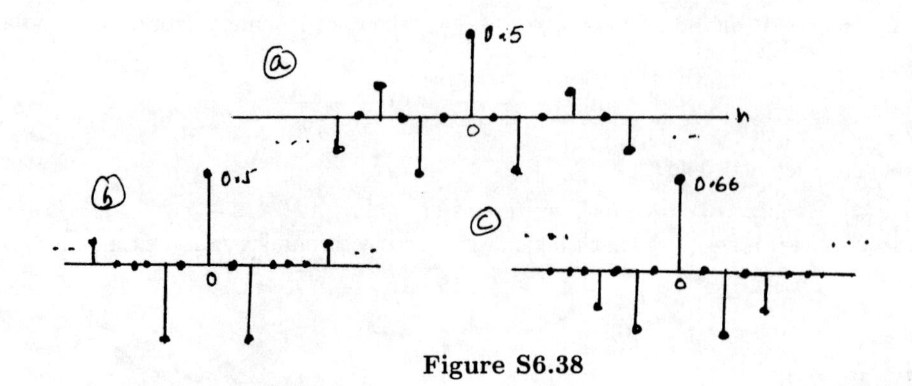

Figure S6.38

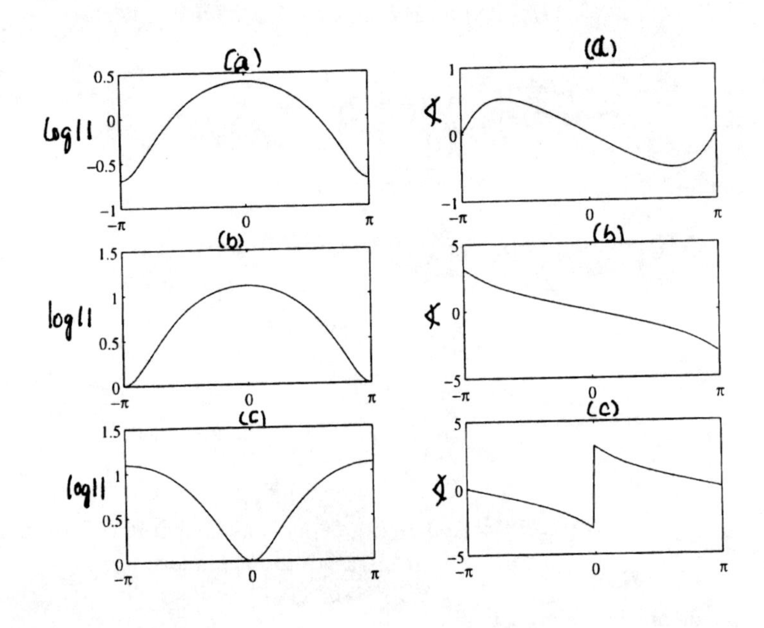

FIGURE S6.39

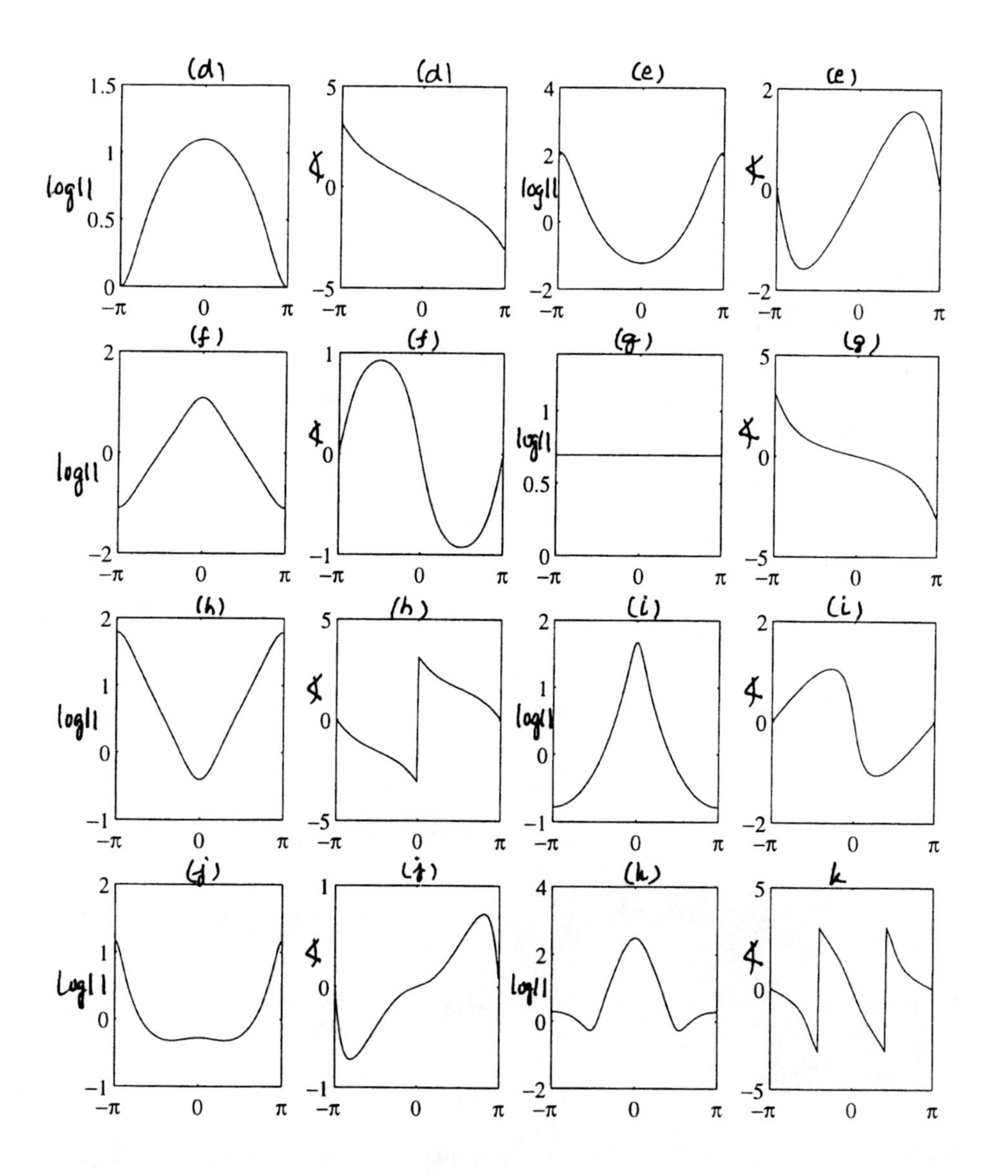

FIGURE S6.39 (CONTD.)

6.40. We may write $h_1[n]$ as

$$\begin{aligned} H_1(e^{j\omega}) &= \sum_{n=-\infty}^{\infty} h_1[n]e^{-j\omega n} \\ &= \sum_{n=-\infty}^{\infty} h_1[2n]e^{-j2\omega n} \\ &= \sum_{n=-\infty}^{\infty} h[n]e^{-j2\omega n} \\ &= H(e^{j2\omega}) \end{aligned}$$

Therefore, $H_1(e^{j\omega})$ is $H(e^{j\omega})$ compressed by a factor of two. This is as shown in Figure S6.40.

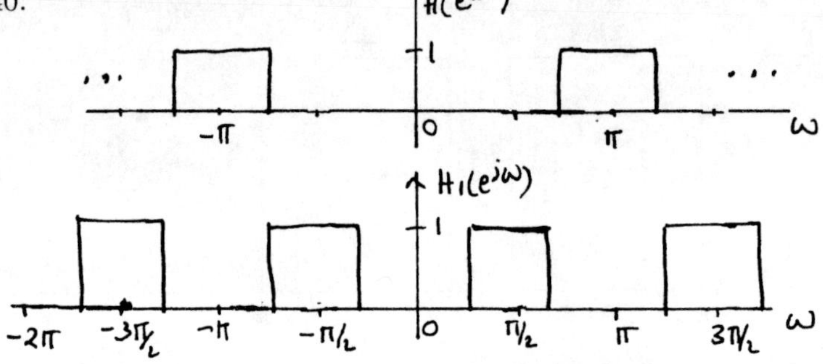

Figure S6.40

Therefore, $H_1(e^{j\omega})$ corresponds to a band-stop filter.

6.41. **(a)** Taking the Fourier transform of both sides of the given difference equation, we obtain

$$H(e^{j\omega}) = \frac{Y(e^{j\omega})}{X(e^{j\omega})} = \frac{1 - e^{-j\omega}}{1 - \frac{1}{\sqrt{2}}e^{-j\omega} + \frac{1}{4}e^{-2j\omega}}.$$

Taking the inverse Fourier transform of $H(e^{j\omega})$ we obtain

$$h[n] = \left(\frac{1}{2}\right)^n \cos(\pi n/4)u[n] - (2\sqrt{2} - 1)\left(\frac{1}{2}\right)^n \sin(\pi n/4)u[n].$$

(b) The log-magnitude and phase of the frequency response are as shown in Figure S6.41.

6.42. **(a)** We get

$$|H_1(e^{j\omega})| = |H_2(e^{j\omega})| = \frac{5/4 + \cos\omega}{17/6 + (1/2)\cos\omega}$$

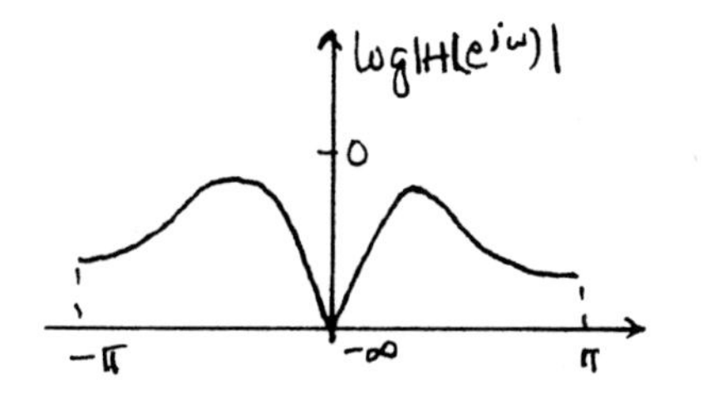

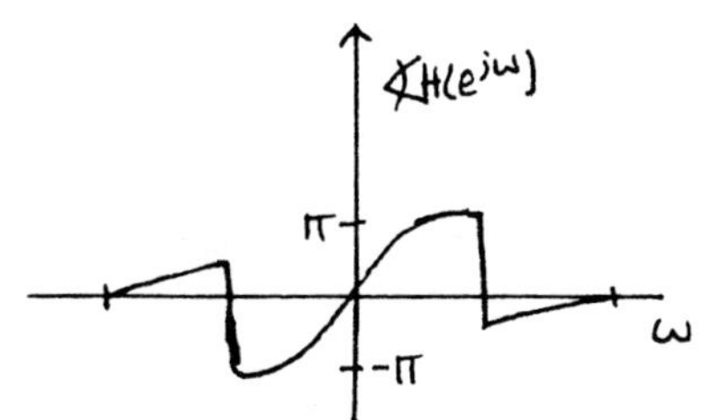

Figure S6.41

and

$$\sphericalangle H_1(e^{j\omega}) = \tan^{-1}\left(\frac{(1/2)\sin\omega}{1+(1/2)\cos(\omega)}\right) \quad \text{and} \quad \sphericalangle H_2(e^{j\omega}) = \tan^{-1}\left(\frac{\sin\omega}{1+(1/2)\cos(\omega)}\right).$$

Comparing tangents of these angle in the range $0 \le \omega \le \pi$, we get

$$\sphericalangle H_2(e^{j\omega}) > \sphericalangle H_1(e^{j\omega}).$$

(b) We get

$$h_1[n] = \left(-\frac{1}{4}\right)^n u[n] + \frac{1}{2}\left(-\frac{1}{4}\right)^{n-1} u[n-1]$$

and

$$h_2[n] = \frac{1}{2}\left(-\frac{1}{4}\right)^n u[n] + \left(-\frac{1}{4}\right)^{n-1} u[n-1].$$

This is as sketched in Figure S6.42.

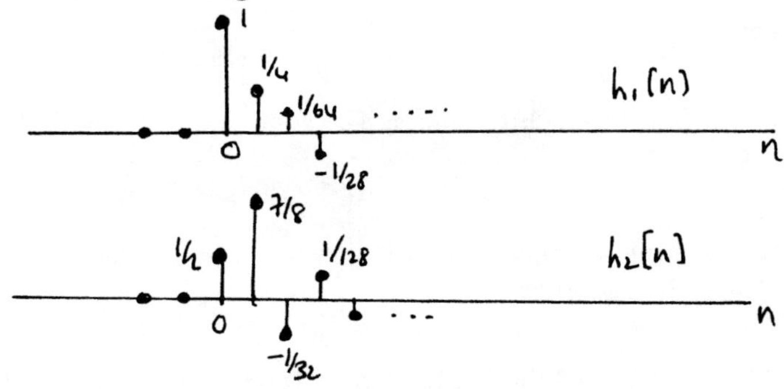

Figure S6.42

(c) We get

$$H_2(e^{j\omega}) = \left(\frac{1/2 + e^{-j\omega}}{1+(1/2)e^{-j\omega}}\right) H_1(e^{j\omega}).$$

Therefore,

$$G(e^{j\omega}) = \left(\frac{1/2 + e^{-j\omega}}{1 + (1/2)e^{-j\omega}}\right)$$

and

$$|G(e^{j\omega})| = \frac{(5/4) + \cos\omega}{(5/4) + \cos\omega} = 1.$$

6.43. **(a)** If $h_{hp}[n] = (-1)^n h_{lp}[n] = e^{j\pi n} h_{lp}[n]$, then

$$H_{hp}(e^{j\omega}) = H_{lp}(e^{j(\omega - \pi)}).$$

Therefore, $H_{hp}(e^{j\omega})$ is as shown in Figure S6.43. Clearly, it corresponds to a highpass filter.

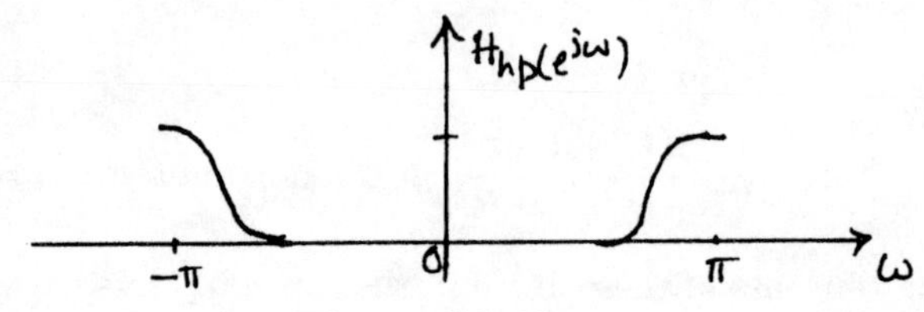

Figure S6.43

(b) Now let us define $h[n] = (-1)^n h_{hp}[n]$, where $h_{hp}[n]$ is the impulse response of a highpass filter. Then

$$H(e^{j\omega}) = H_{hp}(e^{j(\omega - \pi)}).$$

Therefore, if $H_{hp}(e^{j\omega})$ is as shown in Figure S6.43, then $H(e^{j\omega})$ is lowpass.

6.44. **(a)** Note that $(-1)^n = e^{j\pi n}$. From the figure we have

$$y[n] = \left(x[n]e^{j\pi n} * h_{lp}[n]\right) e^{j\pi n}.$$

We may write this as

$$y[n] = a[n]e^{j\pi n},$$

where $a[n] = \left(x[n]e^{j\pi n} * h_{lp}[n]\right)$. Taking the Fourier transform of $a[n]$, we obtain

$$A(e^{j\omega}) = X(e^{j(\omega - \pi)})H_{lp}(e^{j\omega}).$$

Suppose that the input to the system is now $x[n - n_0]$. Let the corresponding output be $y_1[n]$. Then we may write

$$y_1[n] = b[n]e^{j\pi n},$$

where $b[n] = \left(x[n - n_0]e^{j\pi n} * h_{lp}[n]\right)$. Taking the Fourier transform of $b[n]$, we obtain

$$B(e^{j\omega}) = X(e^{j(\omega - \pi)})H_{lp}(e^{j\omega})e^{-j\omega n_0} = A(e^{j\omega})e^{-j\omega n_0}.$$

Therefore,

$$b[n] = a[n - n_0].$$

Consequently, $y_1[n] = y[n - n_0]$. Therefore, the system in time invariant.

(b) Since

$$y[n] = a[n]e^{j\pi n}$$

and

$$A(e^{j\omega}) = X(e^{j(\omega-\pi)})H_{lp}(e^{j\omega}),$$

we obtain

$$Y(e^{j\omega}) = H_{lp}(e^{j(\omega-\pi)})X(e^{j\omega}).$$

Therefore, the frequency response of the overall system is $H_{lp}(e^{j(\omega-\pi)})$. If $H_{lp}(e^{j\omega})$ is lowpass, then $H_{lp}(e^{j(\omega-\pi)})$ is highpass.

6.45. (i) All three first order factors in this frequency response are of the form $\frac{1}{1-\alpha e^{-j\omega}}$, $\alpha > 0$. Therefore, none of these factors contributes an oscillatory component to the step response. Therefore, the step response of the overall system is non oscillatory.

(ii) The factor $\frac{1}{1+\frac{1}{2}e^{-j\omega}}$ contributes an oscillatory component to the step response. Therefore, the step response of the overall system is oscillatory.

(iii) Consider the second order factor $\frac{1}{1-\frac{3}{4}e^{-j\omega}+\frac{9}{16}e^{-2j\omega}}$. For this, we get $r = \frac{3}{4}$ and $\cos\theta = \frac{1}{2}$. Since $\theta \neq 0$, this second order factor contributes an oscillatory component to the step response. Therefore, the step response of the overall system is oscillatory.

6.46. **(a)** We have

$$\begin{aligned} H(e^{j\omega}) &= \sum_{n=-\infty}^{\infty} h[n]e^{-j\omega n} \\ &= h[0] + h[1]e^{-j\omega} + \cdots + h[\frac{N-1}{2}]e^{-j\omega(N-1)/2} + \cdots + h[N-1]e^{-j\omega(N-1)} \end{aligned}$$

Since $h[\frac{N-1}{2}+n] = h[\frac{N-1}{2}-n]$, we may write

$$\begin{aligned} H(e^{j\omega}) &= e^{-j\omega(N-1)/2}\Big[h[0]e^{j\omega(N-1)/2} + h[1]e^{j\omega(\frac{N-1}{2}-1)} + \cdots + h[\frac{N-1}{2}] \\ &\quad + \cdots + h[1]e^{-j\omega(\frac{N-1}{2}-1)} + h[0]e^{-j\omega(\frac{N-1}{2})}\Big] \\ &= e^{-j\omega(N-1)/2}\Big[2h[0]\cos(\omega(N-1)/2) + 2h[1]\cos[\omega(\frac{N-1}{2}-1)] \\ &\quad + \cdots + h[\frac{N-1}{2}]\Big] \\ &= e^{-j\omega(N-1)/2}A(\omega) \end{aligned}$$

where

$$A(\omega) = \Big[2h[0]\cos(\omega(N-1)/2) + 2h[1]\cos[\omega(\frac{N-1}{2}-1)] + \cdots + h[\frac{N-1}{2}]\Big]$$

is a real-valued function.

(b) One such example is $h[n] = \delta[n] + 2\delta[n-1] + 3\delta[n-2] + 2\delta[n-3] + \delta[n-4]$.

(c) We have

$$\begin{aligned} H(e^{j\omega}) &= \sum_{n=-\infty}^{\infty} h[n]e^{-j\omega n} \\ &= h[0] + h[1]e^{-j\omega} + \cdots + h[\frac{N}{2} - 1]e^{-j\omega(\frac{N}{2}-1)} + h[\frac{N}{2}]e^{-j\omega N/2} \\ &\quad + \cdots + h[N-1]e^{-j\omega N-1} \end{aligned}$$

Since $h[\frac{N}{2} + n] = h[\frac{N}{2} - n - 1]$, we may write

$$\begin{aligned} H(e^{j\omega}) &= e^{-j\omega(N-1)/2}\Big[h[0]e^{j\omega(N-1)/2} + h[1]e^{j\omega(\frac{N-1}{2}-1)} \\ &\quad + \cdots + h[\frac{N}{2} - 1]e^{-j\omega/2} + h[\frac{N}{2} - 1]e^{j\omega} \\ &\quad + \cdots + h[1]e^{-j\omega(\frac{N-1}{2}-1)} + h[0]e^{-j\omega(\frac{N-1}{2})}\Big] \\ &= e^{-j\omega(N-1)/2}\Big[2h[0]\cos(\omega(N-1)/2) + 2h[1]\cos[\omega(\frac{N-1}{2} - 1)] \\ &\quad + \cdots + 2h[\frac{N}{2} - 1]\cos(\omega/2)\Big] \\ &= e^{-j\omega(N-1)/2}A(\omega) \end{aligned}$$

where

$$A(\omega) = \Big[2h[0]\cos(\omega(N-1)/2) + 2h[1]\cos[\omega(\frac{N-1}{2} - 1)] + \cdots + 2h[\frac{N}{2} - 1]\cos(\omega/2)\Big]$$

is a real-valued function.

(d) One such example is $h[n] = \delta[n] + 2\delta[n-1] + 2\delta[n-2] + \delta[n-3]$.

6.47. **(a)** Taking the Fourier transform of both sides of the given difference equation, we have

$$H(e^{j\omega}) = \frac{Y(e^{j\omega})}{X(e^{j\omega})} = b[1 + 2a\cos\omega].$$

(b) We want $H(e^{j0}) = b[1 + 2a] = 1$. Therefore, $b = 1/(1 + 2a)$.

(c) If $a = 1/2$, then $b = 1/2$. Therefore, $H(e^{j\omega}) = \frac{1}{2}[1 + \cos\omega]$. This is plotted in Figure S6.47.

6.48. **(a)** Here,

$$H(e^{j\omega}) = b_1 e^{-j\omega} + b_2 e^{-2j\omega} = 2b_1 e^{-j3\omega/2}\cos(\omega/2).$$

Therefore,

$$|H(e^{j\omega})| = 2|b_1||\cos(\omega/2)|.$$

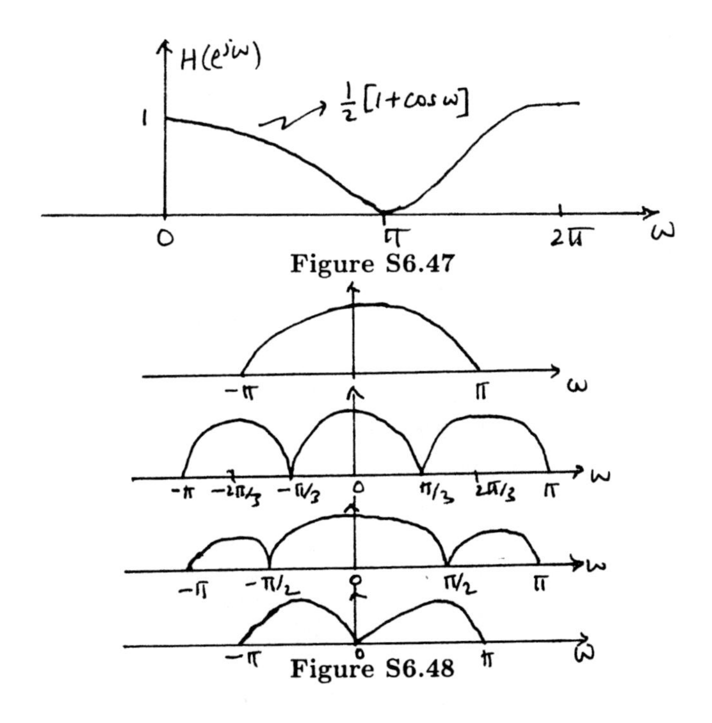

Figure S6.47

Figure S6.48

(b) Here,

$$H(e^{j\omega}) = b_0 + b_3 e^{-3j\omega} = 2b_0 e^{-j3\omega/2}\cos(3\omega/2).$$

Therefore,

$$|H(e^{j\omega})| = 2|b_0||\cos(3\omega/2)|.$$

(c) Here,

$$H(e^{j\omega}) = b_0 + b_1 e^{-j\omega} + b_2 e^{-j2\omega} + b_3 e^{-j3\omega} = 2b_0 e^{-j3\omega/2}\cos(\omega)\cos(\omega/2).$$

Therefore,

$$|H(e^{j\omega})| = 2|b_0||\cos(\omega)||\cos(\omega/2)|.$$

(d) Here,

$$H(e^{j\omega}) = b_0 + b_1 e^{-j\omega} + b_2 e^{-j2\omega} + b_3 e^{-j3\omega} = -2b_0 e^{-j3\omega/2}\sin(\omega)\sin(\omega/2).$$

Therefore,

$$|H(e^{j\omega})| = 2|b_0||\sin(\omega)||\sin(\omega/2)|.$$

The plots for the frequency response magnitudes are shown in Figure S6.48.

6.49. **(a)** Taking the Fourier transform of both sides of the given differential equation, we obtain

$$H(j\omega) = \frac{9}{-\omega^2 + 11j\omega + 10}.$$

Taking the inverse Fourier transform of the partial fraction expansion of $H(j\omega)$, we obtain the impulse response to be

$$h(t) = e^{-t}u(t) - e^{-10t}u(t).$$

Therefore, the step response is

$$s(t) = h(t) * u(t) = \left[1 - e^{-t} - \frac{1}{10} + \frac{1}{10}e^{-10t}\right] u(t).$$

The final value of this response is 9/10. Therefore, the time-constant τ is the time at which the response reaches $9/(10e)$. Therefore,

$$\left[\frac{9}{10} - e^{-\tau} + \frac{1}{10}e^{-10\tau}\right] = \frac{9}{10e}$$

is the equation that we need to solve.

(b) We may write $H(j\omega)$ as

$$H(j\omega) = \frac{1}{1+j\omega} - \frac{1}{10+j\omega} = H_1(j\omega) - H_2(j\omega).$$

Therefore, $H(j\omega)$ may be viewed as the parallel interconnection shown in Figure S6.49.

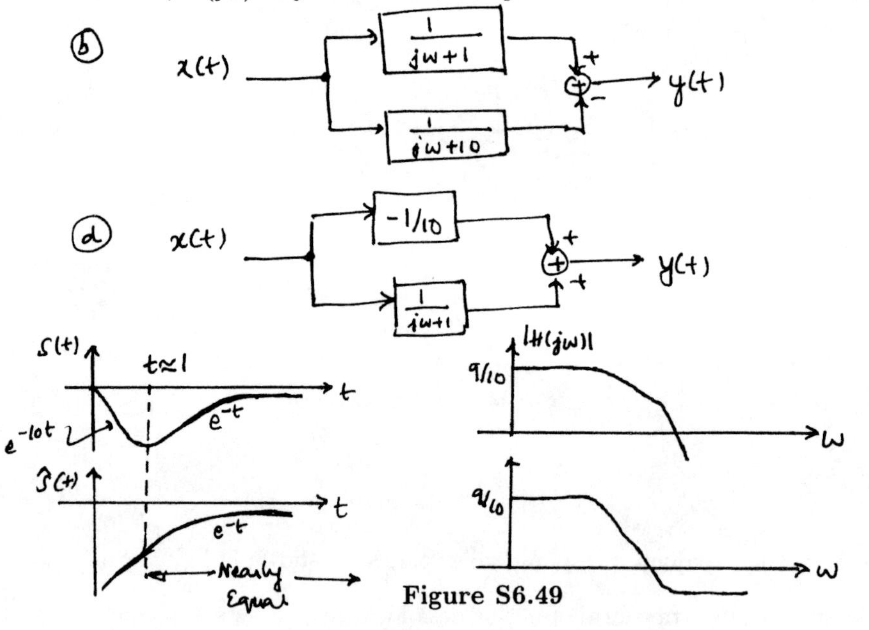

Figure S6.49

The first time constant is $\tau_1 = 1$ and the second time constant is $\tau_2 = \frac{1}{10}$.

(c) Dominant time constant is $\tau = 1$. This approximately satisfies the equation of part (a).

(d) The approximate frequency response may be expressed as

$$\hat{H}(j\omega) = H_1(j\omega) - \hat{H}_2(j\omega) = \frac{1}{1+j\omega} - \frac{1}{10}.$$

The differential equation relating the input and output of the approximate system is

$$\frac{dy(t)}{dt} + y(t) = \frac{1}{10}\frac{dx(t)}{dt} + \frac{9}{10}x(t).$$

The magnitude of the frequency responses of the exact and approximate systems are plotted in Figure S6.49. Clearly, they are identical for low frequencies. The step responses of the exact and approximate systems are also plotted in Figure S6.49. Clearly, they are identical for t approximately greater than 1.

6.50. (a) We have

$$Y(j\omega) = X(j\omega)H(j\omega) = [S(j\omega) + W(j\omega)]\, H(j\omega).$$

Therefore,

$$\epsilon(\omega) = |S(j\omega) - Y(j\omega)|^2 = |S(j\omega) - [S(j\omega) + W(j\omega)]\, H(j\omega)|^2.$$

(b) From part (a), we obtain

$$\begin{aligned}\epsilon(\omega) &= |S(j\omega)|^2 + H^2(j\omega)|S(j\omega) + W(j\omega)|^2 - 2\mathcal{R}e\{S^*(j\omega)\,[S(j\omega) + W(j\omega)]\}H(j\omega) \\ &= |S(j\omega)|^2 + H^2(j\omega)|S(j\omega) + W(j\omega)|^2 - 2H(j\omega)\left[|S(j\omega)|^2 + \mathcal{R}e\{S^*(j\omega)W(j\omega)\}\right]\end{aligned}$$

Therefore,

$$\frac{\partial\epsilon(\omega)}{\partial H(j\omega)} = 2H(j\omega)|S(j\omega) + W(j\omega)|^2 - 2\left[|S(j\omega)|^2 + \mathcal{R}e\{S^*(j\omega)W(j\omega)\}\right].$$

If $\frac{\partial\epsilon(\omega)}{\partial H(j\omega)} = 0$, then

$$H(j\omega) = \frac{\left[|S(j\omega)|^2 + \mathcal{R}e\{S^*(j\omega)W(j\omega)\}\right]}{|S(j\omega) + W(j\omega)|^2}.$$

Note that is $S(j\omega_0) + W(j\omega_0) = 0$, then $X(j\omega_0) = 0$ and $Y(j\omega_0) = 0$ no matter what the value of $H(j\omega_0)$.

(c) If $S(j\omega)$ and $W(j\omega)$ are non-overlapping, then $\mathcal{R}e\{S^*(j\omega)W(j\omega)\} = 0$ for all ω and so

$$H(j\omega) = \begin{cases} \frac{|S(j\omega)|^2}{|S(j\omega)-0|^2} = 1, & \text{for } W(j\omega) = 0, S(j\omega) \neq 0 \\ \frac{0}{|0-W(j\omega)|^2} = 0, & \text{for } W(j\omega) \neq 0, S(j\omega) = 0 \\ 0(\text{arbitrarily}), & \text{for } W(j\omega) = 0, S(j\omega) = 0 \end{cases}$$

Clearly, this is an ideal frequency selective filter.

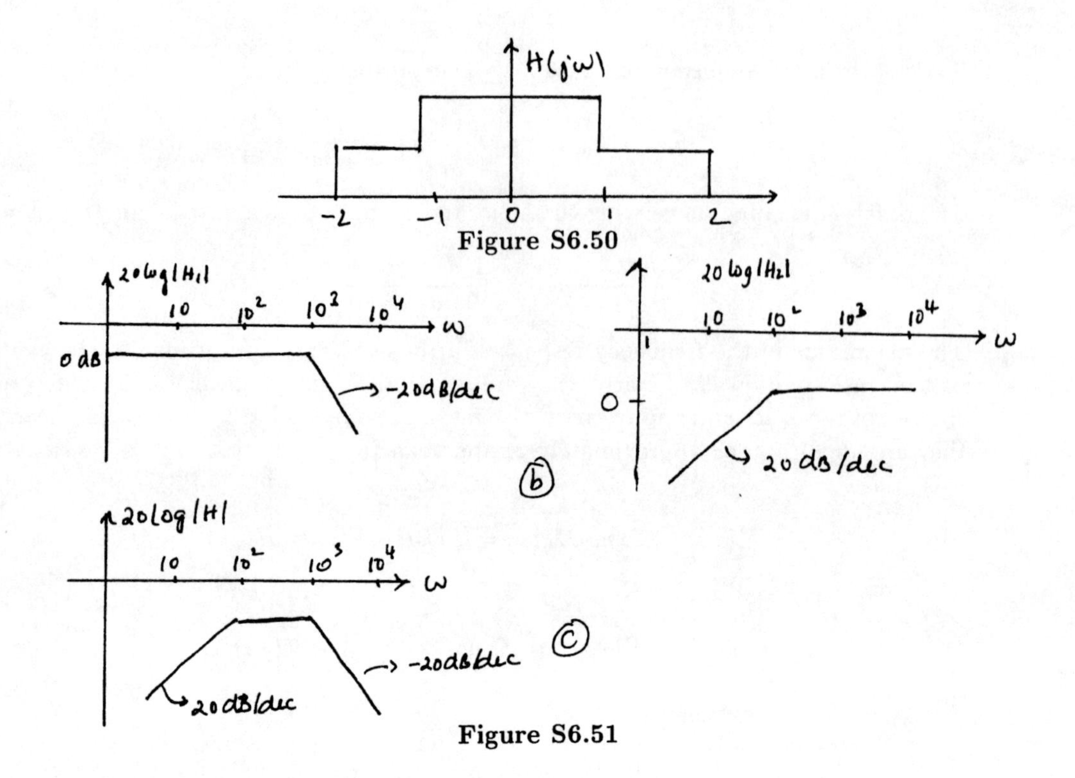

Figure S6.50

Figure S6.51

(d) In this case,

$$H(j\omega) = \begin{cases} 1, & |\omega| \leq 1 \\ \frac{1}{2}, & 1 < |\omega| < 2 \\ 0, & |\omega| \geq 2 \end{cases}$$

This is as shown in Figure S6.50.

6.51. **(a)** We may write $H(j\omega)$ as

$$H(j\omega) = H_{lp}(j\omega) * [\delta(\omega - \omega_0) + \delta(\omega + \omega_0)],$$

where $H_{lp}(j\omega)$ is the frequency response of an ideal lowpass filter with cutoff frequency $\frac{w}{2}$. Therefore,

$$h(t) = 2h_{lp}(t)\cos(\omega_0 t),$$

where

$$h_{lp}(t) = \frac{\sin(wt/2)}{\pi t}.$$

(b) We have

$$H_1(j\omega) = \frac{1}{1 + j\frac{\omega}{10^3}} \quad \text{and} \quad H_2(j\omega) = \frac{j\omega/10^2}{1 + j\frac{\omega}{10^2}}.$$

Therefore the Bode diagrams for these two filters are as shown in Figure S6.51.

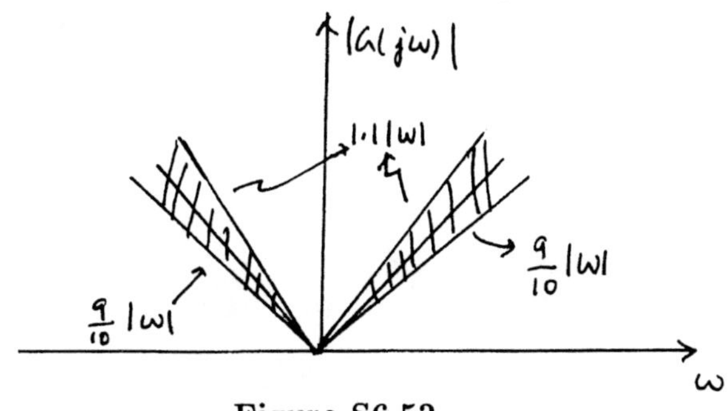

Figure S6.52

(c) Since $H(j\omega) = H_1(j\omega)H_2(j\omega)$,

$$20\log_{10}|H(j\omega)| = 20\log_{10}|H_1(j\omega)| + 20\log_{10}|H_2(j\omega)|.$$

Therefore, the Bode diagram for the bandpass filter is the sum of the two Bode diagrams sketched in part (b).

6.52. (a) Since

$$-0.1|H(j\omega)| \leq [|G(j\omega)| - |H(j\omega)|] \leq 0.1|H(j\omega)|,$$

we have

$$0.9|H(j\omega)| \leq |G(j\omega)| \leq 1.1|H(j\omega)|.$$

Therefore,

$$0.9|\omega| \leq |G(j\omega)| \leq 1.1|\omega|.$$

This is sketched in Figure S6.52.

(b) From Figure P6.52(b) we have

$$y(t) = \frac{1}{T}\left[x(t) - x(t-T)\right].$$

Therefore,

$$Y(j\omega) = \frac{1}{T}\left[X(j\omega) - e^{-j\omega T}X(j\omega)\right]$$

and

$$G(j\omega) = \frac{Y(j\omega)}{X(j\omega)} = \frac{1}{T}[1 - e^{-j\omega T}] = \frac{2}{T}e^{-j\omega T/2}\sin(\omega T/2).$$

Therefore,

$$|G(j\omega)| = \frac{2}{T}|\sin(\omega T/2)|,$$

and

$$\frac{|G(j\omega)|}{|\omega|} = \frac{|\sin(\omega T/2)|}{|\omega T/2|}.$$

For $|G(j\omega)|$ to be within $\pm 10\%$ of $|\omega|$, we require the above ratio to be greater than 0.9. It can be easily shown that for $T = 10^{-2}$, the above ratio falls below 0.9 for $\omega T/2 = \pi/20$, that is, $\omega \approx 31.4$ rad/sec. Therefore, the magnitude of the frequency response of the approximate system remains within $\pm 10\%$ of the ideal differentiator for $|\omega| < 31.4$ rad/sec.

6.53. If $s(t)$ denotes the step response and $h(t)$ the impulse response, then

$$h(t) = \frac{ds(t)}{dt}.$$

If $h(t) \geq 0$, then $\frac{ds(t)}{dt} \geq 0$. This implies that $s(t)$ is a monotonically non-decreasing function.

6.54. **(a)** The cutoff frequency $2\pi \times 10^2$ rad/sec in $H_{lp}(j\omega)$ maps to the frequency $\omega_c = 2\pi \times 10^2/a$ rad/sec in $H_0(j\omega)$. Therefore,

$$a = \frac{2\pi \times 10^2}{\omega_c}.$$

(b) We know from Table 4.1 that

$$x(at) \overset{FT}{\longleftrightarrow} \frac{1}{a} X(j\frac{\omega}{a}).$$

Therefore,

$$h_{lp}(t) = \frac{1}{a} h_0(t/a) = \frac{\omega_c}{2\pi \times 10^2} h_0\left(\frac{\omega_c t}{2\pi \times 10^2}\right).$$

(c) We know that

$$s_0(t) = \int_{-\infty}^{t} h_0(\tau) d\tau.$$

Also,

$$s_{lp}(t) = \int_{-\infty}^{t} h_{lp}(t) d\tau.$$

Therefore,

$$s_{lp}(t) = \frac{1}{a} \int_{-\infty}^{t} h_0(\tau/a) d\tau.$$

Let $\tau' = \tau/a$. Then,

$$s_{lp}(t) = \int_{-\infty}^{t/a} h_0(\tau') d\tau' = s_0(t/a) = s_0(t\omega_c/(2\pi \times 10^2)).$$

(d) Let

$$\lim_{t \to \infty} s_0(t) = A.$$

Then,

$$\tau_r = t_1 - t_0,$$

where $s_0(t_0) = A/10$ and $s_0(t_1) = 9A/10$. Now,

$$\lim_{t\to\infty} s_{lp}(t) = \lim_{t\to\infty} s_{lp}(t/a) = A.$$

We now need to find the times t_2 and t_3 at which $s_{lp}(t)$ is $A/10$ and $9A/10$, respectively. If $s_{lp}(t_2) = A/10$, then $s_0(t_2/a) = A/10$. This implies that $t_2 = at_0$. Also, if $s_{lp}(t_3) = 9A/10$, then $s_0(t_3/a) = 9A/10$. This implies that $t_3 = at_1$. Therefore, the new rise-time is

$$\tau_r' = t_3 - t_2 = a(t_1 - t_0) = a\tau_r = \frac{2\pi}{\omega_c}.$$

τ_r' is sketched in Figure S6.54 as a function of ω_c.

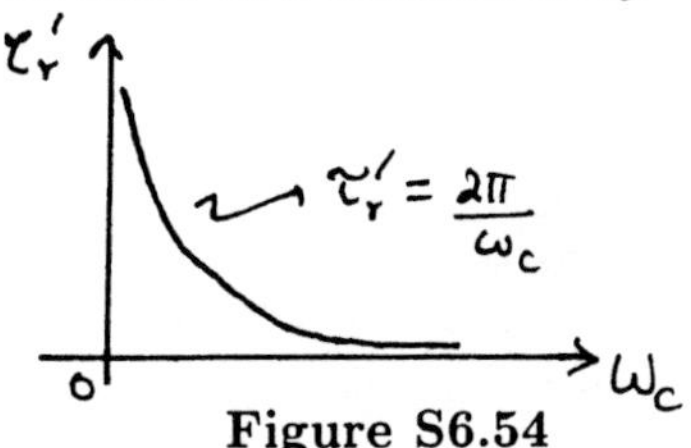

Figure S6.54

6.55. We have

$$|B(j\omega)|^2 = \frac{1}{1 + (\omega/\omega_c)^{2N}} \qquad \text{(S6.55-1)}$$

Also, $|B(j0)|^2 = 1$. Therefore, $|B(j\omega_p)|^2 = 1/2$. From eq.(S6.55-1), we conclude that

$$\left(\frac{\omega_p}{\omega_c}\right)^{2N} = 1 \qquad \Rightarrow \qquad \omega_p = \omega_c.$$

Also, since $|B(j\omega_s)|^2 = 1/100$, we may use eq.(S6.55-1) to conclude that

$$\left(\frac{\omega_s}{\omega_c}\right)^{2N} = 99 \qquad \Rightarrow \qquad \omega_s = (99)^{1/2N}\omega_c.$$

Therefore, the transition ratio is

$$\frac{\omega_s}{\omega_p} = (99)^{1/2N} \approx 10^{1/N}.$$

This is sketched in Figure S6.55.

6.56. **(a)** The conditioning system with frequency response $H_1(j\omega)$ boosts the frequencies that are going to be most affected by the noise. Therefore, its frequency response is chosen to have a magnitude plot as shown in Figure 6.56(a). Therefore,

$$H_1(j\omega) = \frac{\left(1 + \frac{j\omega}{\omega_0}\right)^2}{\left(1 + \frac{j\omega}{\omega_1}\right)^2},$$

where $\omega_0 = 2\pi(5000)$ rad/sec and $\omega_1 = 2\pi(10000)$ rad/sec.

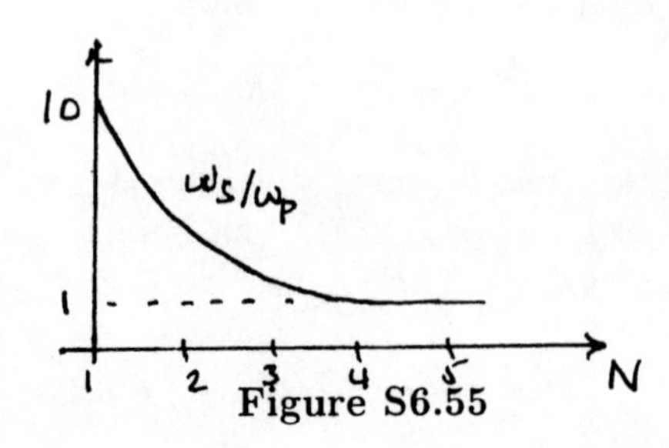

Figure S6.55

(b) The higher frequencies would appear boosted. This would make it sound like the "treble" was higher.

(c) The system with frequency response $H_2(j\omega)$ should undo the effects of $H_1(j\omega)$. Therefore, it has to be the inverse system of $H_1(j\omega)$. The Bode plot for $H_2(j\omega)$ would be as shown in Figure S6.56.

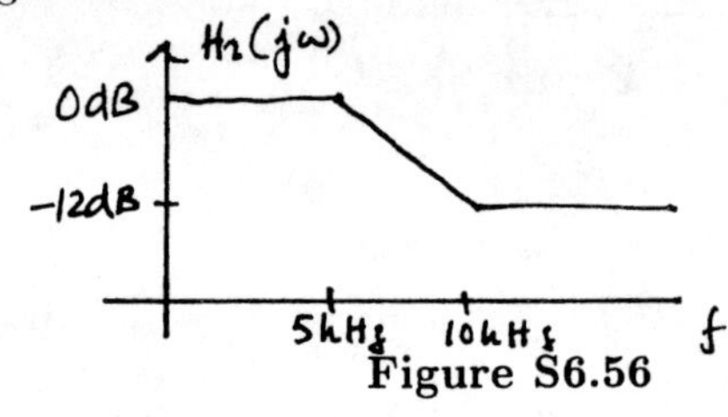

Figure S6.56

Therefore,

$$H_2(j\omega) = \frac{\left(1 + \frac{j\omega}{\omega_1}\right)^2}{\left(1 + \frac{j\omega}{\omega_0}\right)^2},$$

where $\omega_0 = 2\pi(5000)$ rad/sec and $\omega_1 = 2\pi(10000)$ rad/sec. The input $x(t)$ and the output $y(t)$ of $H_2(j\omega)$ are related by the following differential equation

$$\frac{1}{\omega_0^2}\frac{dy^2(t)}{dt^2} + \frac{2}{\omega_0}\frac{dy(t)}{dt} + y(t) = \frac{1}{\omega_1^2}\frac{dx^2(t)}{dt^2} + \frac{2}{\omega_1}\frac{dx(t)}{dt} + x(t).$$

6.57. If $s[n]$ denotes the step response and $h[n]$ the impulse response, then

$$h[n] = s[n] - s[n-1].$$

If $h[n] \geq 0$, then $s[n] \geq s[n-1]$. This implies that $s[n]$ is a monotonically non-decreasing function.

6.58. **(a)** The sequence of operations shown in Figure 6.58(a) may be interpreted as follows:

$$\begin{aligned}
G(e^{j\omega}) &= H(e^{j\omega})X(e^{j\omega}) \\
R(e^{j\omega}) &= G(e^{-j\omega})H(e^{j\omega}) = H(e^{-j\omega})X(e^{-j\omega})H(e^{j\omega}) \\
S(e^{j\omega}) &= R(e^{-j\omega}) = H(e^{j\omega})X(e^{j\omega})H(e^{-j\omega}) = H_1(e^{j\omega})X(e^{j\omega})
\end{aligned}$$

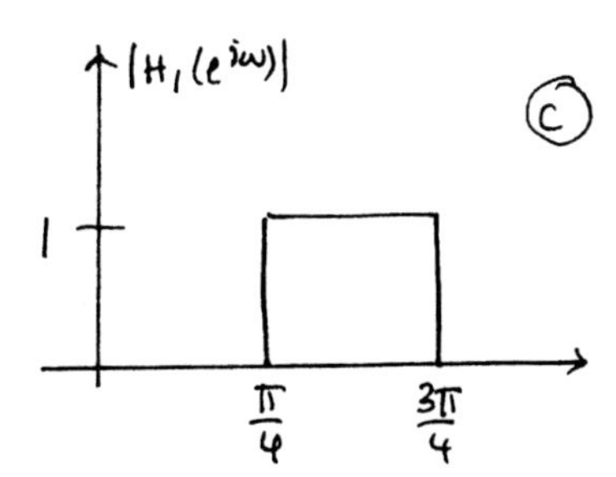

(c)

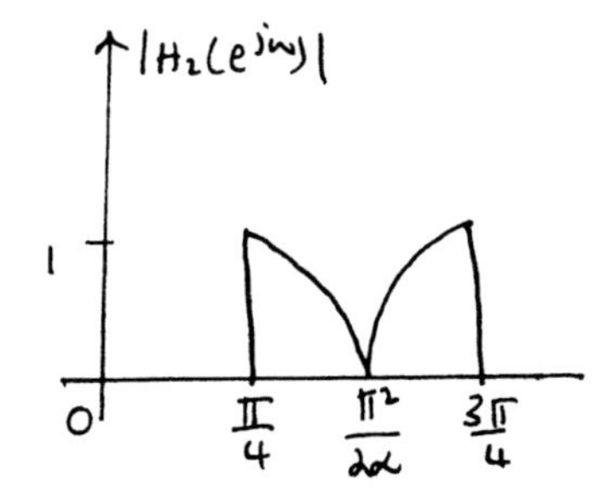

Figure S6.58

Therefore,

$$H_1(e^{j\omega}) = H(e^{j\omega})H(e^{-j\omega}).$$

If $h[n]$ is real, then $H(e^{j\omega}) = H^*(e^{-j\omega})$. Then

$$H_1(e^{j\omega}) = |H(e^{j\omega})|^2.$$

Therefore,

$$h_1[n] = h[n] * h[-n].$$

Also,

$$|H_1(e^{j\omega})| = |H(e^{j\omega})|^2 \quad \text{and} \quad \sphericalangle H_1(e^{j\omega}) = 0.$$

(b) The sequence of operations shown in Figure 6.58(a) may be interpreted as follows:

$$\begin{aligned} G(e^{j\omega}) &= H(e^{j\omega})X(e^{j\omega}) \\ R(e^{j\omega}) &= X(e^{-j\omega})H(e^{j\omega}) \\ Y(e^{j\omega}) &= G(e^{j\omega}) + R(e^{-j\omega}) = X(e^{j\omega})[H(e^{j\omega}) + H(e^{-j\omega})] \end{aligned}$$

Therefore,

$$H_2(e^{j\omega}) = H(e^{j\omega}) + H(e^{-j\omega}).$$

If $h[n]$ is real, then $H(e^{j\omega}) = H^*(e^{-j\omega})$. Then

$$H_1(e^{j\omega}) = 2\mathcal{R}e\{H(e^{j\omega})\} = 2|H(e^{j\omega})|\cos(\sphericalangle H(e^{j\omega})).$$

Therefore,

$$h_2[n] = \frac{h[n] + h[-n]}{2}.$$

Also,

$$|H_2(e^{j\omega}) = 2|H(e^{j\omega})||\cos(\sphericalangle H(e^{j\omega}))|.$$

(c) The plots for $|H_1(e^{j\omega})|$ and $|H_1(e^{j\omega})|$ are shown in Figure S6.58.

Clearly, Method A is preferable because the magnitude of the zero-phase filter does not depend on the phase of $h[n]$.

6.59. **(a)** We have

$$\begin{aligned} E(e^{j\omega}) &= H_d(e^{j\omega}) - H(e^{j\omega}) \\ &= \sum_{n=-\infty}^{\infty} [h_d[n]e^{-j\omega n} - h[n]e^{-j\omega n}] \\ &= \sum_{n=-\infty}^{\infty} (h_d[n] - h[n])e^{-j\omega n} \end{aligned}$$

Therefore, $e[n] = h_d[n] - h[n]$.

(b) Noting that $E(e^{j\omega})$ is the Fourier transform of $e[n]$, we may use Parseval's theorem to obtain

$$\epsilon^2 = \frac{1}{2\pi}\int_{-\pi}^{\pi} |E(e^{j\omega})|^2 d\omega = \sum_{n=-\infty}^{\infty} |e[n]|^2.$$

(c) We have

$$\begin{aligned} \epsilon^2 &= \sum_{n=-\infty}^{\infty} |e[n]|^2 \\ &= \sum_{n=-\infty}^{\infty} |h_d[n] - h[n]|^2 \\ &= \sum_{n=0}^{N-1} |h_d[n] - h[n]|^2 + \sum_{n=-\infty}^{0} |h_d[n]|^2 + \sum_{n=N}^{\infty} |h_d[n]|^2 \end{aligned}$$

The last two terms in the right-hand side of the above equation are constant. The only variable term $\sum_{n=0}^{N-1} |h_d[n] - h[n]|^2$ is minimized when $h_d[n] = h[n]$ in the range $0 \le n \le N-1$.

6.60. The development is identical to that in Problem 6.50. We have

$$\begin{aligned} \epsilon(e^{j\omega}) &= |S(e^{j\omega}) - Y(e^{j\omega})|^2 \\ &= |S(e^{j\omega}) - H(e^{j\omega})[S(e^{j\omega}) + W(e^{j\omega})]|^2 \\ &= |S(e^{j\omega})|^2 + H^2(e^{j\omega})|S(e^{j\omega}) + W(e^{j\omega})|^2 \\ &\quad -2H(e^{j\omega})[|S(e^{j\omega})|^2 + \mathcal{R}e\{S^*(e^{j\omega})W(e^{j\omega})\}] \end{aligned}$$

where $H(e^{j\omega})$ is assumed to be real. With $\partial\epsilon(e^{j\omega})/\partial H(e^{j\omega}) = 0$, we obtain

$$H(e^{j\omega}) = \frac{[|S(e^{j\omega})|^2 + \mathcal{R}e\{S^*(e^{j\omega})W(e^{j\omega})\}]}{|S(e^{j\omega}) + W(e^{j\omega})|^2}.$$

If for some ω_0, $S(e^{j\omega_0}) = W(e^{j\omega_0}) = 0$, then $Y(e^{j\omega_0}) = 0$ regardless of the value of $H(e^{j\omega_0})$.

6.61. **(a)** We have

$$G(e^{j\omega}) = H(e^{j\omega})H(e^{j\omega}) = |H(e^{j\omega})|^2 e^{j2\sphericalangle H(e^{j\omega})}.$$

Therefore,

$$|G(e^{j\omega})| = |H(e^{j\omega})|^2.$$

It follows that the tolerance limits on $|G(e^{j\omega})|$ are given by

$$\begin{aligned}(1-\delta_1)^2 &\leq |G(e^{j\omega})| \leq (1+\delta_1)^2, \qquad 0 \leq \omega \leq \omega_1 \\ 0 &\leq |G(e^{j\omega})| \leq \delta_2^2, \qquad \omega_2 \leq \omega \leq \pi\end{aligned}$$

(b) If $\delta_1 << 1$ and $\delta_2 << 1$, then $(1-\delta_1)^2 \approx 1-2\delta_1$ and $(1+\delta_1)^2 \approx 1+2\delta_1$. Also, $\delta_2^2 < \delta_2$. Therefore, the passband ripple *increases* and the stopband ripple *decreases*.

(c) If N filters are cascaded, then the overall frequency response is

$$G(e^{j\omega}) = |H(e^{j\omega})|^N e^{jN\sphericalangle H(e^{j\omega})}.$$

Therefore,

$$|G(e^{j\omega})| = |H(e^{j\omega})|^N.$$

The tolerance limits are now:

$$\begin{aligned}(1-\delta_1)^N &\leq |G(e^{j\omega})| \leq (1+\delta_1)^N, \qquad 0 \leq \omega \leq \omega_1 \\ 0 &\leq |G(e^{j\omega})| \leq \delta_2^N, \qquad \omega_2 \leq \omega \leq \pi\end{aligned}$$

If $\delta_1 << 1$, then $(1-\delta_1)^N \approx 1-N\delta_1$ and $(1+\delta_1)^N \approx 1+N\delta_1$. Therefore, the tolerance limits on $|G(e^{j\omega})|$ are given by

$$\begin{aligned}1-N\delta_1 &\leq |G(e^{j\omega})| \leq 1+N\delta_1, \qquad 0 \leq \omega \leq \omega_1 \\ 0 &\leq |G(e^{j\omega})| \leq \delta_2^N, \qquad \omega_2 \leq \omega \leq \pi\end{aligned}$$

6.62. **(a)** From Figure P6.62(a) we have

$$W(e^{j\omega}) = \left[2X(e^{j\omega}) - X(e^{j\omega})H(e^{j\omega})\right] H(e^{j\omega}).$$

Therefore,

$$G(e^{j\omega}) = \frac{W(e^{j\omega})}{X(e^{j\omega})} = \left[2 - H(e^{j\omega})\right] H(e^{j\omega}).$$

Let $H(e^{j\omega}) = 1+\delta_1$. Then $G(e^{j\omega}) = [2-1-\delta_1][1+\delta_1] = 1-\delta_1^2$. Let $H(e^{j\omega}) = 1-\delta_1$. Then $G(e^{j\omega}) = [2-1+\delta_1][1-\delta_1] = 1-\delta_1^2$. Therefore,

$$1-\delta_1^2 \leq G(e^{j\omega}) \leq 1, \qquad 0 \leq \omega \leq \omega_p.$$

Therefore, $A = 1-\delta_1^2$ and $B = 1$. Let $H(e^{j\omega}) = -\delta_2$. Then $G(e^{j\omega}) = [2+\delta_2][-\delta_2] = -2\delta_2 - \delta_2^2$. Let $H(e^{j\omega}) = \delta_2$. Then $G(e^{j\omega}) = [2-\delta_2][\delta_2] = 2\delta_2 - \delta_2^2$. Therefore,

$$-2\delta_2 - \delta_2^2 \leq G(e^{j\omega}) \leq 2\delta_2 - \delta_2^2, \qquad \omega_s \leq \omega \leq \pi.$$

Therefore, $C = -2\delta_2 - \delta_2^2$ and $D = 2\delta_2 - \delta_2^2$.

(b) If $\delta_1 << 1$ and $\delta_2 << 1$, then $A \approx 1 - \delta_1^2$, $B \approx 1 + \delta_1^2$, $C \approx -2\delta_2$ and $D \approx 2\delta_2$. Therefore, the passband ripple is smaller and the stopband ripple is larger.

(c) From part (a), we have

$$|G(e^{j\omega})| = |2 - H(e^{j\omega})||H(e^{j\omega})|.$$

Since $|2 - H(e^{j\omega})| \leq 2 + |H(e^{j\omega})|$ and $|2 - H(e^{j\omega})| \geq 2 - |H(e^{j\omega})|$, we may write

$$\left[2 - |H(e^{j\omega})|\right] |H(e^{j\omega})| \leq G(e^{j\omega}) \leq \left[2 + |H(e^{j\omega})|\right] |H(e^{j\omega})| \qquad \text{(S6.62-1)}$$

If $H(e^{j\omega}) \approx 1$, then from the above equation we obtain

$$1 \leq G(e^{j\omega}) \leq 3.$$

If $H(e^{j\omega}) \approx 0$, then from the eq. (S6.62-1) we obtain

$$0 \leq G(e^{j\omega}) \leq 0.$$

Therefore, the filter is a good approximation of a lowpass filter in the stopband. But in the passband, for some $\theta(\omega)$ it is possible to obtain extremely large ripple. Therefore, overall it is not a good approximation for a lowpass filter.

(d) In Figure P6.62(a) if we attach a N point delay to $H(e^{j\omega})$, then the equivalent filter will be a real filter that is a good approximation to a lowpass filter. We have seen that in such a case the overall system is also a good approximation to lowpass.

6.63. **(a)** Let $g[n] = nh[n]$. Then,

$$G(e^{j\omega}) = j\frac{dH(e^{j\omega})}{d\omega}.$$

Using Parseval's theorem (an also noting that $g[n]$ is real)

$$\sum_{n=-\infty}^{\infty} g^2[n] = \frac{1}{2\pi}\int_{-\pi}^{\pi} |G(e^{j\omega})|^2 d\omega.$$

Therefore,

$$D = \sum_{n=-\infty}^{\infty} n^2 h^2[n] = \frac{1}{2\pi} \left|\frac{dH(e^{j\omega})}{d\omega}\right|^2 d\omega.$$

(b) Replacing $H(e^{j\omega})$ by $|H(e^{j\omega})|e^{j\theta(\omega)}$ in the result of part (a),

$$\begin{aligned} D &= \frac{1}{2\pi}\int_{-\pi}^{\pi} \left| e^{j\theta(\omega)}\frac{d|H(e^{j\omega})|}{d\omega} + |H(e^{j\omega})|e^{j\theta(\omega)}\frac{d\theta(\omega)}{d\omega}\right|^2 d\omega \\ &= \frac{1}{2\pi}\int_{-\pi}^{\pi} \left|\frac{d|H(e^{j\omega})|}{d\omega} + |H(e^{j\omega})|\frac{d\theta(\omega)}{d\omega}\right|^2 d\omega \end{aligned}$$

Let $M(\omega) = |H(e^{j\omega})|$ and $\theta'(\omega) = \frac{d\theta(\omega)}{d\omega}$. Also note that $M(\omega) = M(-\omega)$, $M'(\omega) = M'(-\omega)$ and $\theta'(\omega) = \theta'(-\omega)$. Therefore,

$$D = \frac{1}{2\pi} \int_0^{\pi} \left\{ |M'(\omega) + M(\omega)\theta'(\omega)|^2 + |M'(\omega) - M(\omega)\theta'(\omega)|^2 \right\} d\omega.$$

Now since the integrand is positive for all ω, it is sufficient to minimize the integrand to minimize D. Therefore,

$$\frac{d}{d\theta'(\omega)} \left\{ |M'(\omega) + M(\omega)\theta'(\omega)|^2 + |M'(\omega) - M(\omega)\theta'(\omega)|^2 \right\} = 0.$$

Simplifying this, we obtain

$$2M^2(\omega)\theta'(\omega) = 0 \quad \Rightarrow \quad \theta'(\omega) = 0.$$

However, since $\theta(\omega)$ is odd, the only function that satisfies $\theta'(\omega) = 0$ is $\theta(\omega) = 0$.

6.64. **(a)** From Table 5.1 we know that when a signal is real and even, then its Fourier transform is also real and even. Therefore, using duality, we may say that if the Fourier transform of a signal is real and even, then the signal is real and even. Therefore, $h_r[n] = h_r[-n]$.

By using the time shift property, we know that if $H(e^{j\omega}) = H_r(e^{j\omega})e^{-j\omega M}$, then

$$h[n] = h_r[n - M].$$

(b) We have

$$h[M + n] = h_r[M + n - M] = h_r[n].$$

Also,

$$h[M - n] = h_r[M - n - M] = h_r[-n].$$

Since $h_r[n] = h_r[-n]$,

$$h[M + n] = h[M - n].$$

(c) Since $h[n]$ is causal, $h[-k] = 0$ for $k > 0$. But due to the symmetry property,

$$h[-k] = h_r[-k - M] = h_r[k + M] = h[k + 2M].$$

Therefore,

$$h[k + 2M] = 0 \quad \text{for } k > 0.$$

It follows that

$$h[n] = 0 \quad \text{for } n > 2M.$$

6.65. **(a)** We have

$$|[B(e^{j\omega})|^2 = \frac{1}{1 + \tan^2(\omega/2)} = \frac{1}{\sec^2(\omega/2)} = \cos^2(\omega/2).$$

(b) If $B(e^{j\omega}) = a\cos(\omega/2)$, then

$$|[B(e^{j\omega})|^2 = aa^* \cos^2(\omega/2).$$

If we want this to be the same as part (a), then $aa^* = 1$. Therefore,

$$a = e^{j\theta(\omega)}.$$

(c) Taking the Fourier transform of the given difference equation we obtain

$$H(e^{j\omega}) = \frac{Y(e^{j\omega})}{X(e^{j\omega})} = \alpha + \beta e^{-j\omega\gamma} = e^{-j\omega\gamma/2}[\alpha e^{j\omega\gamma/2} + \beta e^{-j\omega\gamma/2}].$$

Comparing with

$$B(e^{j\omega}) = e^{-j\theta(\omega)} \left[\frac{1}{2}e^{j\omega/2} + \frac{1}{2}e^{-j\omega/2}\right],$$

we find that $H(e^{j\omega}) = B(e^{j\omega})$ when

$$\alpha = \beta = \frac{1}{2}, \qquad \gamma = 1.$$

6.66. **(a)** Since $h_k[n] = e^{j2\pi nk/N} h_0[n]$, we have

$$H_k(e^{j\omega}) = H_0(e^{j(\omega - 2\pi k/N)}).$$

Below are shown the sketches of $H_k(e^{j\omega})$ for $N = 16$ in Figure S6.66.

(b) Overall frequency response of the system is $H_{ov}(e^{j\omega}) = \sum_{k=0}^{N-1} H_k(e^{j\omega})$. For this to be an identity system, we require that $H_{ov}(e^{j\omega}) = 1$ for all ω. Therefore, we want the non-zero portions of the $H_k(e^{j\omega})$s to be non-overlapping and yet cover the region from $-\pi$ to π. We see that this is achieved by having $\omega_c = \pi/N$.

(c) Since $H_{ov}(e^{j\omega}) = \sum_{k=0}^{N-1} H_k(e^{j\omega})$, we have

$$h_{ov}[n] = \sum_{k=0}^{N-1} h_k[n] = \sum_{k=0}^{N-1} h_0[n] e^{j2\pi kn/N} = h_0[n] \sum_{k=0}^{N-1} e^{j2\pi kn/N}.$$

Therefore,

$$r[n] = \sum_{k=0}^{N-1} e^{j2\pi kn/N} = \begin{cases} N, & n = 0, \pm N, \pm 2N, \cdots \\ 0, & \text{otherwise} \end{cases}.$$

Therefore, $r[n] = N \sum_{k=-\infty}^{\infty} \delta[n - kN]$ and is as sketched in Figure S6.66.

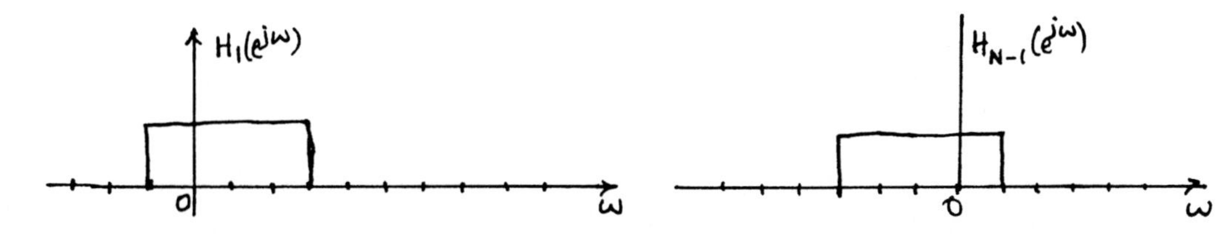

Figure S6.66

(d) In order for $h[n]$ to be the impulse response of an identity system, we require that $h[n] = \delta[n]$. From part (c), we know that

$$h[n] = h_0[n] \sum_{k=-\infty}^{\infty} \delta[n - kN].$$

Therefore, the necessary and sufficient condition for $h[n]$ to be $\delta[n]$ is

$$h_0[0] = \frac{1}{N} \qquad \text{and} \qquad h_0[kN] = 0 \qquad \text{for } k = \pm 1, \pm 2, \cdots .$$

Chapter 7 Answers

7.1. From the Nyquist sampling theorem, we know that only if $X(j\omega) = 0$ for $|\omega| > \omega_s/2$ will be signal be recoverable from its samples. Therefore, $X(j\omega) = 0$ for $|\omega| > 5000\pi$.

7.2. From the Nyquist theorem, we know that the sampling frequency in this case must be at least $\omega_s = 2000\pi$. In other words, the sampling period should be at most $T = 2\pi/(\omega_s) = 1 \times 10^{-3}$. Clearly, only (a) and (c) satisfy this condition.

7.3. **(a)** We can easily show that $X(j\omega) = 0$ for $|\omega| > 4000\pi$. Therefore, the Nyquist rate for this signal is $\omega_N = 2(4000\pi) = 8000\pi$.

(b) From Table 4.2 we know that, $X(j\omega)$ is a rectangular pulse for which $X(j\omega) = 0$ for $|\omega| > 4000\pi$. Therefore, the Nyquist rate for this signal is $\omega_N = 2(4000\pi) = 8000\pi$.

(c) From Tables 4.1 and 4.2, we know that $X(j\omega)$ is the convolution of two rectangular pulses each of which is zero for $|\omega| > 4000\pi$. Therefore, $X(j\omega) = 0$ for $|\omega| > 8000\pi$ and the Nyquist rate for this signal is $\omega_N = 2(8000\pi) = 16000\pi$.

7.4. If the signal $x(t)$ has a Nyquist rate of ω_0, then its Fourier transform $X(j\omega) = 0$ for $|\omega| > \omega_0/2$.

(a) From chapter 4,

$$y(t) = x(t) + x(t-1) \overset{FT}{\longleftrightarrow} Y(j\omega) = X(j\omega) + e^{-j\omega t}X(j\omega).$$

Clearly, we can only guarantee that $Y(j\omega) = 0$ for $|\omega| > \omega_0/2$. Therefore, the Nyquist rate for $y(t)$ is also ω_0.

(b) From chapter 4,

$$y(t) = \frac{dx(t)}{dt} \overset{FT}{\longleftrightarrow} Y(j\omega) = j\omega X(j\omega).$$

Clearly, we can only guarantee that $Y(j\omega) = 0$ for $|\omega| > \omega_0/2$. Therefore, the Nyquist rate for $y(t)$ is also ω_0.

(c) From chapter 4,

$$y(t) = x^2(t) \overset{FT}{\longleftrightarrow} Y(j\omega) = (1/2\pi)[X(j\omega) * X(j\omega).$$

Clearly, we can guarantee that $Y(j\omega) = 0$ for $|\omega| > \omega_0$. Therefore, the Nyquist rate for $y(t)$ is $2\omega_0$.

(d) From chapter 4,

$$y(t) = x(t)\cos(\omega_0 t) \overset{FT}{\longleftrightarrow} Y(j\omega) = (1/2)X(j(\omega-\omega_0)) + (1/2)X(j(\omega+\omega_0)).$$

Clearly, we can guarantee that $Y(j\omega) = 0$ for $|\omega| > \omega_0 + \omega_0/2$. Therefore, the Nyquist rate for $y(t)$ is $3\omega_0$.

7.5. Using Table 4.2

$$p(t) \stackrel{FT}{\longleftrightarrow} \frac{2\pi}{T} \sum_{k=-\infty}^{\infty} \delta(\omega - k2\pi/T).$$

From Table 4.1,

$$p(t-1) \stackrel{FT}{\longleftrightarrow} \frac{2\pi}{T} e^{-j\omega} \sum_{k=-\infty}^{\infty} \delta(\omega - k\frac{2\pi}{T}) = \frac{2\pi}{T} \sum_{k=-\infty}^{\infty} \delta(\omega - k\frac{2\pi}{T}) e^{-jk\frac{2\pi}{T}}.$$

Since $y(t) = x(t)p(t-1)$, we have

$$\begin{aligned} Y(j\omega) &= (1/2\pi)[X(j\omega) * \mathcal{FT}\{p(t-1)\}] \\ &= (1/T) \sum_{k=-\infty}^{\infty} X(j(\omega - k\frac{2\pi}{T}))e^{-jk\frac{2\pi}{T}} \end{aligned}$$

Therefore, $Y(j\omega)$ consists of replicas of $X(j\omega)$ shifted by $k2\pi/T$ and added to each other (see Figure S7.5). In order to recover $x(t)$ from $y(t)$, we need to be able to isolate one replica of $X(j\omega)$ from $Y(j\omega)$.

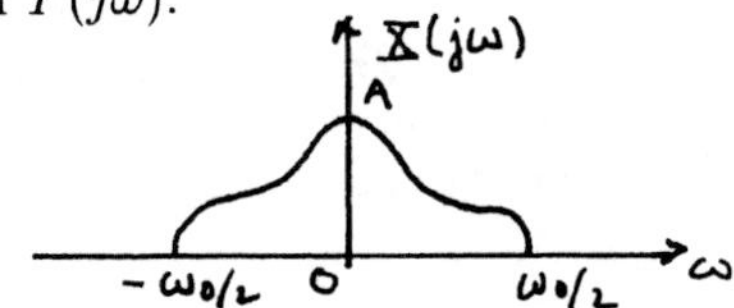

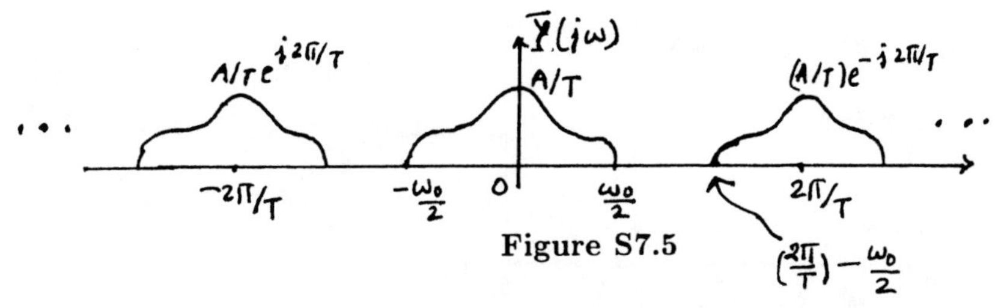

Figure S7.5

From the figure, it is clear that this is possible if we multiply $Y(j\omega)$ with

$$H(j\omega) = \begin{cases} T, & |\omega| \leq \omega_c \\ 0, & \text{otherwise} \end{cases}$$

where $(\omega_0/2) < \omega_c < (2\pi/T) - (\omega_0/2)$.

7.6. Consider the signal $w(t) = x_1(t)x_2(t)$. The Fourier transform $W(j\omega)$ of $w(t)$ is given by

$$W(j\omega) = \frac{1}{2\pi}[X_1(j\omega) * X_2(j\omega)].$$

Since $X_1(j\omega) = 0$ for $|\omega| \geq \omega_1$ and $X_2(j\omega) = 0$ for $|\omega| \geq \omega_2$, we may conclude that $W(j\omega) = 0$ for $|\omega| \geq \omega_1 + \omega_2$. Consequently, the Nyquist rate for $w(t)$ is $\omega_s = 2(\omega_1 + \omega_2)$. Therefore, the *maximum* sampling period which would still allow $w(t)$ to be recovered is $T = 2\pi/(\omega_s) = \pi/(\omega_1 + \omega_2)$.

7.7. We note that

$$x_1(t) = h_1(t) * \left\{ \sum_{n=-\infty}^{\infty} x(nT)\delta(t-nT) \right\}.$$

From Figure 7.7 in the textbook, we know that the output of the zero-order hold may be written as

$$x_0(t) = h_0(t) * \left\{ \sum_{n=-\infty}^{\infty} x(nT)\delta(t-nT) \right\},$$

where $h_0(t)$ is as shown in Figure S7.7. By taking the Fourier transform of the two above equations, we have

$$\begin{aligned} X_1(j\omega) &= H_1(j\omega)X_p(j\omega) \\ X_0(j\omega) &= H_0(j\omega)X_p(j\omega) \end{aligned}$$

We now need to determine a frequency response $H_d(j\omega)$ for a filter which produces $x_1(t)$ at its output when $x_0(t)$ is its input. Therefore, we need

$$X_0(j\omega)H_d(j\omega) = X_1(j\omega).$$

The triangular function $h_1(t)$ may be obtained by convolving two rectangular pulses as shown in Figure S7.7.

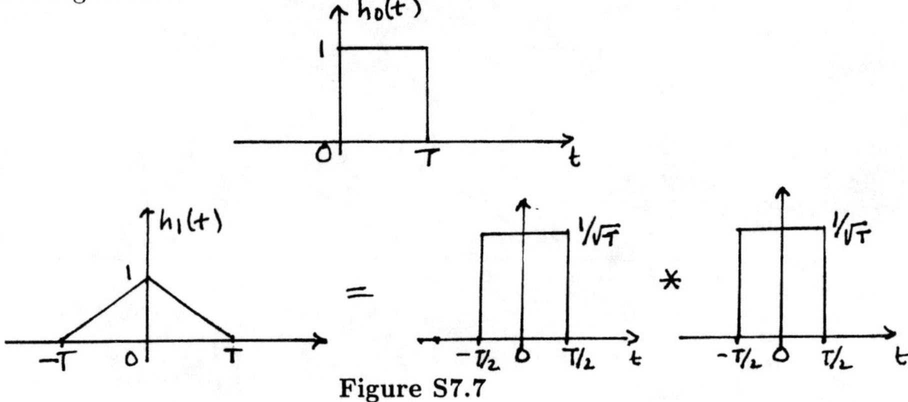

Figure S7.7

Therefore,

$$h_1(t) = \{(1/\sqrt{T})h_0(t+T/2)\} * \{(1/\sqrt{T})h_0(t+T/2)\}.$$

Taking the Fourier transform of both sides of the above equation,

$$H_1(j\omega) = \frac{1}{T}e^{j\omega T}H_0(j\omega)H_0(j\omega).$$

Therefore,

$$\begin{aligned} X_1(j\omega) &= H_1(j\omega)X_p(j\omega) \\ &= \frac{1}{T}e^{j\omega T}H_0(j\omega)H_0(j\omega)X_p(j\omega) \\ &= \frac{1}{T}e^{j\omega T}H_0(j\omega)X_0(j\omega) \end{aligned}$$

Therefore,

$$H_d(j\omega) = \frac{1}{T} e^{j\omega T} H_0(j\omega) = e^{j\omega T/2} \frac{2\sin(\omega T/2)}{\omega T}.$$

7.8. (a) Yes, aliasing does occur in this case. This may be easily shown by considering the sinusoidal term of $x(t)$ for $k = 5$. This term is a signal of the form $y(t) = (1/2)^5 \sin(5\pi t)$. If $x(t)$ is sampled as $T = 0.2$, then we will always be sampling $y(t)$ at exactly its zero-crossings (This is similar to the idea presented in Figure 7.17 of your textbook). Therefore, the signal $y(t)$ appears to be identical to the signal $(1/2)^5 \sin(0\pi t)$ for all time in the sampled signal. Therefore, the sinusoid $y(t)$ of frequency 5π is aliased into a sinusoid of frequency 0 in the sampled signal.

(b) The lowpass filter performs band limited interpolation on the signal $\hat{x}(t)$. But since aliasing has already resulted in the loss of the sinusoid $(1/2)^5 \sin(5\pi t)$, the output will be of the form

$$x_r(t) = \sum_{k=0}^{4} \left(\frac{1}{2}\right)^k \sin(k\pi t).$$

The Fourier series representation of this signal is of the form

$$x_r(t) = \sum_{k=-4}^{4} a_k e^{-j(k\pi/t)}, \qquad \text{where} \qquad a_k = \begin{cases} 0, & k = 0 \\ -j(1/2)^{k+1}, & 1 \le k \le 4 \\ j(1/2)^{-k+1}, & -4 \le k \le -1 \end{cases}.$$

7.9. The Fourier transform $X(j\omega)$ of $x(t)$ is as shown in Figure S7.9.

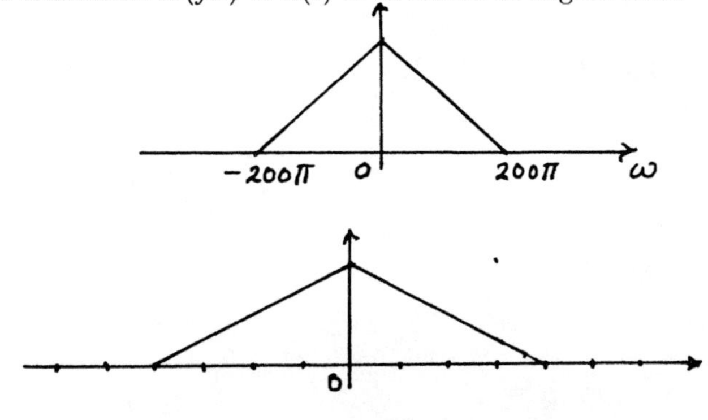

Figure S7.9

We know from the results on impulse-train sampling that

$$G(j\omega) = \frac{1}{T} \sum_{k=-\infty}^{\infty} X(j(\omega - k\omega_s)),$$

where $T = 2\pi/\omega_s = 1/75$. Therefore, $G(j\omega)$ is as shown in Figure S7.9. Clearly, $G(j\omega) = (1/T)X(j\omega) = 75X(j\omega)$ for $|\omega| \le 50\pi$.

7.10. (a) We know that $x(t)$ is not a band-limited signal. Therefore, it *cannot* undergo impulse-train sampling without aliasing.

(b) From the given $X(j\omega)$ it is clear that the signal $x(t)$ which is bandlimited. That is, $X(j\omega) = 0$ for $|\omega| > \omega_0$. Therefore, it must be possible to perform impulse-train sampling on this signal without experiencing aliasing. The minimum sampling rate required would be be $\omega_s = 2\omega_0$. This implies that the sampling period can at most be $T = 2\pi/\omega_s = \pi/\omega_0$.

(c) When $x(t)$ undergoes impulse train sampling with $T = 2\pi/\omega_0$, we would obtain the signal $g(t)$ with Fourier transform

$$G(j\omega) = \frac{1}{T}\sum_{k=-\infty}^{\infty} X(j(\omega - k2\pi/T)).$$

This is as shown in the Figure S7.10.

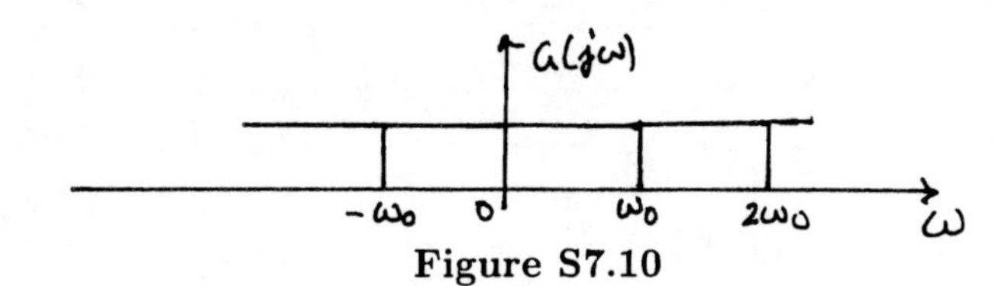

Figure S7.10

It is clear from the figure that no aliasing occurs, and that $X(j\omega)$ can be recovered by using a filter with frequency response

$$H(j\omega) = \begin{cases} T, & 0 \le \omega \le \omega_0 \\ 0, & \text{otherwise} \end{cases}.$$

Therefore, the given statement is true.

7.11. We know from Section 7.4 that

$$X_d(e^{j\omega}) = \frac{1}{T}\sum_{k=-\infty}^{\infty} X_c(j(\omega - 2\pi k)/T).$$

(a) Since $X_d(e^{j\omega})$ is just formed by shifting and summing replicas of $X(j\omega)$, we may argue that if $X_d(e^{j\omega})$ is real, then $X(j\omega)$ must also be real.

(b) $X_d(e^{j\omega})$ consists of replicas of $X(j\omega)$ which are scaled by $1/T$. Therefore, if $X_d(e^{j\omega})$ has a maximum of 1, then $X(j\omega)$ will have a maximum of $T = 0.5 \times 10^{-3}$.

(c) The region $3\pi/4 \le |\omega| \le \pi$ in the discrete-time domain corresponds to the region $3\pi/(4T) \le |\omega| \le \pi/T$ in the continuous-time domain. Therefore, if $X_d(e^{j\omega}) = 0$ for $3\pi/4 \le |\omega| \le \pi$, then $X(j\omega) = 0$ for $1500\pi \le |\omega| \le 2000\pi$. But since we already have $X(j\omega) = 0$ for $|\omega| \ge 2000\pi$, we have $X(j\omega) = 0$ for $|\omega| \ge 1500\pi$.

(d) In this case, since π in discrete-time frequency domain corresponds to 2000π in the continuous-time frequency domain, this condition translates to $X(j\omega) = (j(\omega - 2000\pi))$.

7.12. From Section 7.4, we know that the discrete and continuous-time frequencies Ω and ω are related by $\Omega = \omega T$. Therefore, in this case for $\Omega = \frac{3\pi}{4}$, we find the corresponding value of ω to be $\omega = \frac{3\pi}{4}\frac{1}{T} = 3000\pi/4 = 750\pi$.

7.13. For this problem, we use an approach similar to the one used in Example 7.2. We assume that

$$x_c(t) = \frac{\sin(\pi t/T)}{\pi t}.$$

The overall output is

$$y_c(t) = x_c(t - 2T) = \frac{\sin[(\pi/T)(t - 2T)]}{\pi(t - 2T)}.$$

From $x_c(t)$, we obtain the corresponding discrete-time signal $x_d[n]$ to be

$$x_d[n] = x_c(nT) = \frac{1}{T}\delta[n].$$

Also, we obtain from $y_c(t)$, the corresponding discrete-time signal $y_d[n]$ to be

$$y_d[n] = y_c(nT) = \frac{\sin[\pi(n-2)]}{\pi T(n-2)}.$$

We note that the right-hand side of the above equation is always zero when $n \neq 2$. When $n = 2$, we may evaluate the value of the ratio using L' Hospital's rule to be $1/T$. Therefore,

$$y_d[n] = \frac{1}{T}\delta[n-2].$$

We conclude that the impulse response of the filter is

$$h_d[n] = \delta[n-2].$$

7.14. For this problem, we use an approach similar to the one used in Example 7.2. We assume that

$$x_c(t) = \frac{\sin(\pi t/T)}{\pi t}.$$

The overall output is

$$y_c(t) = \frac{d}{dt}x_c(t - \frac{T}{2}) = \frac{(\pi/T)\cos[(\pi/T)(t - T/2)]}{\pi(t - T/2)} - \frac{\pi \sin[(\pi/T)(t - T/2)]}{(\pi(t - T/2))^2}.$$

From $x_c(t)$, we obtain the corresponding discrete-time signal $x_d[n]$ to be

$$x_d[n] = x_c(nT) = \frac{1}{T}\delta[n].$$

Also, we obtain from $y_c(t)$, the corresponding discrete-time signal $y_d[n]$ to be

$$y_d[n] = y_c(nT) = \frac{(\pi/T)\cos[\pi(n-1/2)]}{\pi T(n-1/2)} - \frac{\sin[\pi(n-1/2)]}{\pi T^2(n-1/2)^2}.$$

The first term in the right-hand side of the above equation is always zero because $\cos[\pi(n-1/2)] = 0$. Therefore,

$$y_d[n] = -\frac{\sin[\pi(n-1/2)]}{\pi T^2(n-1/2)^2}.$$

We conclude that the impulse response of the filter is

$$h_d[n] = -\frac{\sin[\pi(n-1/2)]}{\pi T(n-1/2)^2}.$$

7.15. In this problem we are interested in the lowest rate which $x[n]$ may be sampled without the possibility of aliasing. We use the approach used in Example 7.4 to solve this problem. To find the lowest rate at which $x[n]$ may be sampled while avoiding the possibility of aliasing, we must find an N such that

$$\frac{2\pi}{N} \geq 2\left(\frac{3\pi}{7}\right) \Rightarrow N \leq \frac{7}{3}.$$

Therefore, N can at most be 2.

7.16. Although the signal $x_1[n] = 2\sin(\pi n/2)/(\pi n)$ satisfies the first two conditions, it does not satisfy the third condition. This is because the Fourier transform $X_1(e^{j\omega})$ of this signal is a rectangular pulse which is zero for $\pi/2 < |\omega| < \pi/2$. We also note that the signal $x[n] = 4[\sin(\pi n/2)/(\pi n)]^2$ satisfies the first two conditions. From our numerous encounters with this signal, we know that its Fourier transform $X(e^{j\omega})$ is given by the periodic convolution of $X_1(e^{j\omega})$ with itself. Therefore, $X(e^{j\omega})$ will be a triangular function in the range $0 \leq |\omega| \leq \pi$. This obviously satisfies the third condition as well. Therefore, the desired signal is $x[n] = 4[\sin(\pi n/2)/(\pi n)]^2$.

7.17. In this problem, we wish to determine the effect of decimating the impulse response of the given filter by a factor of 2. As explained in Section 7.5.2, the process of decimation may be broken up into two steps. In the first step we perform impulse train sampling on $h[n]$ to obtain

$$h_p[n] = \sum_{k=-\infty}^{\infty} h[2k]\delta[n-2k].$$

The decimated sequence is then obtained using

$$h_1[n] = h[2n] = h_p[2n].$$

Using eq. (7.37), we obtain the Fourier transform $H_p(e^{j\omega})$ of $h_p[n]$ to be

$$H_p(e^{j\omega}) = (1/2)H(e^{j\omega}) + (1/2)H(e^{j(\omega-\pi)}).$$

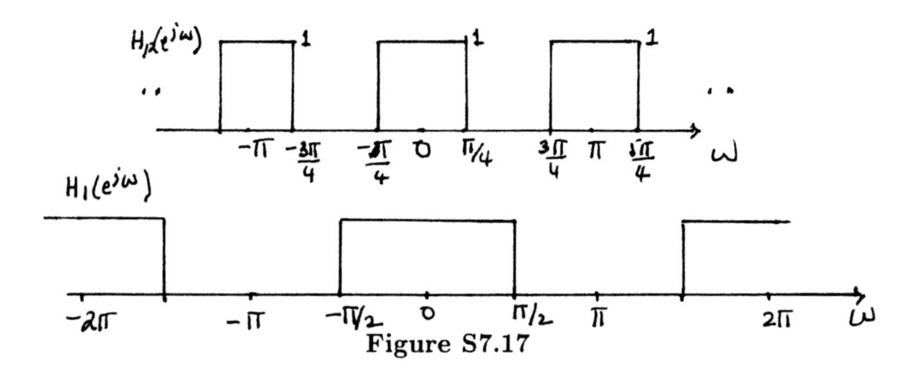

Figure S7.17

This is as shown in Figure S7.17.

From eq. (7.49) we know that the Fourier transform of the decimated impulse response is

$$H_1(e^{j\omega}) = H_p(e^{j\omega/2}).$$

In other words, $H_1(e^{j\omega})$ is $H_p(e^{j\omega})$ expanded by a factor of 2. This is as shown in the figure above. Therefore, $h_1[n] = h[2n]$ is the impulse response of an ideal lowpass filter with a passband gain of unity and a cutoff frequency of $\pi/2$.

7.18. From Figure 7.37, it is clear interpolation by a factor of 2 results in the frequency response getting compressed by a factor of 2. Interpolation also results in a magnitude scaling by a factor of 2. Therefore, in this problem, the interpolated impulse response will correspond to an ideal lowpass filter with cutoff frequency $\pi/$ and a passband gain of 2.

7.19. The Fourier transform of $x[n]$ is given by

$$X(e^{j\omega}) = \begin{cases} 1, & |\omega| \le \omega_1 \\ 0, & \text{otherwise} \end{cases}.$$

This is as shown in Figure S7.19.

(a) When $\omega_1 \le 3\pi/5$, the Fourier transform $X_1(e^{j\omega})$ of the output of the zero-insertion system is as shown in Figure S7.19. The output $W(e^{j\omega})$ of the lowpass filter is as shown in Figure S7.19. The Fourier transform of the output of the decimation system $Y(e^{j\omega})$ is an expanded or stretched out version of $W(e^{j\omega})$. This is as shown in Figure S7.19.

Therefore,

$$y[n] = \frac{1}{5}\frac{\sin(5\omega_1 n/3)}{\pi n}.$$

(b) When $\omega_1 > 3\pi/5$, the Fourier transform $X_1(e^{j\omega})$ of the output of the zero-insertion system is as shown in Figure S7.19. The output $W(e^{j\omega})$ of the lowpass filter is as shown in Figure S7.19.

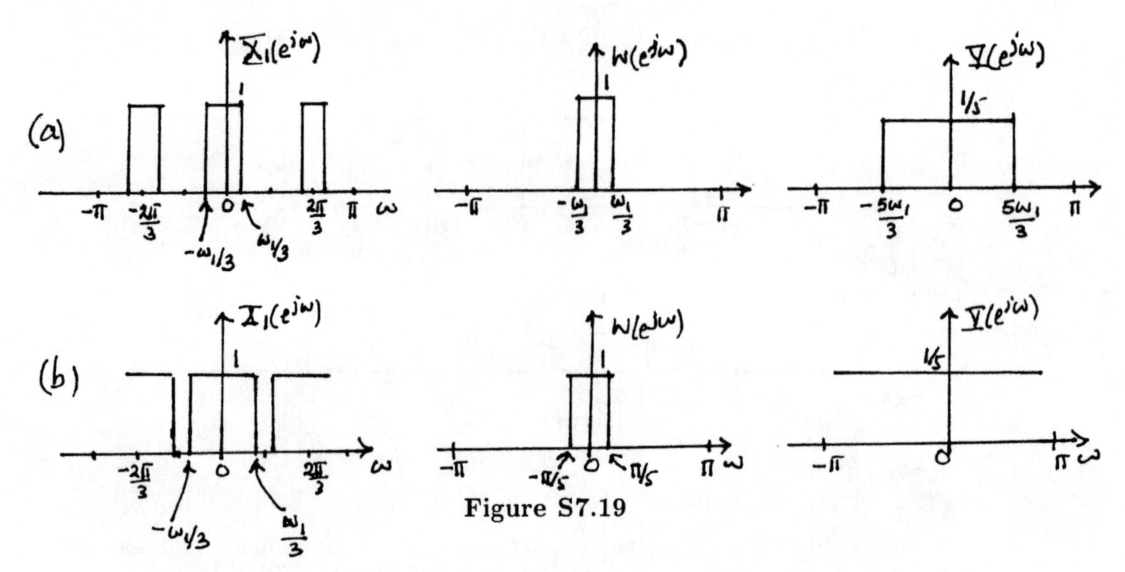

Figure S7.19

The Fourier transform of the output of the decimation system $Y(e^{j\omega})$ is an expanded or stretched out version of $W(e^{j\omega})$. This is as shown in Figure S7.19. Therefore,

$$y[n] = \frac{1}{5}\delta[n].$$

7.20. **(a)** Suppose that $X(e^{j\omega})$ is as shown in Figure S7.20, then the Fourier transform $X_A(e^{j\omega})$ of the output of S_A, the Fourier transform $X_1(e^{j\omega})$ of the output of the lowpass filter, and the Fourier transform $X_B(e^{j\omega})$ of the output of S_B are all shown in the figures below. Clearly this system accomplishes the filtering task.

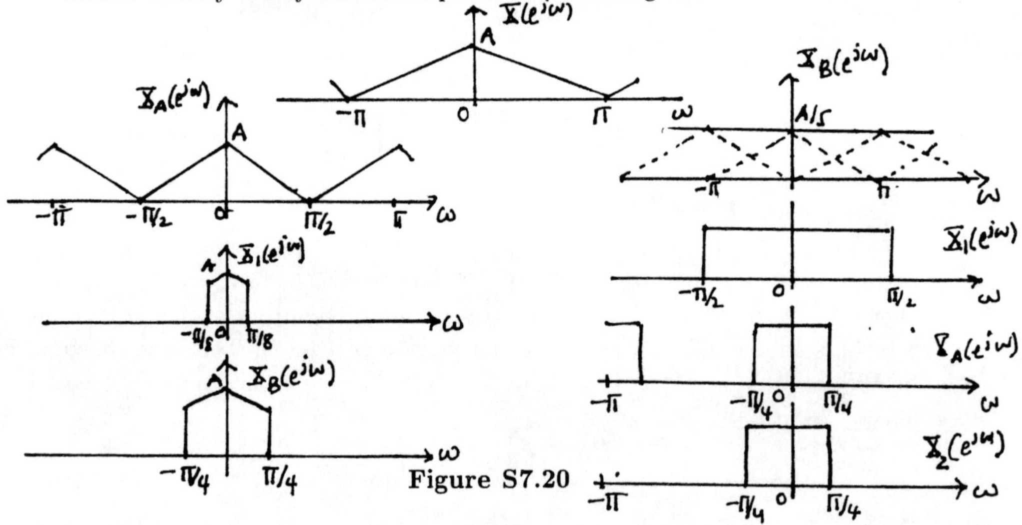

Figure S7.20

(b) Suppose that $X(e^{j\omega})$ is as shown in Figure S7.20, then the Fourier transform $X_B(e^{j\omega})$ of the output of S_B, the Fourier transform $X_1(e^{j\omega})$ of the output of the first lowpass filter, the Fourier transform $X_A(e^{j\omega})$ of the output of S_A, the Fourier transform $X_2(e^{j\omega})$

of the output of the first lowpass filter are all shown in the figures below. Clearly this system does not accomplish the filtering task.

7.21. **(a)** The Nyquist rate for the given signal is $2 \times 5000\pi = 10000\pi$. Therefore, in order to be able to recover $x(t)$ from $x_p(t)$, the sampling period must at most be $T_{max} = \frac{2\pi}{10000\pi} = 2 \times 10^{-4}$ sec. Since the sampling period used is $T = 10^{-4} < T_{max}$, $x(t)$ can be recovered from $x_p(t)$.

(b) The Nyquist rate for the given signal is $2 \times 15000\pi = 30000\pi$. Therefore, in order to be able to recover $x(t)$ from $x_p(t)$, the sampling period must at most be $T_{max} = \frac{2\pi}{30000\pi} = 0.66 \times 10^{-4}$ sec. Since the sampling period used is $T = 10^{-4} > T_{max}$, $x(t)$ cannot be recovered from $x_p(t)$.

(c) Here, $\mathcal{I}m\{X(j\omega)\}$ is not specified. Therefore, the Nyquist rate for the signal $x(t)$ is indeterminate. This implies that one cannot guarantee that $x(t)$ would be recoverable from $x_p(t)$.

(d) Since $x(t)$ is real, we may conclude that $X(j\omega) = 0$ for $|\omega| > 5000$. Therefore, the answer to this part is identical to that of part (a).

(e) Since $x(t)$ is real, $X(j\omega) = 0$ for $|\omega| > 15000\pi$. Therefore, the answer to this part is identical to that of part (b).

(f) If $X(j\omega) = 0$ for $|\omega| > \omega_1$, then $X(j\omega) * X(j\omega) = 0$ for $|\omega| > 2\omega_1$. Therefore, in this part, $X(j\omega) = 0$ for $|\omega| > 7500\pi$. The Nyquist rate for this signal is $2 \times 7500\pi = 15000\pi$. Therefore, in order to be able to recover $x(t)$ from $x_p(t)$, the sampling period must at most be $T_{max} = \frac{2\pi}{15000\pi} = 1.33 \times 10^{-4}$ sec. Since the sampling period used is $T = 10^{-4} < T_{max}$, $x(t)$ can be recovered from $x_p(t)$.

(g) If $|X(j\omega)| = 0$ for $\omega > 5000\pi$, then $X(j\omega) = 0$ for $\omega > 5000\pi$. Therefore, the answer to this part is identical to the answer of part (a).

7.22. Using the properties of the Fourier transform, we obtain

$$Y(j\omega) = X_1(j\omega)X_2(j\omega).$$

Therefore, $Y(j\omega) = 0$ for $|\omega| > 1000\pi$. This implies that the Nyquist rate for $y(t)$ is $2 \times 1000\pi = 2000\pi$. Therefore, the sampling period T can at most be $2\pi/(2000\pi) = 10^{-3}$ sec. Therefore we have to use $T < 10^{-3}$ sec in order to be able to recover $y(t)$ from $y_p(t)$.

7.23. **(a)** We may express $p(t)$ as

$$p(t) = p_1(t) - p_1(t - \Delta),$$

where $p_1(t) = \sum_{k=-\infty}^{\infty} \delta(t - k2\Delta)$. Now,

$$P_1(j\omega) = \frac{\pi}{\Delta} \sum_{k=-\infty}^{\infty} \delta(\omega - \pi/\Delta).$$

Therefore,

$$P(j\omega) = P_1(j\omega) - e^{-j\omega\Delta} P_1(j\omega)$$

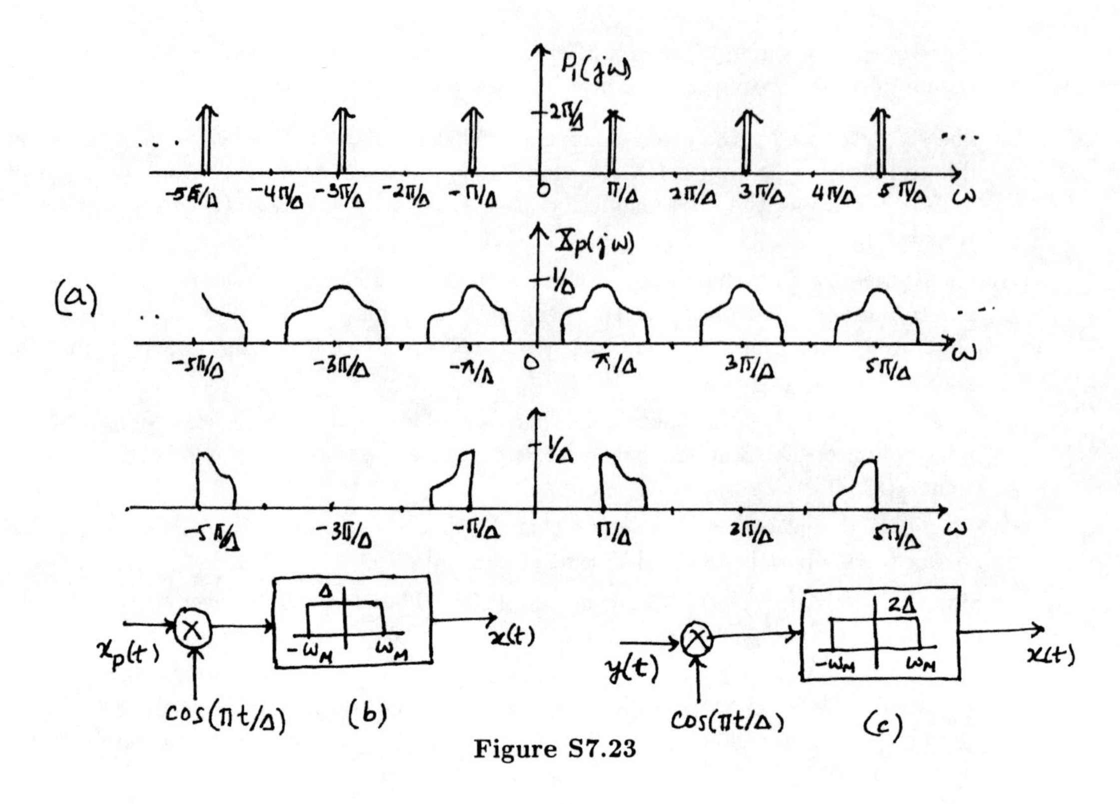

Figure S7.23

is as shown in Figure S7.23.

Now,

$$X_p(j\omega) = \frac{1}{2\pi}\left[X(j\omega) * P(j\omega)\right].$$

Therefore, $X_p(j\omega)$ is as sketched below for $\Delta < \pi/(2\omega_M)$. The corresponding $Y(j\omega)$ is also sketched in Figure S7.23.

(b) The system which can be used to recover $x(t)$ from $x_p(t)$ is as shown in Figure S7.23.

(c) The system which can be used to recover $x(t)$ from $x(t)$ is as shown in Figure S7.23.

(d) We see from the figures sketched in part (a) that aliasing is avoided when $\omega_M \le \pi/\Delta$. Therefore, $\Delta_{max} = \pi/\omega_M$.

7.24. We may express $s(t)$ as $s(t) = \hat{s}(t) - 1$, where $\hat{s}(t)$ is as shown in Figure S7.24.

We may easily show that

$$\hat{S}(j\omega) = \sum_{k=-\infty}^{\infty} \frac{4\sin(2\pi k\Delta/T)}{k}\delta(\omega - k2\pi/T).$$

From this, we obtain

$$S(j\omega) = \hat{S}(j\omega) - 2\pi\delta(\omega) = \sum_{k=-\infty}^{\infty} \frac{4\sin(2\pi k\Delta/T)}{k}\delta(\omega - k2\pi/T) - 2\pi\delta(\omega).$$

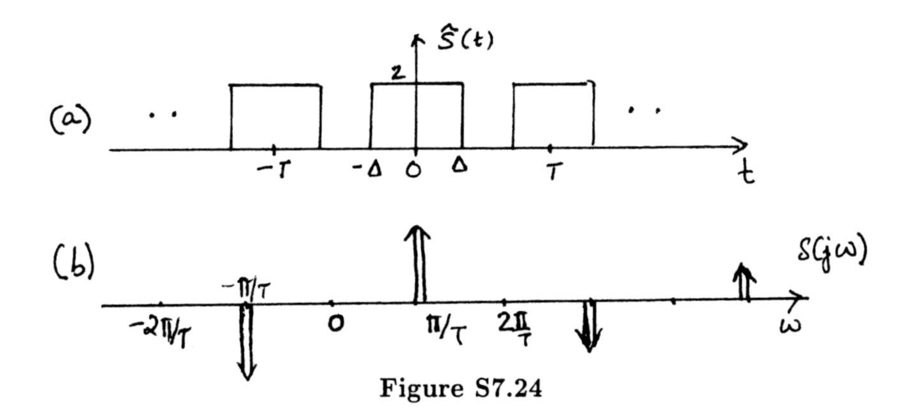

Figure S7.24

Clearly, $S(j\omega)$ consists of impulses spaced every $2\pi/T$.

(a) If $\Delta = T/3$, then

$$S(j\omega) = \sum_{k=-\infty}^{\infty} \frac{4\sin(2\pi k/3)}{k}\delta(\omega - k2\pi/T) - 2\pi\delta(\omega).$$

Now, since $w(t) = s(t)x(t)$,

$$W(j\omega) = \frac{1}{2\pi}\sum_{k=-\infty}^{\infty} \frac{4\sin(2\pi k/3)}{k}X(j(\omega - k2\pi/T)) - 2\pi X(j\omega).$$

Therefore, $W(j\omega)$ consists of replicas of $X(j\omega)$ which are spaced $2\pi/T$ apart. In order to avoid aliasing, ω_M should be less that π/T. Therefore, $T_{max} = \pi/\omega_M$.

(b) If $\Delta = T/3$, then

$$S(j\omega) = \sum_{k=-\infty}^{\infty} \frac{4\sin(2\pi k/4)}{k}\delta(\omega - k2\pi/T) - 2\pi\delta(\omega).$$

We note that $S(j\omega) = 0$ for $k = 0, \pm 2, \pm 4, \cdots$. This is as sketched in Figure S7.24.

Therefore, the replicas of $X(j\omega)$ in $W(j\omega)$ are now spaced $4\pi/T$ apart. In order to avoid aliasing, ω_M should be less that $2\pi/T$. Therefore, $T_{max} = 2\pi/\omega_M$.

7.25. Here, $x_r(kT)$ can be written as

$$x_r(kT) = \sum_{n=-\infty}^{\infty} x(nT)\frac{\sin[\pi(k-n)]}{\pi(k-n)}.$$

Note that when $n \neq k$,

$$\frac{\sin[\pi(k-n)]}{\pi(k-n)} = 0$$

and when $n = k$,

$$\frac{\sin[\pi(k-n)]}{\pi(k-n)} = 1.$$

Therefore,

$$x_r(kT) = x(kT).$$

7.26. We note that

$$P(j\omega) = \frac{2\pi}{T}\delta(\omega - k2\pi/T).$$

Also, since $x_p(t) = x(t)p(t)$,

$$\begin{aligned} X_p(j\omega) &= \frac{1}{2\pi}\{X(j\omega) * P(j\omega)\} \\ &= \frac{1}{T}X(j(\omega - k2\pi/T)). \end{aligned}$$

This is sketched in Figure S7.26.

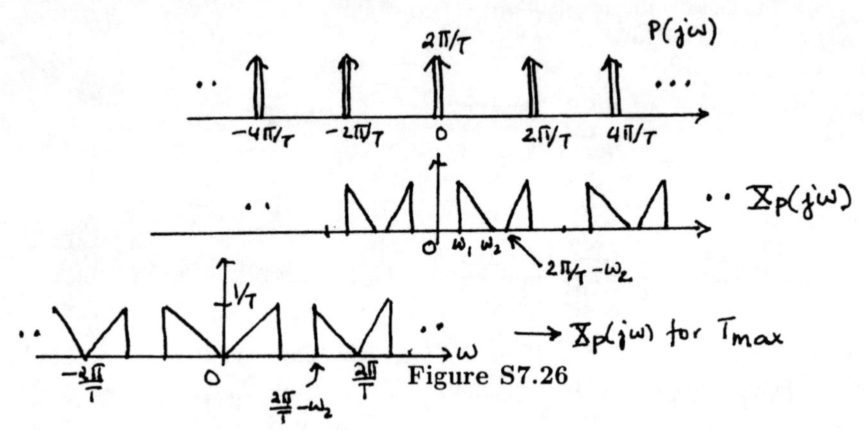

Figure S7.26

Note that as T increases, $\frac{2\pi}{T} - \omega_2$ approaches zero. Also, we note that there is aliasing when

$$2\omega_1 - \omega_2 < \frac{2\pi}{T} - \omega_2 < \omega_2.$$

If $2\omega_1 - \omega_2 \geq 0$ (as given) then it is easy to see that aliasing does not occur when

$$0 \leq \frac{2\pi}{T} - \omega_2 \leq 2\omega_1 - \omega_2.$$

For maximum T, we must choose the minimum allowable value for $\frac{2\pi}{T} - \omega_2$ (which is zero). This implies that $T_{max} = 2\pi/\omega_2$. We plot $X_p(j\omega)$ for this case in Figure S7.26. Therefore, $A = T$, $\omega_b = 2\pi/T$, and $\omega_a = \omega_b - \omega_1$.

7.27. **(a)** Let $X_1(j\omega)$ denote the Fourier transform of the signal $x_1(t)$ obtained by multiplying $x(t)$ with $e^{-j\omega_0 t}$ Let $X_2(j\omega)$ be the Fourier transform of the signal $x_2(t)$ obtained at the output of the lowpass filter. Then, $X_1(j\omega)$, $X_2(j\omega)$, and $X_p(j\omega)$ are as shown in Figure S7.27.

(b) The Nyquist rate for the signal $x_2(t)$ is $2 \times (\omega_2 - \omega_1)/2 = \omega_2 - \omega_1$. Therefore, the sampling period T must be at most $2\pi/(\omega_2 - \omega_1)$ in order to avoid aliasing.

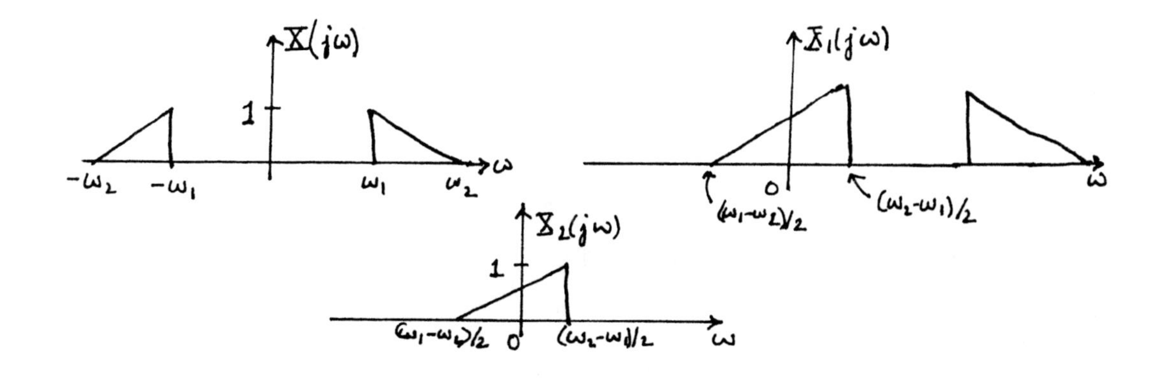

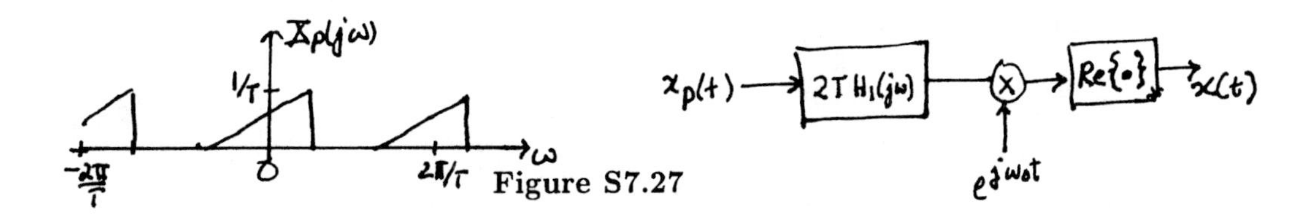

Figure S7.27

(c) A system that can be used to recover $x(t)$ from $x_p(t)$ is shown in Figure S7.27.

7.28. (a) The fundamental frequency of $x(t)$ is 20π rad/sec. From Chapter 4 we know that the Fourier transform of $x(t)$ is given by

$$X(j\omega) = 2\pi \sum_{k=-\infty}^{\infty} a_k \delta(\omega - 20\pi k).$$

This is as sketched below. The Fourier transform $X_c(j\omega)$ of the signal $x_c(t)$ is also sketched in Figure S7.28.

Note that

$$P(j\omega) = \frac{2\pi}{5 \times 10^{-3}} \sum_{k=-\infty}^{\infty} \delta(\omega - 2\pi k/(5 \times 10^{-3}))$$

and

$$X_p(j\omega) = \frac{1}{2\pi}\left[X_c(j\omega) * P(j\omega)\right].$$

Therefore, $X_p(j\omega)$ is as shown in the Figure S7.28. Note that the impulses from adjacent replicas of $X_c(j\omega)$ add up at 200π. Now the Fourier transform $X(e^{j\Omega})$ of the sequence $x[n]$ is given by

$$X(e^{j\Omega}) = X_p(j\omega)\big|_{\omega=\Omega T}.$$

This is as shown in the Figure S7.28.

Since the impulses in $X(e^{j\omega})$ are located at multiples of a 0.1π, the signal $x[n]$ is periodic. The fundamental period is $2\pi/(0.1\pi) = 20$.

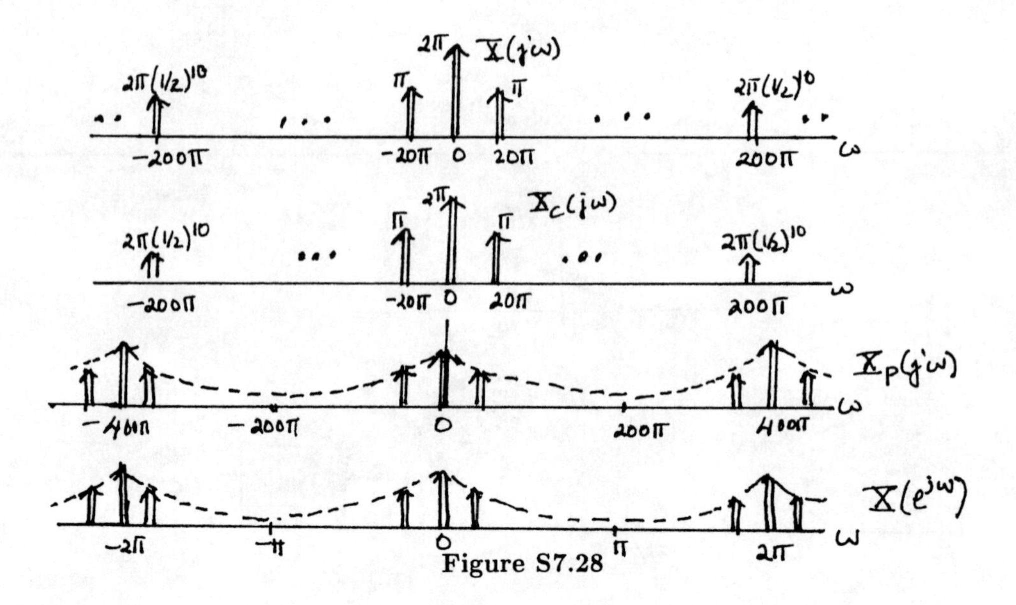

Figure S7.28

(b) The Fourier series coefficients of $x[n]$ are

$$a_k = \begin{cases} \frac{2\pi}{T}\left(\frac{1}{2}\right)^k, & k = 0, \pm 1, \pm 2, \cdots, \pm 9 \\ \frac{4\pi}{T}\left(\frac{1}{2}\right)^{10}, & k = 10 \end{cases}.$$

7.29. From Section 7.1.1 we know that

$$X_p(j\omega) = \frac{1}{T}\sum_{k=-\infty}^{\infty} X(j(\omega - k2\pi/T)).$$

$X(e^{j\omega})$, $Y(e^{j\omega})$, $Y_p(j\omega)$, and $Y_c(j\omega)$ are as shown in Figure S7.29.

7.30. **(a)** Since $x_c(t) = \delta(t)$, we have

$$\frac{dy_c(t)}{dt} + y_c(t) = \delta(t).$$

Taking the Fourier transform we obtain

$$j\omega Y(j\omega) + Y(j\omega) = 1.$$

Therefore,

$$Y_c(j\omega) = \frac{1}{j\omega + 1}, \quad \text{and} \quad y_c(t) = e^{-t}u(t).$$

(b) Since $y_c(t) = e^{-t}u(t)$,

$$y[n] = y_c(nT) = e^{-nT}u[n].$$

Therefore,

$$Y(e^{j\omega}) = \frac{1}{1 - e^{-T}e^{-j\omega}}.$$

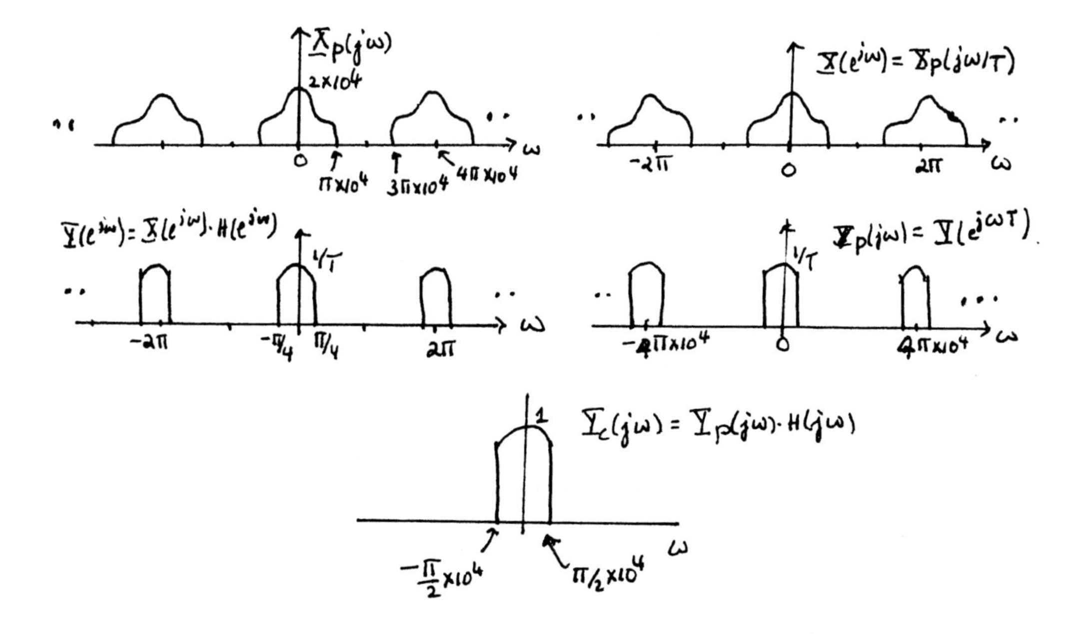

Figure S7.29

Also,

$$H(e^{j\omega}) = \frac{W(e^{j\omega})}{Y(e^{j\omega})} = \frac{1}{1/(1 - e^{-T}e^{-j\omega})} = 1 - e^{-T}e^{-j\omega}.$$

Therefore,

$$h[n] = \delta[n] - e^{-T}\delta[n-1].$$

7.31. In this problem for the sake of clarity we will use the variable Ω to denote discrete frequency.

Taking the Fourier transform of both sides of the given difference equation we obtain

$$H(e^{j\Omega}) = \frac{Y(e^{j\Omega})}{X(e^{j\Omega})} = \frac{1}{1 - \frac{1}{2}e^{-j\Omega}}.$$

Given that the sampling rate is greater than the Nyquist rate, we have

$$X(e^{j\Omega}) = \frac{1}{T}X_c(j\Omega/T), \qquad \text{for } -\pi \leq \Omega \leq \pi.$$

Therefore,

$$Y(e^{j\Omega}) = \frac{\frac{1}{T}X_c(j\Omega/T)}{1 - \frac{1}{2}e^{-j\Omega}}$$

for $-\pi \leq \Omega \leq \pi$. From this we get

$$\tilde{Y}(j\omega) = Y(e^{j\omega T}) == \frac{\frac{1}{T}X_c(j\omega)}{1-\frac{1}{2}e^{-j\omega T}}$$

for $-\pi/T \leq \omega \leq \pi/T$. In this range, $\tilde{Y}(j\omega) = Y_c(j\omega)$. Therefore,

$$H_c(j\omega) = \frac{Y_c(j\omega)}{X_c(j\omega)} = \frac{1/T}{1-\frac{1}{2}e^{-j\omega T}}.$$

7.32. Let $p[n] = \sum_{k=-\infty}^{\infty} \delta[n-1-4k]$. Then from Chapter 5,

$$P(e^{j\omega}) = e^{-j\omega}\frac{2\pi}{4}\sum_{k=-\infty}^{\infty}\delta(\omega - 2\pi k/4) = \frac{\pi}{2}\sum_{k=-\infty}^{\infty} e^{-j2\pi k/4}\delta(\omega-2\pi k/4).$$

Therefore,

$$\begin{aligned} G(e^{j\omega}) &= \frac{1}{2\pi}\int_{-\pi}^{\pi} P(e^{j\theta})X(e^{j(\omega-\theta)})d\theta \\ &= \frac{1}{4}\sum_{k=0}^{3} e^{-j2\pi k/4}X(e^{j(\omega-2\pi k/4)}) \end{aligned}$$

Since $X(e^{j\omega}) = 0$ for $\pi/4 \leq |\omega| \leq \pi$, $G(e^{j\omega})$ is as shown in Figure S7.32.

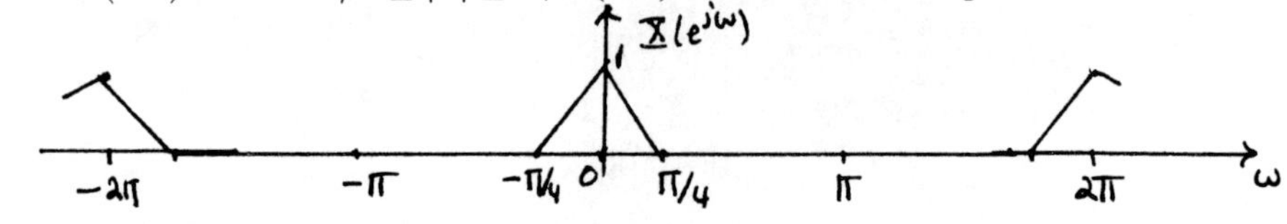

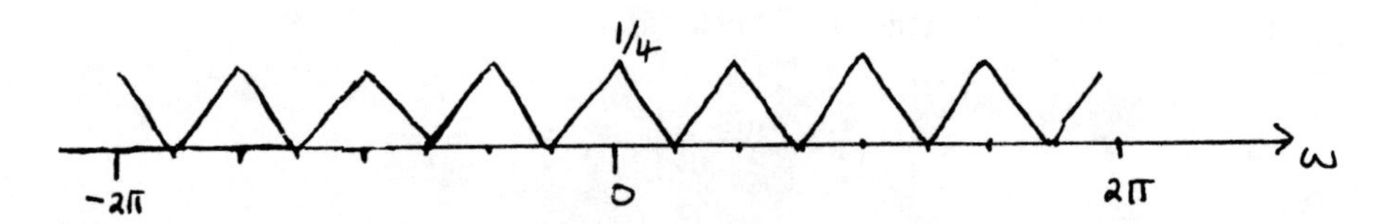

Figure S7.32

Clearly, in order to isolate just $X(e^{j\omega})$ we need to use an ideal lowpass filter with cutoff frequency $\pi/4$ and passband gain of 4. Therefore, in the range $|\omega| < \pi$,

$$H(e^{j\omega}) = \begin{cases} 4, & |\omega| < \pi/4 \\ 0, & \pi/4 \leq |\omega| \leq \pi \end{cases}.$$

7.33. Let $y[n] = x[n] \sum_{k=-\infty}^{\infty} \delta[n-3k]$. Then

$$Y(e^{j\omega}) = \frac{1}{3}\sum_{k=0}^{3} X(e^{j(\omega-2\pi k/3)}).$$

Note that $\sin(\pi n/3)/(\pi n/3)$ is the impulse response of an ideal lowpass filter with cutoff frequency $\pi/3$ and passband gain of 3. Therefore, we now require that $y[n]$ when passed through this filter should yield $x[n]$. Therefore, the replicas of $X(e^{j\omega})$ contained in $Y(e^{j\omega})$ should not overlap with one another. This is possible only if $X(e^{j\omega}) = 0$ for $\pi/3 \leq |\omega| \leq \pi$.

7.34. In order to make $X(e^{j\omega})$ occupy the entire region from $-\pi$ to π, the signal $x[n]$ must be downsampled by a factor of 14/3. Since it is not possible to directly downsample by a non-integer factor, we first upsample the signal by a factor of 3. Therefore, after the upsampling we will need to reduce the sampling rate by $14/3 \times 3 = 14$. Therefore, the overall system for performing the sampling rate conversion is shown in Figure S7.34.

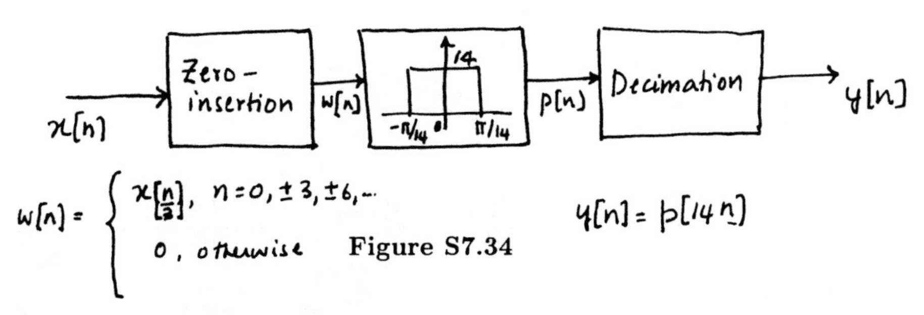

Figure S7.34

7.35. (a) The signals $x_p[n]$ and $x_d[n]$ are sketched in Figure S7.35.

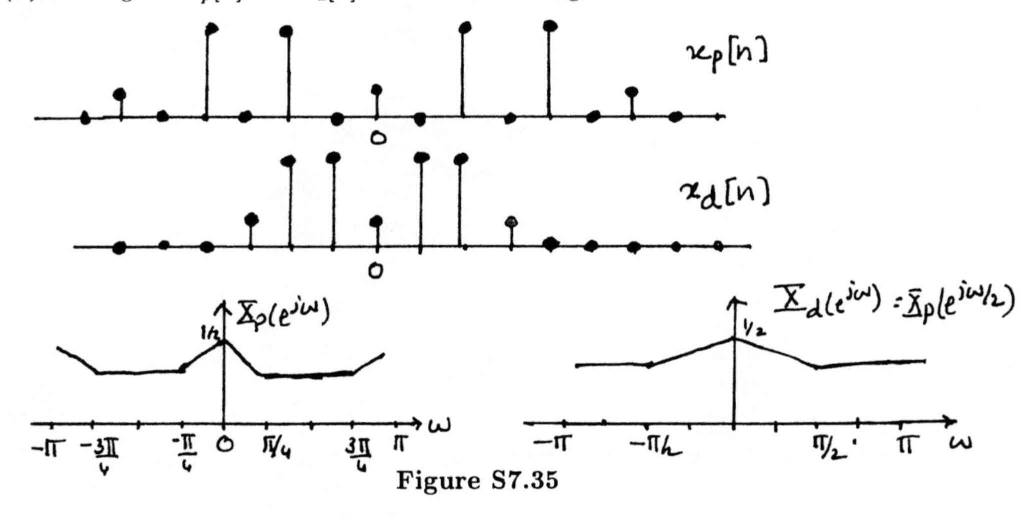

Figure S7.35

(b) $X_p(e^{j\omega})$ and $X_d(e^{j\omega})$ are sketched in Figure S7.35.

7.36. (a) Let us denote the sampled signal by $x_p(t)$. We have

$$x_p(t) = \sum_{n=-\infty}^{\infty} x(nT)\delta(t-nT).$$

Since the Nyquist rate for the signal $x(t)$ is $2\pi/T$, we can reconstruct the signal from $x_p(t)$. From Section 7.2, we know that

$$x(t) = x_p(t) * h(t),$$

where

$$h(t) = \frac{\sin(\pi t/T)}{\pi t/T}.$$

Therefore,

$$\frac{dx(t)}{dt} = x_p(t) * \frac{dh(t)}{dt}.$$

Denoting $\frac{dh(t)}{dt}$ by $g(t)$, we have

$$\frac{dx(t)}{dt} = x_p(t) * g(t) = \sum_{n=-\infty}^{\infty} x(nT)g(t-nT).$$

Therefore,

$$g(t) = \frac{dh(t)}{dt} = \frac{\cos(\pi t/T)}{t} - \frac{T\sin(\pi t/T)}{\pi t^2}.$$

(b) No.

7.37. We may write $p(t)$ as

$$p(t) = p_1(t) + p_1(t-\Delta),$$

where

$$p_1(t) = \sum_{k=-\infty}^{\infty} \delta(t - 2\pi k/W).$$

Therefore,

$$P(j\omega) = (1 + e^{-j\Delta\omega})P_1(j\omega),$$

where

$$P_1(j\omega) = W\sum_{k=-\infty}^{\infty} \delta(\omega - kW). \tag{S7.37–1}$$

Let us denote the product $p(t)f(t)$ by $g(t)$. Then,

$$g(t) = p(t)f(t) = p_1(t)f(t) + p_1(t-\Delta)f(t).$$

This may be written as

$$g(t) = ap_1(t) + bp_1(t - \Delta).$$

Therefore,

$$G(j\omega) = (a + be^{-j\omega\Delta})P_1(j\omega),$$

with $P_1(j\omega)$ is specified in eq. (S7.37-1). Therefore,

$$G(j\omega) = W \sum_{k=-\infty}^{\infty} [a + be^{-jk\Delta W}]\delta(\omega - kW).$$

We now have

$$y_1(t) = x(t)p(t)f(t).$$

Therefore,

$$Y_1(j\omega) = \frac{1}{2\pi}\left[G(j\omega) * X(j\omega)\right].$$

This gives us

$$Y_1(j\omega) = \frac{W}{2\pi} \sum_{k=-\infty}^{\infty} [a + be^{-jk\Delta W}]X(j(\omega - kW)).$$

In the range $0 < \omega < W$, we may specify $Y_1(j\omega)$ as

$$Y_1(j\omega) = \frac{W}{2\pi}\left[(a + b)X(j\omega) + (a + be^{-j\Delta W})X(j(\omega - W))\right].$$

Since $Y_2(j\omega) = Y_1(j\omega)H_1(j\omega)$, in the range $0 < \omega < W$ we may specify $Y_2(j\omega)$ as

$$Y_2(j\omega) = \frac{jW}{2\pi}\left[(a + b)X(j\omega) + (a + be^{-j\Delta W})X(j(\omega - W))\right].$$

Since $y_3(t) = x(t)p(t)$, in the range $0 < \omega < W$ we may specify $Y_3(j\omega)$ as

$$Y_3(j\omega) = \frac{W}{2\pi}\left[2X(j\omega) + (1 + e^{-j\Delta W})X(j(\omega - W))\right].$$

Given that $0 < W\Delta < \pi$, we require that $Y_2(j\omega) + Y_3(j\omega) = KX(j\omega)$ for $0 < \omega < W$. That is,

$$\frac{W}{2\pi}\left[(2 + ja + jb)X(j\omega)\right] + \frac{W}{2\pi}\left[(1 + e^{-j\Delta W} + ja + jbe^{-j\Delta W})X(j(\omega - W))\right] = KX(j\omega).$$

This implies that

$$1 + e^{-j\Delta W} + ja + jbe^{-j\Delta W} = 0.$$

Solving this we obtain

$$a = 1, \qquad b = -1,$$

when $W\Delta = \pi/2$. More generally, we get

$$a = \sin(W\Delta) + \frac{(1 + \cos(W\Delta))}{\tan(W\Delta)} \quad \text{and} \quad b = -\frac{1 + \cos(W\Delta)}{\sin(W\Delta)},$$

except when $W\Delta = \pi/2$. Finally, we also get $K = \frac{2\pi}{W}[1/(2 + ja + jb)]$.

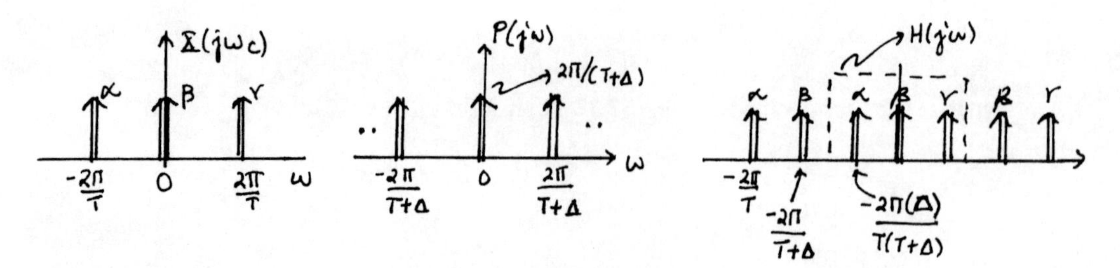

Figure S7.38

7.38. The Fourier transforms $X(j\omega)$, $P(j\omega)$, and $Y(j\omega)$ are as shown in Figure S7.38.

Clearly, we cannot have $\Delta = 0$. Also, from the figures above it is clear that we require

$$\frac{2\pi\Delta}{T(T+\Delta)} \leq \frac{1}{2(T+\Delta)}.$$

This implies that

$$\Delta \leq \frac{T}{4\pi}.$$

Also from the figures, it is clear that

$$a = \frac{\frac{2\pi\Delta}{T(T+\Delta)}}{\frac{2\pi}{T}} = \frac{\Delta}{T+\Delta}.$$

7.39. **(a)** Using Trigonometric identities,

$$\cos\left(\frac{\omega_s}{2}t+\phi\right) = \cos\left(\frac{\omega_s}{2}t\right)\cos(\phi) - \sin\left(\frac{\omega_s}{2}t\right)\sin(\phi).$$

Therefore,

$$g(t) = -\sin\left(\frac{\omega_s}{2}t\right)\sin(\phi).$$

(b) By replacing ω_s with $2\pi/T$, and t by NT in the above equation, we get

$$\begin{aligned} g(nT) &= -\sin\left(\frac{2\pi}{2T}nT\right)\sin(\phi) \\ &= -\sin(n\pi)\sin(\phi). \end{aligned}$$

Clearly, the right-hand side of the above equation is zero for $n = 0, \pm1, \pm2, \cdots$.

(c) From parts (a) and (b), we get

$$\begin{aligned} x_p(t) = \sum_{n=-\infty}^{\infty} x(nT)\delta(t-nT) &= \sum_{n=-\infty}^{\infty} \delta(t-nT)\left\{\cos\left(\frac{\omega_s}{2}nT\right)\cos(\phi) + g(nT)\right\} \\ &= \sum_{n=-\infty}^{\infty} \delta(t-nT)\cos\left(\frac{\omega_s}{2}nT\right)\cos(\phi) \end{aligned}$$

When this signal is passed through a lowpass filter, we are in effect performing band-limited interpolation. This results in the signal

$$y(t) = \cos\left(\frac{\omega_s}{2}t\right)\cos(\phi).$$

7.40. (a) The Fourier transform $V(j\omega)$ is as shown in Figure S7.40.

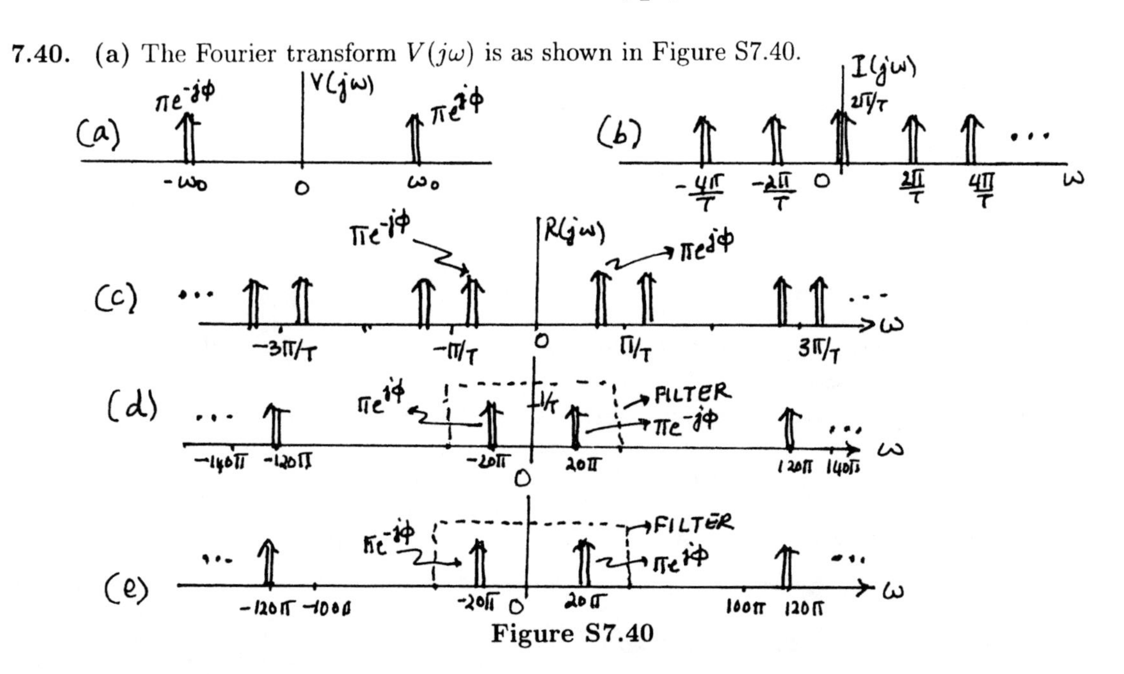

Figure S7.40

(b) The Fourier transform $I(j\omega)$ is

$$I(j\omega) = \frac{2\pi}{T}\sum_{k=-\infty}^{\infty}\delta(\omega - 2\pi k/T).$$

This is as shown in Figure S7.40.

(c) The Nyquist rate for $v(t)$ is $2\omega_0$. Therefore,

$$\frac{2\pi}{T_{max}} = 2\omega_0 \quad \Rightarrow \quad T_{max} = \frac{\pi}{\omega_0}.$$

The cutoff frequency of the lowpass filter has to be ω_0.

(d) Now,

$$R(j\omega) = \frac{1}{2\pi}\frac{1}{T}\sum_{k=-\infty}^{\infty} V(j(\omega - 2\pi k/T)).$$

Since $\omega_0 = 2\pi(60)$ rad/sec, we have $2\pi/T = 120\pi + 20\pi = 140\pi$. Therefore, $R(j\omega)$ is as shown in Figure S7.40.

Therefore, $v_a(t)$ obtained by passing $r(t)$ through a lowpass filter with cutoff frequency $2\pi(20)$ rad/sec is

$$v_a(t) = \frac{1}{T}\cos(20\pi t - \phi).$$

Therefore,

$$\omega_a = 20\pi, \qquad \phi_a = -\phi, \qquad \text{and} \qquad A_a = \frac{1}{T}.$$

(e) Here, $2\pi/T = 120\pi - 20\pi = 100\pi$. Therefore, $R(j\omega)$ is as shown in Figure S7.40.

It follows that

$$v_a(t) = \frac{1}{T}\cos(20\pi t + \phi).$$

and

$$\omega_a = 20\pi, \qquad \phi_a = \phi, \qquad \text{and} \qquad A_a = \frac{1}{T}.$$

7.41. In this problem, to avoid confusion we use the variable Ω to indicate discrete-time frequency.

(a) The Nyquist rate for the signal $x(t)$ is $2\omega_M$. Therefore, the sampling theorem states that $x(t)$ has to be sampled at least every π/ω_M. In this part, $T < \pi/\omega_M$. Therefore, $y_c(t)$ will be equal to $x(t)$ as long as $y[n] = x[n]$. Now,

$$\begin{aligned} s[n] &= x(nT_0) + \alpha x(nT_0 - T_0) \\ &= x[n] + \alpha x[n-1]. \end{aligned}$$

Therefore, if we require $y[n] = x[n]$ then,

$$H(e^{j\Omega}) = \frac{Y(e^{j\Omega})}{S(e^{j\Omega})} = \frac{X(e^{j\Omega})}{X(e^{j\Omega}) + \alpha e^{-j\Omega}X(e^{j\Omega})} = \frac{1}{1+\alpha e^{-j\Omega}}.$$

Therefore, the difference equation for the filter $h[n]$ is

$$y[n] + \alpha y[n-1] = s[n].$$

(b) From Figures P7.41(a) and (b), we have

$$H_{eq}(j\omega) = \frac{A}{T_0}H(e^{j\omega T_0}), \qquad \text{(S7.41–1)}$$

where $H_{eq}(j\omega)$ is the system response of the overall continuous-time system. Since we require that $y_c(t) = x(t)$,

$$H_{eq}(j\omega) = \frac{Y_c(j\omega)}{S_c(j\omega)} = \frac{1}{1+\alpha e^{-j\omega T_0}}. \qquad \text{(S7.41–2)}$$

Comparing this with eq.(S7.41-1), we get $A = T_0$.

(c) We require a T which avoids aliasing. Therefore, $T < \pi/\omega_M$. We also require that

$$H_{eq}(j\omega) = \frac{1}{1+\alpha e^{-j\omega T_0}}, \qquad -\omega_M \le \omega \le \omega_M.$$

But,

$$H_{eq}(j\omega) = \frac{A}{T}H(e^{j\omega T}), \qquad -\frac{\pi}{T} \le \omega \le \frac{\pi}{T}.$$

For these to be consistent, we need $A = T$ and

$$H(e^{j\Omega}) = \frac{1}{1+\alpha e^{-j\Omega T/T_0}}$$

for $-\pi \le \Omega \le \pi$.

7.42. In this problem, to avoid confusion we use the variable Ω to indicate discrete-time frequency.

Using Parseval's theorem and the fact that $X_c(j\omega) = 0$ for $|\omega| \ge \omega_0$, we get

$$E_c = \int_{-\infty}^{\infty} |x_c(t)|^2 dt = \frac{1}{2\pi}\int_{-\omega_0}^{\omega_0} |X_c(j\omega)|^2 d\omega.$$

Also, using Parseval's theorem we have

$$E_d = \sum_{n=-\infty}^{\infty} |x[n]|^2 = \frac{1}{2\pi}\int_{-\pi}^{\pi} |X(e^{j\Omega})|^2 d\Omega.$$

But since $X(e^{j\Omega}) = \frac{1}{T}X_c(j\Omega/T)$ for $-\pi \le \Omega \le \pi$, we may write

$$E_d = \frac{1}{2\pi T^2}\int_{-\pi}^{\pi} |X_c(j\Omega/T)|^2 d\Omega.$$

Replacing Ω/T by ω, we get

$$E_d = \frac{1}{2\pi T}\int_{-\pi/T}^{\pi/T} |X_c(j\omega)|^2 d\omega.$$

Also, since $2\pi/T \ge 2\omega_0$, we may rewrite the above equation as

$$E_d = \frac{1}{2\pi T}\int_{-\omega_0}^{\omega_0} |X_c(j\omega)|^2 d\omega = \frac{E_c}{T}.$$

7.43. Throughout this problem, to avoid confusion we use the variable Ω to indicate discrete-time frequency.

Taking the Fourier transform of both sides of the given differential equation, we get

$$H(j\omega) = \frac{Y_c(j\omega)}{X_c(j\omega)} = \frac{1}{-\omega^2 + 4j\omega + 3}.$$

Taking the inverse Fourier transform of the partial fraction expansion of $H(j\omega)$, we obtain

$$h(t) = \frac{1}{2}e^{-t}u(t) - \frac{1}{2}e^{-3t}u(t).$$

Now, $x_p(t) = \sum_{n=-\infty}^{\infty} x[n]\delta(t-nT)$. Therefore, $X_p(j\omega) = X(e^{j\omega T})$. Also,

$$X_c(j\omega) = TX_p(j\omega) = TX(e^{j\omega T}) \quad \text{for} \quad -\pi/T \le \omega \le \pi/T$$

and 0 otherwise. From this we get

$$Y_c(j\omega) = H(j\omega)TX(e^{j\omega T}) \quad \text{for} \quad -\pi/T \le \omega \le \pi/T$$

and 0 otherwise. Then, one period of $Y_p(j\omega)$ may be specified as

$$Y_p(j\omega) = \frac{1}{T}Y_c(j\omega) = H(j\omega)X(e^{j\omega T}) \quad \text{for} \quad -\pi/T \le \omega \le \pi/T.$$

Therefore, one period of $Y(e^{j\Omega})$ is

$$Y(e^{j\Omega}) = X(e^{j\Omega})H(j\Omega/T), \quad \text{for} \quad -\pi \le \Omega \le \pi.$$

Denoting the frequency response of the equivalent system, by $H(e^{j\omega})$, we have

$$H(e^{j\omega}) = H(j\Omega/T), \quad \text{for } -\pi \le \Omega \le \pi.$$

Note that $H(e^{j\omega})$ represents the Fourier transform of the sequence $h[n]$ obtained by low-pass filtering $h(t)$ (with a filter of cutoff frequency π/T) and sampling the result every T. Therefore,

$$h[n] = \left[h(t) * \frac{\sin(\pi t/T)}{\pi t/T}\right]_{t=nT} = \left[\frac{T}{2}\int_0^{\infty}[e^{-\tau} - e^{-3\tau}]\frac{\sin(\pi(t-\tau)/T)}{\pi(t-\tau)/T}d\tau\right]_{t=nT}.$$

7.44. **(a)** We have

$$y_p(t) = \sum_{k=-\infty}^{\infty} \cos\left(\frac{2\pi k}{N}\right)\delta(t-kT).$$

If $\omega_0 = 2\pi/NT$, then

$$\begin{aligned}
y_p(t) &= \sum_{k=-\infty}^{\infty} \cos(\omega_0 kT)\delta(t-kT) \\
&= \sum_{k=-\infty}^{\infty} \cos(\omega_0 t)\delta(t-kT) \\
&= \cos(\omega_0 t)\sum_{k=-\infty}^{\infty} \delta(t-kT).
\end{aligned}$$

Let the range of T be $T_{min} \leq T \leq T_{max}$. Then with T_{min}, we want to obtain the smallest frequency ω_1 and with T_{max}, we want to obtain the largest frequency ω_2. Therefore,

$$T_{min} = \frac{2\pi}{N\omega_2}, \quad \text{and} \quad T_{max} = \frac{2\pi}{N\omega_1}.$$

(b) Let $c(t) = \cos(\omega_0 t)$ and $p(t) = \sum_{k=-\infty}^{\infty} \delta(t - kT)$. Then

$$Y_p(j\omega) = \frac{1}{2\pi}\left[C(j\omega) * P(j\omega)\right].$$

This is as shown in Figure S7.44.

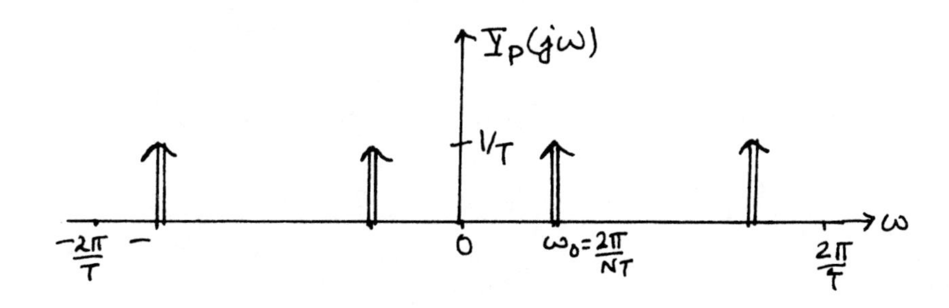

Figure S7.44

(c) To avoid aliasing in $Y(j\omega)$, we require that $2\omega_0 < 2\pi/T$. Therefore, $4\pi/NT < 2\pi/T$. This implies that $N > 2$. Therefore, the minimum value of N is 3. By inspection of $Y(j\omega)$, we obtain $\omega_2 < \omega_c < 4\pi/(3T)$. This keeps the sinusoid at frequency ω_2 while rejecting contributions from cosines centered around $2\pi/T$ and $-2\pi/T$.

(d) We have

$$G(j\omega) = \begin{cases} T, & -\omega_c \leq \omega \leq \omega_c \\ \text{arbitrary}, & \text{otherwise} \end{cases}.$$

7.45. **(a)** The Nyquist rate for the signal $x_c(t)$ is $4\pi \times 10^4$. Therefore, the maximum value of T that can be used to sample $x_c(t)$ is

$$T_{max} = \frac{2\pi}{4\pi \times 10^4} = 5 \times 10^{-5}.$$

(b) We have

$$y[n] = T\sum_{k=-\infty}^{n} x[k] = T\sum_{k=-\infty}^{\infty} x[k]u[n-k] = T\{x[n] * u[n]\}.$$

Therefore,

$$h[n] = Tu[n].$$

(c) We have

$$\lim_{n\to\infty} y[n] = \lim_{n\to\infty} T \sum_{k=-\infty}^{n} x[k] = TX(e^{j0}).$$

Also,

$$\lim_{t\to\infty} x_c(t) = X_c(j0).$$

Therefore, eq. (P7.45-2) requires that

$$TX(e^{j0}) = X_c(j0).$$

Now,

$$X(e^{j\omega}) = X_p(j\omega/T)$$

and

$$X_p(j\omega) = \frac{1}{T}\sum_{k=-\infty}^{\infty} X_c(j(\omega - 2\pi k/T)).$$

To avoid aliasing at $\omega = 0$ in $X_p(\omega)$, we require that $(2\pi/T) > 2\pi \times 10^4$. This implies that $T < 10^{-4}$. With this condition,

$$X(e^{j0}) = (1/T)X_c(j0).$$

7.46. We have

$$\begin{aligned} x_r[mN] &= \sum_{k=-\infty}^{\infty} x[kN]\frac{N\omega_c}{2\pi}\frac{\sin[\omega_c(mN-kN)]}{\omega_c(mN-kN)} \\ &= \sum_{k=-\infty}^{\infty} x[kN]\frac{\sin 2\pi(m-k)}{2\pi(m-k)} \end{aligned}$$

Note that $[\sin[2\pi(m-k)]]/[2\pi(m-k)]$ is 1 when $m = k$, and zero otherwise. Therefore,

$$x_r[mN] = x[mN].$$

7.47. Let us define a signal

$$x_p[n] = x[n]\sum_{k=-\infty}^{\infty}\delta[n-3k] = \sum_{k=-\infty}^{\infty} x[3k]\delta[n-3k].$$

From Section 7.5.1, we know that the Fourier transform of $x_p[n]$ is

$$X_p(e^{j\omega}) = \frac{1}{3}\sum_{k=0}^{2} X(e^{j(\omega-2\pi/3)}).$$

Since $X(e^{j\omega}) = 0$ for $\pi/3 \le |\omega| \le \pi$, there is no aliasing among the replicas of $X(e^{j\omega})$ in $X_p(e^{j\omega})$. This is shown in the Figure S7.47.

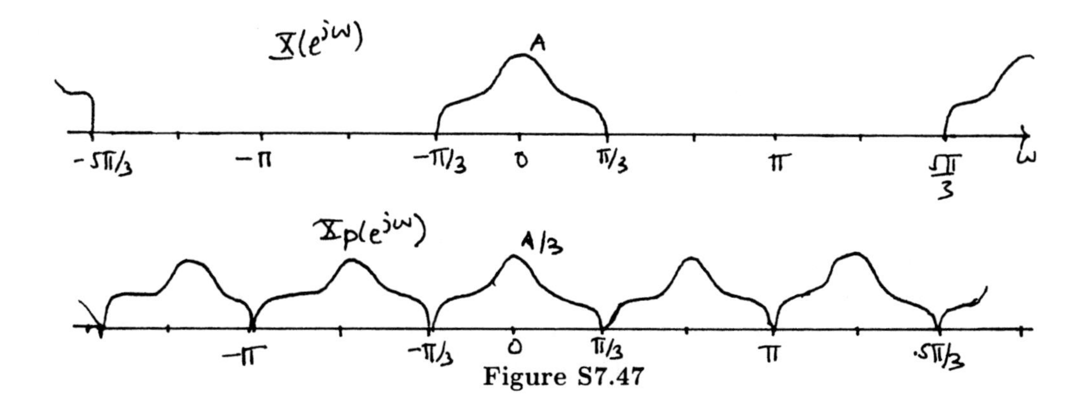

Figure S7.47

In order to be able to recover $x[n]$ from $x_p[n]$, it is clear that we need to pass $x_p[n]$ through a lowpass filter with cutoff frequency $\pi/3$ and passband gain 3. Therefore,

$$\begin{aligned} x[n] &= x_p[n] * \frac{3\sin(\pi n/3)}{\pi n} \\ &= \{\sum_{k=-\infty}^{\infty} x[3k]\delta[n-3k]\} * \frac{3\sin(\pi n/3)}{\pi n} \\ &= \sum_{k=-\infty}^{\infty} x[3k]\frac{\sin[\pi(n-3k)/3]}{\pi(n-3k)/3}. \end{aligned}$$

7.48. In Figure S7.49, we plot the signal $\cos(\pi n/4)$.

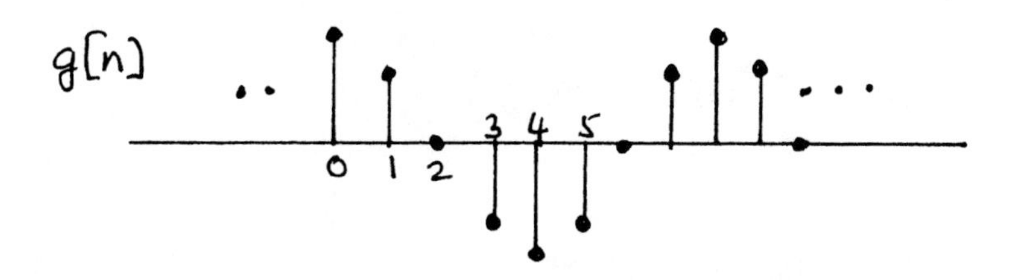

Figure S7.48

Note that the signal $g[n]$ contains every fourth sample of $x[n]$. If the signal $x[n]$ were $\cos[\pi(n+2)/4]$ (see Figure S7.48), then $g[n]$ would be zero for all n. Therefore, there would be no way of recovering $x[n]$ from $g[n]$. Therefore, ϕ_0 should never be $\pi/2$ in order for the given equation to be true.

7.49. **(a)** Let the signals $x_{d_1}[n]$ and $x_{d_2}[n]$ be inputs to system A. Let the corresponding outputs be $x_{p_1}[n]$ and $x_{p_2}[n]$. Now, consider an input of the form $x_{d_3}[n] = \alpha_1 x_{d_1}[n] + \alpha_2 x_{d_2}[n]$.

This gives an output which is

$$x_{p3}[n] = \begin{cases} \alpha_1 x_{d_1}[n/N] + \alpha_2 x_{d_2}[n/N], & n = 0, \pm N, \pm 2N, \cdots \\ 0, & \text{otherwise} \end{cases}.$$

Therefore, $x_{p_3}[n] = \alpha_1 x_{p_1}[n] + \alpha_2 x_{p_2}[n]$. This implies that the system is linear.

(b) Let us consider a signal $x_d[n]$ as shown in Figure S7.49. The output of the system $x_p[n]$ is then as shown in the figure. Let us now define a new input $x_{d_1}[n] = x_d[n-1]$. The corresponding output $x_{p_1}[n]$ is shown in the Figure S7.49. Clearly, $x_{p_1}[n] \neq x_p[n]$. Therefore, the system in *not* time invariant.

(c) We have $X_p(e^{j\omega}) = X_d(e^{j\omega N})$. This is as shown in Figure S7.49.

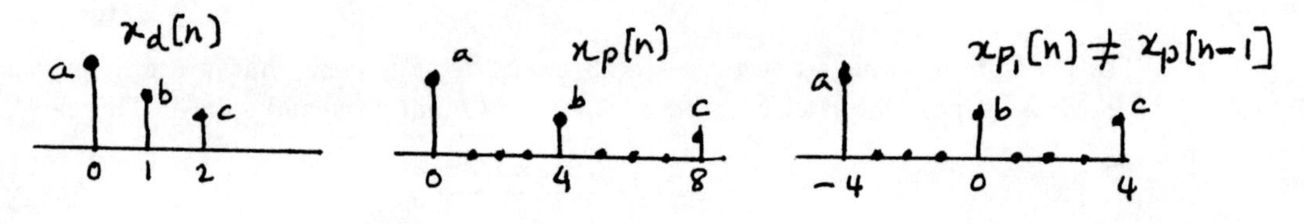

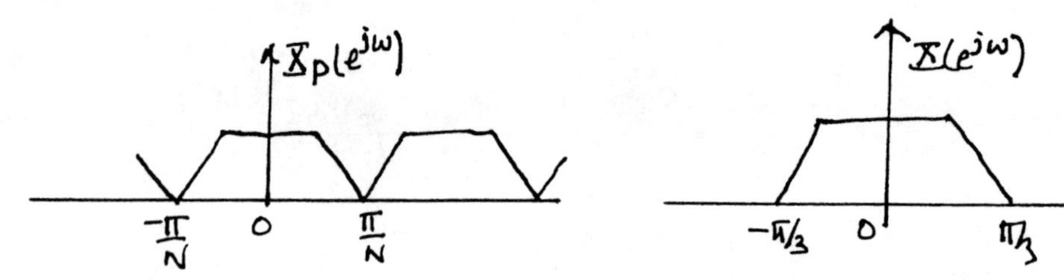

Figure S7.49

(d) $X(e^{j\omega})$ is as sketched in Figure S7.49.

7.50. (a) We have

$$h_0[n] = u[n] - u[n-N].$$

This is as shown in the Figure S7.50.

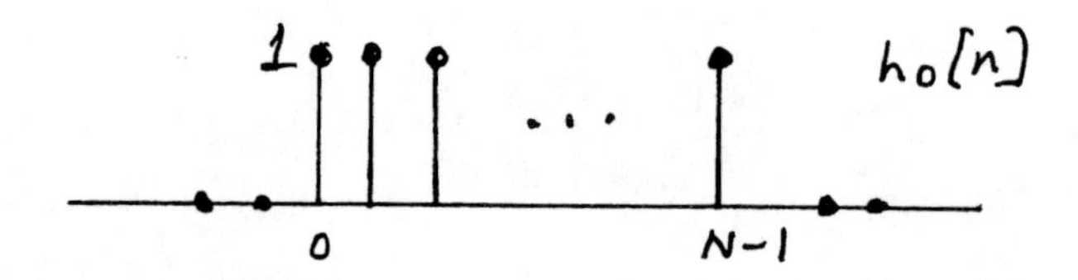

Figure S7.50

(b) We require that $H(e^{j\omega})H_0(e^{j\omega}) = N$ for $|\omega| < \omega_s/2$ and zero otherwise. Here, $\omega_s/2 = \pi/N$. But

$$H_0(e^{j\omega}) = \frac{1 - e^{-j\omega N}}{1 - e^{-j\omega}}.$$

Therefore,

$$H(e^{j\omega}) = \begin{cases} N\frac{1-e^{-j\omega}}{1-e^{-j\omega N}}, & |\omega| < \pi/N \\ 0, & (\pi/N) \le |\omega| \le \pi \end{cases}.$$

(c) We have

$$h_1[n] = \frac{1}{N}\left[h_0[n] * h_0[-n]\right].$$

(d) Again we have $H(e^{j\omega}) = N/H_1(e^{j\omega})$ for $|\omega| < \pi/N$ and zero otherwise. But from part (c),

$$H_1(e^{j\omega}) = (1/N^2)|H_0(e^{j\omega})|^2.$$

Therefore,

$$H(e^{j\omega}) = \begin{cases} N^2\left|\frac{1-e^{-j\omega}}{1-e^{-j\omega N}}\right|^2, & |\omega| < \pi/N \\ 0, & (\pi/N) \le |\omega| \le \pi \end{cases}.$$

7.51. **(a)** This is possible only of $h[kL] = 0$ for $k = \pm 1, \pm 2, \cdots$ and $h[0] = 1$.

(b) N must be odd. In this case, α is an integer. If N is even, α is not an integer. If α were an integer, shifting $h[n]$ by α would make $h[n]$ an even sequence. This is impossible with N even.

(c) N can be odd or even. This time, α is allowed to be fractional. Thus, an even length filter can be designed which is a linear-phase causal symmetric FIR filter.

7.52. **(a)** Since,

$$\tilde{X}(j\omega) = X(j\omega)P(j\omega),$$

we have

$$\tilde{x}(t) = x(t) * p(t).$$

(b) Taking the inverse Fourier transform of $P(j\omega)$, we have

$$p(t) = \frac{1}{\omega_0}\sum_{k=-\infty}^{\infty}\delta\left(t - \frac{2\pi k}{\omega_0}\right).$$

From part (a), we have

$$\begin{aligned} \tilde{x}(t) &= p(t) * x(t) \\ &= \frac{1}{\omega_0}\sum_{k=-\infty}^{\infty} x\left(t - \frac{2\pi k}{\omega_0}\right) \end{aligned}$$

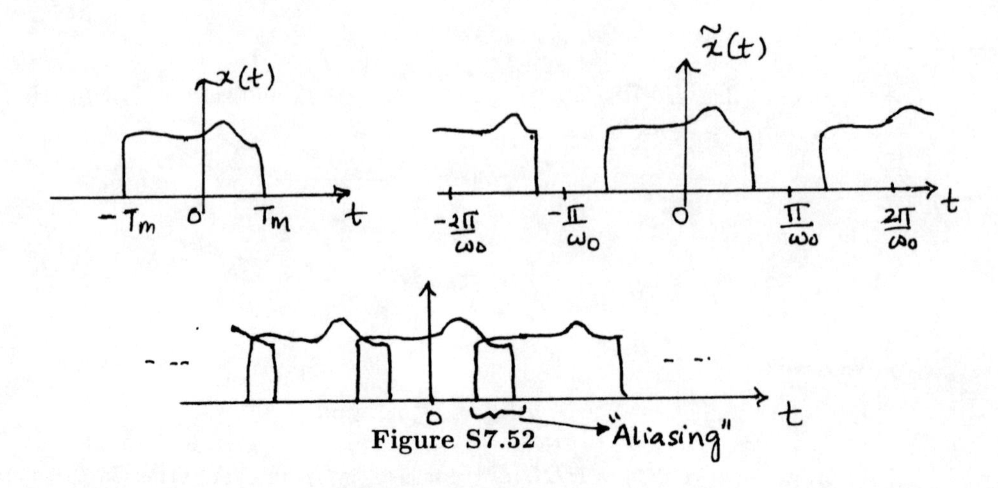

Figure S7.52

Noting that $x(t)$ is time-limited so that $x(t) = 0$ for $|t| > \pi/\omega_0$, we assume that $x(t)$ is as shown in Figure S7.52. Then, $\tilde{x}(t)$ is as shown in the figure below. Clearly, $x(t)$ can be recovered from $\tilde{x}(t)$ by multiplying it with the function

$$w(t) = \begin{cases} \omega_0, & |t| \leq \pi/\omega_0 \\ 0, & \text{otherwise} \end{cases}.$$

(c) If $x(t)$ is not constrained to be zero for $|t| > \pi/\omega_0$, then $\tilde{x}(t)$ is as shown in Figure S7.52. Clearly, there is "time-domain aliasing" between the replicas of $x(t)$ in $\tilde{x}(t)$. Therefore, $x(t)$ cannot be recovered from $\tilde{x}(t)$.

Chapter 8 Answers

8.1. Using Table 4.1, take the inverse Fourier transform of $Y(j(\omega - \omega_c))$. This gives

$$y(t) = 2x(t)e^{j\omega_c t}.$$

Therefore,

$$m(t) = 2e^{j\omega_c t}.$$

8.2. **(a)** The Fourier transform $Y(j\omega)$ of $y(t)$ is given by

$$Y(j\omega) = X(j(\omega - \omega_c)).$$

Clearly $Y(j\omega)$ is just a shifted version of $X(j\omega)$. Therefore, $x(t)$ may be recovered from $y(t)$ simply by multiplying $y(t)$ by $e^{-j\omega_c t}$. There is no constraint that needs to be placed on ω_c to ensure that $x(t)$ is recoverable from $y(t)$.

(b) We know that

$$y_1(t) = \mathcal{R}e\{y(t)\} = x(t)\cos(\omega_c t).$$

The Fourier transform $Y_1(j\omega)$ of $y_1(t)$ is as shown in Figure S8.2

$$Y_1(j\omega) = \frac{1}{2}X(j(\omega - \omega_c)) + \frac{1}{2}X(j(\omega + \omega_c))$$

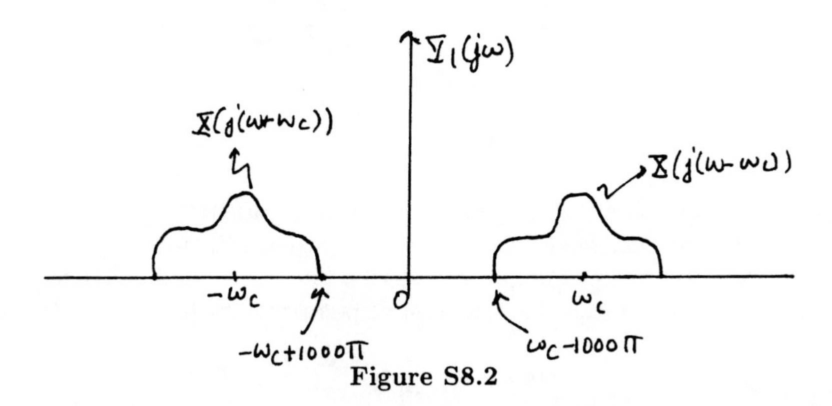

Figure S8.2

If we want to prevent the two shifted replicas of $Y(j\omega)$ from overlapping, then we need to ensure that $|\omega_c| > 1000\pi$.

8.3. When $g(t)$ is multiplied by $\cos(2000\pi t)$, the output will be

$$x_1(t) = g(t)\cos(2000\pi t) = x(t)\sin(2000\pi t)\cos(2000\pi t) = \frac{1}{2}x(t)\sin(4000\pi t).$$

The Fourier transform of this signal is

$$X_1(j\omega) = \frac{1}{4j}X(j(\omega - 4000\pi)) - \frac{1}{4j}X(j(\omega + 4000\pi)).$$

This implies that $X_1(j\omega)$ is zero for $|\omega| \leq 2000\pi$. When $y(t)$ is passed through a lowpass filter with cutoff frequency 2000π, the output will clearly be zero. Therefore $y(t) = 0$.

8.4. Consider the signal

$$\begin{aligned} y(t) &= g(t)\sin(400\pi t) \\ &= \sin(200\pi t)\sin^2(400\pi t) + 2\sin^3(400\pi t) \\ &= \sin(200\pi t)[(1-\cos(800\pi t))/2] + 2\sin(400\pi t)[(1-\cos(800\pi t)/2] \\ &= (1/2)\sin(200\pi t) - (1/4)\{\sin(1000\pi t) - \sin(600\pi t)\} \\ &\quad + \sin(400\pi t) - (1/2)\{\sin(1200\pi t) - \sin(400\pi t)\} \end{aligned}$$

If this signal is passed through a lowpass filter with cutoff frequency 400π, then the output will be

$$y_1(t) = \sin(200\pi t).$$

8.5. The signal $x(t)$ is as shown in the Figure S8.5.

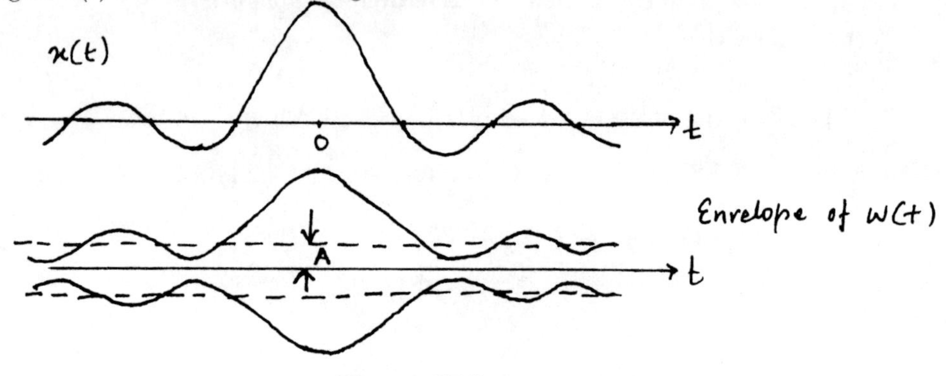

Figure S8.5

The envelope of the signal $w(t)$ is as shown in the Figure S8.5. Clearly. is we want to use asynchronous demodulation to recover the signal $x(t)$, we need to ensure that A is greater than the height h of the highest sidelobe (see Figure S8.5). Let us now determine the height of the highest sidelobe. The first zero-crossing of the signal $x(t)$ occurs at time t_0 such that

$$1000\pi t_0 = \pi, \qquad \Rightarrow \qquad t_0 = 1/1000.$$

Similarly, the second zero-crossing happens at time t_1 such that

$$1000\pi t_1 = 2\pi, \qquad \Rightarrow \qquad t_1 = 2/1000.$$

The highest sidelobe occurs at time $(t_0 + t_1)/2$, that is, at time $t_2 = 3/2000$. At this time, the amplitude of the signal $x(t)$ is

$$x(t_2) = \frac{\sin(3\pi/2)}{\pi 3/2000} = -\frac{2000}{3\pi}.$$

Therefore, A should at least be $\frac{2000}{3\pi}$. The modulation index corresponding to the smallest permissible value of A is

$$m = \frac{\text{Max. value of } x(t)}{\text{Min. possible value of} A} = \frac{1000}{2000/3\pi} = \frac{3\pi}{2}.$$

8.6. Let us denote the Fourier transform of $\sin(\omega_c t)/(\pi t)$ by $H(j\omega)$. This will be a rectangular pulse which is nonzero only in the range $|\omega| \leq \omega_c$. Taking the Fourier transform of the first equation given in the problem, we have

$$\begin{aligned} G(j\omega) &= \mathcal{FT}\{x(t)\cos(\omega_c t)\} - \mathcal{FT}\{x(t)\cos(\omega_c t)\}H(j\omega) \\ &= \mathcal{FT}\{x(t)\cos(\omega_c t)\}\{1 - H(j\omega)\} \\ &= (1/2)\left[X(j(\omega-\omega_c)) + X(j(\omega+\omega_c))\right]\{1 - H(j\omega)\}. \end{aligned}$$

$G(j\omega)$ is as shown in Figure S8.6.

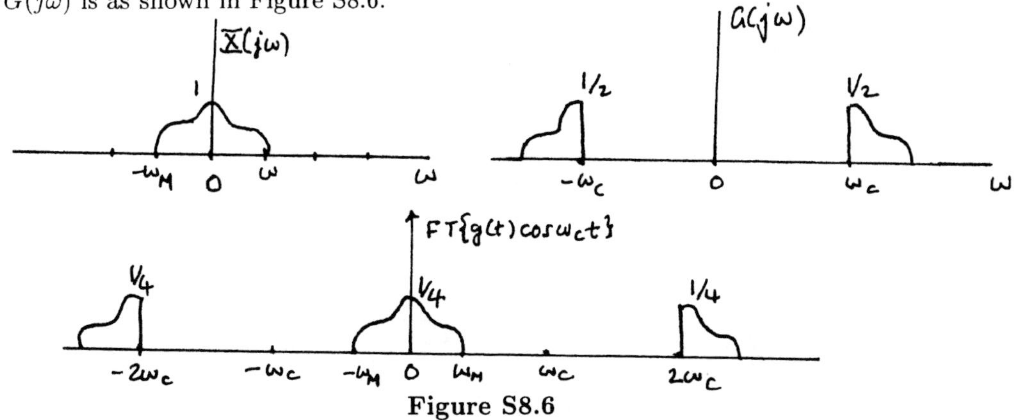

Figure S8.6

The Fourier transform of $g(t)\cos(\omega_c t)$ is also shown in Figure S8.6. Clearly, if we want to recover $x(t)$ from $g(t)\cos(\omega_c t)$, then we have to pass $g(t)\cos(\omega_c t)$ through an ideal lowpass filter with gain 4 and cutoff frequency ω_M. Therefore, $A = 4$.

8.7. In Figure S8.7, we show $X(j\omega)$, $G(j\omega)$, and $Q(j\omega)$. We also show a plot of the Fourier transform of $g(t)\cos(\omega_0 t)$. From this figure, it is clear that if we want to be able to obtain $q(t)$ from $g(t)\cos(\omega_0 t)$, then we need to ensure that (1) $\omega_0 = 2\omega_c$, and (2) an ideal lowpass filter with passband gain of 2 and a cutoff frequency of ω_c is used to filter $g(t)\cos(\omega_0 t)$.

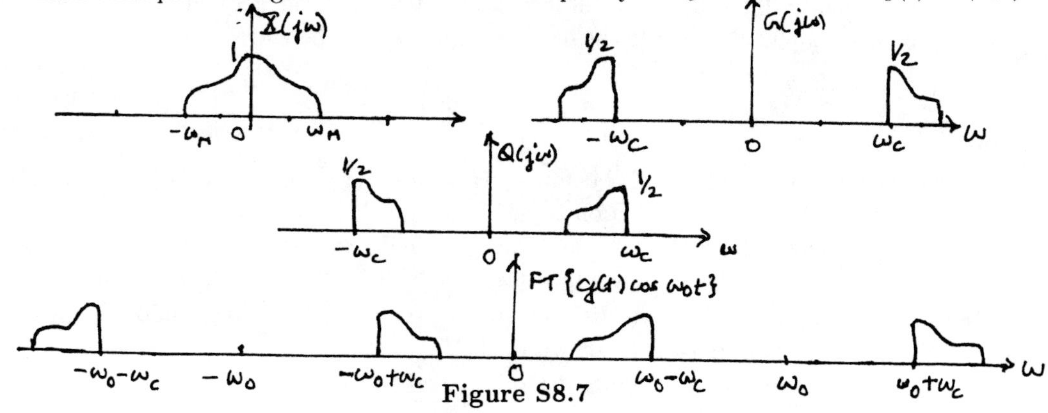

Figure S8.7

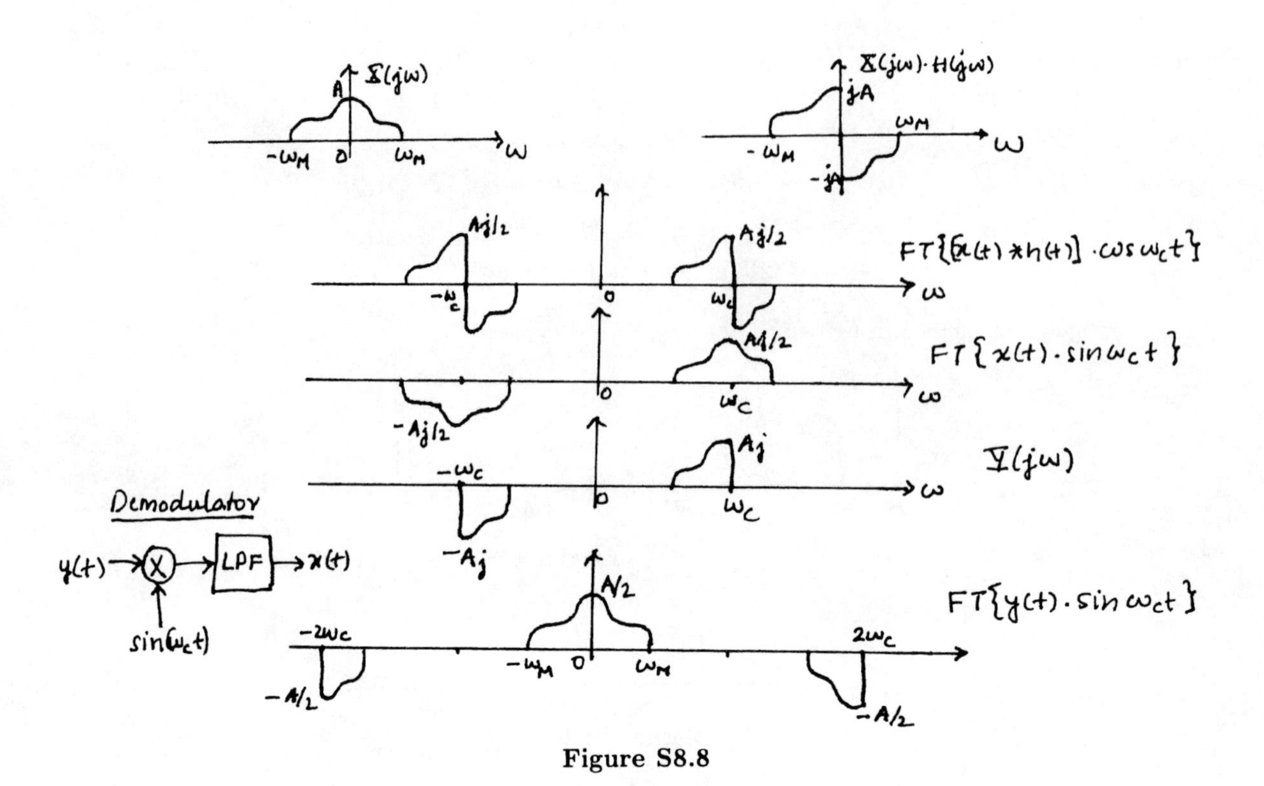

Figure S8.8

8.8. (a) From Figure S8.8, it is clear that $Y(j\omega)$ is conjugate-symmetric. Therefore, $y(t)$ is real.

(b) This part of the problem explores the demodulation of SSB signals through synchronous demodulation. This idea is explored in more detail in Problem 8.29.

Let us assume that we use the synchronous demodulation system shown in the Figure S8.8. The Fourier transform $Y_1(j\omega)$ of the signal $y_1(t)$ is shown in the Figure S8.8. Clearly, if we use an ideal lowpass filter with cutoff frequency ω_c and passband gain of 2, we would recover the original signal $x(t)$. Therefore,

$$x(t) = [y(t)\sin(\omega_c t)] * \left\{\frac{2\sin\omega_c t}{\pi t}\right\}.$$

8.9. Let the signals $x_1(t)$ and $x_2(t)$ have Fourier transforms $X_1(j\omega)$ and $X_2(j\omega)$ as shown in the Figure S8.9. When SSB modulation is performed on the signals $x_1(t)$ and $x_2(t)$, we would obtain the signals $y_1(t)$ and $y_2(t)$, respectively. The Fourier transforms $Y_1(j\omega)$ and $Y_2(j\omega)$ of these signals would be as shown in the Figure S8.9 (see Section 8.4 for details).

(a) From the figure, it is clear that the signal $y(t) = y_1(t) + y_2(t)$ would have a Fourier transform $Y(j\omega)$ which is as shown in the Figure S8.9.

From this figure, it is obvious that $Y(j\omega)$ is zero for $|\omega| > 2\omega_c$.

(b) In order to obtain $x_1(t)$ from $y(t)$, we have to first remove any contribution in $y(t)$ from $x_2(t)$. From the previously drawn figures, it is clear that we can remove all

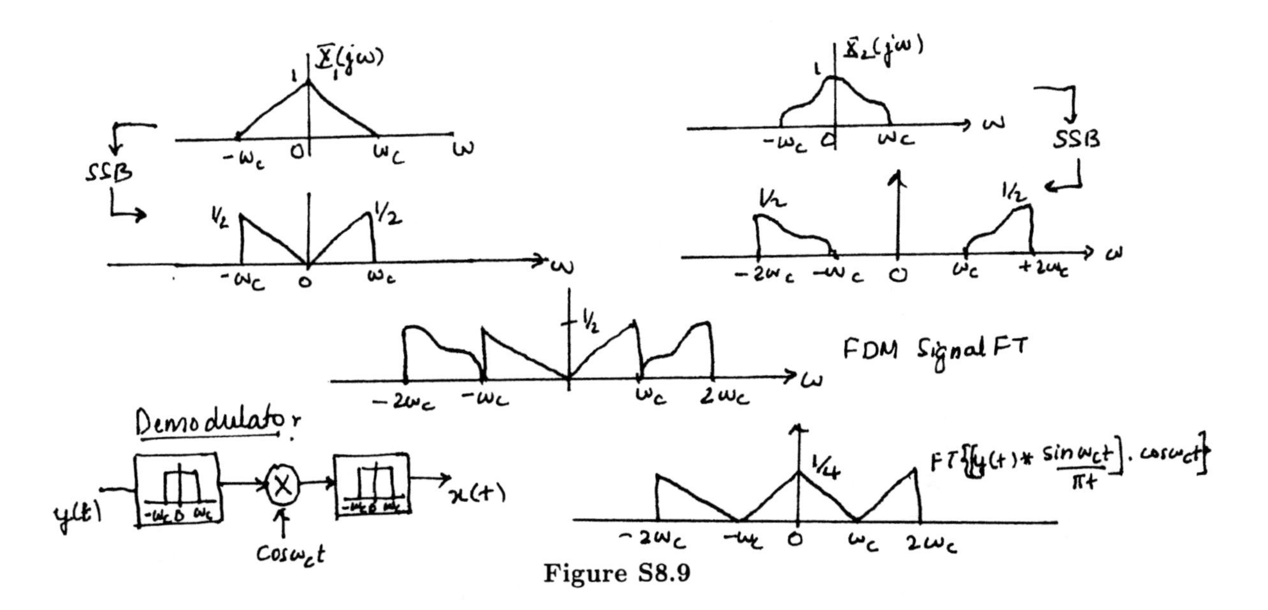

Figure S8.9

contribution to $y(t)$ from $x_2(t)$ by first lowpass filtering $y(t)$ using a lowpass filter with cutoff frequency ω_c. We may then follow this by a synchronous demodulation system. This idea is illustrated in the Figure S8.9. Therefore,

$$x_1(t) = \left[\left\{y(t) * \frac{\sin\omega_c t}{\pi t}\right\}\cos\omega_0 t\right] * \frac{A\sin\omega_c t}{\pi t}.$$

In order to determine the value of the gain A, we first plot the Fourier transform of $\left[\{y(t) * \frac{\sin\omega_c t}{\pi t}\}\cos\omega_0 t\right]$. From this it is clear that $A = 4$.

8.10. (a) From Section 8.5, we know that in order to avoid aliasing, $2\pi/T > 2\omega_M$, where ω_M is the maximum frequency in the original signal and T is the period of $c(t)$. In this case, $T = 10^{-3}$. Therefore $\omega_M < 1000\pi$. Therefore, $X(j\omega) = 0$ for $\omega > 1000\pi$.

(b) From Figure 8.24, we know that the Fourier transform $Y(j\omega)$ of the signal $y(t)$ consists of shifted replicas of $X(j\omega)$. The replica of $X(j\omega)$ centered around $\omega = 0$ is scaled by Δ/T, where Δ is the width of each pulse of $c(t)$. By using a lowpass filter, we may recover $X(j\omega)$ from $Y(j\omega)$. The lowpass filter needs to have a passband gain of T/Δ. In this case, this evaluates to $10^{-3}/(0.25 \times 10^{-3}) = 4$.

8.11. The signal $c(t)$ is

$$c(t) = a_1 e^{j\omega_c t} + a_{-1}e^{-j\omega_c t} + a_2 e^{j2\omega_c t} + a_{-2}e^{-j2\omega_c t} + \cdots$$

Since $c(t)$ is real, $a_k = a_{-k}^*$. The Fourier transform $Y(j\omega)$ of the signal $y(t) = x(t)c(t)$ is

$$Y(j\omega) = a_1 X(j(\omega - \omega_c)) + a_1^* X(j(\omega + \omega_c)) + a_2 X(j(\omega - 2\omega_c)) + a_2^* X(j(\omega + 2\omega_c)) + \cdots$$

This is plotted in Figure S8.11.

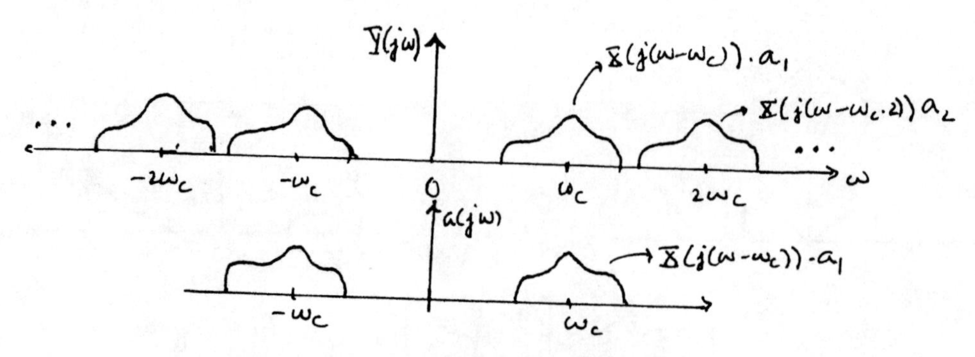

Figure S8.11

(a) The Fourier transform $G(j\omega)$ of $g(t)$ is

$$G(j\omega) = a_1 X(j(\omega - \omega_c)) + a_1^* X(j(\omega + \omega_c)).$$

This is as shown in Figure S8.11. Clearly, by comparing $G(j\omega)$ and $Y(j\omega)$, we know that $g(t)$ may be obtained from $y(t)$ by passing $y(t)$ through an ideal bandpass filter which has a passband gain of unity in the range $(\omega_c/2) \leq |\omega| \leq (3\omega_c/2)$.

(b) If $a_1 = |a_1|e^{j\sphericalangle a}$, then $a_1^* = |a_1|e^{-j\sphericalangle a}$ Also,

$$\begin{aligned} g(t) &= (a_1 e^{j\omega_c t} + a_1^* e^{-j\omega_c t})x(t) \\ &= |a_1|e^{j(\omega_c t + \sphericalangle a_1)}x(t) + |a_1|e^{-j(\omega_c t + j\sphericalangle a_1)}x(t) \\ &= 2|a_1|\cos(\omega_c t + \sphericalangle a_1)x(t) \end{aligned}$$

Therefore, $A = 2|a_1|$, and $\phi = \sphericalangle a_1$.

8.12. We need to first determine the maximum allowable period T. From Section 8.5.1, we know that T should be chosen such that $\frac{2\pi}{T} > 2\omega_M$. In this case, $\omega_M = 2000\pi$. Therefore, $T \leq 0.5 \times 10^{-3}$ sec. We now need to have 10 different pulses within a duration of T. Therefore, each pulse can be at most $\Delta = 0.5 \times 10^{-4}$ sec wide.

8.13. **(a)** We know that

$$p(0) = \frac{1}{2\pi}\int_{-\infty}^{\infty} P(j\omega)d\omega.$$

Therefore,

$$\begin{aligned} p(0) &= \frac{1}{2\pi}\int_{-2\pi/T_1}^{2\pi/T_1} \left(\frac{1}{2} + \cos(\omega T_1/2)\right) d\omega \\ &= 1/T_1 \end{aligned}$$

(b) Since $P(j\omega)$ satisfies eq. (8.28), we know that it must have zero-crossings every T_1. Therefore,

$$p(kT_1) = 0, \qquad \text{for} \qquad k = \pm 1, \pm 2, \cdots.$$

8.14. Given

$$\begin{aligned} y(t) &= \cos(\omega_c t + m\cos\omega_m t) \\ &= \cos(\omega_c t)\cos(m\cos(\omega_m t)) - \sin(\omega_c t)\sin(m\cos(\omega_m t)) \end{aligned}$$

But since $\omega_c >> \omega_m$ and $m << \pi/2$, we may make the following approximations

$$\cos(m\cos(\omega_m t)) \approx 1$$

and

$$\sin(m\cos(\omega_m t)) \approx m\cos(\omega_m t).$$

Therefore,

$$\begin{aligned} y(t) &= \cos(\omega_c t) - \sin(\omega_c t)m\cos(\omega_m t) \\ &= \cos(\omega_c t) - \frac{m}{2}\{\sin[(\omega_c + \omega_m)t] + \sin[(\omega_c - \omega_m)t]\} \end{aligned}$$

Therefore for $\omega > 0$,

$$Y(j\omega) = \pi\delta(\omega - \omega_c) - \frac{m\pi}{2j}\delta(\omega - (\omega_c + \omega_m)) - \frac{m\pi}{2j}\delta(\omega - (\omega_c - \omega_m)).$$

8.15. When a signal $x(t)$ is amplitude modulated with $e^{j\omega_0 n}$, then the Fourier transform of the result is

$$Y_1(e^{j\omega}) = X(e^{j(\omega-\omega_0)}).$$

When a signal $x(t)$ is amplitude modulated with $\cos(\omega_0 n)$, then the Fourier transform of the result is

$$Y_1(e^{j\omega}) = (1/2)X(e^{j(\omega-\omega_0)}) + (1/2)X(e^{j(\omega+\omega_0)}).$$

$Y_1(e^{j\omega}) = Y_2(e^{j\omega})$ only when ω_0 is either 0 or π.

8.16. We know that $c[n] = \sin(5\pi n/2) = \sin(\pi n/2)$.

$$Y(e^{j\omega}) = (1/2j)X(e^{j(\omega-\pi/2)}) - (1/2j)X(e^{j(\omega+\pi/2)}).$$

This is as shown in the Figure S8.16.

From the figure, it is obvious that

$$Y(e^{j\omega}) = 0, \qquad \text{for } 0 \le \omega \le \frac{3\pi}{8} \text{ and } \frac{5\pi}{8} \le \omega \le \pi.$$

8.17. The Fourier transforms $X(e^{j\omega})$, $G(e^{j\omega})$, $Q(e^{j\omega})$, and $Y(e^{j\omega})$ are shown in Figure S8.17.

From the figure, it is obvious that $Y(e^{j\omega}) = 0$ for $0 \le |\omega| \le \frac{\pi}{2}$.

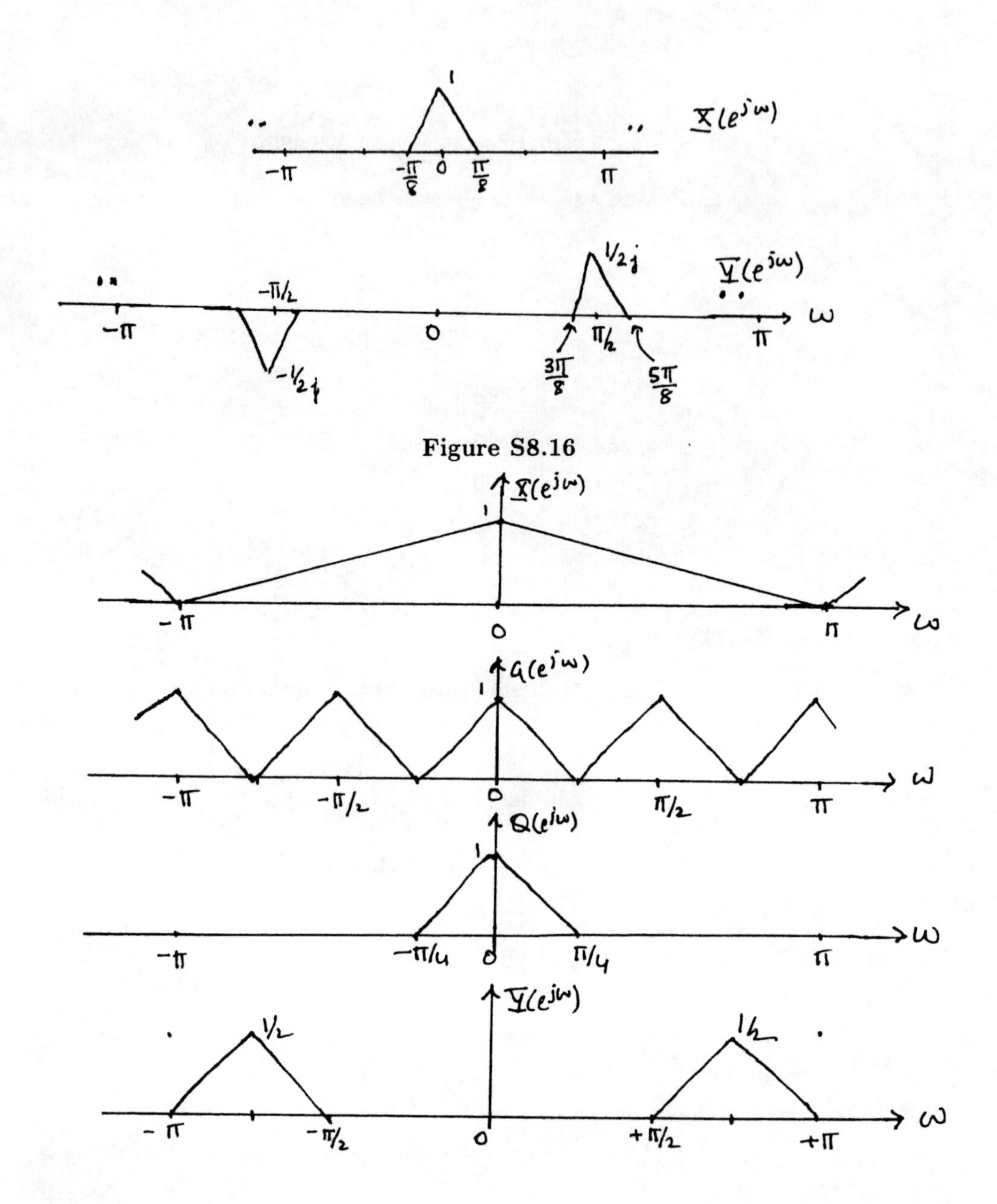

Figure S8.16

Figure S8.17

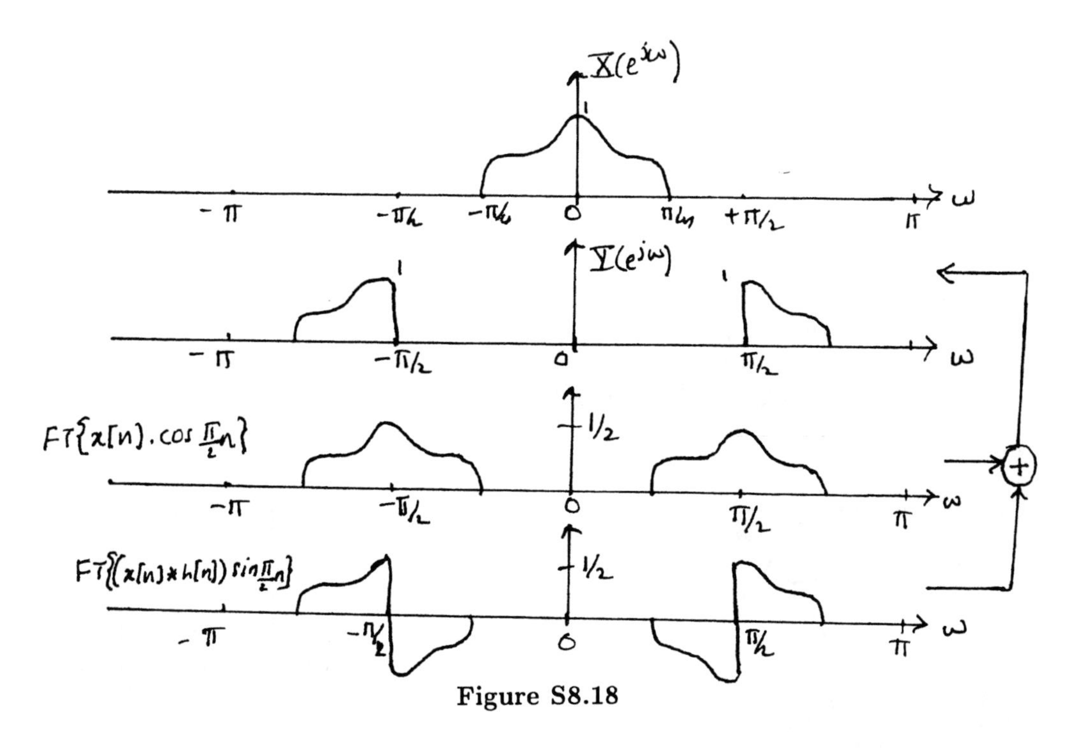

Figure S8.18

8.18. The Fourier transforms $X(e^{j\omega})$ and $Y(e^{j\omega})$ are as shown in Figure S8.18.

From these figures, it is clear that we wish to accomplish single sideband modulation using this system. In particular, we are interested in retaining the upper sidebands of the signal. Note that in Figure 8.21 of Section 8.4, is shown a continuous-time single sideband system for retaining the lower sidebands. In this section, it was also mentioned (see eq. (8.21)) that in order to retain the upper sidebands, the frequency response of the filter used in the system had to be changed to

$$H(j\omega) = \begin{cases} j, & \omega > 0 \\ -j, & \omega < 0 \end{cases}.$$

In this problem, we extend this same idea to discrete-time systems. We assume that the frequency response $H(e^{j\omega})$ of the unknown system is

$$H(e^{j\omega}) = \begin{cases} j, & \omega > 0 \\ -j, & \omega < 0 \end{cases}.$$

Let us now show that this does indeed give us the desired output. We redraw the system given in the problem with appropriate labels for the intermediate outputs. The Fourier transforms of these intermediate outputs are shown in Figure S8.18.

From Figure S8.18, it is clear that the choice of $H(e^{j\omega})$ was appropriate.

8.19. Since 10 different signals have to be squeezed in within a bandwidth of 2π, each signal is allowed to occupy a bandwidth of $\frac{2\pi}{10} = \frac{\pi}{5}$ *after* sinusoidal modulation. Therefore, before sinusoidal modulation, each signal can occupy only a bandwidth of $\frac{\pi}{10}$. The Fourier transform $Y_i(e^{j\omega})$ of the signal obtained by upsampling $x_i[n]$ by a factor of N can be nonzero (in the range $|\omega| < \pi$) only for $|\omega| < \frac{\pi}{20}$. Therefore, N has to be at least 20.

8.20. Note that by choosing $p[n] = \sum_{k=-\infty}^{\infty} \delta[n-2k]$, we would be able to get $\hat{v}_1[n]$ and $\hat{v}_2[n-1]$ at the output of the multipliers. Furthermore, note that $\hat{V}_1(e^{j\omega}) = V_2(e^{2j\omega})$ and $\hat{V}_2(e^{j\omega}) = V_2(e^{2j\omega})e^{-j\omega}$. This is illustrated in Figure S8.20. Therefore, the output of the two branches will be as shown in Figure S8.20. From these figures, it is clear that the sum of the two outputs will be a FDM signal containing both $v_1[n]$ and $v_2[n]$.

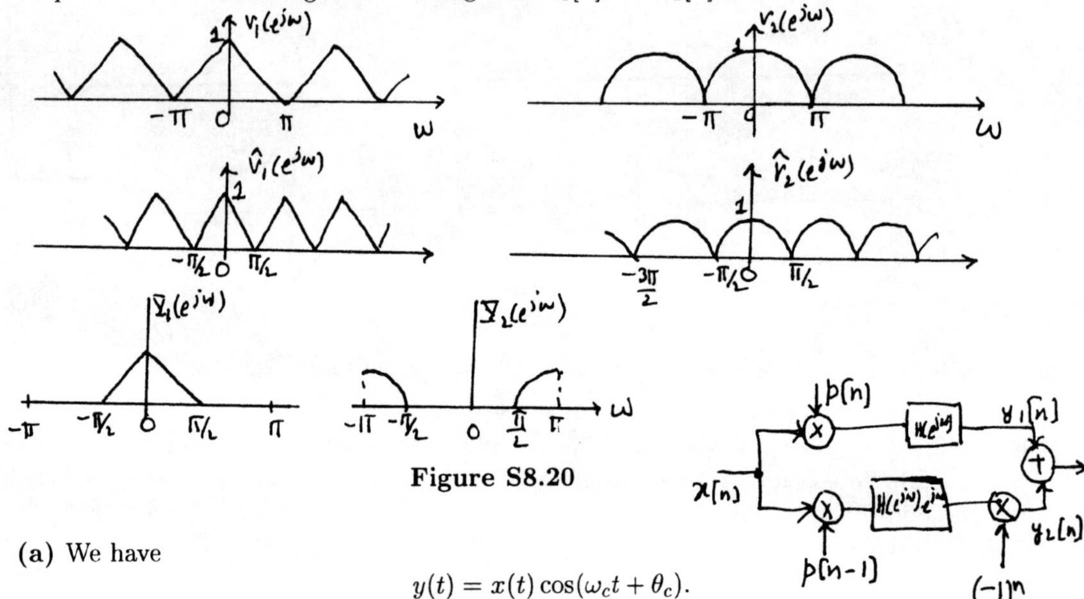

Figure S8.20

8.21. (a) We have

$$y(t) = x(t)\cos(\omega_c t + \theta_c).$$

From this we obtain

$$\begin{aligned} w(t) &= y(t)\cos(\omega_c t + \theta_c) \\ &= x(t)\cos^2(\omega_c t + \theta_c) \\ &= x(t)\left[\frac{1+\cos[2(\omega_c t + \theta_c)]}{2}\right] \\ &= \frac{1}{2}x(t) + \frac{1}{2}x(t)\cos(2\omega_c t + 2\theta_c). \end{aligned}$$

(b) The Fourier transforms $X(j\omega)$ and $Y(j\omega)$ of the signals $x(t)$ and $y(t)$ are sketched in Figure S8.21.

From these figures, it is clear that in order to avoid the any overlap between the sidebands, we must have $\omega_c > \omega_M$. We now sketch $W(j\omega)$ in Figure S8.21. From this figure it is clear that we require to satisfy the following condition in order to have the output be proportional to $x(t)$:

$$\omega_M \leq W \leq (2\omega_c - \omega_M).$$

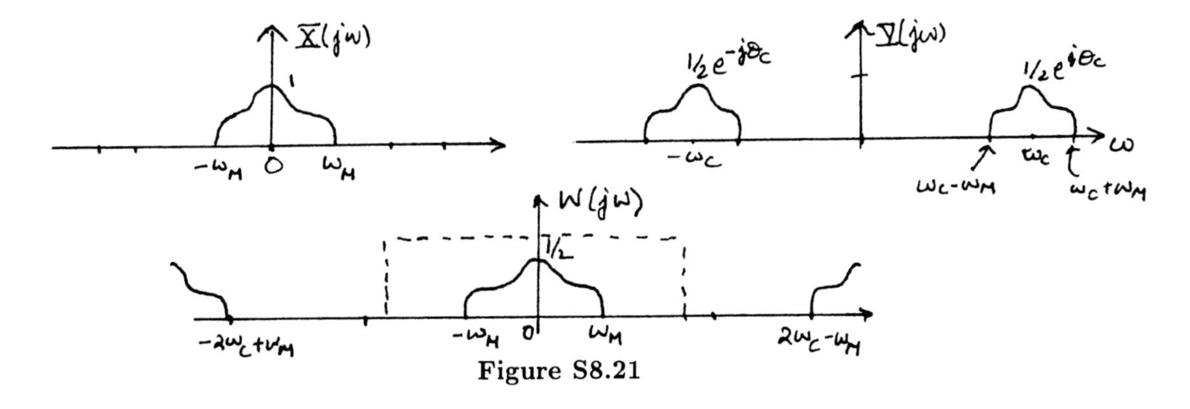

Figure S8.21

8.22. Sketches for (i) the Fourier transforms for each of the intermediate outputs and (ii) the Fourier transform $Y(j\omega)$ of $y(t)$ are shown in Figure S8.22.

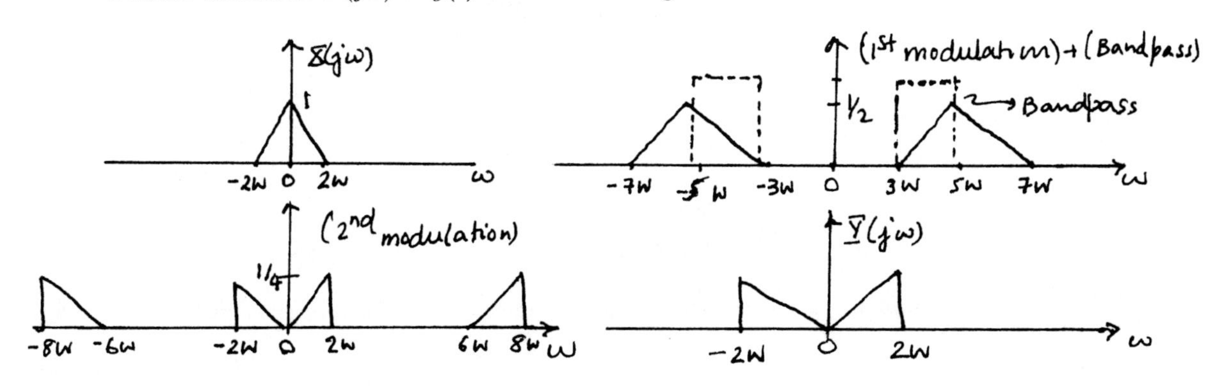

Figure S8.22

8.23. (a) We have

$$w(t) = x(t)\cos(\omega_c t)\cos(\omega_d t).$$

Using Trigonometric identities, we have

$$\begin{aligned} w(t) &= \frac{1}{2}x(t)\left\{\cos[(\omega_c+\omega_d)t] + \cos[(\omega_c-\omega_d)t]\right\} \\ &= \frac{1}{2}x(t)\left\{\cos[(\omega_c+\omega_d)t] + \cos[\Delta\omega t]\right\} \end{aligned}$$

Now, $\frac{1}{2}x(t)\cos[(\omega_c+\omega_d)t]$ has a spectrum in the range $\omega_c+\omega_d-\omega_M \le |\omega| \le \omega_c+\omega_d+\omega_M$. This range may also be expressed as $2\omega_c+\Delta\omega-\omega_M \le |\omega| \le 2\omega_c+\Delta\omega+\omega_M$. Since, we are given that $W < 2\omega_c + \Delta\omega - \omega_M$, lowpass filtering will result in the output $\frac{1}{2}x(t)\cos(\Delta\omega t)$.

(b) We sketch the spectrum of the output for $\Delta\omega = \omega_M/2$ in Figure S8.23.

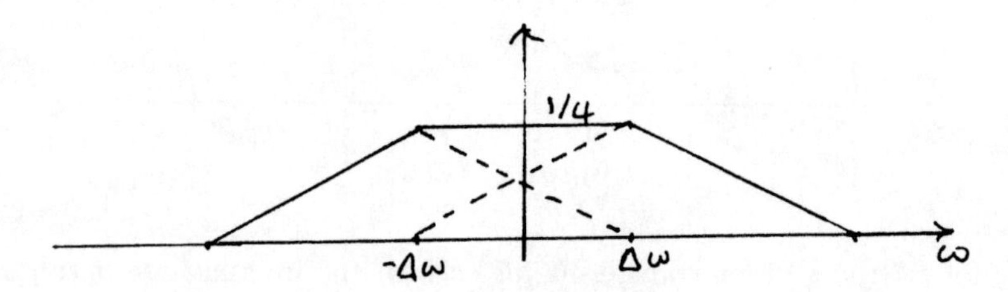

Figure S8.23

8.24. **(a)** Since $s(t) = \sum_{k=-\infty}^{\infty} \delta(t - kT)$, we have

$$S(j\omega) = \frac{2\pi}{T} \sum_{k=-\infty}^{\infty} \delta(\omega - 2k\pi/T) = \omega_c \sum_{k=-\infty}^{\infty} \delta(\omega - k\omega_c).$$

Let us denote $x(t)s(t)$ by $w(t)$. Then the Fourier transform of the signal $w(t)$ is

$$\begin{aligned} W(j\omega) &= \frac{1}{2\pi}[X(j\omega) * S(j\omega)] \\ &= \frac{\omega_c}{2\pi} \sum_{k=-\infty}^{\infty} X(j(\omega - k\omega_c)) \end{aligned}$$

This is as shown in Figure S8.24. Therefore, the Fourier transform $Y(j\omega)$ of the output of the bandpass filter is

$$Y(j\omega) = \frac{A\omega_c}{2\pi} X(j(\omega - \omega_c)) + \frac{A\omega_c}{2\pi} X(j(\omega + \omega_c)),$$

Therefore,

$$y(t) = \frac{A\omega_c}{\pi} x(t) \cos(\omega_c t).$$

(b) If $\Delta \neq 0$, then

$$S(j\omega) = \frac{2\pi}{T} e^{-\Delta\omega} \sum_{k=-\infty}^{\infty} \delta(\omega - 2k\pi/T) = \frac{2\pi}{T} \sum_{k=-\infty}^{\infty} e^{-j\Delta k 2\pi/T} \delta(\omega - 2k\pi/T).$$

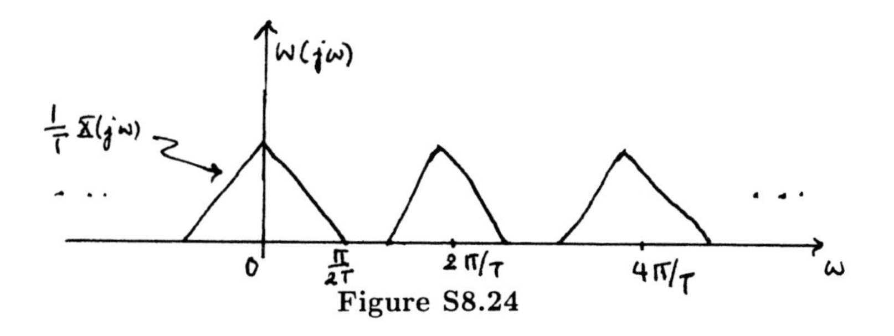

Figure S8.24

Let us denote $x(t)s(t)$ by $w(t)$. Then the Fourier transform of the signal $w(t)$ is

$$\begin{aligned} W(j\omega) &= \frac{1}{2\pi}[X(j\omega) * S(j\omega)] \\ &= \frac{1}{T}\sum_{k=-\infty}^{\infty} e^{-j\Delta k 2\pi/T} X(j(\omega - k2\pi/T)) \end{aligned}$$

Therefore, the Fourier transform $Y(j\omega)$ of the output of the bandpass filter is

$$Y(j\omega) = \frac{A\omega_c}{2\pi} e^{-j2\pi\Delta/T} X(j(\omega - 2\pi/T)) + \frac{A\omega_c}{2\pi} e^{-j2\pi\Delta/T} X(j(\omega + 2\pi/T)),$$

Therefore,

$$y(t) = \frac{A\omega_c}{2\pi} x(t) \cos\left(\frac{2\pi}{T}t - \frac{2\pi}{T}\Delta\right).$$

(c) From the analysis in part (b), it is clear that the maximum allowable value for ω_M is π/T.

8.25. (a) $Y(j\omega)$ is as sketched in Figure S8.25.

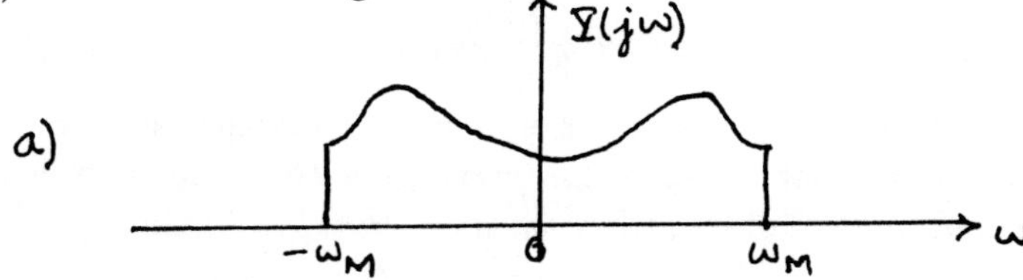

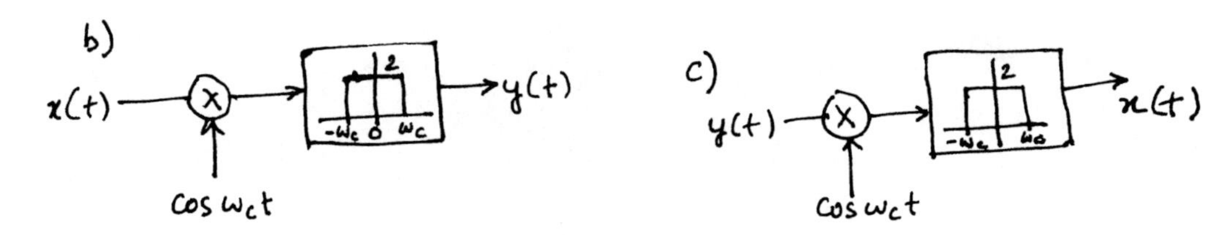

Figure S8.25

(b) The block-diagram of the scrambler is as sketched in Figure S8.25.

(c) The block-diagram of the unscrambler is as sketched in Figure S8.25.

8.26. The Fourier transform of $y(t)$ is as sketched in Figure S8.26. We also sketch the Fourier transforms of $y(t)\cos(\omega_c t)$ and $y(t)\sin(\omega_c t)$ in Figure S8.26.

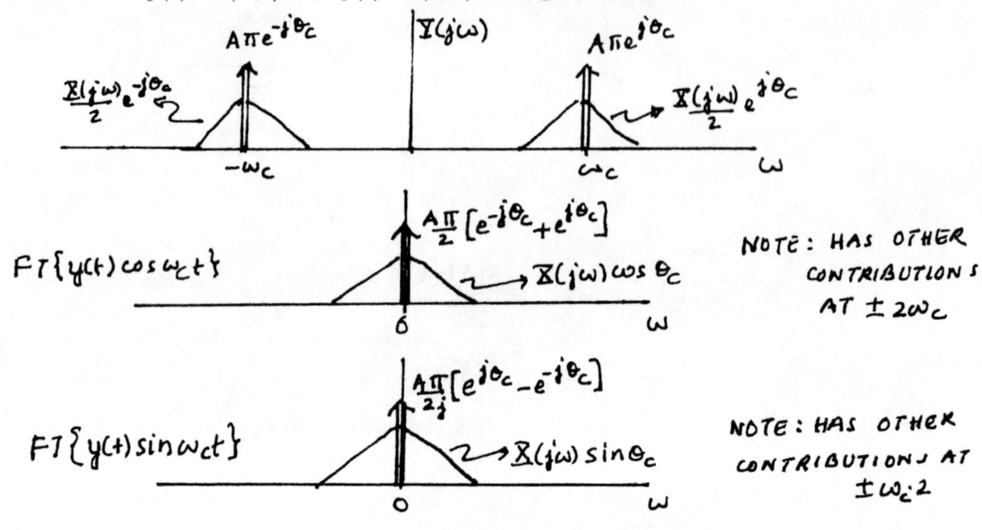

Figure S8.26

From these figures, it is clear that the outputs of the lowpass filters are $[x(t)+A]\cos(\theta_c)$ and $[x(t)+A]\sin(\theta_c)$. Upon squaring and adding, we obtain the signal $[x(t)+A]^2\{\cos^2\theta_c + \sin^2\theta_c\} = [x(t)+A]^2$. Therefore, $r(t) = x(t) + A$.

8.27. **(a)** The maximum value of $x(t)$ is 1. Therefore, $m = 1/A$. Now,

$$y(t) = A\cos(\omega_c t + \theta) + \frac{1}{2}\cos[(\omega_c + \omega_M)t] + \frac{1}{2}\cos[(\omega_c - \omega_M)t].$$

Therefore, $y(t)$ consists of three sinusoids. From Parseval's theorem, we know that the total power in $y(t)$ is the sum of the powers in each of the sinusoids. Now note that the power in a sinusoid of frequency ω_0 is

$$\frac{\omega_0}{2\pi}\int_0^{2\pi/\omega_0} \cos^2(\omega_0 t)dt = \frac{1}{2}.$$

Therefore,

$$P_y = \frac{A^2}{2} + \frac{1}{2} = \frac{1}{2m^2} + \frac{1}{2}.$$

(b) The efficiency is given by

$$E = \frac{\frac{1}{2}}{\frac{1}{2}[m^{-2}+1]} = \frac{m^2}{1+m^2}.$$

This is plotted in Figure S8.27.

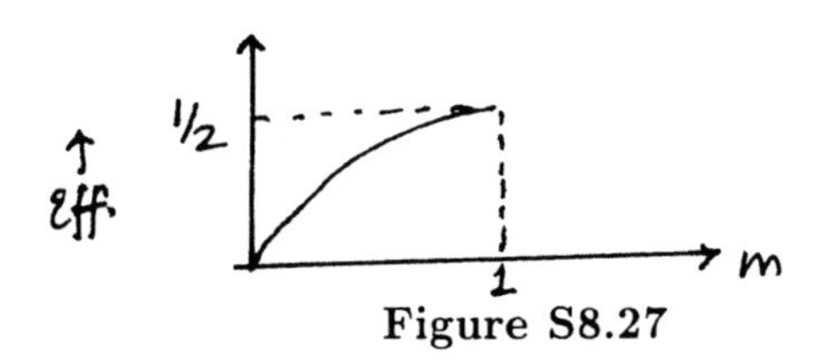

Figure S8.27

8.28. (a) The sketches of $Y_1(j\omega)$, $Y_2(j\omega)$, and $Y(j\omega)$ are as shown in Figure S8.28.

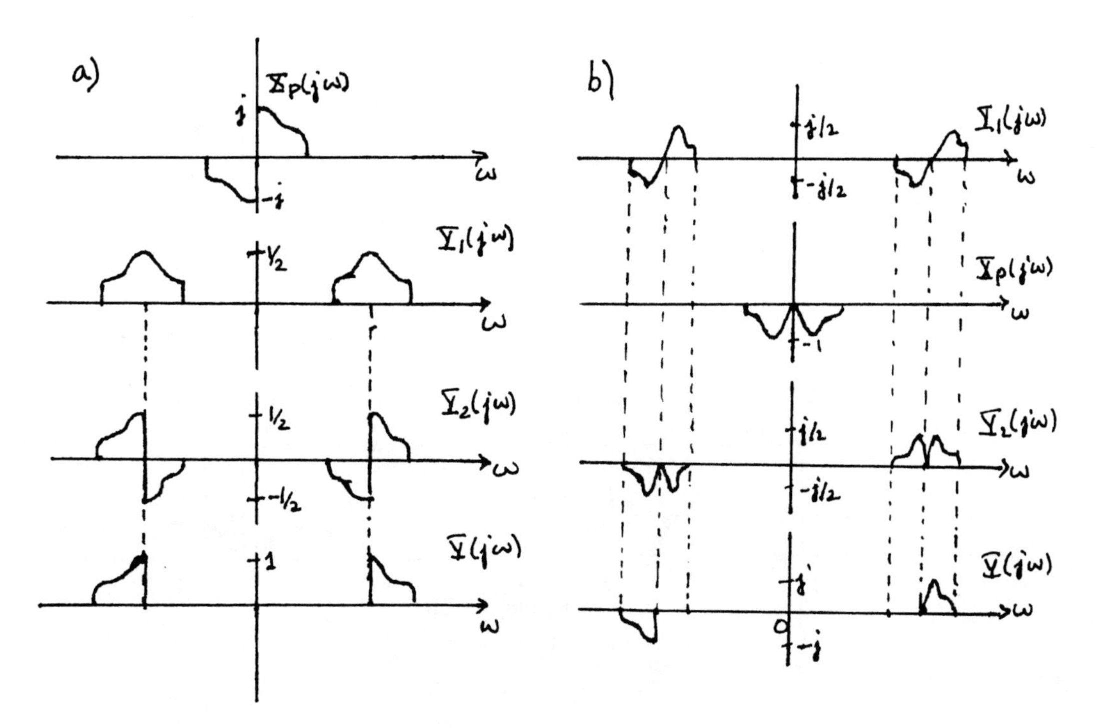

Figure S8.28

(b) The sketches of $Y_1(j\omega)$, $Y_2(j\omega)$, and $Y(j\omega)$ are as shown in Figure S8.28.

8.29. (a) The sketches in Figure S8.29 show $S(j\omega)$ and $R(j\omega)$.

(b) In Figure S8.29 we show how $P(j\omega)$ may be obtained by considering the outputs of the various stages of Figure P8.29(c). From the sketch for $P(j\omega)$, it is clear that $P(j\omega) = 2S(j\omega)$.

(c) In Figure S8.29 we show the result of demodulation on both $s(t)$ and $r(t)$. It is clear that $x(t)$ is recovered in both cases.

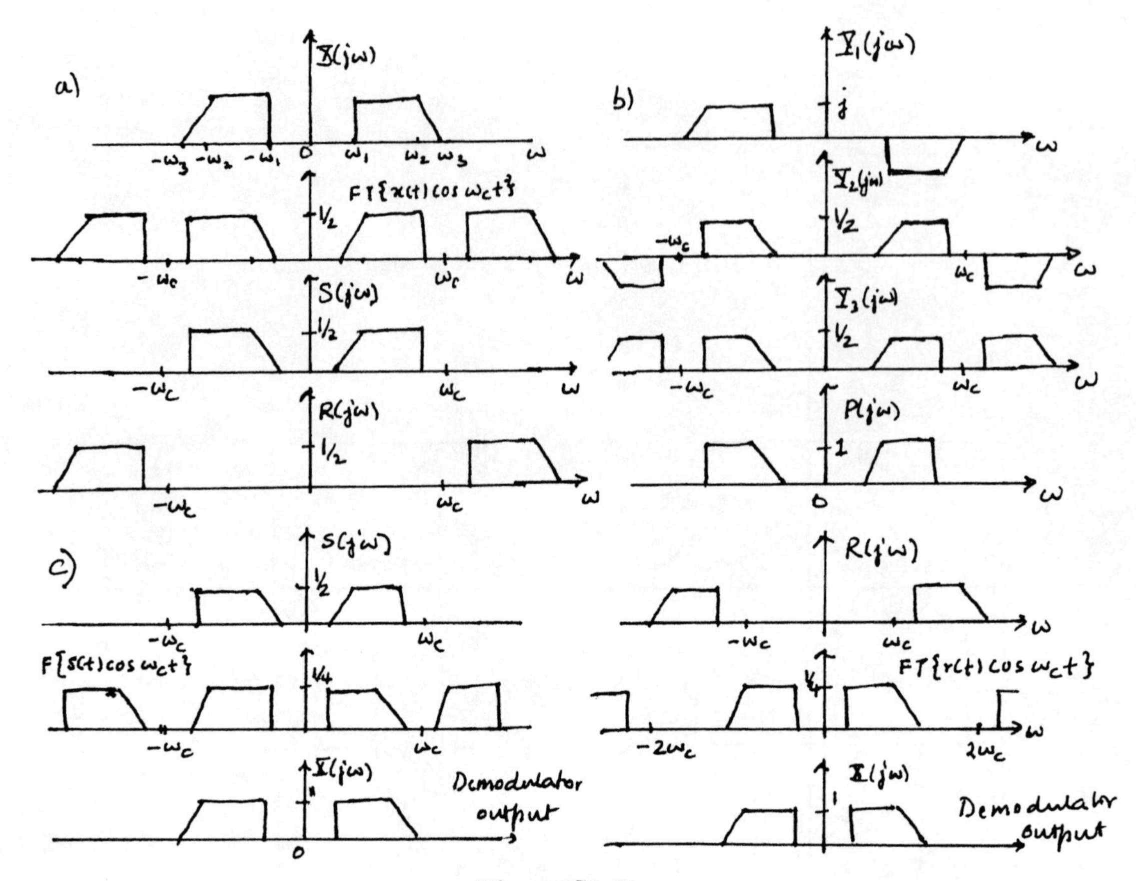

Figure S8.29

8.30. **(a)** We have

$$P(j\omega) = \frac{2\pi}{T} \sum_{k=-\infty}^{\infty} \delta(\omega - 2k\pi/T).$$

Now,

$$\begin{aligned} R(j\omega) &= \frac{1}{2\pi}\left[X(j\omega) * P(j\omega)\right] \\ &= \frac{1}{T}\sum_{k=-\infty}^{\infty} X(j(\omega - 2\pi k/T)). \end{aligned}$$

This is as sketched in Figure S8.30. Now,

$$Q(j\omega) = H(j\omega)R(j\omega),$$

where

$$H(j\omega) = \frac{2\sin(\omega\Delta/2)}{\omega}.$$

Therefore, $Q(j\omega)$ is as sketched in Figure S8.30.

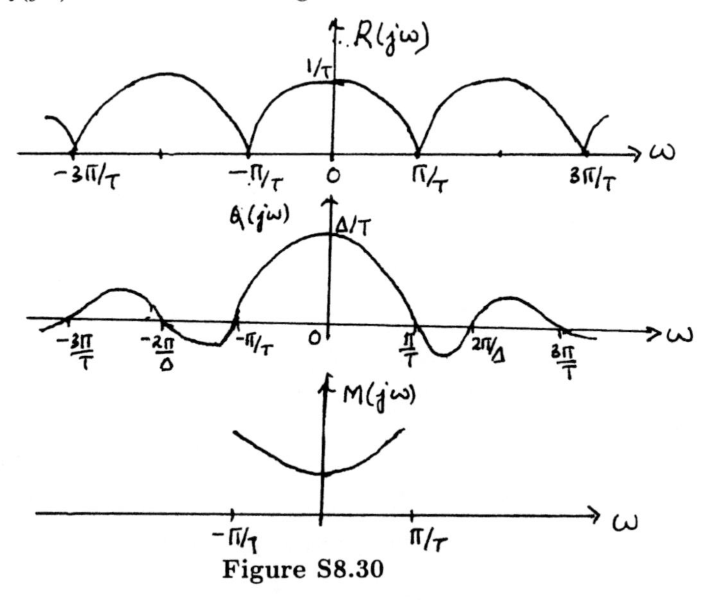

Figure S8.30

(b) We have $Q(j\omega) = H(j\omega)\frac{1}{T}X(j\omega)$ for $|\omega| \leq \pi/T$. To recover $X(j\omega)$ from $Q(j\omega)$, we require that $H(j\omega) \neq 0$ for $|\omega| \leq \pi/T$. This is possible only if $2\pi/\Delta > \pi/T$, that is $\Delta < 2T$.

(c) We require that

$$M(j\omega) = \begin{cases} \frac{T}{H(j\omega)}, & |\omega| \leq \pi/T \\ 0, & \text{otherwise} \end{cases}.$$

This is sketched in Figure S8.30.

8.31. (a) Given

$$y(t) = \sum_{n=-\infty}^{\infty} x[n]p(t-n).$$

Taking the Fourier transform of both sides of this equation, we obtain

$$\begin{aligned} Y(j\omega) &= \sum_{n=-\infty}^{\infty} x[n]P(j\omega)e^{-j\omega n} \\ &= P(j\omega)\sum_{n=-\infty}^{\infty} x[n]e^{-j\omega n} \\ &= P(j\omega)X(e^{j\omega}) \end{aligned}$$

(b) Let us define the signals $c(t)$ and $d(t)$ such that

$$c(t) = \cos(8\pi t)$$

and

$$d(t) = \begin{cases} 1, & 0 \le t \le 1 \\ 0, & \text{otherwise} \end{cases}.$$

Then,

$$C(j\omega) = \pi \left[\delta(\omega - 8\pi) + \delta(\omega + 8\pi)\right]$$

and

$$D(j\omega) = \frac{\sin(\omega/2)}{\omega/2} e^{-j\omega/2}.$$

Noting that $p(t) = c(t)d(t)$, we have

$$\begin{aligned} P(j\omega) &= \frac{1}{2\pi}\left[C(j\omega) * D(j\omega)\right] \\ &= \frac{1}{2}D(j(\omega - 8\pi)) + \frac{1}{2}D(j(\omega - 8\pi)) \\ &= \frac{1}{2}\frac{\sin[(\omega - 8\pi)/2]}{(\omega - 8\pi)/2} e^{-j(\omega - 8\pi)/2} + \frac{1}{2}\frac{\sin[(\omega + 8\pi)/2]}{(\omega + 8\pi)/2} e^{-j(\omega - 8\pi)/2} \\ &= -\frac{1}{2}\frac{\sin(\omega/2)}{(\omega - 8\pi)/2} e^{-j\omega/2} + \frac{1}{2}\frac{\sin(\omega/2)}{(\omega + 8\pi)/2} e^{-j\omega/2} \end{aligned}$$

Therefore,

$$Y(j\omega) = X(e^{j\omega})P(j\omega),$$

where $P(j\omega)$ is as given above.

8.32. **(a)** Let $c[n] = \cos(\omega_c n)$. Noting that $y[n] = x[n]c[n]$, we have

$$Y(e^{j\omega}) = \frac{1}{2\pi}\int_{-\pi}^{\pi} C(e^{j\theta})X(e^{j(\omega - \theta)})d\theta.$$

We now have

$$C(e^{j\omega}) = \pi \sum_{k=-\infty}^{\infty} \left[\delta(\omega - \omega_c + 2k\pi) + \delta(\omega + \omega_c + 2k\pi)\right].$$

Therefore, $Y(e^{j\omega})$ is

$$Y(e^{j\omega}) = \frac{1}{2}\left[X(e^{j(\omega - \omega_c)}) + X(e^{j(\omega + \omega_c)})\right].$$

If we assume that $\omega_c > \omega_{lp}$ and $\omega_c < \pi - \omega_{lp}$, then $Y(e^{j\omega})$ may be sketched as shown in Figure S8.32.

(b) Let $c_1[n] = \cos(\omega_c n + \theta_c)$. Let $q[n] = y[n]c_1[n]$. Then,

$$Q(e^{j\omega}) = \frac{1}{2\pi}\int_{-\pi}^{\pi} C_1(e^{j\theta})Y(e^{j(\omega - \theta)})d\theta.$$

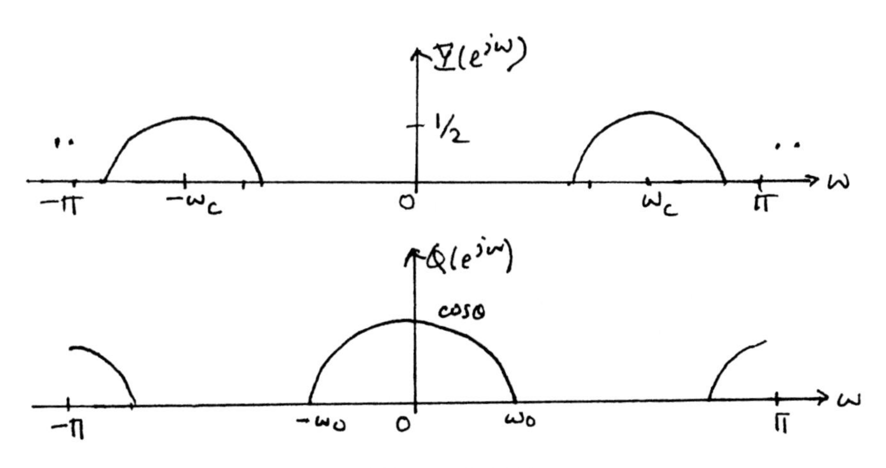

Figure S8.32

We now have

$$C_1(e^{j\omega}) = \pi \sum_{k=-\infty}^{\infty} \left[e^{j\theta_c}\delta(\omega - \omega_c + 2k\pi) + e^{-j\theta_c}\delta(\omega + \omega_c + 2k\pi) \right].$$

Therefore, $Q(e^{j\omega})$ is

$$Q(e^{j\omega}) = \frac{1}{2}\left[e^{j\theta_c} Y(e^{j(\omega-\omega_c)}) + e^{-j\theta_c} Y(e^{j(\omega+\omega_c)}) \right].$$

This is as shown in Figure S8.32.

Thus, $\hat{x}[n] = x[n]$ if $G = 1/\cos(\theta_c)$. We definitely require that $\cos(\theta_c) \neq 0$. This implies that θ_c should not be an odd multiple of $\pi/2$. The restrictions to ω_c and ω_{lp} are the same as the ones mentioned in part (a). That is, $\omega_c > \omega_{lp}$ and $\omega_c < \pi - \omega_{lp}$.

8.33. **(a)** Let us assume that each of the signals is bandlimited to ω_M after upsampling. That is, $X_i(e^{j\omega})$ in the range $|\omega| \leq \pi$ is nonzero only for $|\omega| \leq \omega_M$. Now when one of these upsampled signals is modulated by a sinusoid, the Fourier transform of the resulting signal will occupy a bandwidth of $4\omega_M$ in the range $|\omega| \leq \pi$. If we need to squeeze in four such signals in the range $|\omega| \leq \pi$, then we need to make sure that

$$\frac{2\pi}{4\omega_M} = 4, \qquad \Rightarrow \qquad \omega_M = \frac{\pi}{8}.$$

Therefore, each of the signals has to be upsampled by at least a factor of 8.

(b) We may use to the following procedure to generate the required FDM signal.

(i) For each $x_i[n]$, generate the signal

$$g_i[n] = \begin{cases} x_i[n/4], & n = 0, \pm 4, \pm 8, \cdots \\ 0, & \text{otherwise} \end{cases}.$$

(ii) Pass each $g_i[n]$ through a lowpass filter with cutoff frequency $\pi/4$.

(iii) Perform sinusoidal modulation on each $g_i[n]$ to obtain

$$y_i[n] = g_i[n]\cos(i\pi/4).$$

(iv) Pass each $y_i[n]$ through a lowpass filter with cutoff frequency $i\pi/4$. This gives the signals $r_i[n]$.

(v) Add all $r_i[n]$ together to get the FDM signal.

8.34. The output of the squarer is

$$r(t) = [x(t) + \cos(\omega_c t)]^2 = x^2(t) + \cos^2(\omega_c t) + 2x(t)\cos(\omega_c t).$$

The bandpass filter should reject $x^2(t) + \cos^2(\omega_c t)$ and multiply the remainder by 1/2. Therefore, $A = 1/2$. Since the spectral contribution of $2x(t)\cos(\omega_c t)$ is in the range $\omega_c - \omega_M \le |\omega| \le \omega_c + \omega_M$, we require $\omega_l = \omega_c - \omega_M$ and $\omega_h = \omega_c + \omega_M$.

Note that (i) the spectral contribution of $x^2(t)$ is in the range $|\omega| \le 2\omega_M$ and (ii) the spectral contribution of $\cos^2(\omega_c t)$ is at $\omega = 0$ and $\omega = \pm 2\omega_c$. Therefore, we need to ensure that

$$\omega_l > 2\omega_M \quad \Rightarrow \quad \omega_M < \omega_c/3.$$

We also need to ensure that

$$\omega_h < 2\omega_c \quad \Rightarrow \quad \omega_M < \omega_c.$$

8.35. **(a)** Since $Z(j\omega) = \frac{1}{2}X(j(\omega - \omega_c)) + \frac{1}{2}X(j(\omega + \omega_c))$, it is as shown in Figure S8.35.

The Fourier series coefficients of $p(t)$ can be shown to be $a_k = 4\sin(k\pi/2)/(2\pi k)$ for $k \ne 0$ and zero for $k = 0$. Therefore,

$$P(j\omega) = \sum_{\substack{k=-\infty \\ k\neq 0}}^{\infty} \frac{4\sin(k\pi/2)}{k}\delta(\omega - k\omega_0).$$

This is as shown in Figure S8.35.

Since $y(t) = z(t)p(t)$,

$$Y(j\omega) = \frac{1}{2\pi}\left[Z(j\omega) * P(j\omega)\right].$$

Therefore, $Y(j\omega)$ is as shown in Figure S8.35.

(b) From the last figure in the previous part, it is clear that we require $H(j\omega)$ to be as shown in Figure S8.35 to ensure that $v(t) = x(t)$.

8.36. **(a)** $Z(j\omega)$ is as sketched in Figure S8.36.

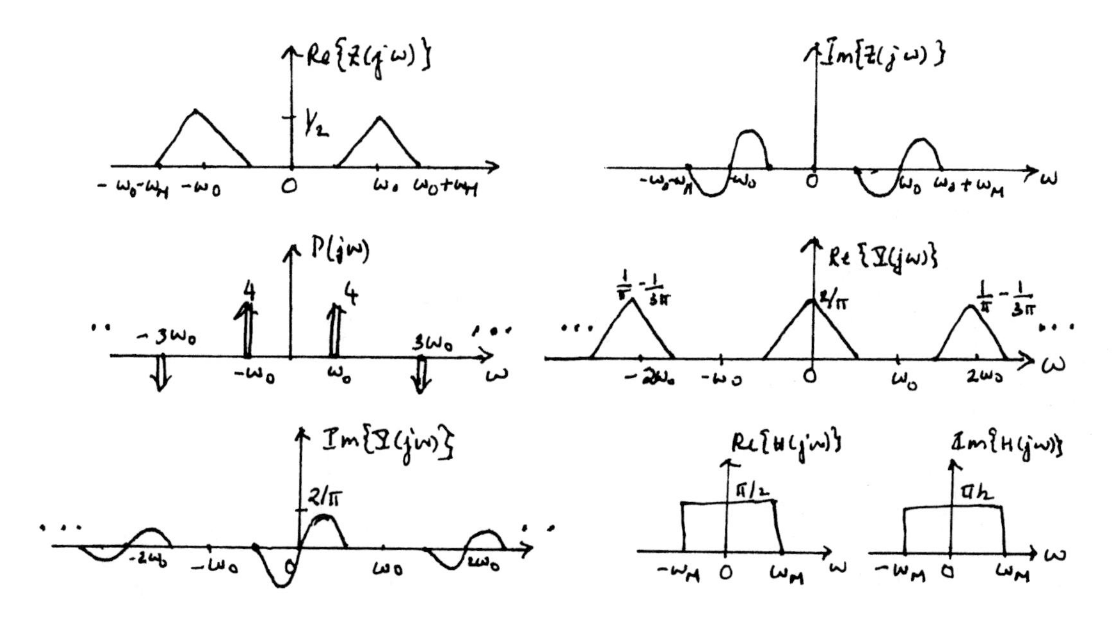

Figure S8.35

(b) From the above figure, it is clear that we require

$$2\omega_c + \omega_f - \omega_T \geq \omega_f + \omega_M \qquad \Rightarrow \qquad \omega_T \leq 2\omega_c - \omega_M.$$

Note that in case $\omega_f - \omega_T$ is negative, then we may additionally require that

$$-\omega_f + \omega_T \leq \omega_f - \omega_M \qquad \Rightarrow \omega_T \leq 2\omega_f - \omega_M.$$

(c) If we want to isolate the replica of $X_1(j\omega)$ centered around ω_f, then we require $G = 2/K$, $\alpha = \omega_f - \omega_M$, and $\beta = \omega_f + \omega_M$.

8.37. **(a)** We have

$$\begin{aligned} z(t) &= y(t) + \frac{1}{2}y^2(t) + \frac{1}{6}y^3(t) \\ &= \cos(\omega_0 t) + \frac{1}{2}\cos^2(\omega_0 t) + \frac{1}{6}\cos^3(\omega_0 t) + x(t) + \frac{1}{2}x^2(t) + \frac{1}{6}x^3(t) \\ &\quad + \left[x(t) + \frac{1}{2}x^2(t)\right]\cos(\omega_0 t). \end{aligned}$$

Therefore, $Z(j\omega)$ is as sketched in Figure S8.37. Note that the overlapping components have to be added together.

(b) From Figure S8.37 it is clear that

$$\omega_c + 2\omega_1 < \alpha < 2\omega_0 - \omega_1$$

and

$$2\omega_0 + \omega_1 < \beta < 3\omega_0.$$

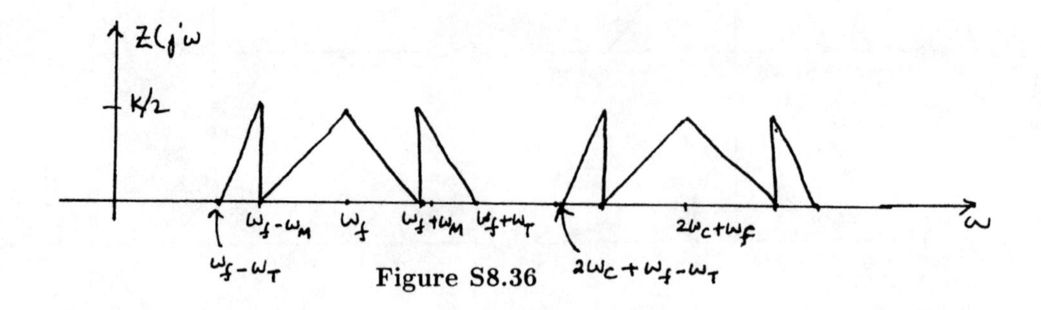

Figure S8.36

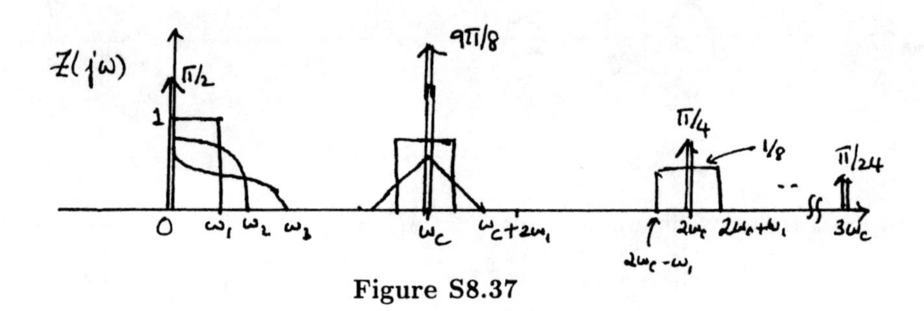

Figure S8.37

8.38. One of the key issues to note in this problem is that the structure of the demodulator is that of a synchronous demodulator. Therefore, the input signal to the demodulator has to have a replica of $X(j\omega)$ centered around ω_c. Only then will the demodulator be successful in recovering $x(t)$.

Case 1:
$P(j\omega)$ is given by

$$P(j\omega) = \frac{\sin[(\omega - \omega_c)D/2]}{\omega} + \frac{\sin[(\omega + \omega_c)D/2]}{\omega}.$$

$M_1(j\omega)$ will consist of impulses which occur at intervals of $2\pi/T$ weighted by $P(j\omega)$. Furthermore, note that if $y_1(t) = x(t)m_1(t)$, then we have

$$Y_1(j\omega) = \frac{1}{2\pi}\left[X(j\omega) * M_1(j\omega)\right].$$

Therefore, $Y_1(j\omega)$ will consist of weighted replicas of $X(j\omega)$ which occur every $2\pi/T$. Note that unless ω_c is a multiple of $2\pi/T$, $M_1(j\omega) = 0$ for $\omega = \pm\omega_c$. If $2\pi/T$ is arbitrary, (i.e., it is not specified to be a multiple of ω_c) $Y_1(j\omega)$ has no replicas of $X(j\omega)$ centered around ω_c. Since $y_1(t)$ constitutes the input to the demodulator, the signal $r(t)$ at the output of the demodulator will not be proportional to $x(t)$.

Case 2:
In this case,

$$M_2(j\omega) = \frac{1}{2}G(j(\omega - \omega_c)) + \frac{1}{2}G(j(\omega + \omega_c),$$

where

$$G(j\omega) = \sum_{k=-\infty}^{\infty} \frac{2\sin(k\pi D/T)}{k}\delta(\omega - 2\pi k/T).$$

Clearly, $M_2(j\omega)$ has equal-valued impulses at $\pm\omega_c$. Therefore, the Fourier transform $Y_2(j\omega)$ of the signal $y_2(t) = x(t)m_2(t)$ has replicas of $X(j\omega)$ at $\pm\omega_c$. These replicas do not alias with other replicas of $X(j\omega)$ in $Y_2(j\omega)$ because $2\pi/T > 2\omega_M$. Thus, when demodulation is performed on $y_2(t)$, then $r(t)$ can be made proportional to $x(t)$ provided $2\omega_M < \omega_{lp} < 2\pi/T$.

8.39. **(a)** The two possible differences between the lines are

$$D_0 = \int_0^T \cos^2(\omega_0 t)dt - \left|\int_0^T \cos(\omega_0 t)\cos(\omega_1 t)dt\right|$$

and

$$D_1 = \int_0^T \cos^2(\omega_1 t)dt - \left|\int_0^T \cos(\omega_0 t)\cos(\omega_1 t)dt\right|.$$

Clearly, D_0 and D_1 are maximum when

$$\left|\int_0^T \cos(\omega_0 t)\cos(\omega_1 t)dt\right| = 0.$$

This condition may also be written as

$$\int_0^T \cos(\omega_0 t)\cos(\omega_1 t)dt = 0.$$

(b) We have

$$\int_0^T \cos(\omega_0 t)\cos(\omega_1 t)dt = \frac{1}{2}\int_0^T \{\cos[(\omega_0 + \omega_1)t] + \cos[(\omega_0 - \omega_1)t]\}dt.$$

Therefore, if we ensure that T is never a common multiple of the periods of both $\cos[(\omega_0 - \omega_1)t]$ and $\cos[(\omega_0 + \omega_1)t]$, then the above integral will be zero.

8.40. Let $X_1(j\omega)$ and $X_2(j\omega)$ be as shown in Figure S8.40.

Then $R(j\omega)$ is as shown in Figure S8.40. The overlapping regions in the figure need to be summed. When $r(t)$ is multiplied by $\cos\omega_c t$, in the vicinity of $\omega = 0$ we get

$$\frac{1}{2}\left\{\frac{1}{2}X_1(j\omega) + \frac{1}{2}jX_2(j\omega) + \frac{1}{2}X_1(j\omega) - \frac{1}{2}jX_2(j\omega)\right\} = \frac{1}{2}X_1(j\omega).$$

Therefore, the first lowpass filter output is equal to $x_1(t)$.

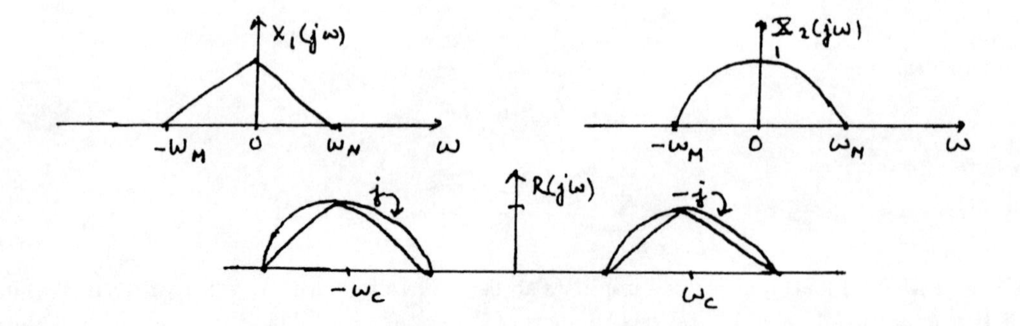

Figure S8.40

When $r(t)$ is multiplied by $\sin \omega_c t$, in the vicinity of $\omega = 0$ we get

$$\frac{1}{2}\left\{-j[\frac{1}{2}jX_2(j\omega) + \frac{1}{2}X_1(j\omega)] + j[-j\frac{1}{2}X_2(j\omega) + \frac{1}{2}X_1(j\omega)]\right\} = \frac{1}{2}X_2(j\omega).$$

Therefore, the second lowpass filter output is equal to $x_2(t)$.

8.41. **(a)** Let $X_1(e^{j\omega})$ and $X_2(e^{j\omega})$ be as shown in Figure S8.41.

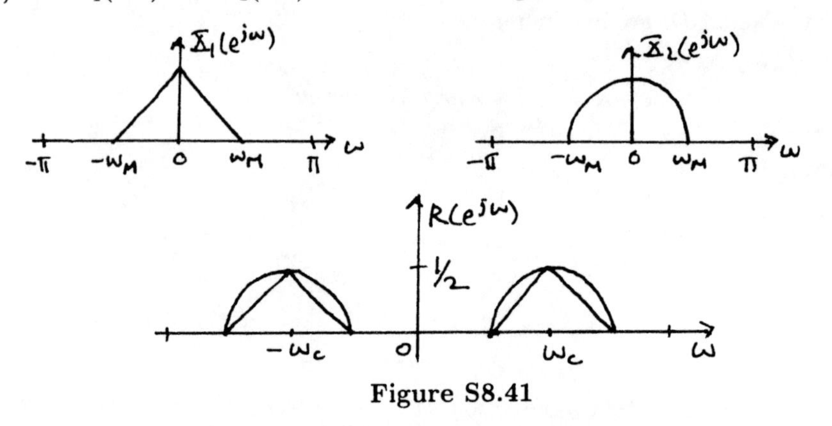

Figure S8.41

Then $R(e^{j\omega})$ is as shown in Figure S8.41. To avoid aliasing, we need to ensure that $\omega_M < \omega_c < \pi - \omega_M$.

(b) When $r[n]$ is multiplied by $\cos \omega_c n$, we get in the vicinity of $\omega = 0$ (i.e. in the range $|\omega| \leq \omega_M$)

$$\frac{1}{2}\left\{\frac{1}{2}X_1(e^{j\omega}) + \frac{1}{2}jX_2(e^{j\omega}) + \frac{1}{2}X_1(e^{j\omega}) - \frac{1}{2}jX_2(e^{j\omega})\right\} = \frac{1}{2}X_1(e^{j\omega}).$$

Therefore, we want $H(e^{j\omega})$ to correspond to a lowpass filter with cutoff frequency ω_M and passband gain 2.

(c) When $r[n]$ is multiplied by $\sin \omega_c n$, we get in the vicinity of $\omega = 0$ (i.e. in the range $|\omega| \leq \omega_M$)

$$-\frac{1}{2}\left\{-j[\frac{1}{2}X_1(e^{j\omega}) + \frac{1}{2}jX_2(e^{j\omega})] + j[\frac{1}{2}X_1(e^{j\omega}) - \frac{1}{2}jX_2(e^{j\omega})]\right\} = \frac{1}{2}X_2(e^{j\omega}).$$

Therefore, we want $H(e^{j\omega})$ to correspond to a lowpass filter with cutoff frequency ω_M and passband gain 2.

8.42. **(a)** From the given information, it is clear that the function $P_1(j\omega)$ shows odd symmetry about π/T_1. Also, $P_1(j\omega)$ is even because $p_1(t)$ is real and even. Therefore, this is a function as shown in Figure S8.42.

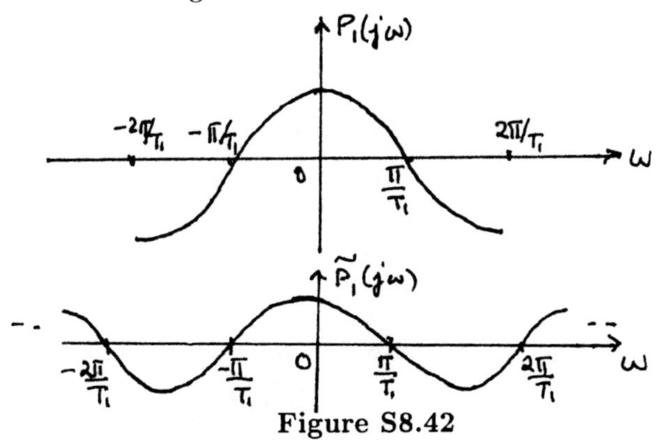

Figure S8.42

If we now define $\tilde{P}_1(j\omega)$ as given in the problem, then $\tilde{P}_1(j\omega)$ is as shown in Figure S8.42. Clearly,

$$\tilde{P}_1(j\omega) = -\tilde{P}_1(j\omega - j2\pi/T_1).$$

(b) Let us define a signal $q(t) = \tilde{P}_1(jt)$. This signal is periodic with a period of $4\pi/T_1$. Let the Fourier series coefficients of $q(t)$ be a_k. We also know from (a) that $q(t)$ satisfies the expression

$$q(t) = -q(t - 2\pi/T_1).$$

Equating the Fourier series of both sides of the above expression, we obtain

$$a_k = -a_k e^{-jk(2\pi/T_0)(2\pi/T_1)}, \qquad \text{where} \qquad T_0 = \frac{4\pi}{T_1}.$$

Therefore,

$$a_k = -a_k e^{-jk\pi}.$$

This implies that $a_k = 0$ for $k = 0, \pm 2, \pm 4, \cdots$. We know that the Fourier transform of $q(t)$ is of the form

$$\begin{aligned} Q(j\omega) &= 2\pi \sum_{k=-\infty}^{\infty} a_k \delta(\omega - k2\pi/T_0) \\ &= 2\pi \sum_{k=-\infty}^{\infty} a_k \delta(\omega - kT_1/2) \end{aligned}$$

Since $a_k = 0$ for $k = 0, \pm 2, \pm 4, \cdots$, $Q(j\omega)$ is zero for $\omega = 0, \pm T_1, \pm 2T_1, \cdots$.

Now note that using the duality property of the Fourier transform, we may infer that the inverse Fourier transform of $\tilde{P}_1(j\omega)$ is $\tilde{p}_1(t) = Q(-jt)/2\pi$. Therefore, $\tilde{p}_1(t) = 0$ zero for $t = 0, \pm T_1, \pm 2T_1, \cdots$.

(c) We may write $\tilde{P}_1(j\omega)$ as

$$\tilde{P}_1(j\omega) = \frac{1}{2\pi}\left[P_1(j\omega) * 2\pi \sum_{k=-\infty}^{\infty} \delta(\omega - k4\pi/T_1)\right].$$

Taking the inverse Fourier transform, we obtain

$$\tilde{p}_1(t) = p_1(t) \sum_{k=-\infty}^{\infty} \delta(t - kT_1/2).$$

Clearly, $\tilde{p}_1(t) = p_1(t)$ for $t = 0, \pm T_1/2, \pm T_1, \pm 3T_1/2, \cdots$. From part (b), it is obvious that $p_1(t) = 0$ for $t = 0, \pm T_1, \pm 2T_1, \cdots$.

(d) Note that $P(j\omega) = P_1(j\omega) + P_2(j\omega)$, where

$$P_2(j\omega) = \begin{cases} 1, & |\omega| \leq \pi/T_1 \\ 0, & \text{otherwise} \end{cases}.$$

Therefore, $p(t) = p_1(t) + p_2(t)$, where

$$p_2(t) = \frac{\sin(\pi t/T_1)}{\pi t}.$$

We can easily see that $p_2(t) = 0$ for $t = 0, \pm T_1, \pm 2T_1, \cdots$. And since we already have shown that $p_1(t) = 0$ for $t = 0, \pm T_1, \pm 2T_1, \cdots$, it is obvious that $p(t) = 0$ for $t = 0, \pm T_1, \pm 2T_1, \cdots$.

8.43. **(a)** We know that

$$H(j\omega) = \frac{10000}{1000 + j\omega}.$$

Therefore, the frequency response of the compensating system has to be

$$G(j\omega) = \frac{1000 + j\omega}{10000}.$$

This implies that $A = 0.1$ and $B = 10^{-4}$.

(b) If the input and output of the compensating systems are denoted by $x(t)$ and $y(t)$, respectively, then

$$\frac{Y(j\omega)}{X(j\omega)} = \frac{1000 + j\omega}{10000}.$$

Cross-multiplying and taking the inverse Fourier transform, we obtain

$$y(t) = 0.1x(t) + 10^{-4}\frac{dx(t)}{dt}.$$

Therefore,

$$\alpha = 10^{-4}, \qquad \beta = 0.1.$$

8.44. **(a)** We may write $y(t)$ as

$$y(t) = x(t) * \sum_{l=-N}^{N} a_l \delta(t - lT_1).$$

Therefore, $y(t)$ is obtained by passing $x(t)$ through a filter with impulse response $h(t) = \sum_{l=-N}^{N} a_l \delta(t - lT_1)$.

(b) Using eq. (P8.44-1), we obtain the following three simultaneous equations

$$y(0) = a_{-1}x(T_1) + a_0 x(0) + a_1 x(-T_1),$$

$$y(T_1) = a_{-1}x(2T_1) + a_0 x(T_1) + a_1 x(0),$$

and

$$y(-T_1) = a_{-1}x(0) + a_0 x(-T_1) + a_1 x(-2T_1).$$

Substituting the given values for $x(t)$ and $y(t)$ and solving, we obtain

$$a_0 = 0, \qquad a_1 = a_{-1}.$$

8.45. **(a)** Since we are given that

$$y(t) = \cos\left(\omega_c t + m\int_{-\infty}^{\infty} x(\tau)d\tau\right),$$

we know that the phase of the carrier is

$$\theta(t) = \omega_c t + m\int_{-\infty}^{\infty} x(\tau)d\tau.$$

Therefore, the instantaneous frequency is

$$\omega_i(t) = \frac{d\theta}{dt} = \omega_c + mx(t).$$

(b) Expanding $y(t)$, we get

$$y(t) = \cos(\omega_c t)\cos\left(m\int_{-\infty}^{\infty} x(\tau)d\tau\right) - \sin(\omega_c t)\sin\left(m\int_{-\infty}^{\infty} x(\tau)d\tau\right).$$

Using the narrowband assumption, we know that $\left(m\int_{-\infty}^{\infty} x(\tau)d\tau\right)$ is very close to zero. Therefore,

$$\cos\left(m\int_{-\infty}^{\infty} x(\tau)d\tau\right) \approx 1$$

and

$$\sin\left(m\int_{-\infty}^{\infty} x(\tau)d\tau\right) \approx \left(m\int_{-\infty}^{\infty} x(\tau)d\tau\right).$$

This implies that

$$y(t) \approx \cos(\omega_c t) - \left(m\int_{-\infty}^{\infty} x(\tau)d\tau\right)\sin(\omega_c t).$$

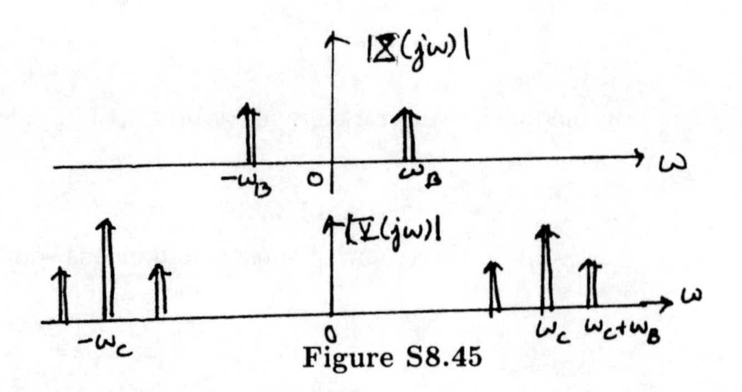

Figure S8.45

(c) We show both $X|(j\omega)|$ and $|Y(j\omega)|$ in Figure S8.45.

Clearly, the bandwidth of $x(t)$ is $2\omega_B$, while the bandwidth of $y(t)$ is $2\omega_c + 2\omega_B$.

8.46. **(a)** The instantaneous frequency is

$$\omega_i(t) = \frac{d\theta(t)}{dt} = \frac{d}{dt}\left[\frac{\omega_0 t^2}{2}\right] = \omega_0 t.$$

(b) We have

$$\begin{aligned} S(j\omega) &= \int_{-\infty}^{\infty} e^{j\omega_0 t^2/2} e^{-j\omega t} dt \\ &= e^{j\omega^2/(2\omega_0)} \int_{-\infty}^{\infty} e^{j[\sqrt{\omega_0/2}(t-\omega/\omega_0)]^2} dt \\ &= \sqrt{\frac{\pi}{\omega_0}}(1+j)e^{j\omega^2/(2\omega_0)} \end{aligned}$$

(c) We have

$$X(j\omega_0 t) = \int_{-\infty}^{\infty} x(\tau)e^{-j\omega_0 t\tau} d\tau.$$

But $t\tau = \frac{1}{2}[t^2 + \tau^2 - (t-\tau)^2]$. Therefore,

$$\begin{aligned} X(j\omega_0 t) &= \int_{-\infty}^{\infty} x(\tau)e^{-j\omega_0\tau^2/2} e^{-j\omega_0 t^2/2} e^{j\omega_0(t-\tau)^2/2} d\tau \\ &= e^{-j\omega_0 t^2/2} \int_{-\infty}^{\infty} x(\tau)e^{-j\omega_0\tau^2/2} e^{j\omega_0(t-\tau)^2/2} d\tau \end{aligned}$$

Let $g(\tau) = x(\tau)e^{-j\omega_0\tau^2/2}$. Then

$$X(j\omega_0 t) = e^{-j\omega_0 t^2/2}[g(t) * e^{j\omega_0 t^2/2}].$$

This is exactly what Figure P8.46 implements.

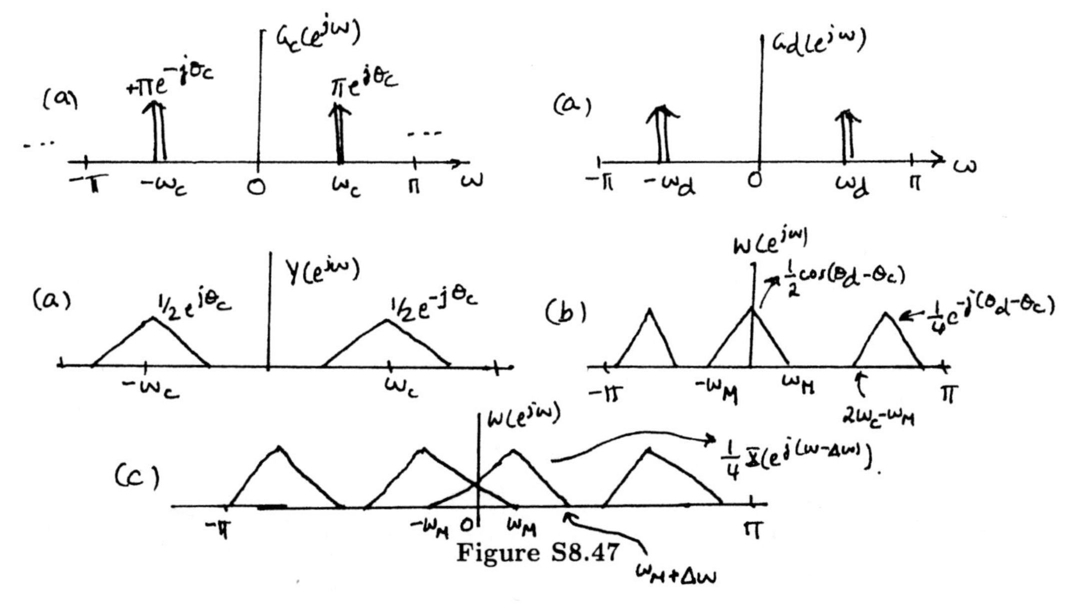

Figure S8.47

8.47. Let $G_c(e^{j\omega})$ represent the Fourier transform of $\cos(\omega_c n + \theta_c)$. This is as shown in Figure S8.47. Let $G_d(e^{j\omega})$ represent the Fourier transform of $\cos(\omega_d n + \theta_d)$. This is as shown in Figure S8.47.

Now,

$$Y(e^{j\omega}) = \frac{1}{2\pi}\int_{-\pi}^{\pi} X(e^{j(\omega-\theta)})G_c(e^{j\theta})d\theta.$$

This is as shown in Figure S8.47.

Here, we assume that $\omega_c > \omega_M$ and $\pi - \omega_M > \omega_c$. Also, we have

$$W(e^{j\omega}) = \frac{1}{2\pi}\int_{-\pi}^{\pi} Y(e^{j(\omega-\theta)})G_d(e^{j\theta})d\theta.$$

(a) If $\Delta\omega = 0$, then $\omega_d = \omega_c$. Therefore, $W(e^{j\omega})$ will be as shown in Figure S8.47.

(b) When $W(e^{j\omega})$ passes through $H(e^{j\omega})$, we obtain $R(e^{j\omega})$ as shown below. Then, $R(e^{j\omega}) = \cos(\theta_d - \theta_c)X(e^{j\omega}) = \cos(\Delta\theta)X(e^{j\omega})$. If $\delta(\theta) = \pi/2$, then $r[n] = 0$.

(c) In this case, $W(e^{j\omega})$ is as shown in Figure S8.47. If $w > \omega_M + \Delta\omega$, then $R(e^{j\omega}) = \frac{1}{2}X(e^{j(\omega-\Delta\omega)}) + \frac{1}{2}X(e^{j(\omega+\Delta\omega)})$. Therefore,

$$r[n] = x[n]\cos(\Delta\omega n).$$

8.48. **(a)** The Fourier series coefficients of $p[n]$ are

$$a_k = \frac{1}{N}\sum_{n=0}^{M} e^{-j2\pi kn/N} = e^{-j2\pi kM/(2N)}\frac{\sin(\frac{2\pi k}{N}\frac{(M+1)}{2})}{N\sin(\frac{2\pi k}{2N})}.$$

Now,

$$P(e^{j\omega}) = 2\pi \sum_{k=-\infty}^{\infty} a_k \delta(\omega - 2\pi k/N).$$

(b) With $M = 1$,

$$a_k = e^{-j2\pi k/N} \frac{\sin(2\pi k/N)}{N \sin(\pi k/N)}.$$

Ignoring the phase factor, and taking $N = 6$, we have $Y(e^{j\omega})$ as shown in Figure S8.48.

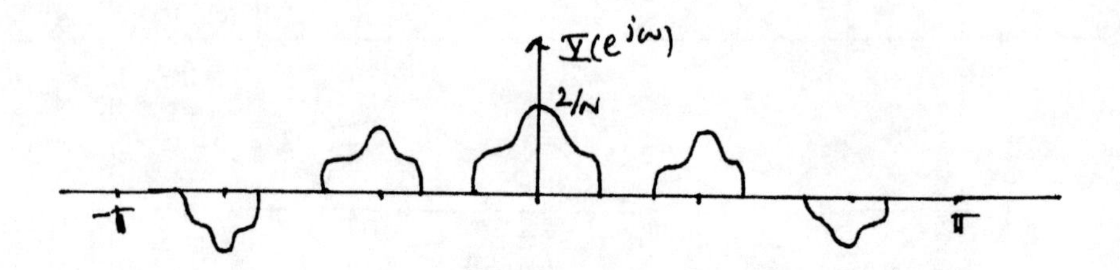

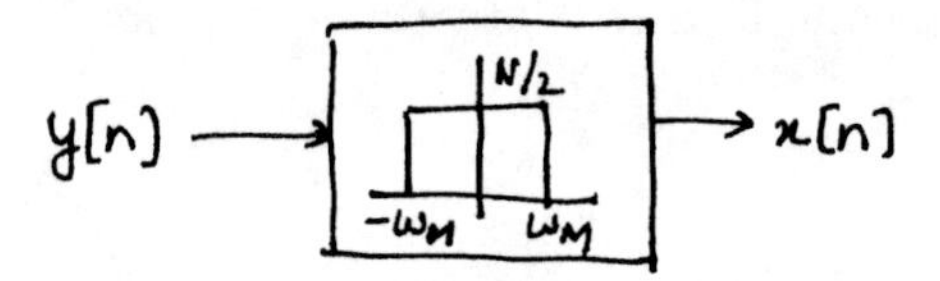

Figure S8.48

(c) We need $\omega_M \leq \pi/N$. The result does not depend on M.

(d) The block-diagram is as shown in Figure S8.48.

8.49. **(a)** The Fourier transform $S(j\omega)$ of $s(t)$ is given by

$$S(j\omega) = \sum_{k=-\infty}^{\infty} \frac{2\sin(k\pi/2)}{k} \delta(\omega - 2\pi k/T).$$

To avoid aliasing $X(j\omega)$ should be 0 for $|\omega| > \pi/T$.

(b) Let $X(j\omega)$ be as shown in Figure S8.49. Then the Fourier transforms of the signals at the outputs of $H_1(j\omega)$ and $H_2(j\omega)$ are as shown in the same figure. Therefore, the gain is $2A/\pi^2$.

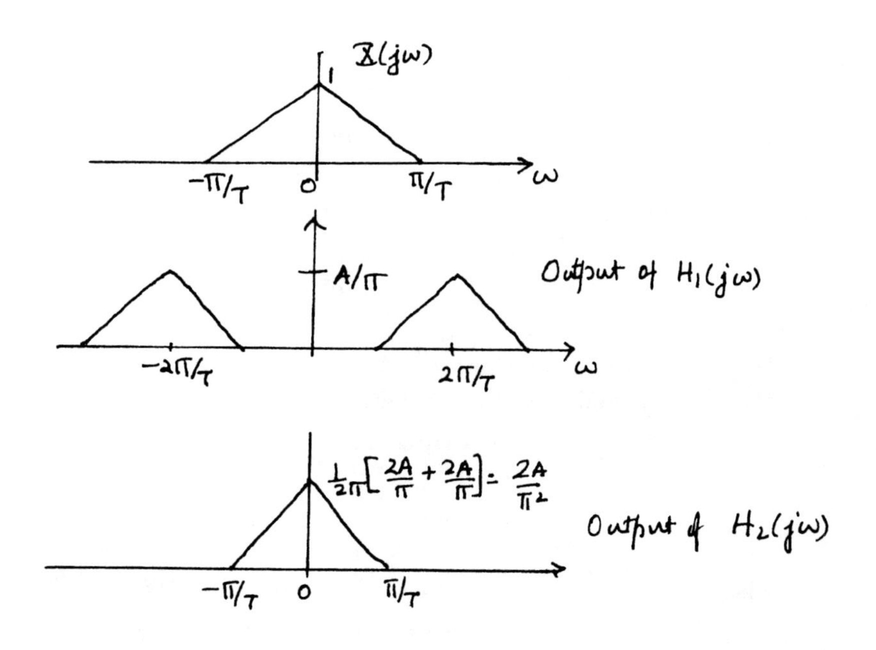

$X(j\omega)$
1
$-\pi/T$
0
π/T
ω
A/π
Output of $H_1(j\omega)$
$-2\pi/T$
$2\pi/T$
ω
$\frac{1}{2\pi}\left[\frac{2A}{\pi}+\frac{2A}{\pi}\right]=\frac{2A}{\pi^2}$
Output of $H_2(j\omega)$
$-\pi/T$
0
π/T

Figure S8.49

Chapter 9 Answers

9.1. (a) The given integral may be written as

$$\int_{0}^{\infty} e^{-(5+\sigma)t} e^{j\omega t} dt.$$

If $\sigma < -5$, then the function $e^{-(5+\sigma)t}$ grows towards ∞ with increasing t and the given integral does not converge. But if $\sigma > -5$, then the integral does converge.

(b) The given integral may be written as

$$\int_{-\infty}^{0} e^{-(5+\sigma)t} e^{j\omega t} dt.$$

If $\sigma > -5$, then the function $e^{-(5+\sigma)t}$ grows towards ∞ as t decreases towards $-\infty$ and the given integral does not converge. But if $\sigma < -5$, then the integral does converge.

(c) The given integral may be written as

$$\int_{-5}^{5} e^{-(5+\sigma)t} e^{j\omega t} dt.$$

Clearly this integral has a finite value for all finite values of σ.

(d) The given integral may be written as

$$\int_{-\infty}^{\infty} e^{-(5+\sigma)t} e^{j\omega t} dt.$$

If $\sigma > -5$, then the function $e^{-(5+\sigma)t}$ grows towards ∞ as t decreases towards $-\infty$ and the given integral does not converge. If $\sigma < -5$, then the function $e^{-(5+\sigma)t}$ grows towards ∞ with increasing t and the given integral does not converge. If $\sigma = 5$, then the integral still does not have a finite value. Therefore, the integral does not converge for any value of σ.

(e) The given integral may be written as

$$\int_{-\infty}^{0} e^{-(-5+\sigma)t} e^{j\omega t} dt + \int_{0}^{\infty} e^{-(5+\sigma)t} e^{j\omega t} dt.$$

The first integral converges for $\sigma < 5$. The second integral converges if $\sigma > -5$. Therefore, the given integral converges when $|\sigma| < 5$.

(f) The given integral may be written as

$$\int_{-\infty}^{0} e^{-(-5+\sigma)t} e^{j\omega t} dt.$$

If $\sigma > 5$, then the function $e^{-(-5+\sigma)t}$ grows towards ∞ as t decreases towards $-\infty$ and the given integral does not converge. But if $\sigma < 5$, then the integral does converge.

9.2. (a)

$$\begin{aligned} X(s) &= \int_{-\infty}^{\infty} e^{-5t} u(t-1) e^{-st} dt \\ &= \int_{1}^{\infty} e^{-(5+s)t} dt \\ &= \frac{e^{-(5+s)}}{s+5} \end{aligned}$$

As shown in Example 9.1, the ROC will be $\mathcal{R}e\{s\} > -5$.

(b) By using eq. (9.3), we can easily show that $g(t) = Ae^{-5t}u(-t-t_0)$ has the Laplace transform

$$G(s) = \frac{Ae^{(s+5)t_0}}{s+5}.$$

The ROC is specified as $\mathcal{R}e\{s\} < -5$. Therefore, $A = 1$ and $t_0 = -1$.

9.3. Using an analysis similar to that used in Example 9.3, we know that the given signal has a Laplace transform of the form

$$X(s) = \frac{1}{s+5} + \frac{1}{s+\beta}.$$

The corresponding ROC is $\mathcal{R}e\{s\} > max(-5, \mathcal{R}e\{\beta\})$. Since we are given that the ROC is $\mathcal{R}e\{s\} > -3$, we know that $\mathcal{R}e\{\beta\} = 3$. There are no constraints on the imaginary part of β.

9.4. We know from Table 9.2 that

$$x_1(t) = -e^{-t}\sin(2t)u(t) \stackrel{\mathcal{L}}{\longleftrightarrow} X_1(s) = -\frac{2}{(s+1)^2+2^2}, \qquad \mathcal{R}e\{s\} > -1.$$

We also know from Table 9.1 that

$$x(t) = x_1(-t) \stackrel{\mathcal{L}}{\longleftrightarrow} X(s) = X_1(-s).$$

The ROC of $X(s)$ is such that if s_0 was in the ROC of $X_1(s)$, then $-s_0$ will be in the ROC of $X(s)$. Putting the two above equations together, we have

$$x(t) = x_1(-t) = e^{-t}\sin(2t)u(-t) \stackrel{\mathcal{L}}{\longleftrightarrow} X(s) = X_1(-s) = -\frac{2}{(s-1)^2+2^2}, \qquad \mathcal{R}e\{s\} < 1.$$

The denominator of $X(s)$ is of the form $s^2 - 2s + 5$. Therefore, the poles of $X(s)$ are $1+2j$ and $1-2j$.

9.5. (a) The given Laplace transform may be written as

$$X(s) = \frac{2s+4}{(s+1)(s+3)}.$$

Clearly, $X(s)$ has a zero at $s = -2$. Since in $X(s)$ the order of the denominator polynomial exceeds the order of the numerator polynomial by 1, $X(s)$ has a zero at ∞. Therefore, $X(s)$ has one zero in the finite s-plane and one zero at infinity.

(b) The given Laplace transform may be written as

$$X(s) = \frac{s+1}{(s-1)(s+1)} = \frac{1}{s-1}.$$

Clearly, $X(s)$ has no zeros in the finite s-plane. Since in $X(s)$ the order of the denominator polynomial exceeds the order of the numerator polynomial by 1, $X(s)$ has a zero at ∞. Therefore, $X(s)$ has no zeros in the finite s-plane and one zero at infinity.

(c) The given Laplace transform may be written as

$$X(s) = \frac{(s-1)(s^2+s+1)}{(s^2+s+1)} = s-1.$$

Clearly, $X(s)$ has a zero at $s = 1$. Since in $X(s)$ the order of the numerator polynomial exceeds the order of the denominator polynomial by 1, $X(s)$ has no zeros at ∞. Therefore, $X(s)$ has one zero in the finite s-plane and no zeros at infinity.

9.6. **(a) No**. From property 3 in Section 9.2 we know that for a finite-length signal, the ROC is the entire s-plane. Therefore, there can be no poles in the finite s-plane for a finite-length signal. Clearly, in this problem this is not the case.

(b) Yes. Since the signal is absolutely integrable, the ROC must include the $j\omega$-axis. Furthermore, $X(s)$ has a pole at $s = 2$. Therefore, one valid ROC for the signal would be $\mathcal{R}e\{s\} < 2$. From property 5 in Section 9.2 we know that this would correspond to a left-sided signal.

(c) No. Since the signal is absolutely integrable, the ROC must include the $j\omega$-axis. Furthermore, $X(s)$ has a pole at $s = 2$. Therefore, we can never have an ROC of the form $\mathcal{R}e\{s\} > \alpha$. From property 4 in Section 9.2 we know that $x(t)$ cannot be a right-sided signal.

(d) Yes. Since the signal is absolutely integrable, the ROC must include the $j\omega$-axis. Furthermore, $X(s)$ has a pole at $s = 2$. Therefore, a valid ROC for the signal could be $\alpha < \mathcal{R}e\{s\} < 2$ such that $\alpha < 0$. From property 6 in Section 9.2, we know that this would correspond to a two-sided signal.

9.7. We may find different signals with the given Laplace transform by choosing different regions of convergence. The poles of the given Laplace transform are

$$s_0 = -2, \qquad s_1 = -3, \qquad s_3 = -\frac{1}{2} + \frac{\sqrt{3}}{2}j, \qquad s_3 = -\frac{1}{2} - \frac{\sqrt{3}}{2}j.$$

Based on the locations of these poles, we my choose from the following regions of convergence:

(i) $\mathcal{R}e\{s\} > -\frac{1}{2}$

(ii) $-2 < \mathcal{R}e\{s\} < -\frac{1}{2}$

(iii) $-3 < \mathcal{R}e\{s\} < -2$

(iv) $\mathcal{R}e\{s\} < -3$

Therefore, we may find four different signals with the given Laplace transform.

9.8. From Table 9.1, we know that

$$g(t) = e^{2t}x(t) \stackrel{\mathcal{L}}{\longleftrightarrow} G(s) = X(s-2).$$

The ROC of $G(s)$ is the ROC of $X(s)$ shifted to the right by 2.

We are also given that $X(s)$ has exactly 2 poles, located at $s = -1$ and $s = -3$. Since $G(s) = X(s-2)$, $G(s)$ also has exactly two poles, located at $s = -1+2 = 1$ and $s = -3+2 = -1$. Since we are given $G(j\omega)$ exists, we may infer that the $j\omega$-axis lies in the ROC of $G(s)$. Given this fact and the locations of the poles, we may conclude that $g(t)$ is a two sided sequence. Obviously $x(t) = e^{-2t}g(t)$ will also be two sided.

9.9. Using partial fraction expansion

$$X(s) = \frac{4}{s+4} - \frac{2}{s+3}.$$

Taking the inverse Laplace transform,

$$x(t) = 4e^{-4t}u(t) - 2e^{-3t}u(t).$$

9.10. The pole-zero plots for each of the three Laplace transforms is as shown in Figure S9.10.

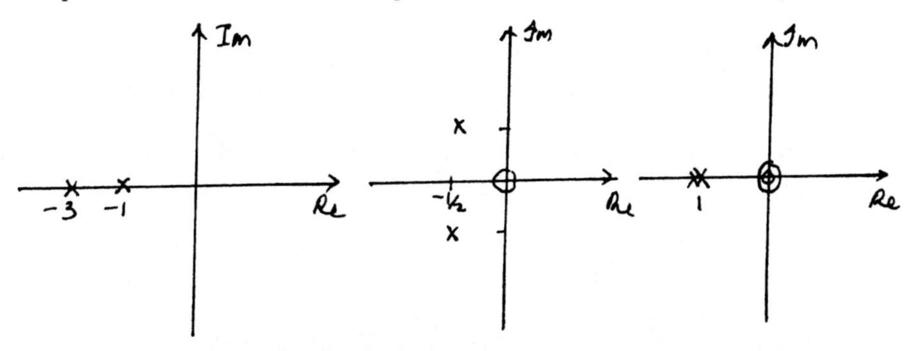

Figure S9.10

(a) From Section 9.4 we know that the magnitude of the Fourier transform may be expressed as

$$\frac{1}{(\text{Length of vector from } \omega \text{ to } -1)(\text{Length of vector from } \omega \text{ to } -2)}.$$

We see that the right-hand side of the above expression is maximum for $\omega = 0$ and decreases as ω becomes increasingly more positive or more negative. Therefore $|H_1(j\omega)|$ is approximately lowpass.

(b) From Section 9.4 we know that the magnitude of the Fourier transform may be expressed as

$$\frac{\text{(Length of vector from } \omega \text{ to } 0)}{\text{(Length of vector from } \omega \text{ to } -\frac{1}{2}+j\frac{\sqrt{3}}{2})\text{(Length of vector from } \omega \text{ to } -\frac{1}{2}-j\frac{\sqrt{3}}{2})}.$$

We see that the right-hand side of the above expression is zero for $\omega = 0$. It then increases with increasing $|\omega|$ until $|\omega|$ reaches $\frac{1}{2}$. Then it starts decreasing as $|\omega|$ increases even further. Therefore $|H_2(j\omega)|$ is approximately bandpass.

(c) From Section 9.4 we know that the magnitude of the Fourier transform may be expressed as

$$\frac{\text{(Length of vector from } \omega \text{ to } 0)^2}{\text{(Length of vector from } \omega \text{ to } -\frac{1}{2}+j\frac{\sqrt{3}}{2})\text{(Length of vector from } \omega \text{ to } -\frac{1}{2}-j\frac{\sqrt{3}}{2})}.$$

We see that the right-hand side of the above expression is zero for $\omega = 0$. It then increases with increasing $|\omega|$ until $|\omega|$ reaches $\frac{1}{2}$. Then $|\omega|$ increases, $|H_3(j\omega)|$ decreases towards a value of 1 (because all the vector lengths become almost identical and the ratio becomes 1). Therefore $|H_3(j\omega)|$ is approximately highpass.

9.11. $X(s)$ has poles at $s = -\frac{1}{2}+j\frac{\sqrt{3}}{2}$ and $-\frac{1}{2}-j\frac{\sqrt{3}}{2}$. $X(s)$ has zeros at $s = \frac{1}{2}+j\frac{\sqrt{3}}{2}$ and $\frac{1}{2}-j\frac{\sqrt{3}}{2}$. From Section 9.4, we know that $|X(j\omega)|$ is

$$\frac{\text{(Length of vector from } \omega \text{ to } \frac{1}{2}+j\frac{\sqrt{3}}{2})\text{(Length of vector from } \omega \text{ to } \frac{1}{2}-j\frac{\sqrt{3}}{2})}{\text{(Length of vector from } \omega \text{ to } -\frac{1}{2}+j\frac{\sqrt{3}}{2})\text{(Length of vector from } \omega \text{ to } -\frac{1}{2}-j\frac{\sqrt{3}}{2})}.$$

The terms in the numerator and denominator of the right-hand side of the above expression cancel out giving us $|X(j\omega)| = 1$

9.12. **(a)** If $X(s)$ has only one pole, then $x(t)$ would be of the form Ae^{-at}. Clearly such a signal violates condition 2. Therefore, this statement is inconsistent with the given information.

(b) If $X(s)$ has only two poles, then $x(t)$ would be of the form $Ae^{-at}sin(\omega_o t)$. Clearly such a signal could be made to satisfy all three conditions (Example: $\omega_0 = 80\pi$, $a = 19200$). Therefore, this statement is consistent with the given information.

(c) If $X(s)$ has more than two poles (say 4 poles), then $x(t)$ could be assumed to be of the form $Ae^{-at}sin(\omega_o t) + Be^{-bt}sin(\omega_o t)$. Clearly such a signal could still be made to satisfy all three conditions. Therefore, this statement is consistent with the given information.

9.13. We have

$$X(s) = \frac{\beta}{s+1}, \qquad \mathcal{R}e\{s\} > -1.$$

Also,

$$G(s) = X(s) + \alpha X(-s), \qquad -1 < \mathcal{R}e\{s\} < 1.$$

Therefore,

$$G(s) = \beta\left[\frac{1 - s + \alpha s + \alpha}{1 - s^2}\right].$$

Comparing with the given equation for $G(s)$,

$$\alpha = -1, \qquad \beta = \frac{1}{2}.$$

9.14. Since $X(s)$ has 4 poles and no zeros in the finite s-plane, we may assume that $X(s)$ is of the form

$$X(s) = \frac{A}{(s-a)(s-b)(s-c)(s-d)}.$$

Since $x(t)$ is real, the poles of $X(s)$ must occur in conjugate reciprocal pairs. Therefore, we may assume that $b = a^*$ and $d = c^*$. This results in

$$X(s) = \frac{A}{(s-a)(s-a^*)(s-c)(s-c^*)}.$$

Since the signal $x(t)$ is also even, the Laplace transform $X(s)$ must also be even. This implies that the poles have to be symmetric about the $j\omega$-axis. Therefore, we may assume that $c = -a^*$. This results in

$$X(s) = \frac{A}{(s-a)(s-a^*)(s+a^*)(s+a)}.$$

We are given that the location of one of the poles is $(1/2)e^{j\pi/4}$. If we assume that this pole is a, we have

$$X(s) = \frac{A}{(s - \frac{1}{2}e^{j\frac{\pi}{4}})(s - \frac{1}{2}e^{-j\frac{\pi}{4}})(s + \frac{1}{2}e^{-j\frac{\pi}{4}})(s + \frac{1}{2}e^{j\frac{\pi}{4}})}.$$

This gives us

$$X(s) = \frac{A}{(s^2 - \frac{s}{\sqrt{2}} + \frac{1}{4})(s^2 + \frac{s}{\sqrt{2}} + \frac{1}{4})}.$$

Also, we are given that

$$\int_{-\infty}^{\infty} x(t)dt = X(0) = 4.$$

Substituting in the above expression for $X(s)$, we have $A = 1/4$. Therefore,

$$X(s) = \frac{(1/4)}{(s^2 - \frac{s}{\sqrt{2}} + \frac{1}{4})(s^2 + \frac{s}{\sqrt{2}} + \frac{1}{4})}.$$

9.15. Taking the Laplace transforms of both sides of the two differential equations, we have

$$sX(s) = -2Y(s) + 1 \qquad \text{and} \qquad sY(s) = 2X(s).$$

Solving for $X(s)$ and $Y(s)$, we obtain

$$X(s) = \frac{s}{s^2+4} \qquad \text{and} \qquad Y(s) = 2s^2 + 4.$$

The region of convergence for both $X(s)$ and $Y(s)$ is $\mathcal{R}e\{s\} > 0$ because both are right-sided signals.

9.16. Taking the Laplace transform of both sides of the given differential equation, we obtain

$$Y(s)[s^3 + (1+\alpha)s^2 + \alpha(\alpha+1)s + \alpha^2] = X(s).$$

Therefore,

$$H(s) = \frac{Y(s)}{X(s)} = \frac{1}{s^3 + (1+\alpha)s^2 + \alpha(\alpha+1)s + \alpha^2}.$$

(a) Taking the Laplace transform of both sides of the given equation, we have

$$G(s) = sH(s) + H(s).$$

Substituting for $H(s)$ from above,

$$G(s) = \frac{(s+1)}{s^3 + (1+\alpha)s^2 + \alpha(\alpha+1)s + \alpha^2} = \frac{1}{s^2 + \alpha s + \alpha^2}.$$

Therefore, $G(s)$ has 2 poles.

(b) We know that

$$H(s) = \frac{1}{(s+1)(s^2 + \alpha s + \alpha^2)}.$$

Therefore, $H(s)$ has poles at -1, $\alpha(-\frac{1}{2} + j\frac{\sqrt{3}}{2})$, and $\alpha(-\frac{1}{2} - j\frac{\sqrt{3}}{2})$. If the system has to be stable, then the real part of the poles has to be less than zero. For this to be true, we require that $-\alpha/2 < 0$, i.e., $\alpha > 0$.

9.17. The overall system shown in Figure P9.17 may be treated as two feedback systems of the form shown in Figure 9.31 connected in parallel. By carrying out an analysis similar to that described in in Section 9.8.1, we find the system function of the upper feedback system to be

$$H_1(s) = \frac{2/s}{1 + 4(2/s)} = \frac{2}{s+8}.$$

Similarly, the system function of the lower feedback system is

$$H_2(s) = \frac{1/s}{1 + 2(1/2)} = \frac{1}{s+2}.$$

The system function of the overall system is now

$$H(s) = H_1(s) + H_2(s) = \frac{3s + 12}{s^2 + 10s + 16}.$$

Since $H(s) = Y(s)/X(s)$, we may write

$$Y(s)[s^2 + 10s + 16] = X(s)[3s + 12].$$

Taking the inverse Laplace transform, we obtain

$$\frac{d^2y(t)}{dt} + 10\frac{dy(t)}{dt} + 16y(t) = 12x(t) + 3\frac{dx(t)}{dt}.$$

9.18. (a) From Problem 3.20, we know that the differential equation relating the input and output of the RLC circuit is

$$\frac{d^2y(t)}{dt} + \frac{dy(t)}{dt} + y(t) = x(t).$$

Taking the Laplace transform of this (while noting that the system is causal and stable), we obtain

$$Y(s)[s^2 + s + 1] = X(s).$$

Therefore,

$$H(s) = \frac{Y(s)}{X(s)} = \frac{1}{s^2 + s + 1}, \qquad \mathcal{R}e\{s\} > -\frac{1}{2}.$$

(b) We note that $H(s)$ has two poles at $s = -\frac{1}{2} - j\frac{\sqrt{3}}{2}$ and $s = -\frac{1}{2} + j\frac{\sqrt{3}}{2}$. It has no zeros in the finite s-plane. From Section 9.4 we know that the magnitude of the Fourier transform may be expressed as

$$\frac{1}{(\text{Length of vector from } \omega \text{ to } -\frac{1}{2} + j\frac{\sqrt{3}}{2})(\text{Length of vector from } \omega \text{ to } -\frac{1}{2} - j\frac{\sqrt{3}}{2}))}.$$

We see that the right-hand side of the above expression increases with increasing $|\omega|$ until $|\omega|$ reaches $\frac{1}{2}$. Then it starts decreasing as $|\omega|$ increases even further. It finally reaches 0 for $|\omega| = \infty$. Therefore $H_2(j\omega)|$ is approximately lowpass.

(c) By repeating the analysis carried out in Problem 3.20 and part (a) of this problem with $R = 10^{-3}\Omega$, we can show that

$$H(s) = \frac{Y(s)}{X(s)} = \frac{1}{s^2 + 10^{-3}s + 1}, \qquad \mathcal{R}e\{s\} > -0.0005.$$

(d) We have

$$\frac{1}{(\text{Vect. Len. from } \omega \text{ to } -0.0005 + j\frac{\sqrt{3}}{2})(\text{Vect. Len. from } \omega \text{ to } -0.0005 - j\frac{\sqrt{3}}{2}))}.$$

We see that when $|\omega|$ is in he vicinity 0.0005, the right-hand side of the above equation takes on extremely large values. On either side of this value of $|\omega|$ the value of $|H(j\omega)|$ rolls off rapidly. Therefore, $H(s)$ may be considered to be approximately bandpass.

9.19. (a) The unilateral Laplace transform is

$$\begin{aligned}
\mathcal{X}(s) &= \int_{0^-}^{\infty} e^{-2t}u(t+1)e^{-st}dt \\
&= \int_{0^-}^{\infty} e^{-2t}e^{-st}dt \\
&= \frac{1}{s+2}
\end{aligned}$$

(b) The unilateral Laplace transform is

$$\begin{aligned}\mathcal{X}(s) &= \int_{0^-}^{\infty} [\delta(t+1) + \delta(t) + e^{-2(t+3)}u(t+1)]e^{-st}dt \\ &= \int_{0^-}^{\infty} [\delta(t) + e^{-2(t+3)}]e^{-st}dt \\ &= 1 + \frac{e^{-6}}{s+2}\end{aligned}$$

(c) The unilateral Laplace transform is

$$\begin{aligned}\mathcal{X}(s) &= \int_{0^-}^{\infty} [e^{-2t}u(t)e^{-4t}u(t)]e^{-st}dt \\ &= \int_{0^-}^{\infty} [e^{-2t} + e^{-4t}]e^{-st}dt \\ &= \frac{1}{s+2} + \frac{1}{s+4}\end{aligned}$$

9.20. In Problem 3.19, we showed that the input and output of the RL circuit are related by

$$\frac{dy(t)}{dt} + y(t) = x(t).$$

Applying the unilateral Laplace transform to this equation, we have

$$s\mathcal{Y}(s) - y(0^-) + \mathcal{Y}(s) = \mathcal{X}(s).$$

(a) For the zero-state response, set $y(0^-) = 0$. Also we have

$$\mathcal{X}(s) = \mathcal{UL}\{e^{-2t}u(t)\} = \frac{1}{s+2}.$$

Therefore,

$$\mathcal{Y}(s)(s+1) = \frac{1}{s+2}.$$

Computing the partial fraction expansion of the right-hand side of the above equation and then taking its inverse unilateral Laplace transform, we have

$$y(t) = e^{-t}u(t) - e^{-2t}u(t).$$

(b) For the zero-input response, assume that $x(t) = 0$. Since we are given that $y(0^-) = 1$,

$$s\mathcal{Y}(s) - 1 + \mathcal{Y}(s) = 0 \quad \Rightarrow \mathcal{Y}(s) = \frac{1}{s+1}.$$

Taking the inverse unilateral Laplace transform we have

$$y(t) = e^{-t}u(t).$$

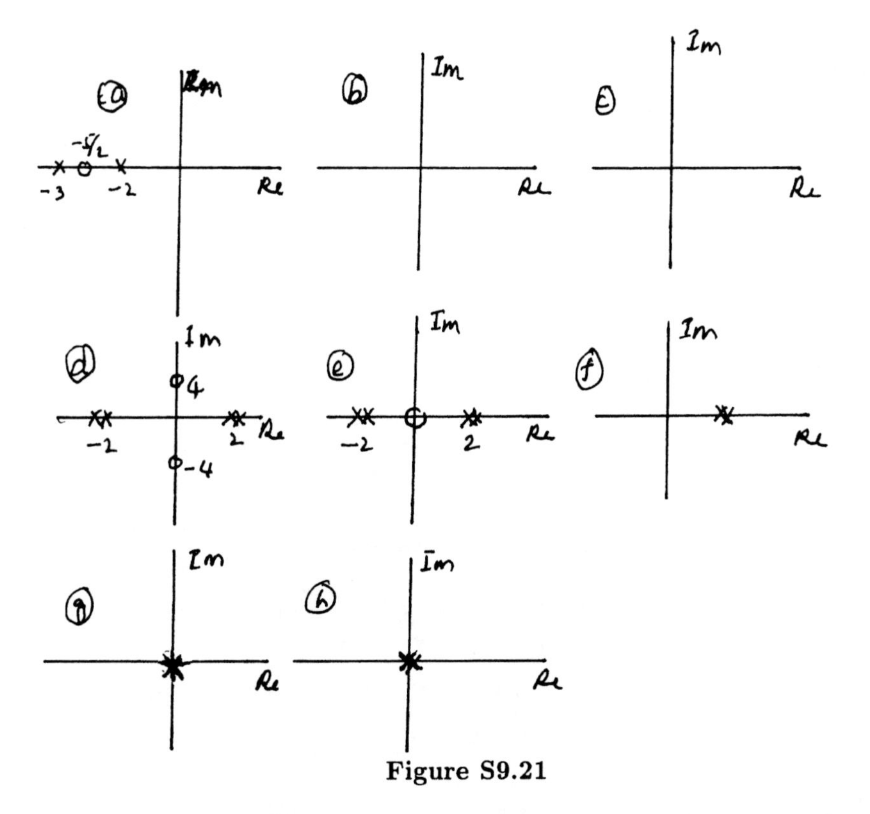

Figure S9.21

(c) The total response is the sum of the zero-state and zero-input responses. This is

$$y(t) = 2e^{-t}u(t) - e^{-2t}u(t).$$

9.21. The pole zero plots for all the subparts are shown in Figure S9.21.

(a) The Laplace transform of $x(t)$ is

$$\begin{aligned} X(s) &= \int_0^{\infty} (e^{-2t} + e^{-3t})e^{-st}dt \\ &= [-e^{-(s+2)t}/(s+2)]|_0^{\infty} + [-e^{-(s+3)t}/(s+3)]|_0^{\infty} \\ &= \frac{1}{s+2} + \frac{1}{s+3} = \frac{2s+5}{s^2+5s+6} \end{aligned}$$

The region of convergence (ROC) is $\mathcal{Re}\{s\} > -2$.

(b) Using an approach similar to that shown in part (a), we have

$$e^{-4t}u(t) \overset{\mathcal{L}}{\longleftrightarrow} \frac{1}{s+4}, \qquad \mathcal{Re}\{s\} > -4.$$

Also,

$$e^{-5t}e^{j5t}u(t) \overset{\mathcal{L}}{\longleftrightarrow} \frac{1}{s+5-j5}, \qquad \mathcal{Re}\{s\} > -5.$$

and

$$e^{-5t}e^{-j5t}u(t) \stackrel{\mathcal{L}}{\longleftrightarrow} \frac{1}{s+5+j5}, \qquad \mathcal{R}e\{s\} > -5.$$

From this we obtain

$$e^{-5t}\sin(5t)u(t) = \frac{1}{2j}[e^{-5t}e^{j5t} - e^{-5t}e^{-j5t}]u(t) \stackrel{\mathcal{L}}{\longleftrightarrow} \frac{5}{(s+5)^2+25},$$

where $\mathcal{R}e\{s\} > -5$. Therefore,

$$e^{-4t}u(t) + e^{-5t}\sin(5t)u(t) \stackrel{\mathcal{L}}{\longleftrightarrow} \frac{s^2+15s+70}{s^3+14s^2+90s+100}, \qquad \mathcal{R}e\{s\} > -5.$$

(c) The Laplace transform of $x(t)$ is

$$\begin{aligned} X(s) &= \int_{-\infty}^{0} (e^{2t} + e^{3t})e^{-st}dt \\ &= [-e^{(s-2)t}/(s-2)]|_{-\infty}^{0} + [-e^{-(s-3)t}/(s-3)]|_{-\infty}^{0} \\ &= \frac{1}{s-2} + \frac{1}{s-3} = \frac{2s-5}{s^2-5s+6} \end{aligned}$$

The region of convergence (ROC) is $\mathcal{R}e\{s\} < 2$.

(d) Using an approach along the lines of part (a), we obtain

$$e^{-2t}u(t) \stackrel{\mathcal{L}}{\longleftrightarrow} \frac{1}{s+2}, \qquad \mathcal{R}e\{s\} > -2. \qquad \text{(S9.21–1)}$$

Using an approach along the lines of part (c), we obtain

$$e^{2t}u(-t) \stackrel{\mathcal{L}}{\longleftrightarrow} \frac{1}{s-2}, \qquad \mathcal{R}e\{s\} < 2. \qquad \text{(S9.21–2)}$$

From these we obtain

$$e^{-2|t|} = e^{-2t}u(t) + e^{2t}u(-t) \stackrel{\mathcal{L}}{\longleftrightarrow} \frac{2s}{s^2-4}, \qquad -2 < \mathcal{R}e\{s\} < 2.$$

Using the differentiation in the s-domain property, we obtain

$$te^{-2|t|} \stackrel{\mathcal{L}}{\longleftrightarrow} -\frac{d}{ds}\left[\frac{2s}{s^2-4}\right] = -\frac{2s^2+8}{(s^2-4)^2}, \qquad -2 < \mathcal{R}e\{s\} < 2.$$

(e) Using the differentiation in the s-domain property on eq. (S9.21-1), we get

$$te^{-2t}u(t) \stackrel{\mathcal{L}}{\longleftrightarrow} -\frac{d}{ds}\left[\frac{1}{s+2}\right] = \frac{1}{(s+2)^2}, \qquad \mathcal{R}e\{s\} > -2.$$

Using the differentiation in the s-domain property on eq. (S9.21-2), we get

$$-te^{2t}u(-t) \stackrel{\mathcal{L}}{\longleftrightarrow} \frac{d}{ds}\left[\frac{1}{s-2}\right] = -\frac{1}{(s-2)^2}, \qquad \mathcal{R}e\{s\} < 2.$$

Therefore,

$$|t|e^{-2|t|} = te^{-2t}u(t) + -te^{2t}u(-t) \stackrel{\mathcal{L}}{\longleftrightarrow} \frac{-4s}{(s+2)^2(s-2)^2}, \qquad -2 < \mathcal{R}e\{s\} < 2.$$

(f) From the previous part, we have

$$|t|e^{2t}u(-t) = -te^{2t}u(-t) \stackrel{\mathcal{L}}{\longleftrightarrow} -\frac{1}{(s-2)^2}, \qquad \mathcal{R}e\{s\} < 2.$$

(g) Note that the given signal may be written as $x(t) = u(t) - u(t-1)$. Note that

$$u(t) \stackrel{\mathcal{L}}{\longleftrightarrow} \frac{1}{s}, \qquad \mathcal{R}e\{s\} > 0.$$

Using the time shifting property, we get

$$u(t-1) \stackrel{\mathcal{L}}{\longleftrightarrow} \frac{e^{-s}}{s}, \qquad \mathcal{R}e\{s\} > 0.$$

Therefore,

$$u(t) - u(t-1) \stackrel{\mathcal{L}}{\longleftrightarrow} \frac{1-e^{-s}}{s}, \qquad \text{All } s.$$

Note that in this case, since the signal is finite duration, the ROC is the entire s-plane.

(h) Consider the signal $x_1(t) = t[u(t)-u(t-1)]$. Note that the signal $x(t)$ may be expressed as $x(t) = x_1(t) + x_1(-t+2)$. We have from the previous part

$$u(t) - u(t-1) \stackrel{\mathcal{L}}{\longleftrightarrow} \frac{1-e^{-s}}{s}, \qquad \text{All } s.$$

Using the differentiation in s-domain property, we have

$$x_1(t) = t[u(t) - u(t-1)] \stackrel{\mathcal{L}}{\longleftrightarrow} \frac{d}{ds}\left[\frac{1-e^{-s}}{s}\right] = \frac{se^{-s} - 1 + e^{-s}}{s^2}, \qquad \text{All } s.$$

Using the time-scaling property, we obtain

$$x_1(-t) \stackrel{\mathcal{L}}{\longleftrightarrow} \frac{-se^{s} - 1 + e^{s}}{s^2}, \qquad \text{All } s.$$

Then, using the shift property, we have

$$x_1(-t+2) \stackrel{\mathcal{L}}{\longleftrightarrow} e^{-2s}\frac{-se^{s} - 1 + e^{s}}{s^2}, \qquad \text{All } s.$$

Therefore,

$$x(t) = x_1(t) + x_1(-t+2) \stackrel{\mathcal{L}}{\longleftrightarrow} \frac{se^{-s} - 1 + e^{-s}}{s^2} + e^{-2s}\frac{-se^{s} - 1 + e^{s}}{s^2}, \qquad \text{All } s.$$

(i) The Laplace transform of $x(t) = \delta(t) + u(t)$ is $X(s) = 1 + 1/s$, $\mathcal{R}e\{s\} > 0$.

(j) Note that $\delta(3t) + u(3t) = \delta(t) + u(t)$. Therefore, the Laplace transform is the same as the result of the previous part.

9.22. **(a)** From Table 9.2, we have

$$x(t) = \frac{1}{3}\sin(3t)u(t).$$

(b) From Table 9.2 we know that

$$\cos(3t)u(t) \overset{\mathcal{L}}{\longleftrightarrow} \frac{s}{s^2+9}, \qquad \mathcal{R}e\{s\} > 0.$$

Using the time scaling property, we obtain

$$\cos(3t)u(-t) \overset{\mathcal{L}}{\longleftrightarrow} -\frac{s}{s^2+9}, \qquad \mathcal{R}e\{s\} < 0.$$

Therefore, the inverse Laplace transform of $X(s)$ is

$$x(t) = -\cos(3t)u(-t).$$

(c) From Table 9.2 we know that

$$e^t\cos(3t)u(t) \overset{\mathcal{L}}{\longleftrightarrow} \frac{s-1}{(s-1)^2+9}, \qquad \mathcal{R}e\{s\} > 1.$$

Using the time scaling property, we obtain

$$e^{-t}\cos(3t)u(-t) \overset{\mathcal{L}}{\longleftrightarrow} -\frac{s+1}{(s+1)^2+9}, \qquad \mathcal{R}e\{s\} < -1.$$

Therefore, the inverse Laplace transform of $X(s)$ is

$$x(t) = -e^{-t}\cos(3t)u(-t).$$

(d) Using partial fraction expansion on $X(s)$, we obtain

$$X(s) = \frac{2}{s+4} - \frac{1}{s+3}.$$

From the given ROC, we know that $x(t)$ must be a two-sided signal. Therefore,

$$x(t) = 2e^{-4t}u(t) + e^{-3t}u(-t).$$

(e) Using partial fraction expansion on $X(s)$, we obtain

$$X(s) = \frac{2}{s+3} - \frac{1}{s+2}.$$

From the given ROC, we know that $x(t)$ must be a two-sided signal. Therefore,

$$x(t) = 2e^{-3t}u(t) + e^{-2t}u(-t).$$

(f) We may rewrite $X(s)$ as

$$\begin{aligned} X(s) &= 1 + \frac{3s}{s^2-s+1} \\ &= 1 + \frac{3s}{(s-1/2)^2+(\sqrt{3}/2)^2} \\ &= 1 + 3\frac{s-1/2}{(s-1/2)^2+(\sqrt{3}/2)^2} + \frac{3/2}{(s-1/2)^2+(\sqrt{3}/2)^2} \end{aligned}$$

Using Table 9.2, we obtain

$$x(t) = \delta(t) + 3e^{-t/2}\cos(\sqrt{3}t/2)u(t) + \sqrt{3}e^{-t/2}\sin(\sqrt{3}t/2)u(t).$$

(g) We may rewrite $X(s)$ as

$$X(s) = 1 - \frac{3s}{(s+1)^2}.$$

From Table 9.2, we know that

$$tu(t) \overset{\mathcal{L}}{\longleftrightarrow} \frac{1}{s^2}, \qquad \mathcal{R}e\{s\} > 0.$$

Using the shifting property, we obtain

$$e^{-t}tu(t) \overset{\mathcal{L}}{\longleftrightarrow} \frac{1}{(s+1)^2}, \qquad \mathcal{R}e\{s\} > -1.$$

Using the differentiation property,

$$\frac{d}{dt}[e^{-t}tu(t)] = e^{-t}u(t) - te^{-t}u(t) \overset{\mathcal{L}}{\longleftrightarrow} \frac{s}{(s+1)^2}, \qquad \mathcal{R}e\{s\} > -1.$$

Therefore,

$$x(t) = \delta(t) - 3e^{-t}u(t) - 3te^{-t}u(t).$$

9.23. The four pole-zero plots shown may have the following possible ROCs:

- Plot (a): $\mathcal{R}e\{s\} < -2$ or $-2 < \mathcal{R}e\{s\} < 2$ or $\mathcal{R}e\{s\} > 2$.
- Plot (b): $\mathcal{R}e\{s\} < -2$ or $\mathcal{R}e\{s\} > -2$.
- Plot (c): $\mathcal{R}e\{s\} < 2$ or $\mathcal{R}e\{s\} > 2$.
- Plot (d): Entire s-plane.

Also, suppose that the signal $x(t)$ has a Laplace transform $X(s)$ with ROC R.

(1) We know from Table 9.1 that

$$e^{-3t}x(t) \overset{\mathcal{L}}{\longleftrightarrow} X(s+3).$$

The ROC R_1 of this new Laplace transform is R shifted by 3 to the left. If $x(t)e^{-3t}$ is absolutely integrable, then R_1 must include the $j\omega$ axis.

- For plot (a), this is possible only if R was $\mathcal{R}e\{s\} > 2$.
- For plot (b), this is possible only if R was $\mathcal{R}e\{s\} > -2$.
- For plot (c), this is possible only if R was $\mathcal{R}e\{s\} > 2$.
- For plot (d), R is the entire s-plane.

(2) We know from Table 9.2 that

$$e^{-t}u(t) \overset{\mathcal{L}}{\longleftrightarrow} \frac{1}{s+1}, \qquad \mathcal{R}e\{s\} > -1.$$

Also, from Table 9.1 we obtain

$$x(t) * [e^{-t}u(t)] \overset{\mathcal{L}}{\longleftrightarrow} \frac{X(s)}{s+1}, \qquad R_2 = R \cap [\mathcal{R}e\{s\} > -1].$$

If $e^{-t}u(t) * x(t)$ is absolutely integrable, then R_2 must include the $j\omega$-axis.

•For plot (a), this is possible only if R was $-2 < \mathcal{R}e\{s\} < 2$.
•For plot (b), this is possible only if R was $\mathcal{R}e\{s\} > -2$.
•For plot (c), this is possible only if R was $\mathcal{R}e\{s\} < 2$.
•For plot (d), R is the entire s-plane.

(3) If $x(t) = 0$ for $t > 1$, then the signal is a left-sided signal or a finite-duration signal.
•For plot (a), this is possible only if R was $\mathcal{R}e\{s\} < -2$.
•For plot (b), this is possible only if R was $\mathcal{R}e\{s\} < -2$.
•For plot (c), this is possible only if R was $\mathcal{R}e\{s\} < 2$.
•For plot (d), R is the entire s-plane.

(4) If $x(t) = 0$ for $t < -1$, then the signal is a right-sided signal or a finite-duration signal.
•For plot (a), this is possible only if R was $\mathcal{R}e\{s\} > 2$.
•For plot (b), this is possible only if R was $\mathcal{R}e\{s\} > -2$.
•For plot (c), this is possible only if R was $\mathcal{R}e\{s\} > 2$.
•For plot (d), R is the entire s-plane.

9.24. **(a)** The pole-zero diagram with the appropriate markings is shown in Figure S9.24.

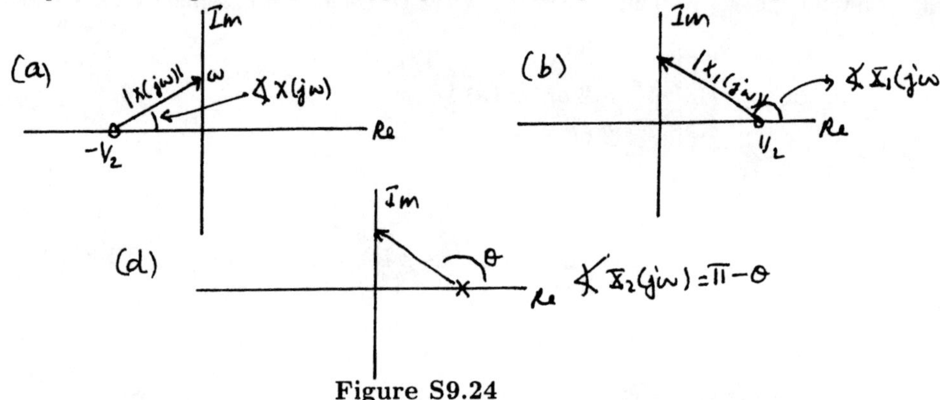

Figure S9.24

(b) By inspecting the pole-zero diagram of part (a), it is clear that the pole-zero diagram shown in Figure S9.24 will also result in the same $|X(j\omega)|$. This would correspond to the Laplace transform

$$X_1(s) = s - \frac{1}{2}, \quad \mathcal{R}e\{s\} < \frac{1}{2}.$$

(c) $\sphericalangle X(j\omega) = \pi - \sphericalangle X_1(j\omega)$.

(d) $X_2(s)$ with the pole-zero diagram shown below in Figure S9.24 would have the property that $\sphericalangle X_2(j\omega) = \sphericalangle X(j\omega)$. Here, $X_2(s) = \frac{-1}{s-1/2}$.

(e) $|X_2(j\omega)| = 1/|X(j\omega)|$.

(f) From the result of part (b), it is clear that $X_1(s)$ may be obtained by reflecting the poles and zeros in the right-half of the s-plane to the left-half of the s-plane. Therefore,

$$X_1(s) = \frac{s + 1/2}{s + 2}.$$

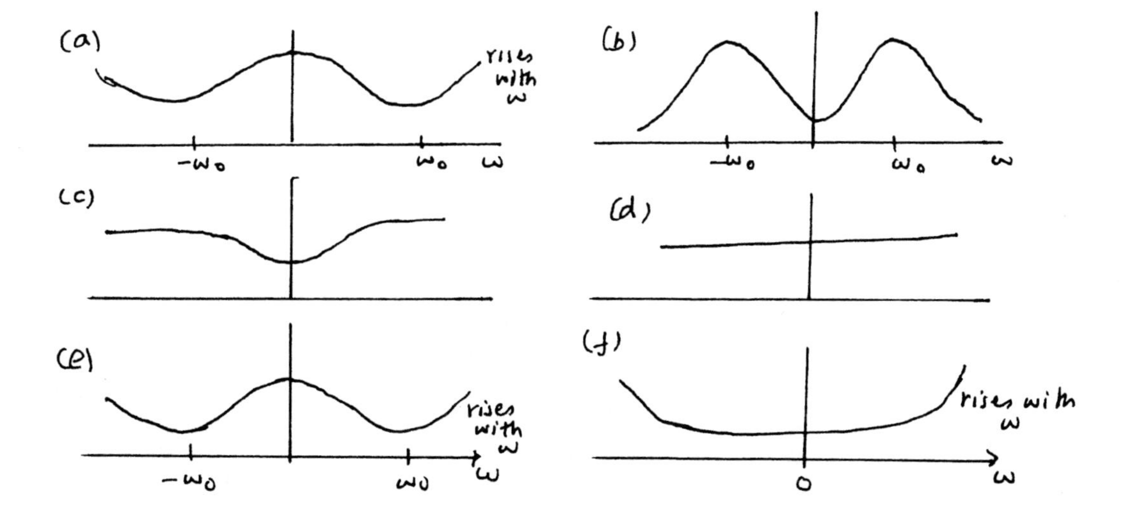

Figure S9.25

From part (d), it is clear that $X_2(s)$ may be obtained by reflecting the poles (zeros) in the right-half of the s-plane to the left-half and simultaneously changing them to zeros (poles). Therefore,

$$X_2(s) = \frac{(s+1)^2}{(s+1/2)(s+2)}.$$

9.25. The plots are as shown in Figure S9.25.

9.26. From Table 9.2 we have

$$x_1(t) = e^{-2t}u(t) \stackrel{\mathcal{L}}{\longleftrightarrow} X_1(s) = \frac{1}{s+2}, \qquad \mathcal{R}e\{s\} > -2$$

and

$$x_1(t) = e^{-3t}u(t) \stackrel{\mathcal{L}}{\longleftrightarrow} X_1(s) = \frac{1}{s+3}, \qquad \mathcal{R}e\{s\} > -3.$$

Using the time-shifting time-scaling properties from Table 9.1, we obtain

$$x_1(t-2) \stackrel{\mathcal{L}}{\longleftrightarrow} e^{-2s}X_1(s) = \frac{e^{-2s}}{s+2}, \qquad \mathcal{R}e\{s\} > -2$$

and

$$x_2(-t+3) \stackrel{\mathcal{L}}{\longleftrightarrow} e^{-3s}X_2(-s) = \frac{e^{-3s}}{3-s}, \qquad \mathcal{R}e\{s\} > -3.$$

Therefore, using the convolution property we obtain

$$y(t) = x_1(t-2) * x_2(-t+3) \stackrel{\mathcal{L}}{\longleftrightarrow} Y(s) = \left[\frac{e^{-2s}}{s+2}\right]\left[\frac{e^{-3s}}{3-s}\right].$$

9.27. From clues 1 and 2, we know that $X(s)$ is of the form

$$X(s) = \frac{A}{(s+a)(s+b)}.$$

Furthermore, we are given that one of the poles of $X(s)$ is $-1+j$. Since $x(t)$ is real, the poles of $X(s)$ must occur in conjugate reciprocal pairs. Therefore, $a = 1-j$ and $b = 1+j$ and

$$H(s) = \frac{A}{(s+1-j)(s+1+j)}.$$

From clue 5, we know that $X(0) = 8$. Therefore, we may deduce that $A = 16$ and

$$H(s) = \frac{16}{s^2+2s+2}.$$

Let R denote the ROC of $X(s)$. From the pole locations we know that there are two possible choices of R. R may either be $\mathcal{R}e\{s\} < -1$ or $\mathcal{R}e\{s\} > -1$. We will now use clue 4 to pick one. Note that

$$y(t) = e^{2t}x(t) \stackrel{\mathcal{L}}{\longleftrightarrow} Y(s) = X(s-2).$$

The ROC of $Y(s)$ is R shifted by 2 to the right. Since it is given that $y(t)$ is not absolutely integrable, the ROC of $Y(s)$ should not include the $j\omega$-axis. This is possible only of R is $\mathcal{R}e\{s\} > -1$.

9.28. **(a)** The possible ROCs are
(i) $\mathcal{R}e\{s\} < -2$.
(ii) $-2 < \mathcal{R}e\{s\} < -1$.
(iii) $-1 < \mathcal{R}e\{s\} < 1$.
(iv) $\mathcal{R}e\{s\} > 1$.

(b) (i) Unstable and anticausal.
(ii) Unstable and non causal.
(iii) Stable and non causal.
(iv) Unstable and causal.

9.29. **(a)** Using Table 9.2, we obtain

$$X(s) = \frac{1}{s+1}, \qquad \mathcal{R}e\{s\} > -1$$

and

$$H(s) = \frac{1}{s+2}, \qquad \mathcal{R}e\{s\} > -2.$$

(b) Since $y(t) = x(t) * h(t)$, we may use the convolution property to obtain

$$Y(s) = X(s)H(s) = \frac{1}{(s+1)(s+2)}.$$

The ROC of $Y(s)$ is $\mathcal{R}e\{s\} > -1$.

(c) Performing partial fraction expansion on $Y(s)$, we obtain

$$Y(s) = \frac{1}{s+1} - \frac{1}{s+2}.$$

Taking the inverse Laplace transform, we get

$$y(t) = e^{-t}u(t) - e^{-2t}u(t).$$

(d) Explicit convolution of $x(t)$ and $h(t)$ gives us

$$\begin{aligned} y(t) &= \int_{-\infty}^{\infty} h(\tau)x(t-\tau)d\tau \\ &= \int_{0}^{\infty} e^{-2\tau}e^{-(t-\tau)}u(t-\tau)d\tau \\ &= e^{-t}\int_{0}^{t} e^{-\tau}d\tau \qquad \text{for } t > 0 \\ &= [e^{-t} - e^{-2t}]u(t). \end{aligned}$$

9.30. For the input $x(t) = u(t)$, the Laplace transform is

$$X(s) = \frac{1}{s}, \qquad \mathcal{R}e\{s\} > 0.$$

The corresponding output $y(t) = [1 - e^{-t} - te^{-t}]u(t)$ has the Laplace transform

$$Y(s) = \frac{1}{s} - \frac{1}{s+1} - \frac{1}{(s+1)^2} = \frac{1}{s(s+1)^2}, \qquad \mathcal{R}e\{s\} > 0.$$

Therefore,

$$H(s) = \frac{Y(s)}{X(s)} = \frac{1}{(s+1)^2}, \qquad \mathcal{R}e\{s\} > 0.$$

Now, the output $y_1(t) = [2 - 3e^{-t} + e^{-3t}]u(t)$ has the Laplace transform

$$Y_1(s) = \frac{2}{s} - \frac{3}{s+1} + \frac{1}{s+3} = \frac{6}{s(s+1)(s+3)}, \qquad \mathcal{R}e\{s\} > 0.$$

Therefore, the Laplace transform of the corresponding input will be

$$X_1(s) = \frac{Y_1(s)}{H(s)} = \frac{6(s+1)}{s(s+3)}, \qquad \mathcal{R}e\{s\} > 0.$$

Taking the inverse Laplace transform of the partial fraction expansion of $X_1(s)$, we obtain

$$x_1(t) = 2u(t) + 4e^{-3t}u(t).$$

9.31. (a) Taking the Laplace transform of both sides of the given differential equation and simplifying, we obtain

$$H(s) = \frac{Y(s)}{X(s)} = \frac{1}{s^2 - s - 2}.$$

The pole-zero plot for $H(s)$ is as shown in Figure S9.31.

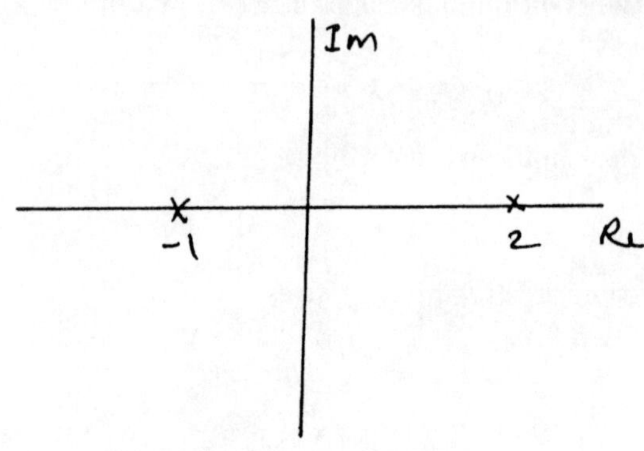

Figure S9.31

(b) The partial fraction expansion of $H(s)$ is

$$H(s) = \frac{1/3}{s-2} - \frac{1/3}{s+1}.$$

(i) If the system is stable, the ROC for $H(s)$ has to be $-1 < \mathcal{R}e\{s\} < 2$. Therefore,

$$h(t) = -\frac{1}{3}e^{2t}u(-t) - \frac{1}{3}e^{-t}u(t).$$

(ii) If the system is causal, the ROC for $H(s)$ has to be $\mathcal{R}e\{s\} > 2$. Therefore,

$$h(t) = \frac{1}{3}e^{2t}u(t) - \frac{1}{3}e^{-t}u(t).$$

(iii) If the system is neither stable nor causal, the ROC for $H(s)$ has to be $\mathcal{R}e\{s\} < -1$. Therefore,

$$h(t) = -\frac{1}{3}e^{2t}u(-t) + \frac{1}{3}e^{-t}u(-t).$$

9.32. If $x(t) = e^{2t}$ produces $y(t) = (1/6)e^{2t}$, then $H(2) = 1/6$. Also, by taking the Laplace transform of both sides of the given differential equation we get

$$H(s) = \frac{s + b(s+4)}{s(s+4)(s+2)}.$$

Since $H(2) = 1/6$, we may deduce that $b = 1$. Therefore,

$$H(s) = \frac{2(s+2)}{s(s+4)(s+2)} = \frac{2}{s(s+4)}.$$

9.33. Since $x(t) = e^{-|t|} = e^{-t}u(t) + e^{t}u(-t)$,

$$X(s) = \frac{1}{s+1} - \frac{1}{s-1} = \frac{-2}{(s+1)(s-1)}, \qquad -1 < \mathcal{R}e\{s\} < 1.$$

We are also given that

$$H(s) = \frac{s+1}{s^2+2s+2}.$$

Since the poles of $H(s)$ are at $-1 \pm j$, and since $h(t)$ is causal, we may conclude that the ROC of $H(s)$ is $\mathcal{R}e\{s\} > -1$. Now,

$$Y(s) = H(s)X(s) = \frac{-2}{(s^2+2s+2)(s-1)}.$$

The ROC of $Y(s)$ will be the intersection of the ROCs of $X(s)$ and $H(s)$. This is $-1 < \mathcal{R}e\{s\} < 1$.

We may obtain the following partial fraction expansion for $Y(s)$:

$$Y(s) = -\frac{2/5}{s-1} + \frac{2s/5 + 6/5}{s^2+2s+2}.$$

We may rewrite this as

$$Y(s) = -\frac{2/5}{s-1} + \frac{2}{5}\left[\frac{s+1}{(s+1)^2+1}\right] + \frac{4}{5}\left[\frac{1}{(s+1)^2+1}\right].$$

Noting that the ROC of $Y(s)$ is $-1 < \mathcal{R}e\{s\} < 1$ and using Table 9.2, we obtain

$$y(t) = \frac{2}{5}e^{t}u(-t) + \frac{2}{5}e^{-t}\cos tu(t) + \frac{4}{5}e^{-t}\sin tu(t).$$

9.34. We know that

$$x_1(t) = u(t) \stackrel{\mathcal{L}}{\longleftrightarrow} X_1(s) = \frac{1}{s}, \qquad \mathcal{R}e\{s\} > 0.$$

Therefore, $X_1(s)$ has a pole at $s = 0$. Now, the Laplace transform of the output $y_1(t)$ of the system with $x_1(t)$ as the input is

$$Y_1(s) = H(s)X_1(s).$$

Since in clue 2, $Y_1(s)$ is given to be absolutely integrable, $H(s)$ must have a zero at $s = 0$ which cancels out the pole of $X_1(s)$ at $s = 0$.

We also know that

$$x_2(t) = tu(t) \stackrel{\mathcal{L}}{\longleftrightarrow} X_2(s) = \frac{1}{s^2}, \qquad \mathcal{R}e\{s\} > 0.$$

Therefore, $X_2(s)$ has two poles at $s = 0$. Now, the Laplace transform of the output $y_2(t)$ of the system with $x_2(t)$ as the input is

$$Y_2(s) = H(s)X_2(s).$$

Since in clue 3, $Y_2(s)$ is given to be *not* absolutely integrable, $H(s)$ does not have two zeros at $s = 0$. Therefore, we conclude that $H(s)$ has exactly one zero at $s = 0$.

From Clue 4 we know that the signal

$$p(t) = \frac{d^2h(t)}{dt^2} + 2\frac{dh(t)}{dt} + 2h(t)$$

is finite duration. Taking the Laplace transform of both sides of the above equation, we get

$$P(s) = s^2H(s) + 2sH(s) + 2H(s).$$

Therefore,

$$H(s) = \frac{P(s)}{s^2 + 2s + 2}.$$

Since $p(t)$ is of finite duration, we know that $P(s)$ will have no poles in the finite s-plane. Therefore, $H(s)$ is of the form

$$H(s) = \frac{A\prod_{i=1}^{N}(s - z_i)}{s^2 + 2s + 2},$$

where z_i, $i = 1, 2, \cdots, N$ represent the zeros of $P(s)$. Here, A is some constant.

From Clue 5 we know that the denominator polynomial of $H(s)$ has to have a degree which is *exactly* one greater than the degree of the numerator polynomial. Therefore,

$$H(s) = \frac{A(s - s_1)}{s^2 + 2s + 2}.$$

Since we already know that $H(s)$ has a zero at $s = 0$, we may rewrite this as

$$H(s) = \frac{As}{s^2 + 2s + 2}.$$

From Clue 1 we know that $H(1)$ is 0.2. From this, we may easily show that $A = 1$. Therefore,

$$H(s) = \frac{s}{s^2 + 2s + 2}.$$

Since the poles of $H(s)$ are at $-1 \pm j$ and since $h(t)$ is causal and stable, the ROC of $H(s)$ is $\mathcal{R}e\{s\} > -1$.

9.35. (a) We may redraw the given block diagram as shown in Figure S9.35.

From the figure, it is clear that

$$\frac{F(s)}{s} = Y_1(s).$$

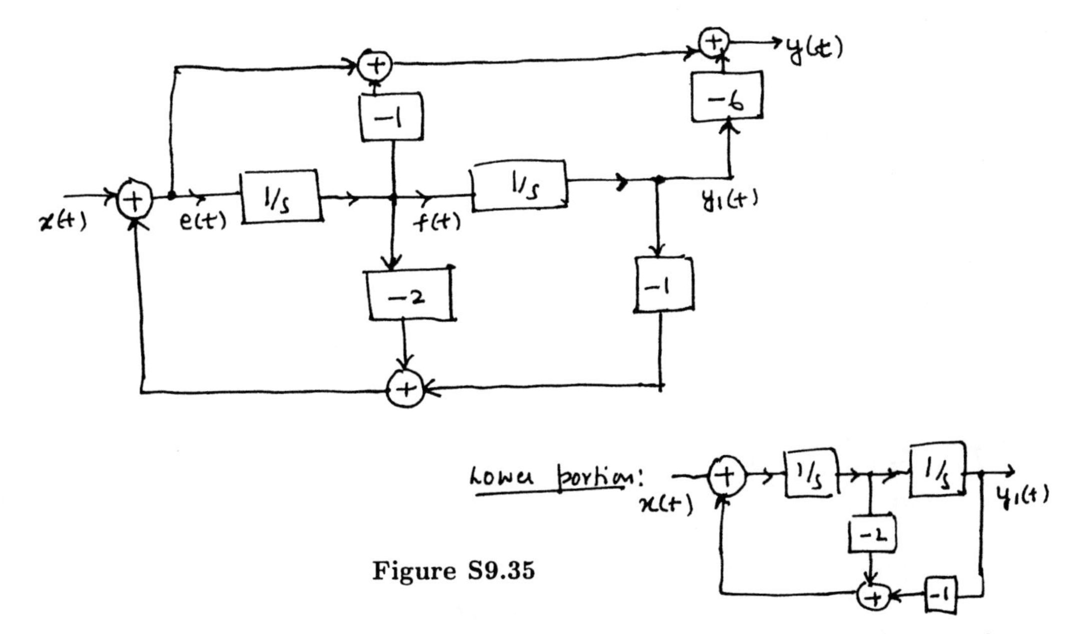

Figure S9.35

Therefore, $f(t) = dy_1(t)/dt$. Similarly, $e(t) = df(t)/dt$. Therefore, $e(t) = d^2y_1(t)/dt^2$. From the block diagram it is clear that

$$y(t) = e(t) - f(t) - 6y_1(t) = \frac{d^2y_1(t)}{dt^2} - \frac{dy_1(t)}{dt} - 6y_1(t).$$

Therefore,

$$Y(s) = s^2Y_1(s) - sY_1(s) - 6Y_1(s). \qquad \text{(S9.35−1)}$$

Now, let us determine the relationship between $y_1(t)$ and $x(t)$. This may be done by concentrating on the lower half of the above figure. We redraw this in Figure S9.35.

From Example 9.30, it is clear that $y_1(t)$ and $x(t)$ must be related by the following differential equation:

$$\frac{d^2y_1(t)}{dt^2} + 2\frac{dy_1(t)}{dt} + y_1(t) = x(t).$$

Therefore,

$$Y_1(s) = \frac{X(s)}{s^2 + 2s + 1}.$$

Using this in conjunction with eq (S9.35-1), we get

$$Y(s) = \frac{s^2 - s - 6}{s^2 + 2s + 1}X(s).$$

Taking the inverse Laplace transform, we obtain

$$\frac{d^2y(t)}{dt^2} + 2\frac{dy(t)}{dt} + y(t) = \frac{d^2x(t)}{dt^2} - \frac{dx(t)}{dt} - 6x(t).$$

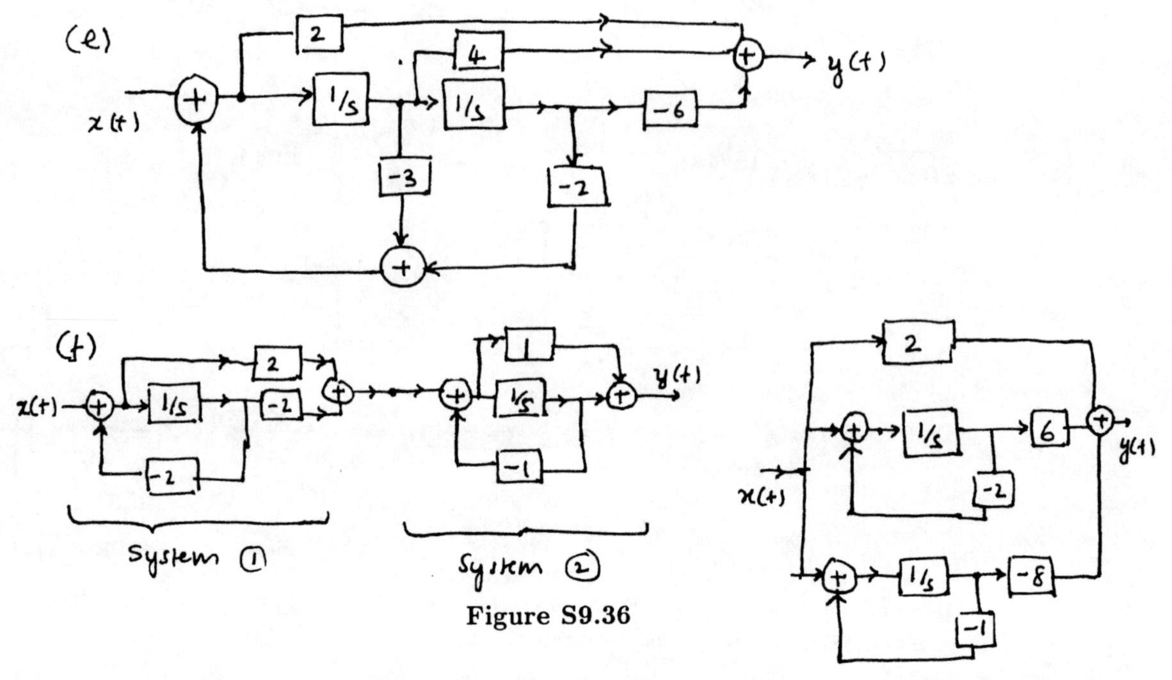

Figure S9.36

(b) The two poles of the system are at -1. Since the system is causal, the ROC must be to the right of $s = -1$. Therefore, the ROC must include the $j\omega$-axis. Hence, the system is stable.

9.36. **(a)** We know that $Y_1(s)$ and $Y(s)$ are related by

$$Y(s) = (2s^2 + 4s - 6)Y_1(s).$$

Taking the inverse Laplace transform, we get

$$y(t) = 2\frac{d^2y_1(t)}{dt^2} + 4\frac{dy_1(t)}{dt} - 6y_1(t).$$

(b) Since $Y_1(s) = F(s)/s$, $f(t) = dy_1(t)/dt$.

(c) Since $F(s) = E(s)/s$, $e(t) = df(t)/dt = d^2y_1(t)/dt^2$.

(d) From part (a), $y(t) = 2e(t) + 4f(t) - 6y_1(t)$.

(e) The extended block diagram is as shown in Figure S9.36.

(f) The block diagram is as shown in Figure S9.36.

(g) The block diagram is as shown in Figure S9.36.

The three subsystems may be connected in parallel as shown in the figure above to obtain the overall system

9.37. The block diagrams are shown in Figure S9.37.

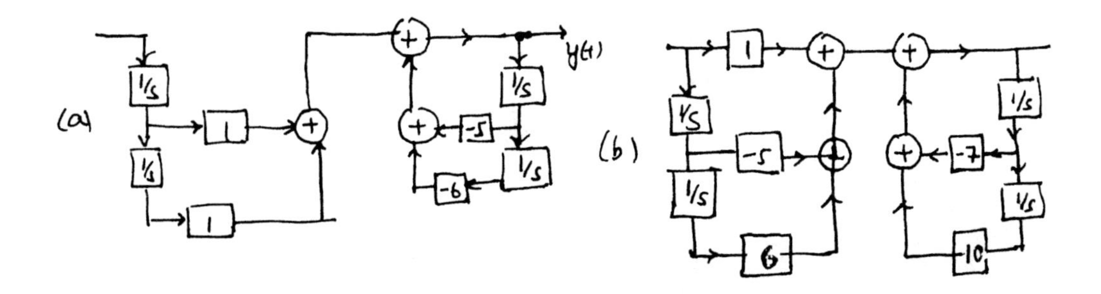

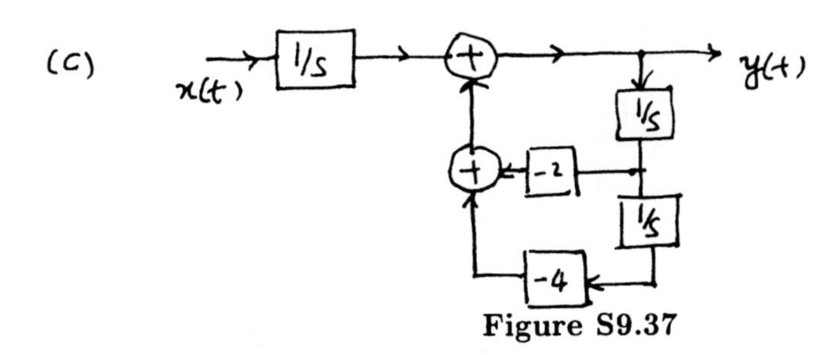

Figure S9.37

9.38. **(a)** We may rewrite $H(s)$ as

$$H(s) = \left[\frac{1}{s+1}\right]\left[\frac{1}{s+1}\right]\left[\frac{1}{s-\frac{1}{2}+\frac{j\sqrt{3}}{2}}\right]\left[\frac{1}{s-\frac{1}{2}-\frac{j\sqrt{3}}{2}}\right].$$

$H(s)$ clearly may be treated as the cascade combination of four first order subsystems. Consider one of these subsystems with the system function

$$H_1(s) = \left[\frac{1}{s-\frac{1}{2}-\frac{j\sqrt{3}}{2}}\right].$$

The block diagram for this is as shown in Figure S9.38. Clearly, it contains multiplications with coefficients that are not real.

(b) We may write $H(s)$ as

$$H(s) = \left[\frac{1}{s^2+2s+1}\right]\left[\frac{1}{s^2-s+1}\right] = H_1(s)H_2(s).$$

The block diagram for $H(s)$ may be constructed as a cascade of the block diagrams of $H_1(s)$ and $H_2(s)$ as shown in Figure S9.38.

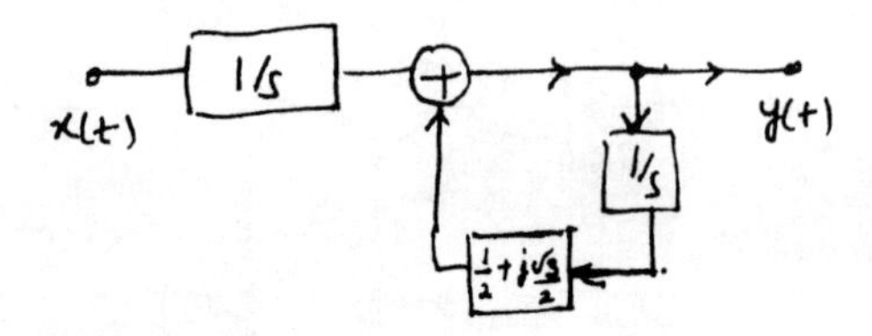

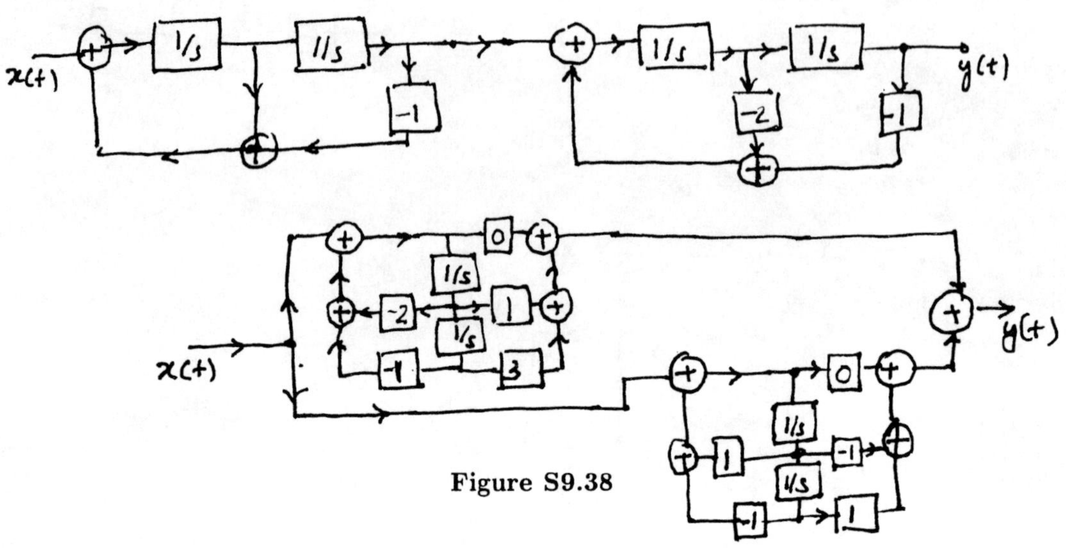

Figure S9.38

(c) We may rewrite $H(s)$ as

$$H(s) = \frac{1}{3}\left[\frac{s+3}{s^2+2s+1}\right] + \frac{1}{3}\left[\frac{1-s}{s^2-s+1}\right] = H_3(s) + H_4(s).$$

The block diagram for $H(s)$ may be constructed as a parallel combination of the block diagrams of $H_3(s)$ and $H_4(s)$ as shown in Figure S9.38.

9.39. **(a)** For $x_1(t)$, the unilateral and bilateral Laplace transforms are identical.

$$X_1(s) = \mathcal{X}_1(s) = \frac{1}{s+2}, \qquad \mathcal{R}e\{s\} > -2.$$

(b) Here, using Table 9.2 and the time shifting property we get

$$X_2(s) = \frac{e^s}{s+3}, \qquad \mathcal{R}e\{s\} > -3.$$

The unilateral Laplace transform is

$$\mathcal{X}_2(s) = e^{-3}\frac{1}{s+3}, \qquad \mathcal{R}e\{s\} > -3.$$

(c) We have

$$
\begin{aligned}
G(s) &= X_1(s)X_2(s) = \frac{e^s}{(s+2)(s+3)} \\
&= e^s\left[\frac{1}{s+2} - \frac{1}{s+3}\right]
\end{aligned}
$$

Taking the inverse Laplace transform, we obtain

$$g(t) = e^{-2(t+1)}u(t+1) - e^{-3(t+1)}u(t+1).$$

(d) We have

$$
\begin{aligned}
\mathcal{R}(s) &= \mathcal{X}_1(s)\mathcal{X}_2(s) = \frac{e^{-3}}{(s+2)(s+3)} \\
&= e^{-3}\left[\frac{1}{s+2} - \frac{1}{s+3}\right]
\end{aligned}
$$

Taking the inverse unilateral Laplace transform, we obtain

$$r(t) = e^{-2t-3}u(t) - e^{-3(t+1)}u(t).$$

Clearly, $r(t) \neq g(t)$ for $t > 0^-$.

9.40. Taking the unilateral Laplace transform of both sides of the given differential equation, we get

$$
\begin{aligned}
s^3\mathcal{Y}(s) &- s^2y(0^-) - sy'(0^-) - y''(0^-) + 6s^2\mathcal{Y}(s) - 6sy(0^-) \\
&-6y(0^-) + 11s\mathcal{Y}(s) - 11y(0^-) + 6\mathcal{Y}(s) = \mathcal{X}(s).
\end{aligned} \tag{S9.40–1}
$$

(a) For the zero state response, assume that all the initial conditions are zero. Furthermore, from the given $x(t)$ we may determine

$$\mathcal{X}(s) = \frac{1}{s+4}, \qquad \mathcal{R}e\{s\} > -4.$$

From eq. (S9.40-1), we get

$$\mathcal{Y}(s)[s^3 + 6s^2 + 11s + 6] = \frac{1}{s+4}.$$

Therefore,

$$\mathcal{Y}(s) = \frac{1}{(s+4)(s^3 + 6s^2 + 11s + 6)}.$$

Taking the inverse unilateral Laplace transform of the partial fraction expansion of the above equation, we get

$$y(t) = \frac{1}{6}e^{-t}u(t) - \frac{1}{6}e^{-4t}u(t) + \frac{1}{2}e^{-2t}u(t) - \frac{1}{2}e^{-3t}u(t).$$

(b) For the zero-input response, we assume that $\mathcal{X}(s) = 0$. Assuming that the initial conditions are as given, we obtain from (S9.40-1)

$$\mathcal{Y}(s) = \frac{s^2 + 5s + 6}{s^3 + 6s^2 + 11s + 6} = \frac{1}{s+1}.$$

Taking the inverse unilateral Laplace transform of the above equation, we get

$$y(t) = e^{-t}u(t).$$

(c) The total response is the sum of the zero-state and zero-input responses.

$$y(t) = \frac{7}{6}e^{-t}u(t) - \frac{1}{6}e^{-4t}u(t) + \frac{1}{2}e^{-2t}u(t) - \frac{1}{2}e^{-3t}u(t).$$

9.41. Let us first find the Laplace transform of the signal $y(t) = x(-t)$. We have

$$\begin{aligned} Y(s) &= \int_{-\infty}^{\infty} x(-t)e^{-st}dt \\ &= \int_{-\infty}^{\infty} x(t)e^{st}dt \\ &= X(-s). \end{aligned}$$

(a) Since $x(t) = x(-t)$ for an even signal, we can conclude that $\mathcal{L}\{x(t)\} = \mathcal{L}\{x(-t)\}$. Therefore, $X(s) = X(-s)$.

(b) Since $x(t) = -x(-t)$ for an odd signal, we can conclude that $\mathcal{L}\{x(t)\} = -\mathcal{L}\{x(-t)\}$. Therefore, $X(s) = -X(-s)$.

(c) First of all note that for a signal to be even, it must be either two-sided or finite duration. Therefore, if $X(s)$ has poles, the ROC must be a strip in the s-plane.

From plot (a), we get

$$X(s) = \frac{As}{(s+1)(s-1)}.$$

Therefore,

$$X(-s) = \frac{-As}{(s-1)(s+1)} = -X(s).$$

Therefore, $x(t)$ is not even (in fact it is odd).

For plot (b), we note that the ROC cannot be chosen to correspond to a two-sided function $x(t)$. Therefore, this signal is not even.

From plot (c), we get

$$X(s) = \frac{A(s-j)(s+j)}{(s+1)(s-1)} = \frac{A(s^2+1)}{s^2-1}.$$

Therefore,

$$X(-s) = \frac{A(s^2+1)}{s^2-1} = X(s).$$

Therefore, $x(t)$ is even provided the ROC is chosen to be $-1 < \mathcal{R}e\{s\} < 1$.

For plot (d), we note that the ROC cannot be chosen to correspond to a two-sided function $x(t)$. Therefore, this signal is not even.

9.42. (a) From table 9.2 we know that the Laplace transform of $t^2u(t)$ is $1/s^3$ with the ROC $\mathcal{R}e\{s\} > 0$. Therefore, the given statement is false.

(b) We know that the Laplace transform of a signal $x(t)$ is the same as the Fourier transform of the signal $x(t)e^{-\sigma t}$. The ROC is given by the range of σ for which this Fourier transform exists.

Now, if $x(t) = e^{t^2}u(t)$, then we note that as $t \to \infty$, the signal $x(t)$ becomes unbounded. Therefore, for the Fourier transform of of $e^{-\sigma t}x(t)$ to exist, we need to find a range of σ which ensures that $e^{-\sigma t}x(t)$ is bounded as $t \to \infty$. Clearly, this is not possible. Therefore, the given statement is true.

(c) This statement is true. Consider the signal $x(t) = e^{j\omega_0 t}$. Then

$$X(s) = \int_{-\infty}^{\infty} e^{j\omega_0 t}e^{-st}dt = \left.\frac{e^{t(j\omega_0 - s)}}{j\omega_0 - s}\right|_{-\infty}^{\infty}.$$

This integral does not converge for any value of s.

(d) This statement is false. Consider the signal $x(t) = e^{j\omega_0 t}u(t)$. Then

$$X(s) = \int_{0}^{\infty} e^{j\omega_0 t}e^{-st}dt = \left.\frac{e^{t(j\omega_0 - s)}}{j\omega_0 - s}\right|_{0}^{\infty}.$$

This integral converges for any value of $s > 0$.

(e) This statement is false. Consider the signal $x(t) = |t|$. Then

$$X(s) = \int_{0}^{\infty} te^{-st}dt + \int_{\infty}^{0} -te^{-st}dt.$$

Both integrals on the right-hand side converge for any value of $s > 0$.

9.43. We are given that $h(t)$ is causal and stable. Therefore, all poles are in the left half of the s-plane.

(a) Note that

$$g(t) = \frac{dh(t)}{dt} \stackrel{\mathcal{L}}{\longleftrightarrow} G(s) = sH(s).$$

Now, $G(s)$ has the same poles as $H(s)$ and hence the ROC for $G(s)$ remains the same. Therefore, $g(t)$ is also guaranteed to be causal and stable.

(b) Note that

$$r(t) = \int_{-\infty}^{t} h(\tau)d\tau \stackrel{\mathcal{L}}{\longleftrightarrow} R(s) = \frac{H(s)}{s}.$$

Note that $R(s)$ does not have a pole at $s = 0$ *only if* $H(s)$ has a zero at $s = 0$. Therefore, we cannot guarantee that $r(t)$ is always causal and unstable.

9.44. (a) Note that

$$\delta(t - nT) \stackrel{\mathcal{L}}{\longleftrightarrow} e^{-snT}, \qquad \text{All } s.$$

Therefore,

$$X(s) = \sum_{n=0}^{\infty} e^{-nT} e^{-snT} = \frac{1}{1 - e^{-T(1+s)}}.$$

In order to determine the ROC, let us first find the poles of $X(s)$. Clearly, the poles occur when $e^{-T(1+s)} = 1$. This implies that the poles s_k satisfy the following equation:

$$e^{-T(1+s_k)} = e^{jk2\pi}, k = 0, \pm1, \pm2, \cdots.$$

Taking the logarithm of both sides of the above equation and simplifying , we get

$$s_k = -1 + \frac{jk2\pi}{T}, k = 0, \pm1, \pm2, \cdots.$$

Therefore, the poles all lie on a vertical line (parallel to the $j\omega$-axis) passing through $s = -1$. Since the signal is right-sided, the ROC is $\mathcal{R}e\{s\} > -1$.

(b) The pole-zero plot is as shown in Figure S9.44.

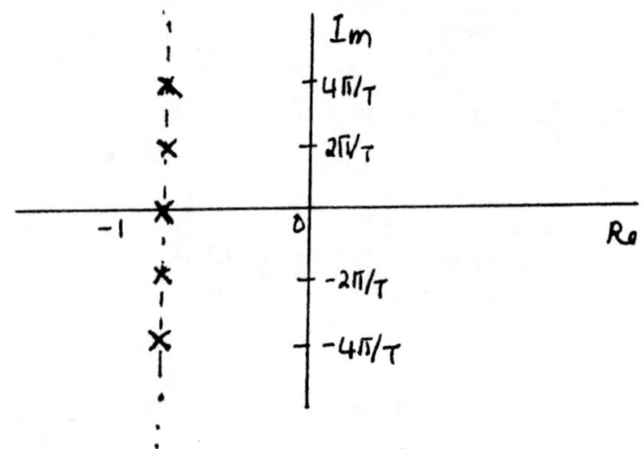

Figure S9.44

(c) The magnitude of the Fourier transform $X(j\omega)$ is given by the product of the reciprocals of the lengths of the vectors from the poles to the point $j\omega$. The phase of $X(j\omega)$ is given by the negative of the sum of the angles of these vectors. Clearly from the pole-zero plot above it is clear that both the magnitude and phase have to vary periodically with a period of $2\pi/T$.

9.45. (a) Taking the Laplace transform of the signal $x(t)$, we get

$$Y(s) = \frac{2/3}{s-2} + \frac{1/3}{s+1} = \frac{s}{(s-2)(s-1)}.$$

The ROC is $-1 < \mathcal{R}e\{s\} < 2$. Also, note that since $x(t)$ is a left-sided signal, the ROC for $X(s)$ is $\mathcal{R}e\{s\} < 2$.

Now,

$$H(s) = \frac{Y(s)}{X(s)} = \frac{s}{(s+2)(s+1)}.$$

We know that the ROC of $Y(s)$ has to be the intersection of the ROCs of $X(s)$ and $H(s)$. This leads us to conclude that the ROC of $H(s)$ is $\mathcal{R}e\{s\} > -1$.

(b) The partial fraction expansion of $H(s)$ is

$$H(s) = \frac{2}{s+2} - \frac{1}{s+1}.$$

Therefore,

$$h(t) = 2e^{-2t}u(t) - e^{-t}u(t).$$

(c) e^{3t} is an Eigen function of the LTI system. Therefore,

$$y(t) = H(3)e^{3t} = \frac{3}{20}e^{3t}.$$

9.46. Since $y(t)$ is real, the third input must be of the form $e^{s_0^* t}$. Since $x(t)$ is of the form $\delta(t) + e^{s_0 t} + e^{s_0^* t}$ and the output is $y(t) = -6e^{-t}u(t) + \frac{4}{34}e^{4t}\cos(3t) + \frac{18}{34}e^{4t}\sin(3t)$, we may conclude that $H(4 \pm 3j) = \frac{4}{34} \pm j\frac{18}{34}$.

Let us try $h(t) = \delta(t) - 6e^{-t}u(t)$. Then

$$H(s) = \frac{s-5}{s+1}.$$

We may easily show that $H(4 \pm 3j) = \frac{4}{34} \pm j\frac{18}{34}$. Therefore, $H(s)$ as given above is consistent with the given information.

9.47. **(a)** Taking the Laplace transform of $y(t)$, we obtain

$$Y(s) = \frac{1}{s+2}, \qquad \mathcal{R}e\{s\} > -2.$$

Therefore,

$$X(s) = \frac{Y(s)}{H(s)} = \frac{s+1}{(s-1)(s+2)}.$$

The pole-zero diagram for $X(s)$ is as shown in Figure S9.47. Now, the ROC of $H(s)$ is $\mathcal{R}e\{s\} > -1$. We know that the ROC of $Y(s)$ is at least the intersection of the ROCs of $X(s)$ and $H(s)$. Note that the ROC can be larger if some poles are canceled out by zeros at the same location. In this case, we can choose the ROC of $X(s)$ to be either $-2 < \mathcal{R}e\{s\} < 1$ or $\mathcal{R}e\{s\} > 1$. In both cases, we get the same ROC for $Y(s)$ because the poles at $s = -1$ and $s = 1$ in $H(s)$ and $X(s)$, respectively are canceled out by zeros.

The partial fraction expansion of $X(s)$ is

$$X(s) = \frac{2/3}{s-1} + \frac{1/3}{s-2}.$$

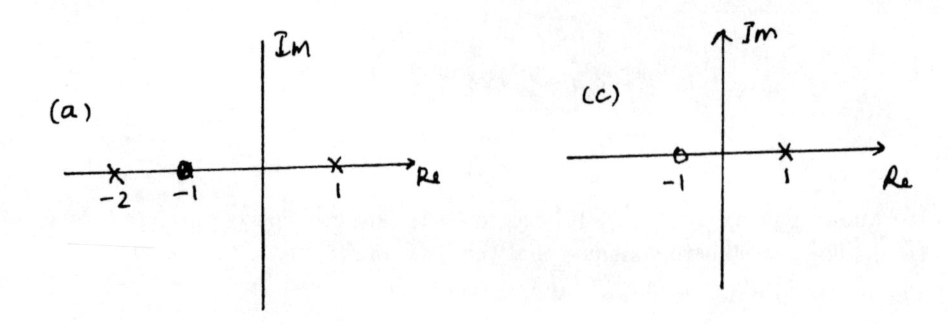

Figure S9.47

Taking the ROC of $X(s)$ to be $-2 < \mathcal{R}e\{s\} < 1$, we get

$$x(t) = -\frac{2}{3}e^{t}u(-t) + \frac{1}{3}e^{-2t}u(t).$$

Taking the ROC of $X(s)$ to be $\mathcal{R}e\{s\} > 1$, we get

$$x(t) = \frac{2}{3}e^{t}u(t) + \frac{1}{3}e^{-2t}u(t).$$

(b) Since it is given that $x(t)$ is absolutely integrable, we can conclude that the ROC of $X(s)$ must include the $j\omega$-axis. Therefore, the first choice of $x(t)$ given above is the one we want.

(c) We need to first find a $H(s)$ such that $H(s)Y(s) = X(s)$. Clearly,

$$H(s) = \frac{X(s)}{Y(s)} = \frac{s+1}{s-1}.$$

The pole-zero plot for $H(s)$ is as shown in Figure S9.47. Since $h(t)$ is given to be stable, the ROC of $H(s)$ has to be $\mathcal{R}e\{s\} < 1$. The partial fraction expansion of $H(s)$ is

$$H(s) = 1 + \frac{2}{s-1}.$$

Therefore,

$$h(t) = \delta(t) - 2e^{-t}u(-t).$$

Also, $Y(s)$ has the ROC $\mathcal{R}e\{s\} > -2$. Therefore, $X(s)$ must have the ROC $-2 < \mathcal{R}e\{s\} < 1$ (the intersection of the ROCs of $Y(s)$ and $H(s)$. From this we get (as shown in part (a))

$$x(t) = -\frac{2}{3}e^{t}u(-t) + \frac{1}{3}e^{-2t}u(t).$$

Verification: Now,

$$\begin{aligned} h(t) * y(t) &= [\delta(t) - 2e^{-t}u(-t)] * [e^{-2t}u(t)] \\ &= e^{-2t}u(t) - 2\int_0^{\infty} e^{-2\tau}e^{t-\tau}u(\tau - t)d\tau \end{aligned}$$

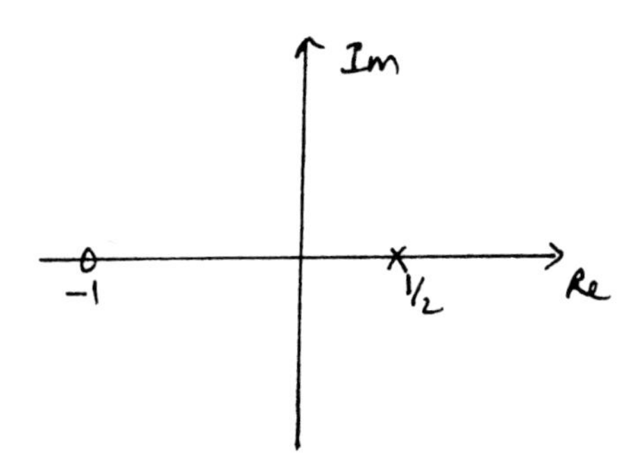

Figure S9.48

For $t > 0$, the integral in the above equation is

$$e^t \int_t^{\infty} e^{-3\tau} d\tau = \frac{1}{3} e^{-2t}.$$

For $t < 0$, the integral in the above equation is

$$e^t \int_0^{\infty} e^{-3\tau} d\tau = \frac{1}{3} e^{t}.$$

Therefore,

$$h(t) * y(t) = -\frac{2}{3} e^t u(-t) + \frac{1}{3} e^{-2t} u(t) = x(t).$$

9.48. **(a)** $H_1(s) = 1/H(s)$.

(b) From the above relationship it is clear that the poles of the inverse system will be the zeros of original system. Also, the zeros of the inverse system will be the poles of the original system. Therefore, the pole-zero plot for $H_1(s)$ is as sketched in Figure S9.48.

9.49. If a system is causal and stable, then the poles of its transfer function must all be in the left half of the s-plane. This is because the ROC of a causal system is to the right of the right-most pole. For the ROC to contain the $j\omega$-axis, the right-most pole must be in the left-half of the s-plane.

Now, if the inverse system is also causal and stable, then its poles must also all lie in the left half of the s-plane. But we know that the poles of the inverse system are the zeros of the original system. Therefore, the zeros of the original system must also lie in the left-half of the s-plane.

9.50. **(a)** False. Counter-example: $H(s) = 1/(s-2), \mathcal{R}e\{s\} < 2$.

(b) True. If the system function has more poles than zeros, then $h(t)$ does not have an impulse at $t = 0$. Since we know that $h(t)$ is the derivative of the step response, we may conclude that the step response has no discontinuities at $t = 0$.

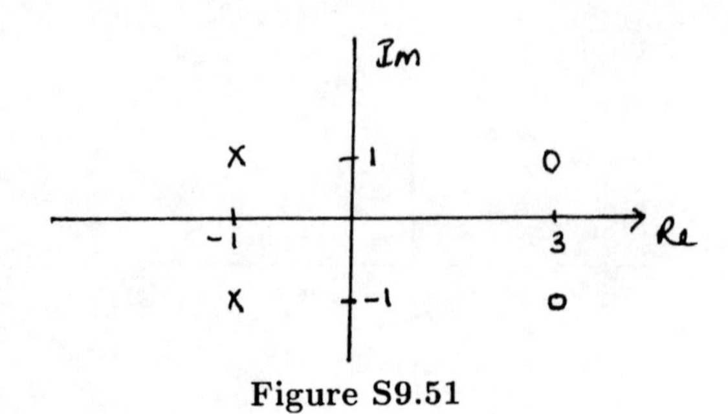

Figure S9.51

(c) False. Causality plays no part in the argument of part (b).

(d) False. Counter-example: $H(s) = (s-1)/(s+2), \mathcal{R}e\{s\} > -2$.

9.51. Since $h(t)$ is real, its poles and zeros must occur in complex conjugate pairs. Therefore, the known poles and zeros of $H(s)$ are as shown in Figure S9.51. Since $H(s)$ has exactly 2 zeros at infinity, $H(s)$ has *at least* two more unknown finite poles. In case $H(s)$ has more than 4 poles, then it will have a zero at some location for every additional pole. Furthermore, since $h(t)$ is causal and stable, all poles of $H(s)$ must lie in the left half of the s-plane and the ROC must include the $j\omega$-axis.

(a) True. Consider

$$g(t) = h(t)e^{-3t} \overset{\mathcal{L}}{\longleftrightarrow} G(s) = H(s+3).$$

The ROC of $G(s)$ will be the ROC of $H(s)$ shifted by 3 to the left. Clearly this ROC will still include the $j\omega$-axis. Therefore, $g(t)$ has to be stable.

(b) Insufficient information. As mentioned earlier, $H(s)$ has some unknown poles. So we do not know which the rightmost pole is in $H(s)$. Therefore, we cannot determine what its exact ROC is.

(c) True. Since $H(s)$ is rational, $H(s)$ may be expressed as a ratio of two polynomials in s. Furthermore, since $h(t)$ is real, the coefficients of these polynomials will be real. Now,

$$H(s) = \frac{Y(s)}{X(s)} = \frac{P(s)}{Q(s)}.$$

Here, $P(s)$ and $Q(s)$ are polynomials in s. The differential equation relating $x(t)$ and $y(t)$ is obtained by taking the inverse Laplace transform of $Y(s)Q(s) = X(s)P(s)$. Clearly, this differential equation has to have only real coefficients.

(d) False. We are given that $H(s)$ has 2 zeros at $s = \infty$. Therefore, $\lim_{s\to\infty} H(s) = 0$.

(e) True. See the reasoning at the beginning of the problem.

(f) Insufficient information. $H(s)$ may have other zeros. See reasoning at the beginning of the problem.

(g) False. We know that $e^{3t}\sin(t) = (1/2j)e^{(3+j)t} - (1/2j)e^{(3-j)t}$. Both $e^{(3+j)t}$ and $e^{(3-j)t}$ are Eigen functions of the LTI system. Therefore, the response of the system to these exponentials is $H(3+j)e^{(3+j)t}$ and $H(3-j)e^{(3-j)t}$, respectively. Since $H(s)$ has zeros at $3 \pm j$, we know that the output of the system to the two exponentials has to be zero. Hence, the response of the system to $e^{3t}\sin(t)$ has to be zero.

9.52. (a) Consider the signal $y(t) = x(t - t_0)$. Now,

$$Y(s) = \int_{-\infty}^{\infty} x(t - t_0)e^{-st}dt.$$

Replacing $t - t_0$ by τ, we get

$$\begin{aligned} Y(s) &= \int_{-\infty}^{\infty} x(\tau)e^{-s(\tau+t_0)}d\tau \\ &= e^{-st_0}\int_{-\infty}^{\infty} x(\tau)e^{-s\tau}d\tau \\ &= e^{-st_0}X(s) \end{aligned}$$

This obviously converges when $X(s)$ converges because e^{-st_0} has no poles. Therefore, the ROC of $Y(s)$ is the same as the ROC of $X(s)$.

(b) Consider the signal $y(t) = e^{s_0 t}x(t)$. Now,

$$\begin{aligned} Y(s) &= \int_{-\infty}^{\infty} x(t)e^{s_0 t}e^{-st}dt \\ &= \int_{-\infty}^{\infty} x(t)e^{-(s-s_0)t}dt \\ &= X(s - s_0) \end{aligned}$$

If $X(s)$ converges in the range $a < \mathcal{R}e\{s\} < b$, then $X(s - s_0)$ converges in the range $a + s_0 < s < b + s_0$. This is the ROC of $Y(s)$.

(c) Consider the signal $y(t) = x(at)$. Now,

$$Y(s) = \int_{-\infty}^{\infty} x(at)e^{-st}dt.$$

Replacing at by τ and assuming that $a > 1$, we get

$$\begin{aligned} Y(s) &= (1/a)\int_{-\infty}^{\infty} x(\tau)e^{-s(\tau/a)}d\tau \\ &= (1/a)X(s/a). \end{aligned}$$

If $a < 0$, then

$$\begin{aligned} Y(s) &= -(1/a)\int_{-\infty}^{\infty} x(\tau)e^{-s(\tau/a)}d\tau \\ &= -(1/a)X(s/a). \end{aligned}$$

Therefore,

$$Y(s) = \frac{1}{|a|}X\left(\frac{s}{a}\right).$$

If $X(s)$ converges in the range $\alpha < \mathcal{R}e\{s\} < \beta$, then $X(s/a)$ converges in the range $\alpha/a < s < \beta/a$ when $a > 0$. When $a < 0$, then $X(s/a)$ converges in the range $\beta/a < s < \alpha/a$.

(d) Consider the signal $y(t) = x(t) * h(t)$. Now,

$$\begin{aligned} Y(s) &= \int_{-\infty}^{\infty} [x(t) * h(t)] e^{-st} dt \\ &= \int_{-\infty}^{\infty} \int_{-\infty}^{\infty} x(\tau) h(t-\tau) d\tau e^{-st} dt \\ &= \int_{-\infty}^{\infty} x(\tau) \left[\int_{-\infty}^{\infty} h(t-\tau) e^{-st} dt \right] d\tau \end{aligned}$$

Using the time-shifting property, we get

$$\begin{aligned} Y(s) &= \int_{-\infty}^{\infty} x(\tau) H(s) e^{-s\tau} d\tau \\ &= H(s) \int_{-\infty}^{\infty} x(\tau) e^{-s\tau} d\tau \\ &= H(s) X(s) \end{aligned}$$

Clearly, $Y(s)$ converges at least in the region where both $X(s)$ and $H(s)$ converge. Its ROC may be larger depending on whether some of the poles of either $H(s)$ or $X(s)$ get cancelled out by the zeros of $X(s)$ or $H(s)$, respectively.

9.53. **(a)** From the example worked out in the text we have

$$e^{-at}\left(\frac{t^n}{n!}\right) u(t) \stackrel{\mathcal{L}}{\longleftrightarrow} \frac{1}{(s+a)^{n+1}}, \qquad \mathcal{R}e\{s\} > a.$$

With $a = 0$, we get

$$x^n(0+)\left(\frac{t^n}{n!}\right) u(t) \stackrel{\mathcal{L}}{\longleftrightarrow} \frac{x^n(0+)}{s^{n+1}}, \qquad \mathcal{R}e\{s\} > 0.$$

(b) We may rewrite eq. (P9.53-1) as

$$x(t) = \sum_{-\infty}^{\infty} x^n(0+)\left(\frac{t^n}{n!}\right).$$

Taking the Laplace transform of both sides of this equation and using the result of part (a), we get

$$X(s) = \sum_{n=0}^{\infty} \frac{x^n(0+)}{s^{n+1}}. \qquad \text{(S9.53–1)}$$

(c) From the result of part (b), we have

$$sX(s) = x^{(0)}(0+) + x^{(1)}(0+)/s + \cdots .$$

Therefore,

$$\lim_{s\to\infty} sX(s) = x^{(0)}(0+) = x(0+).$$

(d) (1) Assuming that the ROC is $s > -2$, we get

$$x(t) = e^{-2t}u(t).$$

Therefore, $x(0+) = 1$. Now,

$$\lim_{s\to\infty} sX(s) = \lim_{s\to\infty} \frac{s}{s+2} = 1.$$

(2) The partial fraction expansion of $X(s)$ is

$$X(s) = \frac{2}{(s+3)} - \frac{1}{s+2}.$$

Assuming that the ROC is $s > -2$, we get

$$x(t) = 2e^{-3t}u(t) - e^{-2t}u(t).$$

Therefore, $x(0+) = 1$. Now,

$$\lim_{s\to\infty} sX(s) = \lim_{s\to\infty} \frac{s^2+s}{s^2+5s+6} = 1.$$

(e) Assuming that $x^{(n)}(0+) = 0$ for $n < N$, eq.(S9.53-1) may be written as

$$X(s) = \sum_{n=N}^{\infty} \frac{x^n(0+)}{s^{n+1}}.$$

Now,

$$s^{N+1}X(s) = x^{(N)}(0+) + \frac{x^{N+1}(0+)}{s} + \frac{x^{N+2}(0+)}{s^2} + \cdots .$$

Therefore,

$$\lim_{s\to\infty} s^{N+1}X(s) = x^{(N)}(0+).$$

9.54. **(a)** We have

$$x(t) = \frac{1}{2\pi j}\int_{\sigma-j\infty}^{\sigma+j\infty} X(s)e^{-st}ds.$$

Conjugating both sides, we get

$$x^*(t) = -\frac{1}{2\pi j}\int_{\sigma+j\infty}^{\sigma-j\infty} X^*(s)e^{s^*t}ds.$$

For a real signal $x(t) = x^*(t)$. Therefore,

$$x(t) = -\frac{1}{2\pi j}\int_{\sigma+j\infty}^{\sigma-j\infty} X^*(s)e^{s^*t}ds.$$

Replacing s^* by p and noting that $dp = -ds$ for a fixed σ, we get

$$\begin{aligned} x(t) &= -\frac{1}{2\pi j}\int_{\sigma+j\infty}^{\sigma-j\infty} X^*(p^*)e^{pt}dp \\ &= \frac{1}{2\pi j}\int_{\sigma-j\infty}^{\sigma+j\infty} X^*(p^*)e^{pt}dp \end{aligned}$$

Therefore, $\mathcal{L}\{x(t)\} = X^*(s^*)$. This implies that $X(s) = X^*(s^*)$.

(b) Let $X(s)$ have a zero at $s = s_1$. Then $X(s_1) = 0$. From the result of part (a), we know that $X^*(s_1^*) = 0$. This implies that $X(s_1^*) = 0$, which in turn implies that $X(s)$ has a zero at s_1^*. The same approach may be used to show that poles occur in conjugate pairs.

9.55 Pair 10:From pair 1 of Table 9.2, we have

$$\delta(t) \overset{\mathcal{L}}{\longleftrightarrow} 1, \qquad \text{All } s.$$

Using the time-shifting property, we get

$$\delta(t-T) \overset{\mathcal{L}}{\longleftrightarrow} e^{-sT}, \qquad \text{All } s.$$

Pair 11:From pair 6 of Table 9.2, we have

$$e^{j\omega_0 t}u(t) \overset{\mathcal{L}}{\longleftrightarrow} \frac{1}{s-j\omega_0}, \qquad \mathcal{R}e\{s\} > 0 \qquad \text{(S9.55–1)}$$

and

$$e^{-j\omega_0 t}u(t) \overset{\mathcal{L}}{\longleftrightarrow} \frac{1}{s+j\omega_0}, \qquad \mathcal{R}e\{s\} > 0. \qquad \text{(S9.55–2)}$$

Note that $\cos(\omega_0 t) = (1/2)e^{j\omega_0 t} + (1/2)e^{-j\omega_0 t}$. Now using eqs. (S9.55-1) and (S9.55-2) with the linearity property, we get

$$\cos(\omega_0 t)u(t) \overset{\mathcal{L}}{\longleftrightarrow} \frac{1}{2}\left[\frac{1}{s-j\omega_0}\right] + \frac{1}{2}\left[\frac{1}{s+j\omega_0}\right] = \frac{s}{s^2+\omega_0^2}.$$

The ROC will be $\mathcal{R}e\{s\} > 0$.

Pair 12:Note that $\sin(\omega_0 t) = (1/2j)e^{j\omega_0 t} - (1/2j)e^{-j\omega_0 t}$. Now using eqs. (S9.55-1) and (S9.55-2) with the linearity property, we get

$$\sin(\omega_0 t)u(t) \overset{\mathcal{L}}{\longleftrightarrow} \frac{1}{2j}\left[\frac{1}{s-j\omega_0}\right] - \frac{1}{2j}\left[\frac{1}{s+j\omega_0}\right] = \frac{\omega_0}{s^2+\omega_0^2}.$$

The ROC will be $\mathcal{R}e\{s\} > 0$.

Pair 13:Using the shifting in the s-domain property on pair 11, we get

$$e^{-at}\cos(\omega_0 t)u(t) \overset{\mathcal{L}}{\longleftrightarrow} \frac{s+a}{(s+a)^2+\omega_0^2}, \qquad \mathcal{R}e\{s\} > -a.$$

Pair 14:Using the shifting in the s-domain property on pair 12, we get

$$e^{-at}\sin(\omega_0 t)u(t) \overset{\mathcal{L}}{\longleftrightarrow} \frac{\omega_0}{(s+a)^2+\omega_0^2}, \qquad \mathcal{R}e\{s\} > -a.$$

Pair 15:From pair 1 of Table 9.2, we have

$$u_0(t) = \delta(t) \overset{\mathcal{L}}{\longleftrightarrow} 1, \qquad \text{All } s.$$

Using the differentiation in time-domain property on this signal, we get

$$u_1(t) = \frac{d\delta(t)}{dt} \overset{\mathcal{L}}{\longleftrightarrow} s, \qquad \text{All} s.$$

Continuing along these lines and differentiating $\delta(t)$ n times, we get

$$u_n(t) = \frac{d^n\delta(t)}{dt^n} \overset{\mathcal{L}}{\longleftrightarrow} s^n, \qquad \text{All} s.$$

Pair 16:From pair 2 of Table 9.2, we have

$$u(t) \overset{\mathcal{L}}{\longleftrightarrow} \frac{1}{s}, \qquad \mathcal{R}e\{s\} > 0.$$

By applying the convolution property, we get

$$u_{-2}(t) = u(t) * u(t) \overset{\mathcal{L}}{\longleftrightarrow} \frac{1}{s^2}, \qquad \mathcal{R}e\{s\} > 0.$$

Continuing along these lines and convolving $u(t)$ with itself n times, we get

$$u_{-n}(t) \overset{\mathcal{L}}{\longleftrightarrow} \frac{1}{s^n}, \qquad \mathcal{R}e\{s\} > 0.$$

9.56. Given that

$$\int_{-\infty}^{\infty} |x(t)|e^{-\sigma_0 t}dt < \infty,$$

we need to prove that $|X(s_0)| = 0$, where $s_0 = \sigma_0 + j\omega_0$. We have

$$|X(s_0)| = \left|\int_{-\infty}^{\infty} x(t)e^{-s_0 t}dt\right| = \left|\int_{-\infty}^{\infty} x(t)e^{-\sigma_0 t}e^{-j\omega_0 t}dt\right|.$$

Using eq. (P9.56-1), we get

$$\begin{aligned} |X(s_0)| &\leq \int_{-\infty}^{\infty} |x(t)e^{-\sigma_0 t}e^{-j\omega_0 t}dt| \\ &\leq \int_{-\infty}^{\infty} |x(t)|e^{-\sigma_0 t}dt \\ &\leq \infty \end{aligned}$$

Plausibility of eq.(P9.56-1): Integration is akin to the addition of an infinite number of complex numbers. For any two complex numbers A and B, we know that $|A+B| \leq |A|+|B|$. Using this, we may argue that the same should hold for a infinite sum of complex numbers or the integration of a complex function.

9.57. Since $x(t)$ has an impulse at $t = 0$, the numerator polynomial of $X(s)$ must be of the same/larger degree than the denominator polynomial of $X(s)$. This implies that $X(s)$ has at least 4 zeros.

9.58. Since $g(t) = \mathcal{R}e\{h(t)\}$,

$$g(t) = \frac{h(t) + h^*(t)}{2}.$$

Using the linearity and conjugation properties, we get

$$G(s) = \frac{H(s) + H^*(s^*)}{2}.$$

The ROC of $G(s)$ will be at least the intersection of the ROCs of $H(s)$ and $H^*(s^*)$. This means that the ROC of $G(s)$ will be at least as much as the ROC of $H(s)$. Therefore, if $H(s)$ is causal and stable, then $G(s)$ also has to be causal and stable.

9.59. **(a)** Let $y(t) = x(t-1)$. Then,

$$\mathcal{Y}(s) = e^{-s}\mathcal{X}(s) + e^{-s}\int_{-1}^{0} x(t)e^{-st}dt.$$

(b) Let $y(t) = x(t+1)$. Then,

$$\mathcal{Y}(s) = e^{s}\mathcal{X}(s) - e^{s}\int_{0}^{1} x(t)e^{-st}dt.$$

(c) Let $y(t) = \int_{-\infty}^{t} x(\tau)d\tau$. Then,

$$\mathcal{Y}(s) = \frac{\mathcal{X}(s) + \int_{-\infty}^{0} x(t)e^{-st}dt}{s}.$$

(d) Let $y(t) = d^3x(t)/dt^3$. Then,

$$\mathcal{Y}(s) = s^3\mathcal{X}(s) - s^2x(0^-) - sx'(0^-) - x''(0^-).$$

9.60. **(a)** We have

$$h(t) = \alpha\delta(t-T) + \alpha^3\delta(t-3T).$$

From Tables 9.1 and 9.2,

$$H(s) = \alpha e^{-sT} + \alpha^3 e^{-3sT}, \qquad \text{All } s.$$

(b) To determine the zeros of $H(s)$, note that we require

$$\alpha e^{-sT} + \alpha^3 e^{-3sT} = \alpha e^{-sT}[1 + \alpha^2 e^{-2sT}] = 0.$$

Therefore, at the zeros

$$1 + \alpha^2 e^{-2sT} = 0 \qquad \Rightarrow \qquad \alpha e^{-sT} = \pm j.$$

This implies that the zeros occur at

$$s = \frac{1}{T}\log_e \alpha \pm \left[\frac{\pi}{2T} \pm \frac{2k\pi}{T}\right], \qquad k = 0, \pm 1, \pm 2, \cdots.$$

At the poles, $H(s) = \infty$. Therefore, at the poles we require that

$$\alpha e^{-sT} + \alpha^3 e^{-3sT} = \alpha e^{-sT}[1 + \alpha^2 e^{-2sT}] = \infty.$$

This is not possible at any finite s. Therefore, there are no poles in the finite s-plane.

(c) The pole-zero plot is as shown in Figure S9.60.

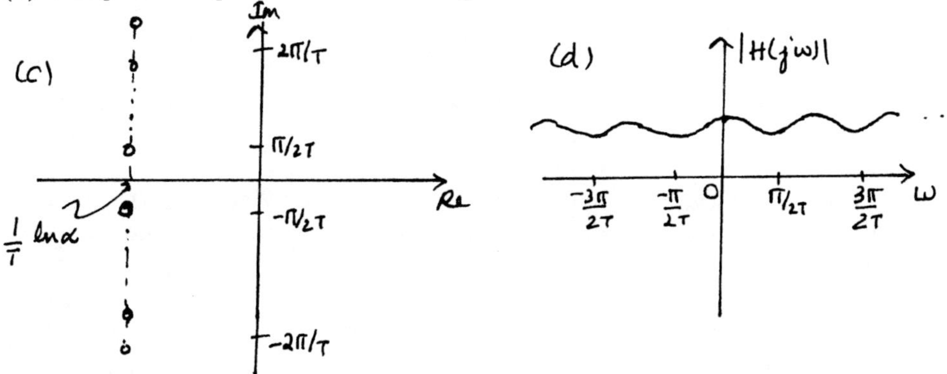

Figure S9.60

(d) From the figure it is clear that $H(j\omega)$ will be periodic and will be as shown in Figure S9.60.

9.61. **(a)** If we want $\phi_{xx}(t)$ to be the output of the system when $x(t)$ is the input, then

$$\phi_{xx}(t) = \int_{-\infty}^{\infty} x(\tau)h(t-\tau)d\tau.$$

Also we are given that

$$\phi_{xx}(t) = \int_{-\infty}^{\infty} x(\tau)x(t+\tau)d\tau.$$

Therefore,

$$x(t+\tau) = h(t-\tau) \qquad \Rightarrow \qquad h(t) = x(-t).$$

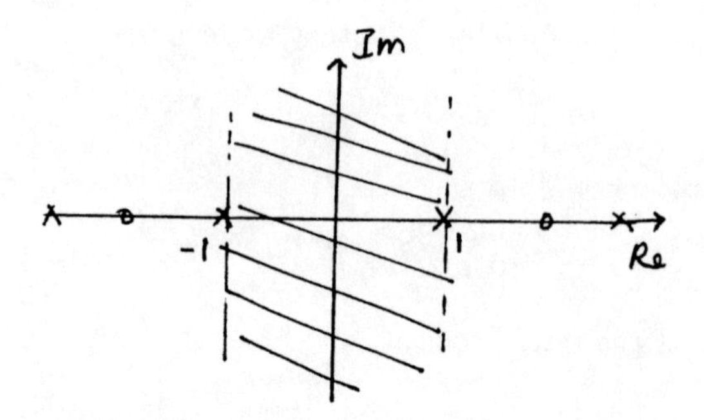

Figure S9.61

(b) Since $\phi_{xx}(t) = x(t) * x(-t)$,

$$\Phi_{xx}(s) = X(s)X(-s)$$

and

$$\Phi_{xx}(j\omega) = X(j\omega)X(-j\omega).$$

If $x(t)$ is real, $X^*(j\omega) = X(-j\omega)$ and

$$\Phi_{xx}(j\omega) = |X(j\omega)|^2.$$

(c) If $X(s)$ has a pole-zero pattern as shown in Figure P9.61, then $X(-s)$ has a pole-zero pattern as shown in Figure S9.61. The corresponding ROC is also shown in Figure S9.61.

Now, $\Phi_{xx}(s)$ will include the poles of both $X(s)$ and $X(-s)$. Furthermore, its ROC will be the intersection of the ROCs of $X(s)$ and $X(-s)$. (See Figure S9.61)

9.62. (a) We have

$$L_0(t) = e^t e^{-t} = 1,$$

$$L_1(t) = e^t \frac{d(te^{-t}))}{dt} = e^t[e^{-t} - te^{-t}] = 1 - t$$

and

$$\begin{aligned} L_2(t) &= \frac{e^t}{2}\frac{d^2(t^2e^{-t})}{dt^2} \\ &= \frac{e^t}{2}[2e^{-t} - 2te^{-t} - 2te^{-t} + t^2e^{-t}] \\ &= 1 - 2t + \frac{1}{2}t^2. \end{aligned}$$

(b) We have

$$\begin{aligned} \phi_n(t) &= \frac{1}{n!}e^{t/2}\frac{d^n[t^ne^{-t}]}{dt^n}u(t) \\ &= \frac{1}{n!}e^{t/2}\frac{d^n[t^ne^{-t}u(t) + t^ne^{-t}u(-t)]}{dt^n}u(t) \\ &= \frac{1}{n!}e^{t/2}\frac{d^n[t^ne^{-t}u(t)]}{dt^n}. \end{aligned}$$

But

$$\frac{t^n e^{-t}}{n!}u(t) \stackrel{\mathcal{L}}{\longleftrightarrow} \frac{1}{(s+1)^{n+1}}, \qquad s > -1.$$

Therefore,

$$\frac{1}{n!}\frac{d^n[t^n e^{-t}]}{dt^n}u(t) \stackrel{\mathcal{L}}{\longleftrightarrow} \frac{s^n}{(s+1)^{n+1}}, \qquad s > -1.$$

It follows that

$$e^{t/2}\frac{1}{n!}\frac{d^n[t^n e^{-t}]}{dt^n}u(t) \stackrel{\mathcal{L}}{\longleftrightarrow} \frac{(s-1/2)^n}{(s+1/2)^{n+1}}, \qquad s > -1/2.$$

Therefore,

$$\Phi_n(s) = \frac{(s-1/2)^n}{(s+1/2)^{n+1}}, \qquad s > -1/2.$$

(c) Choose

$$H_1(s) = \frac{1}{s+1/2}$$

and

$$H_2(s) = \frac{(s-1/2)}{(s+1/2)}.$$

9.63. (a) We have

$$H(s) = \frac{1}{s+1/2}.$$

The pole-zero plot for $H(s)$ is as shown in the Figure S9.63. Using the geometric method for evaluating the magnitude of the Fourier transform, we may sketch $|H(j\omega)|$ as shown in Figure S9.63.

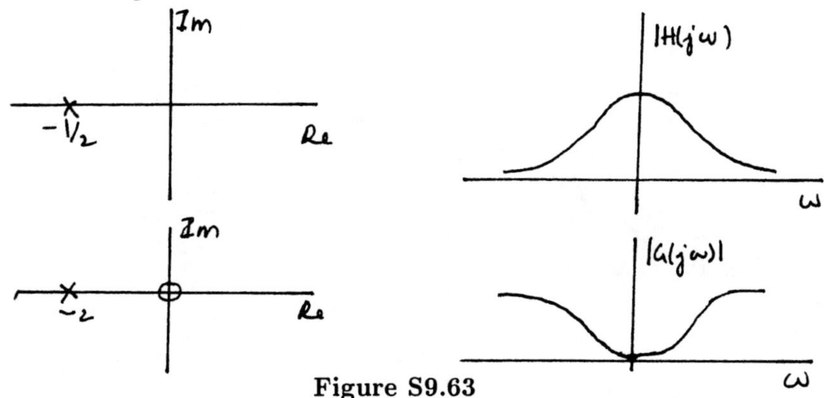

Figure S9.63

Also,

$$G(s) = H(1/s) = \frac{2s}{s+2}.$$

The pole-zero plot for $G(s)$ is as shown in the Figure S9.63. Using the geometric method for evaluating the magnitude of the Fourier transform, we may sketch $|G(j\omega)|$ as shown in Figure S9.63.

(b) LCCDE associated with $H(s)$:

Consider

$$H(s) = \frac{Y(s)}{X(s)} = \frac{1}{s + 1/2}.$$

Cross-multiplying and taking the inverse Laplace transform, we obtain

$$\frac{dy(t)}{dt} + \frac{1}{2}y(t) = x(t).$$

LCCDE associated with $G(s)$:

Consider

$$G(s) = \frac{Y(s)}{X(s)} = \frac{2s}{s+2}.$$

Cross-multiplying and taking the inverse Laplace transform, we obtain

$$\frac{dy(t)}{dt} + 2y(t) = 2\frac{dx(t)}{dt}.$$

(c) Taking the Laplace transform of eq.(P9.63-1), we obtain

$$\sum_{k=0}^{N} a_k s^k Y(s) = \sum_{k=0}^{N} b_k s^k X(s).$$

Therefore,

$$H(s) = \frac{Y(s)}{X(s)} = \frac{\sum_{k=0}^{N} b_k s^k}{\sum_{k=0}^{N} a_k s^k}.$$

Now,

$$G(s) = H(1/s) = \frac{\sum_{k=0}^{N} b_k s^{-k}}{\sum_{k=0}^{N} a_k s^{-k}} = \frac{\sum_{k=0}^{N} b_k s^{N-k}}{\sum_{k=0}^{N} a_k s^{N-k}}.$$

(d) Now from the previous part, we have

$$G(s) = \frac{Y(s)}{X(s)} = \frac{\sum_{k=0}^{N} b_k s^{N-k}}{\sum_{k=0}^{N} a_k s^{N-k}}.$$

Cross-multiplying and taking the inverse Laplace transform, we obtain

$$\sum_{k=0}^{N} a_k \frac{d^{N-k}y(t)}{dt^{N-k}} = \sum_{k=0}^{N} b_k \frac{d^{N-k}x(t)}{dt^{N-k}}.$$

9.64. For the circuit, we know that the differential equation relating the input $x(t)$ and output $y(t)$ is

$$LC\frac{d^2y(t)}{dt^2} + RC\frac{dy(t)}{dt} + y(t) = x(t).$$

Taking the Laplace transform of both sides and simplifying, we get

$$H(s) = \frac{Y(s)}{X(s)} = \frac{1/LC}{s^2 + (R/L)s + (1/LC)}.$$

(a) Note that the poles of $H(s)$ are at

$$\frac{-RC \pm \sqrt{R^2C^2 - 4LC}}{2}.$$

If R, L, and C are always positive, then the poles are always in the left half of the s-plane (because the real part of the numerator of the above equation is always negative). Since the system is causal, the ROC is to the right of the right-most pole. Therefore, the ROC includes the $j\omega$-axis and the system is stable.

(b) From $H(s)$ we obtain

$$\begin{aligned} H(s)H(-s) &= \frac{1}{L^2C^2s^4 + (RLC^2 - RLC^2)s^3 + (2LC - R^2C^2)s^2 + (RC - RC)s + 1} \\ &= \frac{1}{L^2C^2s^4 + (2LC - R^2C^2)s^2 + 1}. \end{aligned}$$

For this to represent a second order Butterworth filter, we require

$$2LC - R^2C^2 = 0 \quad \Rightarrow \quad R = 2\sqrt{\frac{L}{C}}.$$

9.65. **(a)** The differential equation relating $v_i(t)$ and $v_o(t)$ may be obtained by putting $x(t) = v_i(t)$ and $y(t) = v_o(t)$ in the differential equation given in the previous problem. Therefore,

$$LC\frac{d^2v_o(t)}{dt^2} + RC\frac{dv_o(t)}{dt} + v_o(t) = v_i(t)$$

or

$$\frac{d^2v_o(t)}{dt^2} + \frac{R}{L}\frac{dv_o(t)}{dt} + \frac{1}{LC}v_o(t) = \frac{1}{LC}v_i(t)$$

(b) Taking the unilateral Laplace transform of the above differential equation, we get

$$s^2\mathcal{V}_o(s) - s\mathcal{V}_o(0^-) - \mathcal{V}_o'(0^-) + \frac{R}{L}s\mathcal{V}_o(s) - \mathcal{V}_o(0^-) + \frac{1}{LC}\mathcal{V}_o(s) = \frac{1}{LC}\mathcal{V}_i(s). \tag{S9.65–1}$$

Now, since $v_i(t) = e^{-3t}u(t)$,

$$\mathcal{V}_i(s) = \frac{1}{s+3}, \qquad \mathcal{R}e\{s\} > -3.$$

Substituting this along with the values of R, L, and C in eq. (S9.65-1), we get

$$\mathcal{V}_o(S) = \frac{2(s^2+5s+7)}{(s+1)(s+2)(s+3)}.$$

The partial fraction expansion of $\mathcal{V}_o(s)$ is

$$\mathcal{V}_o(s) = \frac{3}{s+1} - \frac{2}{s+2} + \frac{1}{s+3}.$$

Taking the inverse Laplace transform, we get

$$v_o(t) = 3e^{-t}u(t) - 2e^{-2t}u(t) + e^{-3t}u(t).$$

9.66. **(a)** The differential equation relating $i(t)$ and v_2 is

$$\frac{di(t)}{dt} + \frac{R}{L}i(t) = \frac{v_2}{L}u(t).$$

Also, $i(0^-) = v_1/R$.

(b) Taking the unilateral Laplace transform of the above differential equation, we get

$$sI(s) - i(0^-) + \frac{R}{L}I(s) = \frac{v_2}{Ls}.$$

(i) This corresponds to the zero state response of the circuit. Here,

$$I(s) = \frac{v_2}{s(s+1)} = v_2\left[\frac{1}{s} - \frac{1}{s+1}\right].$$

Therefore,

$$i(t) = 2u(t) - 2e^{-t}u(t).$$

(ii) This corresponds to the zero state response of the circuit. Here, $i(0^-) = 4$ and

$$I(s) = \frac{4}{s+1}.$$

Therefore,

$$i(t) = 4e^{-t}u(t).$$

(iii) This corresponds to the total response of the system. It will be the sum of the results of the previous two parts.

$$i(t) = 2u(t) + 2e^{-t}u(t).$$

Chapter 10 Answers

10.1. (a) The given summation may be written as

$$\sum_{n=-1}^{\infty} \frac{1}{2}\left(\frac{1}{2}r^{-1}\right)^n e^{-j\omega n},$$

by replacing z with $re^{j\omega}$. If $r < \frac{1}{2}$, then $\frac{1}{2}r^{-1} > 1$ and the function within the summation grows towards infinity with increasing n. Also, the summation does not converge. But if $r > \frac{1}{2}$, then the summation converges.

(b) The given summation may be written as

$$\sum_{n=1}^{\infty} \frac{1}{2}(2r)^n e^{j\omega n},$$

by replacing z with $re^{j\omega}$. If $r > (1/2)$, then $2r > 1$ and the function within the summation grows towards infinity with increasing n. Also, the summation does not converge. But if $r < \frac{1}{2}$, then the summation converges.

(c) The summation may be written as

$$\sum_{n=0}^{\infty} \frac{r^{-n} + (-r)^{-n}}{2} e^{-j\omega n}$$

by replacing z with $re^{j\omega}$. If $r > 1$, then the function inside the summation grows towards infinity with increasing n. Also, the summation does not converge. But if $r < 1$, then the summation converges.

(d) The summation may be written as

$$\sum_{n=0}^{\infty} (\frac{1}{2}r^{-1})^n \cos(\pi n/4) e^{-j\omega n} + \sum_{n=-\infty}^{0} (\frac{1}{2}r)^{-n} \cos(\pi n/4) e^{-j\omega n}$$

by replacing z with $re^{j\omega}$. The first summation converges for $r > \frac{1}{2}$. The second summation converges for $r < 2$. Therefore, the sum of these two summations converges for $\frac{1}{2} < r < 2$.

10.2. Using eq. (10.3),

$$\begin{aligned}
X(z) &= \sum_{n=-\infty}^{\infty} \left(\frac{1}{5}\right)^n u[n-3] z^{-n} \\
&= \sum_{n=3}^{\infty} \left(\frac{1}{5}\right)^n z^{-n} \\
&= \left[\frac{z^{-3}}{125}\right] \sum_{n=0}^{\infty} \left(\frac{1}{5}\right)^n z^{-n} \\
&= \left[\frac{z^{-3}}{125}\right] \frac{1}{1 - \frac{1}{5}z^{-1}} \qquad |z| > \frac{1}{5}
\end{aligned}$$

10.3. By using eq. (9.3), we can easily show that

$$\alpha^n u[-n-n_0] \stackrel{z}{\longleftrightarrow} \frac{-z^{-n_0}}{1-\alpha z^{-1}}, \qquad |z| < |\alpha|.$$

We then obtain

$$X(z) = \frac{1}{1+z^{-1}} + \frac{-z^{-n_0-1}}{1-\alpha z^{-1}}, \qquad 1 < |z| < |\alpha|.$$

Therefore, $|\alpha|$ has to be 2. n_0 can take on any value.

10.4. Using eq. (9.3), we have

$$\begin{aligned} X(z) &= \sum_{n=-\infty}^{0} (\frac{1}{3})^n \cos(\frac{\pi}{4}n) z^{-n} \\ &= (1/2) \sum_{n=-\infty}^{0} (\frac{1}{3})^n e^{j\pi n/4} z^{-n} + (1/2) \sum_{n=-\infty}^{0} (\frac{1}{3})^n e^{-j\pi n/4} z^{-n} \\ &= (1/2)\sum_{n=0}^{\infty}(\frac{1}{3})^{-n} e^{-j\pi n/4} z^n + (1/2)\sum_{n=0}^{\infty}(\frac{1}{3})^{-n} e^{j\pi n/4} z^n \\ &= (1/2)\frac{1}{1-3e^{-j\pi/4}z} + (1/2)\frac{1}{1-3e^{j\pi/4}z}, \qquad |z| < \frac{1}{3} \end{aligned}$$

The poles are at $z = \frac{1}{3}e^{j\pi/4}$ and $z = \frac{1}{3}e^{-j\pi/4}$.

10.5. **(a)** The given z-transform may be written as

$$X(z) = \frac{z - \frac{1}{2}}{(z-\frac{1}{3})(z-\frac{1}{4})}.$$

Clearly, $X(z)$ has a zero at $z = \frac{1}{2}$. Since in $X(z)$ the order of the denominator polynomial exceeds the order of the numerator polynomial by 1, $X(z)$ has a zero at ∞. Therefore, $X(z)$ has one zero in the finite z-plane and one zero at infinity.

(b) The given z-transform may be written as

$$X(z) = \frac{(z-1)(z-2)}{(z-3)(z-4)}.$$

Clearly, $X(z)$ has zeros at $z = 1$ and $z = 2$. Since in $X(z)$, the orders of the numerator and denominator polynomials are identical, $X(z)$ has no zeros at infinity. Therefore, $X(z)$ has two zeros in the finite z-plane and no zeros at infinity.

(c) The given z-transform may be written as

$$X(z) = \frac{(z-1)}{z(z-\frac{1}{4})(z+\frac{1}{4})}.$$

Clearly, $X(z)$ has a zero at $z = 1$. Since in $X(z)$ the order of the denominator polynomial exceeds the order of the numerator polynomial by 2, $X(z)$ has two zeros at ∞. Therefore, $X(z)$ has one zero in the finite z-plane and two zeros at infinity.

10.6. (a) **No.** From property 3 in Section 10.2, we know that for a finite-length signal, the ROC is the entire z-plane. Therefore, there can be no poles in the finite z-plane for a finite-length signal. Clearly, in this problem this is not the case.

(b) **No.** Since the signal is absolutely summable, the ROC must include the unit circle. Also, since the signal has a pole at $z = 1/2$, the ROC can never be of the form $0 < |z| < r_0$. From property 5 in Section 10.2, we know that the signal cannot be left sided.

(c) **Yes.** Since the signal is absolutely summable, the ROC must include the unit circle. Since it is given that the signal has a pole at $z = 1/2$, a valid ROC for this signal would be $|z| > 1/2$. From property 4 in Section 10.2 we know that this would correspond to a right-sided signal.

(d) **Yes.** Since the signal is absolutely summable, the ROC must include the unit circle. Clearly, we can define an ROC which is a ring in the z-plane and includes the unit circle. From property 6 in Section 10.2, we know we ma conclude that the signal could be two sided.

10.7. We may find different signals with the given z-transform by choosing different regions of convergence. The poles of the z-transform are

$$z_0 = \frac{1}{2}j, \qquad z_1 = -\frac{1}{2}j, \qquad z_2 = -\frac{1}{2}, \qquad z_4 = \frac{3}{4}.$$

Based on these pole locations, we may choose from the following regions of convergence:
(i) $0 < |z| < \frac{1}{2}$
(ii) $\frac{1}{2} < |z| < \frac{3}{4}$
(iii) $|z| > \frac{3}{4}$
Therefore, we may have 3 different signals with the given z-transform.

10.8. If

$$x[n] \stackrel{Z}{\longleftrightarrow} X(z), \qquad R,$$

then from Table 10.1 we have

$$\left(\frac{1}{4}\right)^n x[n] \stackrel{Z}{\longleftrightarrow} X(4z), \qquad \frac{1}{4}R$$

and

$$\left(\frac{1}{8}\right)^n x[n] \stackrel{Z}{\longleftrightarrow} X(8z), \qquad \frac{1}{8}R.$$

Since $\frac{1}{4}R$ includes the unit circle, and $X(z)$ has a pole at $z = 1/2$, we may conclude that R is definitely outside the circle with radius $1/2$. The only question we now have to answer is whether R extends to infinity outside this circle of radius $1/2$. Since $\frac{1}{8}R$ does not include the unit circle, it is clear that this is not the case. Therefore, R is a ring in the z-plane. From property 6 in Section 10.2 we know that $x[n]$ must be a two-sided signal.

10.9. Using partial-fraction expansion,

$$X(z) = \frac{2/9}{1 - z^{-1}} + \frac{7/9}{1 + 2z^{-1}}, \qquad |z| > 2.$$

Taking the inverse z-transform,

$$x[n] = \frac{2}{9}u[n] + \frac{7}{9}(-2)^n u[n].$$

10.10. We use the approach developed in Example 10.11 to solve this problem.

(a) Since $|z| > \frac{1}{3}$, we may use long division to obtain the power-series expansion of $X(z)$ as shown below.

$$1 + \tfrac{1}{3}z^{-1} \,\big)\, 1 + z^{-1} \,\big(\, 1 + \tfrac{2}{3}z^{-1} + \tfrac{2}{9}z^{-2} + \cdots\cdots$$

$$1 + \tfrac{1}{3}z^{-1}$$

$$\tfrac{2}{3}z^{-1} + \tfrac{2}{9}z^{-2}$$

$$\tfrac{2}{9}z^{-2} + \tfrac{2}{27}z^{-3}$$

$$\vdots$$

Comparing $X(z)$ with the definition of the z-transform in eq. (10.3), we see that

$$x[0] = 1, \qquad x[1] = \frac{2}{3}, \qquad x[2] = -\frac{2}{9}.$$

(b) Since $|z| < \frac{1}{3}$, we may use long division to obtain the power-series expansion of $X(z)$ as shown below.

$$\tfrac{1}{3}z^{-1} + 1 \,\big)\, z^{-1} + 1 \,\big(\, 3 - 6z + 18z^2 \cdots$$

$$z^{-1} + 3$$

$$-2$$

$$-2 - 6z^{-1}$$

$$6z^{-1}$$

$$\vdots$$

Comparing $X(z)$ with the definition of the z-transform in eq. (10.3), we see that

$$x[0] = 3, \qquad x[-1] = -6, \qquad x[-2] = 18.$$

10.11. Since the ROC includes the entire z-plane, we know that the signal must be finite length. From the finite-sum formula, we have

$$\frac{1}{1024}\left[\frac{1024 - z^{10}}{1 - \frac{1}{2}z^{-1}}\right] = \sum_{n=0}^{9}\left(\frac{1}{2}\right)^n z^{-n}.$$

Comparing this with the definition of the z-transform in eq. (10.3), we obtain

$$x[n] = \begin{cases} (\frac{1}{2})^n, & 0 \le n \le 9 \\ 0, & \text{otherwise} \end{cases}$$

10.12. The pole-zero plots for each of the three z-transforms is as shown in Figure S10.12.

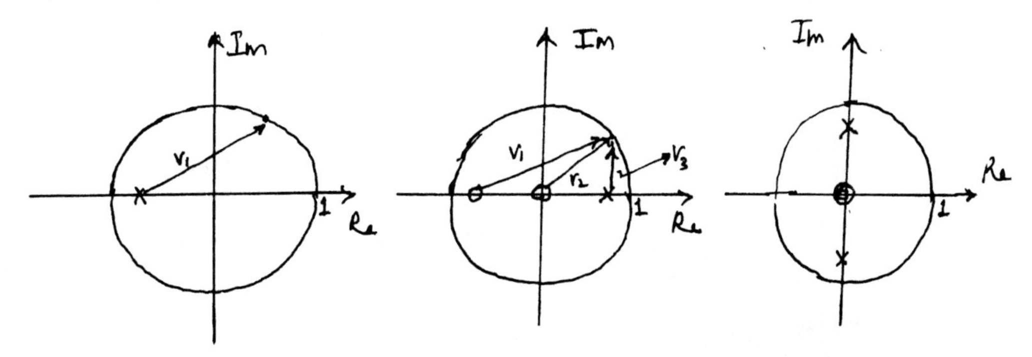

Figure S10.12

(a) From Section 10.4, we know that the magnitude of the Fourier transform may be expressed as

$$|H_1(e^{j\omega})| = \frac{1}{\text{Length of } \vec{v_1}},$$

where $\vec{v_1}$ is as shown in the figure above. Clearly, for small values of ω (ω near zero), the right-hand side of the above equation is small. But as ω approaches π, the right-hand side of the above equation becomes large. Therefore, $H_1(e^{j\omega})$ is approximately highpass.

(b) From Section 10.4, we know that the magnitude of the Fourier transform may be expressed as

$$|H_2(e^{j\omega})| = \frac{(\text{Length of } \vec{v_1})(\text{Length of } \vec{v_2})}{(\text{Length of } \vec{v_3})^2},$$

where $\vec{v_1}$, $\vec{v_1}$, and $\vec{v_3}$ are as shown in the figure above. Clearly, for small values of ω (ω near zero), the numerator of the right-hand side of the above equation is much larger than the denominator. Therefore, $H_1(e^{j\omega})$ is large near $\omega = 0$. But as ω approaches π, the denominator of the right-hand side of the above equation is much larger than the numerator. Therefore, $H_2(e^{j\omega})$ is small near $\omega = \pi$. Therefore, $H_2(e^{j\omega})$ is approximately lowpass.

(c) From Section 10.4, we know that the magnitude of the Fourier transform may be expressed as

$$|H_3e^{j\omega})| = \frac{(\text{Length of } \vec{v_1})^2}{(\text{Length of } \vec{v_2})(\text{Length of } \vec{v_3})},$$

where $\vec{v_1}$, $\vec{v_1}$, and $\vec{v_3}$ are as shown in the figure above. Clearly, for small values of ω (ω near zero), and for values of ω near π the numerator of the right-hand side of the above equation is almost the same as the denominator. But when $|\omega|$ is near $\pi/2$, the numerator of the right-hand side of the above equation is much larger than the denominator. Therefore, $H_1(e^{j\omega})$ is large near $\omega = \pi/2$. Therefore, $H_1(e^{j\omega})$ is approximately bandpass.

10.13. (a) The signal $g[n]$ is

$$g[n] = \delta[n] - \delta[n-6].$$

Using the definition of the z-transform in eq. (10.3), we obtain

$$G(z) = 1 - z^{-6}, \qquad |z| > 0.$$

(b) From Table 10.1, we have

$$x[n] = \sum_{k=-\infty}^{n} g[k] \stackrel{Z}{\longleftrightarrow} X(z) = \frac{1}{1-z^{-1}} G(z), \qquad \text{At least} |z| > 1.$$

Therefore,

$$X(z) = \frac{1-z^{-6}}{1-z^{-1}}, \qquad |z| > 0.$$

The ROC is $|z| > 0$ because $x[n]$ is a finite-length signal.

10.14. (a) We know that $x[n] * x[n]$ will be triangular signal whose first non-zero value occurs at $n = 0$. Furthermore, we also know that $x[n] * x[n - n_0]$ has its first nonzero value at $n = n_0$. Therefore, $n_0 = 2$.

(b) From Problem 10.13 we have

$$X(z) = \frac{1-z^{-6}}{1-z^{-1}}, \qquad |z| > 0.$$

Using the shift property,

$$x[n-2] \stackrel{Z}{\longleftrightarrow} z^{-2}\frac{1-z^{-6}}{1-z^{-1}}, \qquad |z| > 0.$$

Using the convolution property,

$$g[n] = x[n] * x[n-1] \stackrel{Z}{\longleftrightarrow} z^{-2}\left(\frac{1-z^{-6}}{1-z^{-1}}\right)^2, \qquad |z| > 0.$$

Since

$$\lim_{z\to\infty} G(z) = 0 = g[0],$$

$G(z)$ does satisfy the initial value theorem.

10.15. Taking the z-transform of $y[n]$, we have

$$Y(z) = \frac{1}{1 - \frac{1}{9}z^{-1}}, \qquad |z| > \frac{1}{9}.$$

Now from Table 10.1, we have

$$y_1[n] = y_{(2)}[n] = \begin{cases} y[r], & ,n = 2r \\ 0, & ,n \neq 2r \end{cases} \quad \xleftrightarrow{z} Y_1(z) = Y(z^2), \qquad |z| > \frac{1}{3}.$$

Therefore,

$$y_1[0] = 1,\ y_1[1] = 0,\ y_1[2] = \frac{1}{9},\ y_1[3] = 0,\ y_1[4] = \frac{1}{81}, \cdots$$

This may be written as

$$y_1[n] = \frac{1}{2}\left[\left(\frac{1}{3}\right)^n u[n] + (-1)^n \left(\frac{1}{3}\right)^n u[n]\right].$$

If we now choose $x[n]$ to be $\frac{1}{2}\left[\left(\frac{1}{3}\right)^n u[n]\right]$, then

$$Y_1(z) = Y(z^2) = (1/2)[X(z) + X(-z)], \qquad |z| > \frac{1}{3}.$$

Furthermore, since $X(z)$ has only one pole and one zero, this choice of $x[n]$ satisfies both the given conditions.

We may also choose $x[n]$ to be $\frac{1}{2}\left[(-1)^n \left(\frac{1}{3}\right)^n u[n]\right]$. This would still satisfy both given conditions.

10.16. For a system to be both causal and stable, the corresponding z-transform must not have any poles outside the unit circle.

(a) The given z-transform has a pole at infinity. Therefore, it is **not causal**.

(b) The poles of this z-transform are at $z = \frac{1}{4}$ and $z = -\frac{3}{4}$. Therefore, it is **causal**.

(c) This z-transform has a pole at $-\frac{4}{3}$. Therefore, it is **not causal**.

10.17. **(a)** Since $\lim_{z\to\infty} = 1$, $H(z)$ has no poles at infinity. Furthermore, since $h[n]$ is given to be right-sided, $h[n]$ has to be causal.

(b) Since $h[n]$ is causal, the numerator and denominator polynomials of $H(z)$ have the same order. Since $H(z)$ is given to have two zeros, we may conclude that it also has two poles.

Since $h[n]$ is real, the poles must occur in conjugate pairs. Also, it is given that one of the poles lies on the circle defined by $|z| = \frac{3}{4}$. Therefore, the other pole also lies on the same circle.

Clearly, the ROC for $H(z)$ will be of the form $|z| > \frac{3}{4}$. and will include the unit circle. Therefore, we may conclude that the system is stable.

10.18. **(a)** Using the analysis of Example 10.28, we may show that

$$H(z) = \frac{1 - 6z^{-1} + 8z^{-2}}{1 - \frac{2}{3}z^{-2} + \frac{1}{9}z^{-2}}.$$

Since $H(z) = Y(z)/X(z)$, we may write

$$Y(z)[1 - \frac{2}{3}z^{-1} + \frac{1}{9}z^{-2}] = X(z)[1 - 6z^{-1} + 8z^{-2}].$$

Taking the inverse z-transform we obtain

$$y[n] - \frac{2}{3}y[n-1] + \frac{1}{9}y[n-2] = x[n] - 6x[n-1] + 8x[n-2].$$

(b) $H(z)$ has only two poles. These are both at $z = \frac{1}{3}$. Since the system is causal, the ROC of $H(z)$ will be of the form $|z| > \frac{1}{3}$. Since the ROC includes the unit circle, the system is stable.

10.19. **(a)** The unilateral z-transform is

$$\begin{aligned} \mathcal{X}(z) &= \sum_{n=0}^{\infty}(\frac{1}{4})^n u[n+5]z^{-n} \\ &= \sum_{n=0}^{\infty}(\frac{1}{4})^n z^{-n} \\ &= \frac{1}{1-(1/4)z^{-1}}, \qquad |z| > \frac{1}{4} \end{aligned}$$

(b) The unilateral z-transform is

$$\begin{aligned} \mathcal{X}(z) &= \sum_{n=0}^{\infty}(\delta[n+3] + \delta[n] + 2^n[-n])z^{-n} \\ &= \sum_{n=0}^{\infty}(0 + \delta[n] + \delta[n])z^{-n} \\ &= 2, \qquad \text{All } z \end{aligned}$$

The unilateral z-transform is

$$\begin{aligned} \mathcal{X}(z) &= \sum_{n=0}^{\infty}(\frac{1}{2})^{|n|} z^{-n} \\ &= \sum_{n=0}^{\infty}(\frac{1}{2})^n z^{-n} \\ &= \frac{1}{1-(1/2)z^{-1}}, \qquad |z| > \frac{1}{2} \end{aligned}$$

10.20. Applying the unilateral z-transform to the given difference equation, we have

$$z^{-1}\mathcal{Y}(z) + y[-1] + 2\mathcal{Y}(z) = \mathcal{X}(z).$$

(a) For the zero-input response, assume that $x[n] = 0$. Since we are given that $y[-1] = 2$,

$$z^{-1}\mathcal{Y}(z) + y[-1] + 2\mathcal{Y}(z) = 0 \Rightarrow \mathcal{Y}(z) = \frac{-1}{1 + (1/2)z^{-1}}.$$

Taking the inverse unilateral z-transform,

$$y[n] = -\left(-\frac{1}{2}\right)^n u[n].$$

(b) For the zero-state response, set $y[-1] = 0$. Also, we have

$$\mathcal{X}(z) = \mathcal{UZ}\{(1/2)^n u[n]\} = \frac{1}{1 - \frac{1}{2}z^{-1}}, \qquad |z| > \frac{1}{2}.$$

Therefore,

$$\mathcal{Y}(z) = \left(\frac{1}{1 - \frac{1}{4}z^{-1}}\right)\left(\frac{2}{2 + z^{-1}}\right).$$

We use partial fraction expansion followed by the inverse unilateral z-transform to obtain

$$y[n] = \frac{1}{3}\left(-\frac{1}{2}\right)^n u[n] + \frac{1}{6}\left(\frac{1}{4}\right)^n u[n].$$

(c) The total response is the sum of the zero-state and zero-input responses. This is

$$y[n] = -\frac{2}{3}\left(-\frac{1}{2}\right)^n u[n] + \frac{1}{6}\left(\frac{1}{4}\right)^n u[n].$$

10.21. The pole-zero plots are all shown in Figure S10.21.

(a) For $x[n] = \delta[n+5]$,

$$X(z) = z^5, \qquad \text{All } z.$$

The Fourier transform exists because the ROC includes the unit circle.

(b) For $x[n] = \delta[n-5]$,

$$X(z) = z^{-5}, \qquad \text{All } z \text{ except } 0.$$

The Fourier transform exists because the ROC includes the unit circle.

(c) For $x[n] = (-1)^n u[n]$,

$$\begin{aligned} X(z) &= \sum_{n=-\infty}^{\infty} x[n] z^{-n} \\ &= \sum_{n=0}^{\infty} (-1)^n z^{-n} \\ &= 1/(1 + z^{-1}), \qquad |z| > 1 \end{aligned}$$

The Fourier transform does not exist because the ROC does not include the unit circle.

(d) For $x[n] = (1/2)^{n+1}u[n+3]$,

$$\begin{aligned} X(z) &= \sum_{n=-\infty}^{\infty} x[n]z^{-n} \\ &= \sum_{n=-3}^{\infty} (1/2)^{n+1} z^{-n} \\ &= \sum_{n=0}^{\infty} (1/2)^{n-2} z^{-n+3} \\ &= 4z^3/(1-(1/2)z^{-1}), \qquad |z| > 1/2 \end{aligned}$$

The Fourier transform exists because the ROC includes the unit circle.

(e) For $x[n] = (-1/3)^n u[-n-2]$,

$$\begin{aligned} X(z) &= \sum_{n=-\infty}^{\infty} x[n]z^{-n} \\ &= \sum_{n=-\infty}^{-2} (-1/3)^n z^{-n} \\ &= \sum_{n=2}^{\infty} (-1/3)^{-n} z^{n} \\ &= \sum_{n=0}^{\infty} (-1/3)^{-n-2} z^{n+2} \\ &= 9z^2/(1+3z), \qquad |z| < 1/3 \\ &= 3z/(1+(1/3)z^{-1}), \qquad |z| < 1/3 \end{aligned}$$

The Fourier transform does not exist because the ROC does not include the unit circle.

(f) For $x[n] = (1/4)^n u[-n+3]$,

$$\begin{aligned} X(z) &= \sum_{n=-\infty}^{\infty} x[n]z^{-n} \\ &= \sum_{n=-\infty}^{3} (1/4)^n z^{-n} \\ &= \sum_{n=-3}^{\infty} (1/4)^{-n} z^{n} \\ &= \sum_{n=0}^{\infty} (1/4)^{-n+3} z^{n-3} \\ &= (1/64)z^{-3}/(1-4z), \qquad |z| < 1/4 \\ &= (1/16)z^{-4}/(1-(1/4)z^{-1}), \qquad |z| < 1/4 \end{aligned}$$

The Fourier transform does not exist because the ROC does not include the unit circle.

(g) Consider $x_1[n] = 2^n u[-n]$.

$$\begin{aligned}
X_1(z) &= \sum_{n=-\infty}^{\infty} x_1[n]z^{-n} \\
&= \sum_{n=-\infty}^{0} (2)^n z^{-n} \\
&= \sum_{n=0}^{\infty} (2)^{-n} z^n \\
&= 1/(1-(1/2)z), \qquad |z| < 2 \\
&= -2z^{-1}/(1-2z^{-1}), \qquad |z| < 2
\end{aligned}$$

Consider $x_2[n] = (1/4)^n u[n-1]$.

$$\begin{aligned}
X_2(z) &= \sum_{n=-\infty}^{\infty} x_2[n]z^{-n} \\
&= \sum_{n=1}^{\infty} (1/4)^n z^{-n} \\
&= \sum_{n=0}^{\infty} (1/4)^{n+1} z^{-n-1} \\
&= (z^{-1}/4)[1/(1-(1/4)z^{-1})], \qquad |z| > 1/4
\end{aligned}$$

The z-transform of the overall sequence $x[n] = x_1[n] + x_2[n]$ is

$$X(z) = -\frac{2z^{-1}}{(1-2z^{-1})} + \frac{z^{-1}/4}{1-(1/4)z^{-1}}, \qquad (1/4) < |z| < 2.$$

The Fourier transform exists because the ROC includes the unit circle.

(h) Consider $x[n] = (1/3)^{n-2}u[n-2]$.

$$\begin{aligned}
X(z) &= \sum_{n=-\infty}^{\infty} x[n]z^{-n} \\
&= \sum_{n=2}^{\infty} (1/3)^{n-2} z^{-n} \\
&= \sum_{n=0}^{\infty} (1/3)^n z^{-n-2} \\
&= z^{-2}[1/(1-(1/3)z^{-1})], \qquad |z| > 1/3
\end{aligned}$$

The Fourier transform exists because the ROC includes the unit circle.

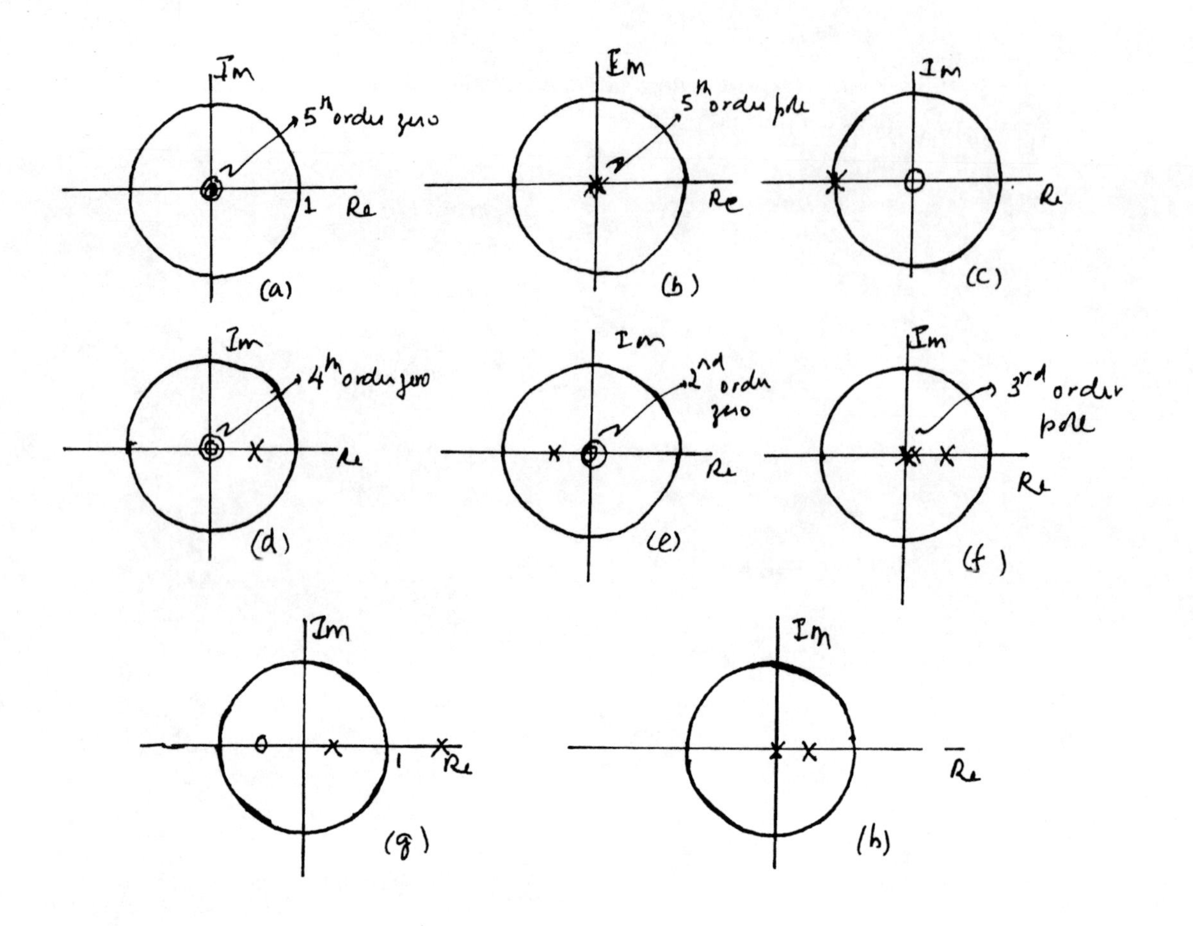

Figure S10.21

10.22. (a) Using the z-transform analysis equation,

$$\begin{aligned} X(z) &= (1/2)^{-4}z^4 + (1/2)^{-3}z^3 + (1/2)^{-2}z^2 + (1/2)^{-1}z^1 + (1/2)^0 z^0 \\ &+ (1/2)^1 z^{-1} + (1/2)^2 z^{-2} + (1/2)^3 z^{-3} + (1/2)^4 z^{-4} \end{aligned}$$

This may be expressed as

$$X(z) = (1/2)^{-4}z^4 \left[\frac{1 - (1/2)^9 z^{-9}}{1 - (1/2)z^{-1}} \right].$$

This has four zeros at $z = 0$ and 8 more zeros distributed on a circle of radius $1/2$. The ROC is the entire z plane. (Although from an inspection of the expression for $X(z)$ it seems like there is a pole at $1/2$, note that there is also a zero at $1/2$ which cancels with this pole.) Since the ROC includes the unit circle, the Fourier transform exists.

(b) Consider the sequence $x_1[n] = (1/2)^{|n|}$. This may be written as

$$x_1[n] = (1/2)^n u[n] + 2^n u[-n-1].$$

Now,

$$(1/2)^n u[n] \stackrel{z}{\longleftrightarrow} \frac{1}{1-(1/2)z^{-1}}, \qquad |z| > (1/2)$$

and

$$(2)^n u[-n-1] \stackrel{z}{\longleftrightarrow} -\frac{1}{1-2z^{-1}}, \qquad |z| < 2.$$

Therefore,

$$X_1(z) = \frac{1}{1-(1/2)z^{-1}} - \frac{1}{1-2z^{-1}}, \qquad (1/2) < |z| < 2.$$

Note that $x[n] = nx_1[n]$. Therefore,

$$X(z) = -z\frac{d}{dz}X_1(z) = -\frac{(1/2)z^{-1}}{(1-(1/2)z^{-1})^2} + \frac{2z^{-1}}{(1-2z^{-1})^2}.$$

The ROC is $(1/2) < |z| < 2$. Therefore, the Fourier transform exists.

(c) Write $x[n]$ as

$$x[n] = n(1/2)^n u[n] - n2^n u[-n-1] = nx_1[n] - nx_2[n]$$

where

$$x_1[n] = (1/2)^n u[n] \stackrel{z}{\longleftrightarrow} X_1(z) = \frac{1}{1-(1/2)z^{-1}}, \qquad |z| > (1/2)$$

and

$$x_2[n] = (2)^n u[-n-1] \stackrel{z}{\longleftrightarrow} X_2(z) = -\frac{1}{1-2z^{-1}}, \qquad |z| < 2.$$

Using the differentiation property, we get

$$X(z) = -z\frac{d}{dz}X_1(z) + z\frac{d}{dz}X_2(z) = -\frac{(1/2)z^{-1}}{(1-(1/2)z^{-1})^2} - \frac{2z^{-1}}{(1-2z^{-1})^2}.$$

The ROC is $(1/2) < |z| < 2$. Therefore, the Fourier transform exists.

(d) The sequence may be written as

$$x[n] = 4^n \left\{ \frac{e^{j[(2\pi n/6)+(\pi/4)]} + e^{-j[(2\pi n/6)+(\pi/4)]}}{2} \right\} u[-n-1].$$

Now,

$$4^n e^{j[(2\pi n/6)+(\pi/4)]} u[-n-1] \stackrel{z}{\longleftrightarrow} \frac{e^{j\pi/4}}{2} \frac{1}{1-4e^{j2\pi/6}z^{-1}}, \qquad |z| < 4$$

and

$$4^n e^{-j[(2\pi n/6)+(\pi/4)]} u[-n-1] \stackrel{z}{\longleftrightarrow} \frac{e^{-j\pi/4}}{2} \frac{1}{1-4e^{-j2\pi/6}z^{-1}}, \qquad |z| < 4$$

Therefore,

$$X(z) = \frac{e^{j\pi/4}}{2} \frac{1}{1-4e^{j2\pi/6}z^{-1}} + \frac{e^{-j\pi/4}}{2} \frac{1}{1-4e^{-j2\pi/6}z^{-1}}, \qquad |z| < 4.$$

The ROC is $|z| < 4$. Therefore, the Fourier transform exists.

10.23. (i) The partial fraction expansion of the given $X(z)$ is

$$X(z) = \frac{-1/2}{1-\frac{1}{2}z^{-1}} + \frac{3/2}{1+\frac{1}{2}z^{-1}}.$$

Since the ROC is $|z| > 1/2$,

$$x[n] = -\frac{1}{2}\left(\frac{1}{2}\right)^n u[n] + \frac{3}{2}\left(-\frac{1}{2}\right)^n u[n].$$

Performing long-division in order to get a right-sided sequence, we obtain

$$X(z) = 1 - z^{-1} + \frac{1}{4}z^{-2} - \frac{1}{4}z^{-3} + \frac{1}{16}z^{-4} - \frac{1}{16}z^{-5} + - \cdots.$$

This may be rewritten as

$$\begin{aligned} X(z) &= \frac{3}{2}[1 - \frac{1}{2}z^{-1} + \frac{1}{4}z^{-2} - \frac{1}{8}z^{-3} + - \cdots] \\ &- \frac{1}{2}[1 + \frac{1}{2}z^{-1} + \frac{1}{4}z^{-2} + \frac{1}{8}z^{-3} + \cdots]. \end{aligned}$$

Therefore,

$$x[n] = -\frac{1}{2}\left(\frac{1}{2}\right)^n u[n] + \frac{3}{2}\left(-\frac{1}{2}\right)^n u[n].$$

(ii) The partial fraction expansion of the given $X(z)$ is

$$X(z) = \frac{-1/2}{1-\frac{1}{2}z^{-1}} + \frac{3/2}{1+\frac{1}{2}z^{-1}}.$$

Since the ROC is $|z| < 1/2$,

$$x[n] = \frac{1}{2}\left(\frac{1}{2}\right)^n u[-n-1] - \frac{3}{2}\left(-\frac{1}{2}\right)^n u[-n-1].$$

Performing long-division in order to get a left-sided sequence, we obtain

$$X(z) = 4z - 4z^2 + 16z^3 - 16z^4 + 64z^5 - 64z^6 + - \cdots.$$

This may be rewritten as

$$\begin{aligned} X(z) &= \frac{3}{2}[2z - 4z^2 + 8z^3 - 16z^4 + - \cdots] \\ &+ \frac{1}{2}[2z + 4z^2 + 8z^3 + 16z^4 + \cdots]. \end{aligned}$$

Therefore,

$$x[n] = \frac{1}{2}\left(\frac{1}{2}\right)^n u[-n-1] - \frac{3}{2}\left(-\frac{1}{2}\right)^n u[-n-1].$$

(iii) The partial fraction expansion of the given $X(z)$ is

$$X(z) = -2 + \frac{3/2}{1-(1/2)z^{-1}}.$$

Since the ROC is $|z| > 1/2$,

$$x[n] = -2\delta[n] + \frac{3}{2}\left(\frac{1}{2}\right)^n u[n].$$

Performing long-division in order to get a right-sided sequence, we obtain

$$X(z) = -\frac{1}{2} + \frac{3}{4}z^{-1} + \frac{3}{8}z^{-2} + \frac{3}{16}z^{-3} + \cdots.$$

This may be rewritten as

$$X(z) = -2 + \frac{3}{2}[1 + \frac{1}{2}z^{-1} + \frac{1}{4}z^{-2} + \cdots].$$

Therefore,

$$x[n] = -2\delta[n] + \frac{3}{2}\left(\frac{1}{2}\right)^n u[n].$$

(iv) The partial fraction expansion of the given $X(z)$ is

$$X(z) = -2 + \frac{3/2}{1-(1/2)z^{-1}}.$$

Since the ROC is $|z| < 1/2$,

$$x[n] = -2\delta[n] - \frac{3}{2}\left(\frac{1}{2}\right)^n u[-n-1].$$

Performing long-division in order to get a left-sided sequence, we obtain

$$X(z) = -2 - 3z - 6z^2 - 12z^3 - 24z^4 - \cdots.$$

This may be rewritten as

$$X(z) = -2 - \frac{3}{2}[2z + 4z^2 + 8z^3 + 16z^4 + \cdots].$$

Therefore,

$$x[n] = -2\delta[n] - \frac{3}{2}\left(\frac{1}{2}\right)^n u[-n-1].$$

(v) We may similarly show that in this case,

$$x[n] = 2n(1/2)^n u[n] - n(1/2)^{n+1} u[n+1].$$

(vi) We may similarly show that in this case,

$$x[n] = -2n(1/2)^n u[-n-1] + n(1/2)^{n+1} u[-n-2].$$

10.24. **(a)** We may write $X(z)$ as

$$X(z) = \frac{1 - 2z^{-1}}{(1 - \frac{1}{2}z^{-1})(1 - 2z^{-1})}.$$

Therefore,

$$X(z) = \frac{1}{1 - \frac{1}{2}z^{-1}}.$$

If $x[n]$ is absolutely summable, then the ROC of $X(z)$ has to include the unit circle. Therefore, the ROC is $|z| > 1/2$. It follow that

$$x[n] = \left(\frac{1}{2}\right)^2 u[n].$$

(b) Carrying out long division on $X(z)$, we get

$$X(z) = 1 - z^{-1} + \frac{1}{2}z^{-2} - \frac{1}{4}z^{-2} + - \cdots.$$

Using the analysis equation (10.3), we get

$$x[n] = \delta[n] - \left(-\frac{1}{2}\right)^{n-1} u[n-1].$$

(c) We may write $X(z)$ as

$$X(z) = \frac{3z^{-1}}{1 - \frac{1}{4}z^{-1} - \frac{1}{8}z^{-2}} = \frac{3z^{-1}}{(1 - \frac{1}{2}z^{-1})(1 + \frac{1}{4}z^{-1})}.$$

The partial fraction expansion of $X(z)$ is

$$X(z) = \frac{4}{1 - \frac{1}{2}z^{-1}} - \frac{4}{1 + \frac{1}{4}z^{-1}}.$$

Since $x[n]$ is absolutely summable, the ROC must be $|z| > 1/2$ in order to include the unit circle. It follows that

$$x[n] = 4\left(\frac{1}{2}\right)^2 u[n] - 4\left(-\frac{1}{4}\right)^2 u[n].$$

10.25. **(a)** The partial fraction expansion of $X(z)$ is

$$X(z) = -\frac{1}{1 - \frac{1}{2}z^{-1}} + \frac{2}{1 - z^{-1}}.$$

Since $x[n]$ is right-sided, the ROC has to be $|z| > 1$. Therefore, it follows that

$$x[n] = -\left(\frac{1}{2}\right)^2 u[n] + 2u[n].$$

(b) $X(z)$ may be rewritten as

$$X(z) = \frac{z^2}{(z-\frac{1}{2})(z-1)}.$$

Using partial fraction expansion, we may rewrite this as

$$\begin{aligned} X(z) &= 2z^2\left[-\frac{1}{z-\frac{1}{2}} + \frac{1}{z-1}\right] \\ &= 2z\left[-\frac{z}{z-\frac{1}{2}} + \frac{z}{z-1}\right] \end{aligned}$$

If $x[n]$ is right-sided, then the ROC for this signal is $|z| > 1$. Using this fact, we may find the inverse z-transform of the term within square brackets above to be $y[n] = -(1/2)^n u[n] + u[n]$. Note that $X(z) = 2zX(z)$. Therefore, $x[n] = 2y[n+1]$. This gives

$$x[n] = -2\left(\frac{1}{2}\right)^{n+1} u[n+1] + 2u[n+1].$$

Noting that $x[-1] = 0$, we may rewrite this as

$$x[n] = -\left(\frac{1}{2}\right)^n u[n] + 2u[n].$$

This is the answer that we obtained in part (a).

10.26. (a) From part (b) of the previous problem,

$$X(z) = \frac{z^2}{(z-\frac{1}{2})(z-1)}.$$

(b) From part (b) of the previous problem,

$$X(z) = 2z\left[-\frac{z}{z-\frac{1}{2}} + \frac{z}{z-1}\right].$$

(c) If $x[n]$ is left-sided, then the ROC for this signal is $|z| < 1/2$. Using this fact, we may find the inverse z-transform of the term within square brackets above to be $y[n] = (1/2)^n u[-n-1] - u[-n-1]$. Note that $X(z) = 2zX(z)$. Therefore, $x[n] = 2y[n+1]$. This gives

$$x[n] = 2\left(\frac{1}{2}\right)^{n+1} u[-n-2] - 2u[-n-2].$$

10.27. We perform long-division on $X(z)$ so as to obtain a right-sided sequence. This gives us

$$X(z) = z^3 + 4z^2 + 5z + \cdots.$$

Therefore, comparing this with eq. (10.3) we get

$$x[-3] = 1, \qquad x[-2] = 4, \qquad x[-1] = 5,$$

and $x[n] = 0$ for $n < -3$.

10.28. **(a)** Using eq. (10.3), we get

$$X(z) = 1 - 0.95z^{-6} = \frac{z^6 - 0.95}{z^6}.$$

(b) Therefore, $X(z)$ has six zeros lying on a circle of radius 0.95 (as shown in Figure S10.28) and 6 poles at $z = 0$.

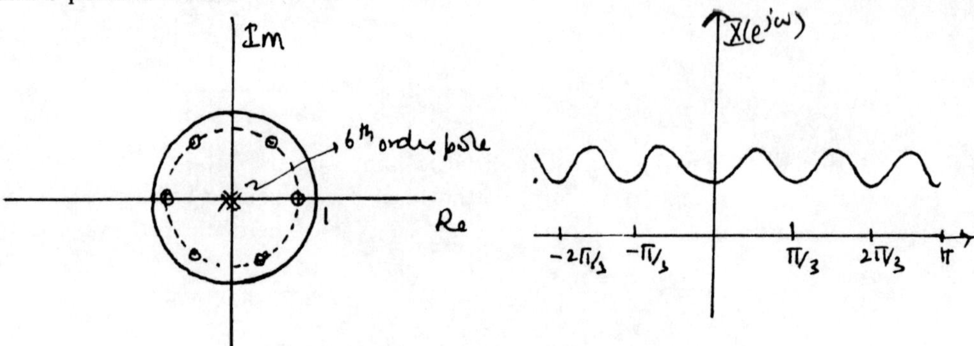

Figure S10.28

(c) The magnitude of the Fourier transform is as shown in Figure S10.28.

10.29. The plots are as shown in Figure S10.29.

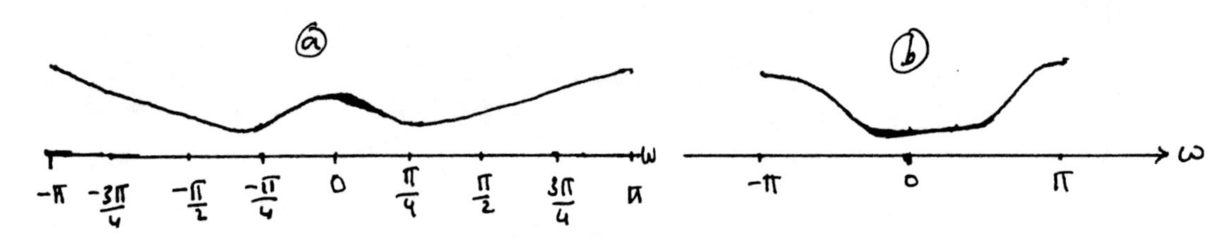

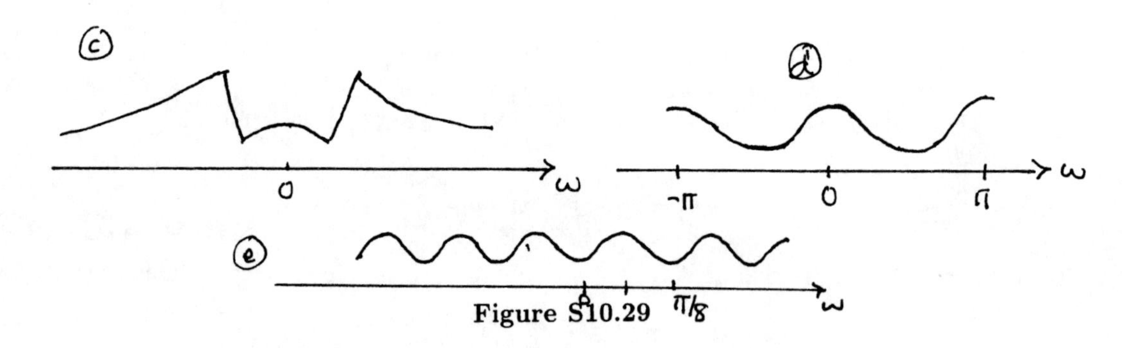

Figure S10.29

10.30. From the given information, we have

$$x_1[n] \overset{z}{\longleftrightarrow} X_1(z) = \frac{1}{1 - \frac{1}{2}z^{-1}}, \qquad |z| > \frac{1}{2}$$

and

$$x_2[n] \stackrel{\mathcal{Z}}{\longleftrightarrow} X_2(z) = \frac{1}{1 - \frac{1}{3}z^{-1}}, \qquad |z| > \frac{1}{3}.$$

Using the time shifting property, we get

$$x_1[n+3] \stackrel{\mathcal{Z}}{\longleftrightarrow} z^3 X_1(z), \qquad |z| > \frac{1}{2}.$$

Using the time reversal and shift properties, we get

$$x_2[-n+1] \stackrel{\mathcal{Z}}{\longleftrightarrow} z^{-1} X_2(z^{-1}), \qquad |z| < 3.$$

Now, using the convolution property, we get

$$y[n] = x_1[n+3] * x_2[-n+1] \stackrel{\mathcal{Z}}{\longleftrightarrow} Y(z) = z^2 X_1(z) X_2(z^{-1}), \qquad \frac{1}{2} < |z| < 3.$$

Therefore,

$$Y(z) = \frac{z^2}{(1 - \frac{1}{2}z^{-1})(1 - \frac{1}{3}z)}.$$

10.31. From Clue 1, we know that $x[n]$ is real. Therefore, the poles and zeros of $X(z)$ have to occur in conjugate pairs. Since Clue 4 tells us that $X(z)$ has a pole at $z = (1/2)e^{j\pi/3}$, we can conclude that $X(z)$ must have another pole at $z = (1/2)e^{-j\pi/3}$. Now, since $X(z)$ has no more poles, we have to assume that $X(z)$ has 2 or less zeros. If $X(z)$ has more than 2 zeros, then $X(z)$ must have poles at infinity. Since Clue 3 tells us that $X(z)$ has 2 zeros at the origin, we know that $X(z)$ must be of the form

$$X(z) = \frac{Az^2}{(z - \frac{1}{2}e^{j\pi/3})(z - \frac{1}{2}e^{-j\pi/3})}.$$

Since Clue 5 tells us that $X(1) = 8/3$, we may conclude that $A = 2$. Therefore,

$$X(z) = \frac{2z^2}{(z - \frac{1}{2}e^{j\pi/3})(z - \frac{1}{2}e^{-j\pi/3})}.$$

Since $x[n]$ is right-sided, the ROC must be $|z| > 1/3$.

10.32. **(a)** We are given that $h[n] = a^n u[n]$ and $x[n] = u[n] - u[n-N]$. Therefore,

$$\begin{aligned} y[n] &= x[n] * h[n] \\ &= \sum_{k=-\infty}^{\infty} h[n-k]x[k] \\ &= \sum_{k=0}^{N-1} a^{n-k} u[n-k] \end{aligned}$$

Now, $y[n]$ may be evaluated to be

$$y[n] = \begin{cases} 0, & n < 0 \\ \sum_{k=0}^{n} a^n a^{-k}, & 0 \le n \le N-1 \\ \sum_{k=0}^{N-1} a^n a^{-k}, & n > N-1 \end{cases}.$$

Simplifying,

$$y[n] = \begin{cases} 0, & n < 0 \\ (a^n - a^{-1})/(1 - a^{-1}), & 0 \le n \le N-1 \\ a^n(1 - a^{-N})/(1 - a^{-1}), & n > N-1 \end{cases}.$$

(b) Using Table 10.2, we get

$$H(z) = \frac{1}{1 - az^{-1}}, \qquad |z| > |a|$$

and

$$X(z) = \frac{1 - z^{-N}}{1 - z^{-1}}, \qquad \text{All} z.$$

Therefore,

$$Y(z) = X(z)H(z) = \frac{1}{(1 - z^{-1})(1 - az^{-1})} - \frac{z^{-N}}{(1 - z^{-1})(1 - az^{-1})}.$$

The ROC is $|z| > |a|$. Consider

$$P(z) = \frac{1}{(1 - z^{-1})(1 - az^{-1})}$$

with ROC $|z| > |a|$. The partial fraction expansion of $P(z)$ is

$$P(z) = \frac{1/(1-a)}{1 - z^{-1}} + \frac{1/(1 - a^{-1})}{1 - az^{-1}}.$$

Therefore,

$$p[n] = \frac{1}{1-a} u[n] + \frac{1}{1 - a^{-1}} a^n u[n].$$

Now, note that

$$Y(z) = P(z)[1 - z^{-N}].$$

Therefore,

$$y[n] = p[n] - p[n-N] = \frac{1}{1-a}\{u[n] - u[n-N]\} + \frac{1}{1-a^{-1}}\{a^n u[n] - a^{n-N} u[n-N]\}.$$

This may be written as

$$y[n] = \begin{cases} 0, & n < 0 \\ (a^n - a^{-1})/(1 - a^{-1}), & 0 \le n \le N-1 \\ a^n(1 - a^{-N})/(1 - a^{-1}), & n > N-1 \end{cases}.$$

This is the same as the result of part (a).

10.33. (a) Taking the z-transform of both sides of the given difference equation and simplifying, we get

$$H(z) = \frac{Y(z)}{X(z)} = \frac{1}{1 - \frac{1}{2}z^{-1} + \frac{1}{4}z^{-2}}.$$

The poles of $H(z)$ are at $(1/4) \pm j(\sqrt{3}/4)$. Since $h[n]$ is causal, the ROC has to be $|z| > |(1/4) + j(\sqrt{3}/4)| = (1/2)$.

(b) We have

$$X(z) = \frac{1}{1 - \frac{1}{2}z^{-1}}, \qquad |z| > \frac{1}{2}.$$

Therefore,

$$Y(z) = H(z)X(z) = \frac{1}{(1 - \frac{1}{2}z^{-1})(1 - \frac{1}{2}z^{-1} + \frac{1}{4}z^{-2})}.$$

The ROC of $Y(z)$ will be the intersection of the ROCs of $X(z)$ and $H(z)$. This implies that the ROC of $Y(z)$ is $|z| > 1/2$. The partial fraction expansion of $Y(z)$ is

$$Y(z) = \frac{1}{1 - \frac{1}{2}z^{-1}} + \frac{z^{-1}/2}{1 - \frac{1}{2}z^{-1} + \frac{1}{4}z^{-2}}.$$

Using Table 10.2 we get

$$y[n] = \left(\frac{1}{2}\right)^n u[n] + \frac{2}{\sqrt{3}}\left(\frac{1}{2}\right)^n \sin\left(\frac{\pi n}{3}\right) u[n].$$

10.34. (a) Taking the z-transform of both sides of the given difference equation and simplifying, we get

$$H(z) = \frac{Y(z)}{X(z)} = \frac{z^{-1}}{1 - z^{-1} - z^{-2}}.$$

The poles of $H(z)$ are at $z = (1/2) \pm (\sqrt{5}/2)$. $H(z)$ has a zero at $z = 0$. The pole-zero plot for $H(z)$ is as shown in Figure S10.34. Since $h[n]$ is causal, the ROC for $H(z)$ has to be $|z| > (1/2) + (\sqrt{5}/2)$.

(b) The partial fraction expansion of $H(z)$ is

$$H(z) = -\frac{1/\sqrt{5}}{1 - (\frac{1+\sqrt{5}}{2})z^{-1}} + \frac{1/\sqrt{5}}{1 - (\frac{1-\sqrt{5}}{2})z^{-1}}.$$

Therefore,

$$h[n] = -\frac{1}{\sqrt{5}}\left(\frac{1+\sqrt{5}}{2}\right)^n u[n] + \frac{1}{\sqrt{5}}\left(\frac{1-\sqrt{5}}{2}\right)^n u[n].$$

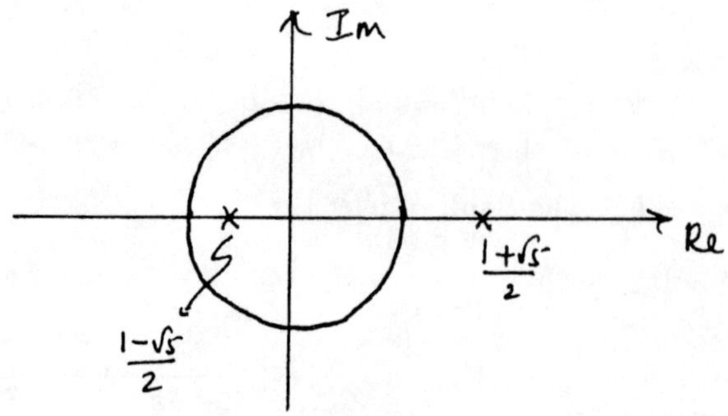

Figure S10.34

(c) Now assuming that the ROC is $(\sqrt{5}/2) - (1/2) < |z| < (1/2) + (\sqrt{5}/2)$, we get

$$h[n] = \frac{1}{\sqrt{5}}\left(\frac{1+\sqrt{5}}{2}\right)^n u[-n-1] + \frac{1}{\sqrt{5}}\left(\frac{1-\sqrt{5}}{2}\right)^n u[n].$$

10.35. Taking the z-transform of both sides of the given difference equation and simplifying, we get

$$H(z) = \frac{Y(z)}{X(z)} = \frac{1}{z - \frac{5}{2} + z^{-1}} = \frac{z^{-1}}{1 - \frac{5}{2}z^{-1} + z^{-2}}.$$

The partial fraction expansion of $H(z)$ is

$$H(z) = \frac{-2/3}{1 - \frac{1}{2}z^{-1}} + \frac{2/3}{1 - 2z^{-1}}.$$

If the ROC is $|z| > 2$, then

$$h_1[n] = -\frac{2}{3}\left(\frac{1}{2}\right)^n u[n] + \frac{2}{3}(2)^n u[n].$$

If the ROC is $1/2 < |z| < 2$, then

$$h_2[n] = -\frac{2}{3}\left(\frac{1}{2}\right)^n u[n] - \frac{2}{3}(2)^n u[-n-1].$$

If the ROC is $|z| < 1/2$, then

$$h_3[n] = \frac{2}{3}\left(\frac{1}{2}\right)^n u[-n-1] - \frac{2}{3}(2)^n u[-n-1].$$

For each $h_i[n]$, we now need to show that if $y[n] = h_i[n]$ in the difference equation, then $x[n] = \delta[n]$. Consider substituting $h_1[n]$ into the difference equation. This yields

$$\begin{array}{rl}\frac{2}{3} & \left(\frac{1}{2}\right)^{n-1} u[n-1] - \frac{2}{3}(2)^{n-1}u[n-1] - \frac{5}{3}\left(\frac{1}{2}\right)^n u[n] \\ & +\frac{5}{3}(2)^n u[n] + \frac{2}{3}\left(\frac{1}{2}\right)^{n+1} u[n+1] - \frac{2}{3}(2)^{n+1}u[n+1] = x[n]\end{array}.$$

Then,

$$x[n] = 0, \qquad \text{for } n < -1,$$

$$x[-1] = 2/3 - 2/3 = 0,$$

$$x[n] = 0, \qquad \text{for } n > 0.$$

It follows that $x[n] = \delta[n]$. It can similarly be shown that $h_2[n]$ and $h_3[n]$ satisfy the difference equation.

10.36. Taking the z-transform of both sides of the given difference equation and simplifying, we get

$$H(z) = \frac{Y(z)}{X(z)} = \frac{1}{z^{-1} - \frac{10}{3} + z} = \frac{z^{-1}}{1 - \frac{10}{3}z^{-1} + z^{-2}}.$$

The partial fraction expansion of $H(z)$ is

$$H(z) = -\frac{3/8}{1 - \frac{1}{3}z^{-1}} + \frac{3/8}{1 - 3z^{-1}}.$$

Since $H(z)$ corresponds to a stable system, the ROC has to be $(1/3) < |z| < 3$. Therefore,

$$h[n] = -\frac{3}{8}\left(\frac{1}{3}\right)^n u[n] - \frac{3}{8}(3)^n u[-n-1].$$

10.37. (a) The block-diagram may be redrawn as shown in part (a) of the figure below. This may be treated as a cascade of the two systems shown within the dotted lines in Figure S10.37. These two systems may be interchanged as shown in part (b) of the figure Figure S10.37 without changing the system function of the overall system. From the figure below, it is clear that

$$y[n] = x[n] + \frac{9}{8}x[n-1] - \frac{1}{3}y[n-1] + \frac{2}{9}y[n-2].$$

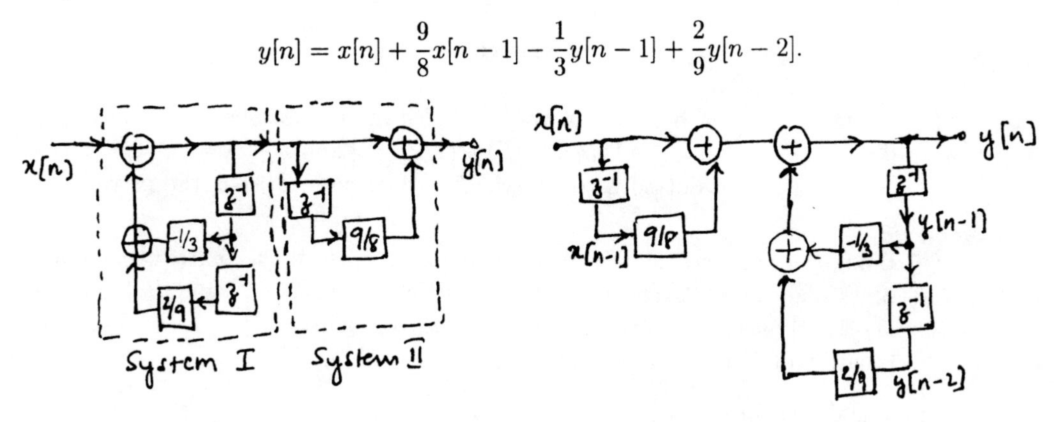

Figure S10.37

(b) Taking the z-transform of the above difference equation and simplifying, we get

$$H(z) = \frac{Y(z)}{X(z)} = \frac{1+\frac{9}{8}z^{-1}}{1+\frac{1}{3}z^{-1}-\frac{2}{9}z^{-2}} = \frac{1+\frac{9}{8}z^{-1}}{(1+\frac{2}{3}z^{-1})(1-\frac{1}{3}z^{-1})}.$$

$H(z)$ has poles at $z = 1/3$ and $z = -2/3$. Since the system is causal, the ROC has to be $|z| > 2/3$. The ROC includes the unit circle and hence the system is stable.

10.38. **(a)** $e_1[n] = f_1[n]$.

(b) $e_2[n] = f_2[n]$.

(c) Using the results of parts (a) and (b), we may redraw the block-diagram as shown in Figure S10.38.

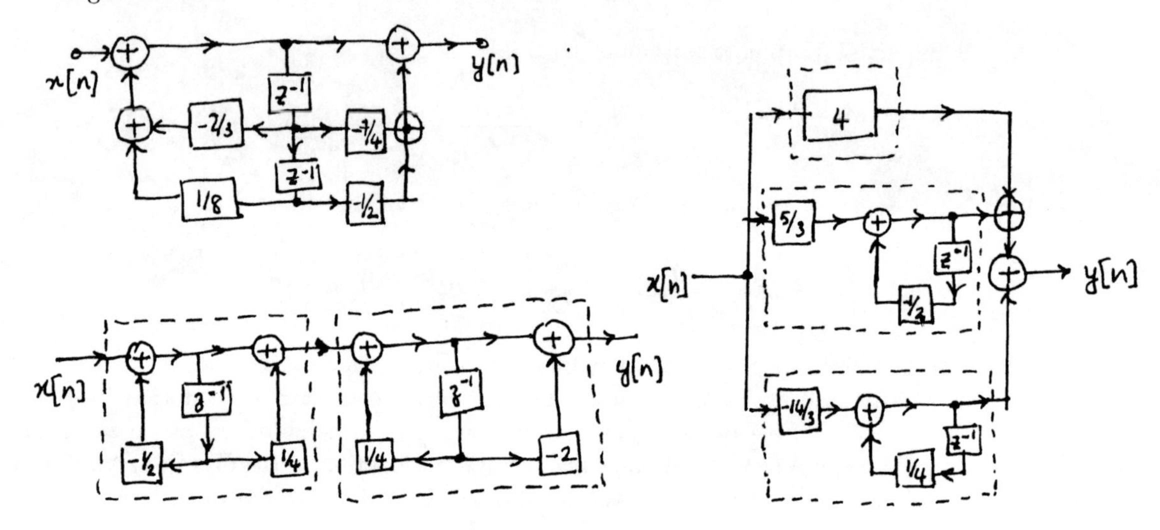

Figure S10.38

(d) Using the approach shown in the examples in the textbook we may draw the block-diagram of $H_1(z) = [1+(1/4)z^{-1}]/[1+(1/2)z^{-1}]$ and $H_2(z) = [1-2z^{-1}]/[1-(1/4)z^{-1}]$ as shown in the dotted boxes in the figure below. $H(z)$ is the cascade of these two systems.

(e) Using the approach shown in the examples shown in the textbook, we may draw the block-diagram of $H_1(z) = 4$, $H_2(z) = [5/3]/[1+(1/2)z^{-1}]$ and $H_3(z) = [-14/3]/[1-(1/4)z^{-1}]$ as shown in the dotted boxes in the figure below. $H(z)$ is the parallel combination of $H_1(z)$, $H_z(z)$, and $H_3(z)$.

10.39. **(a)** The direct form block diagram may be drawn as shown in part (a-i) of Figure S10.39 by noting that

$$H_1(z) = \frac{1}{1-\frac{5}{3}z^{-1}-\frac{11}{36}z^{-2}-\frac{5}{18}z^{-3}+\frac{1}{36}z^{-4}}.$$

The cascade block-diagram is as shown in part (a-ii) of Figure S10.39.

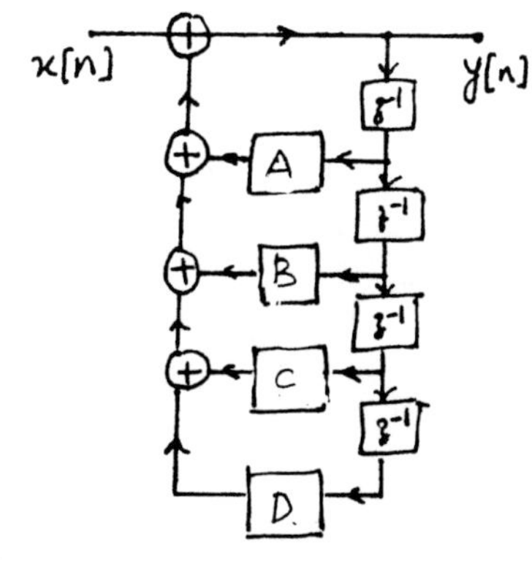

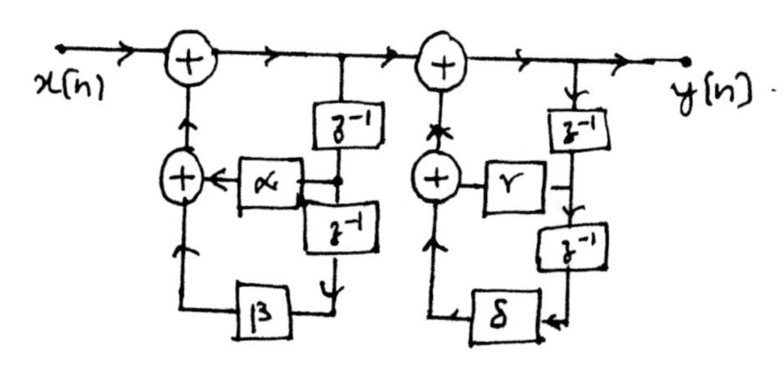

Figure S10.39

Note that

$$H_1(z) = \left[\frac{1}{1-\frac{1}{2}z^{-1}}\right]\left[\frac{1}{1-\frac{1}{2}z^{-1}}\right]\left[\frac{1}{1-\frac{2}{3}z^{-1}}\right]\left[\frac{1}{1-\frac{2}{3}z^{-1}}\right].$$

Therefore, $H_1(z)$ may be drawn as a cascade of four systems for which the coefficient multipliers are all real.

(b) The direct form block diagram may be drawn as shown in part (b-i) of Figure S10.39 by noting that

$$H_2(z) = \frac{1}{1-\frac{3}{2}z^{-1}+2z^{-2}-\frac{5}{4}z^{-3}+\frac{1}{2}z^{-4}}.$$

The cascade block-diagram is as shown in part (b-ii) of Figure S10.39.

Note that

$$H_2(z) = \left[\frac{1}{1-\frac{(1+j)}{2}z^{-1}}\right]\left[\frac{1}{1-\frac{1-j}{2}z^{-1}}\right]\left[\frac{1}{1-\frac{1+j\sqrt{15}}{4}z^{-1}}\right]\left[\frac{1}{1-\frac{1-j\sqrt{15}}{4}z^{-1}}\right].$$

Therefore, $H_1(z)$ cannot be drawn as a cascade of four systems for which the coefficient multipliers are all real.

(c) The direct form block diagram may be drawn as shown in part (c-i) of the Figure S10.39 by noting that

$$H_3(z) = \frac{1}{1-2z^{-1}+\frac{7}{4}z^{-2}-\frac{3}{4}z^{-3}+\frac{1}{8}z^{-4}}.$$

The cascade block-diagram is as shown in part (c-ii) of the Figure S10.39.

Note that

$$H_3(z) = \left[\frac{1}{1 - \frac{(1+j)}{2} z^{-1}}\right] \left[\frac{1}{1 - \frac{1-j}{2} z^{-1}}\right] \left[\frac{1}{1 - \frac{1}{2} z^{-1}}\right] \left[\frac{1}{1 - \frac{1}{2} z^{-1}}\right].$$

Therefore, $H_1(z)$ cannot be drawn as a cascade of four systems for which the coefficient multipliers are all real.

10.40. The definition of the unilateral z-transform is

$$\mathcal{X}(z) = \sum_{n=0}^{\infty} x[n] z^{-n}.$$

(a) Since $x[n] = \delta[n+5]$ is zero in the range $0 \le n \le \infty$, $\mathcal{X}(z) = 0$.

(b) The unilateral Laplace transform of $x[n] = \delta[n-5]$ is

$$\mathcal{X}(z) = \sum_{n=0}^{\infty} \delta[n-5] z^{-n} = e^{-5\omega}.$$

(c) The unilateral Laplace transform of $x[n] = (-1)^n u[n]$ is

$$\mathcal{X}(z) = \sum_{n=0}^{\infty} (-1)^n u[n] z^{-n} = \frac{1}{1 + z^{-1}}, \qquad |z| > 1.$$

(d) The unilateral Laplace transform of $x[n] = (1/2)^n u[n+3]$ is

$$\begin{aligned} \mathcal{X}(z) &= \sum_{n=0}^{\infty} (1/2)^n u[n+3] z^{-n} \\ &= \sum_{n=0}^{\infty} (1/2)^n z^{-n} \\ &= \frac{1}{1 - (1/2) z^{-1}}, \qquad |z| > 1/2. \end{aligned}$$

(e) Since $x[n] = (-1/3)^n u[-n-2]$ is zero in the range $0 \le n \le \infty$, $\mathcal{X}(z) = 0$.

(f) The unilateral Laplace transform of $x[n] = (1/4)^n u[-n+3]$ is

$$\begin{aligned} \mathcal{X}(z) &= \sum_{n=0}^{\infty} (1/4)^n u[-n+3] z^{-n} \\ &= \sum_{n=0}^{3} (1/4)^n z^{-n} \\ &= 1 + \frac{1}{4} z^{-1} + \frac{1}{16} z^{-2} + \frac{1}{64} z^{-3}, \qquad \text{All } z. \end{aligned}$$

(g) The unilateral Laplace transform of $x[n] = 2^n u[-n] + (1/4)^n u[n-1]$ is

$$\begin{aligned}\mathcal{X}(z) &= \sum_{n=0}^{\infty} 2^n u[-n] + (1/4)^n u[n-1] z^{-n} \\ &= \sum_{n=0}^{\infty} (1/4)^n z^{-n} \\ &= \frac{1}{1-\frac{1}{4}z^{-1}}, \qquad \text{All } z.\end{aligned}$$

(h) The unilateral Laplace transform of $x[n] = (1/3)^{n-2} u[n-2]$ is

$$\begin{aligned}\mathcal{X}(z) &= \sum_{n=0}^{\infty} (1/3)^{n-2} u[n-2] z^{-n} \\ &= z^{-2} \sum_{n=0}^{\infty} (1/3)^n z^{-n} \\ &= \frac{z^{-2}}{1-(1/2)z^{-1}}, \qquad |z| > 1/2.\end{aligned}$$

10.41. From the given information,

$$\begin{aligned}\mathcal{X}_1(z) &= \sum_{n=0}^{\infty} (1/2)^{n+1} u[n+1] z^{-n} \\ &= (1/2) \sum_{n=0}^{\infty} (1/2)^n z^{-n} \\ &= \frac{1/2}{1-(1/2)z^{-1}}, \qquad |z| > 1/2\end{aligned}$$

and

$$\begin{aligned}\mathcal{X}_2(z) &= \sum_{n=0}^{\infty} (1/4)^n u[n] z^{-n} \\ &= \sum_{n=0}^{\infty} (1/4)^n z^{-n} \\ &= \frac{1}{1-(1/4)z^{-1}}, \qquad |z| > 1/4.\end{aligned}$$

Using Table 10.2 and the time shift property we get

$$X_1(z) = \frac{z}{1-\frac{1}{2}z^{-1}}, \qquad |z| > 1/2.$$

and

$$X_2(z) = \frac{1}{1-\frac{1}{4}z^{-1}}, \qquad |z| > 1/4.$$

(a) We have

$$G(z) = X_1(z)X_2(z) = \frac{z}{(1-\frac{1}{2}z^{-1})(1-\frac{1}{4}z^{-1})}.$$

The ROC is $|z| > (1/2)$. The partial fraction expansion of $G(z)$ is

$$G(z) = z\left[\frac{2}{1-\frac{1}{2}z^{-1}} - \frac{1}{1-\frac{1}{4}z^{-1}}\right].$$

Using Table 10.2 and the time shift property, we get

$$g[n] = 2\left(\frac{1}{2}\right)^{n+1} u[n+1] - \left(\frac{1}{4}\right)^{n+1} u[n+1].$$

(b) We have

$$\mathcal{Q}(z) = \mathcal{X}_1(z)\mathcal{X}_2(z) = \frac{1/2}{(1-\frac{1}{2}z^{-1})(1-\frac{1}{4}z^{-1})}.$$

The ROC of $\mathcal{Q}(z)$ is $|z| > (1/2)$. The partial fraction expansion of $\mathcal{Y}(z)$ is

$$\mathcal{Q}(z) = \frac{1}{2}\left[\frac{2}{1-\frac{1}{2}z^{-1}} - \frac{1}{1-\frac{1}{4}z^{-1}}\right].$$

Therefore,

$$q[n] = \left(\frac{1}{2}\right)^{n} u[n] - \frac{1}{2}\left(\frac{1}{4}\right)^{n} u[n].$$

Clearly, $q[n] \neq g[n]$ for $n > 0$.

10.42. (a) Taking the unilateral z-transform of both sides of the given difference equation, we get

$$\mathcal{Y}(z) + 3z^{-1}\mathcal{Y}(z) + 3y[-1] = \mathcal{X}(z).$$

Setting $\mathcal{X}(z) = 0$, we get

$$\mathcal{Y}(z) = \frac{-3}{1+3z^{-1}}.$$

The inverse unilateral z-transform gives the zero-input response

$$y_{zi}[n] = -3(-3)^n u[n] = (-3)^{n+1}u[n].$$

Now, since it is given that $x[n] = (1/2)^n u[n]$, we have

$$\mathcal{X}(z) = \frac{1}{1-\frac{1}{2}z^{-1}}, \qquad |z| > 1/2.$$

Setting $y[-1]$ to be zero, we get

$$\mathcal{Y}(z) + 3z^{-1}\mathcal{Y}(z) = \frac{1}{1-\frac{1}{2}z^{-1}}.$$

Therefore,

$$\mathcal{Y}(z) = \frac{1}{(1 - \frac{1}{2}z^{-1})(1 + 3z^{-1})}.$$

The partial fraction expansion of $\mathcal{Y}(z)$ is

$$\mathcal{Y}(z) = \frac{1/7}{1 - \frac{1}{2}z^{-1}} + \frac{6/7}{1 + 3z^{-1}}.$$

The inverse unilateral z-transform gives the zero-state response

$$y_{zs}[n] = \frac{1}{7}\left(\frac{1}{2}\right)^n u[n] + \frac{6}{7}(-3)^n u[n].$$

(b) Taking the unilateral z-transform of both sides of the given difference equation, we get

$$\mathcal{Y}(z) - \frac{1}{2}z^{-1}\mathcal{Y}(z) - \frac{1}{2}y[-1] = \mathcal{X}(z) - \frac{1}{2}z^{-1}\mathcal{X}(z).$$

Setting $\mathcal{X}(z) = 0$, we get

$$\mathcal{Y}(z) = 0.$$

The inverse unilateral z-transform gives the zero-input response

$$y_{zi}[n] = 0.$$

Now, since it is given that $x[n] = u[n]$, we have

$$\mathcal{X}(z) = \frac{1}{1 - z^{-1}}, \qquad |z| > 1.$$

Setting $y[-1]$ to be zero, we get

$$\mathcal{Y}(z) - \frac{1}{2}z^{-1}\mathcal{Y}(z) = \frac{1}{1 - z^{-1}} - \frac{(1/2)z^{-1}}{1 - z^{-1}}.$$

Therefore,

$$\mathcal{Y}(z) = \frac{1}{1 - z^{-1}}.$$

The inverse unilateral z-transform gives the zero-state response

$$y_{zs}[n] = u[n].$$

(c) Taking the unilateral z-transform of both sides of the given difference equation, we get

$$\mathcal{Y}(z) - \frac{1}{2}z^{-1}\mathcal{Y}(z) - \frac{1}{2}y[-1] = \mathcal{X}(z) - \frac{1}{2}z^{-1}\mathcal{X}(z).$$

Setting $\mathcal{X}(z) = 0$, we get

$$\mathcal{Y}(z) = \frac{1/2}{1 - \frac{1}{2}z^{-1}}.$$

The inverse unilateral z-transform gives the zero-input response

$$y_{zi}[n] = \left(\frac{1}{2}\right)^{n+1} u[n].$$

Since the input $x[n]$ is the same as the one used in the part (b), the zero-state response is still

$$y_{zs}[n] = u[n].$$

10.43. (a) First let us determine the z-transform $X_1(z)$ of the sequence $x_1[n] = x[-n]$ in terms of $X(z)$:

$$\begin{aligned} X_1(z) &= \sum_{n=-\infty}^{\infty} x[-n]z^{-n} \\ &= \sum_{n=-\infty}^{\infty} x[n]z^{n} \\ &= X(1/z) \end{aligned}$$

Therefore, if $x[n] = x[-n]$, then $X(z) = X(1/z)$.

(b) If z_0 is a pole, then $1/X(z_0) = 0$. From the result of part (a), we know that $X(z_0) = X(1/z_0)$. Therefore, $1/X(z_0) = 1/X(1/z_0) = 0$. This implies that there is a pole at $1/z_0$.

If z_0 is a zero, then $X(z_0) = 0$. From the result of part (a), we know that $X(z_0) = X(1/z_0) = 0$. This implies that there is a zero at $1/z_0$.

(c) (1) In this case,

$$X(z) = z + z^{-1} = \frac{1+z^2}{z}, \qquad |z| > 0.$$

$X(z)$ has zeros $z_1 = j$ and $z_2 = -j$. Also, $X(z)$ has the poles $p_1 = 0$ and $p_2 = \infty$. Clearly, $z_2 = 1/z_1$ and $p_1 = 1/p_2$, which proves that the statement of (b) is true.

(2) In this case,

$$X(z) = z - \frac{5}{2} + z^{-1} = \frac{1 - \frac{5}{2}z + z^2}{z}, \qquad |z| > 0.$$

$X(z)$ has zeros $z_1 = -1/2$ and $z_2 = -2$. Also, $X(z)$ has the poles $p_1 = 0$ and $p_2 = \infty$. Clearly, $z_2 = 1/z_1$ and $p_1 = 1/p_2$, which proves that the statement of (b) is true.

10.44. (a) Using the shift property, we get

$$\mathcal{Z}\{\Delta x[n]\} = X(z) - z^{-1}X(z) = (1 - z^{-1})X(z).$$

(b) The z-transform $X_1(z)$ is given by

$$\begin{aligned} X_1(z) &= \sum_{n=-\infty}^{\infty} x_1[n]z^{-n} \\ &= \sum_{n=-\infty}^{\infty} x[n]z^{-2n} \\ &= X(2z). \end{aligned}$$

(c) Let us define a signal $g[n] = \{x[n]+(-1)^n x[n]\}/2$. Note that $g[2n] = x[2n]$ and $g[n] = 0$ for n odd. Also, using Table 10.1, we get

$$G(z) = \frac{1}{2}X(z) + \frac{1}{2}X(-z).$$

The z-transform $X_1(z)$ is given by

$$\begin{aligned} X_1(z) &= \sum_{n=-\infty}^{\infty} x_1[n]z^{-n} \\ &= \sum_{n=-\infty}^{\infty} g[2n]z^{-n} \\ &= \sum_{\substack{n=-\infty \\ \text{even}}}^{\infty} g[n]z^{-n/2} \\ &= \sum_{n=-\infty}^{\infty} g[n]z^{-n/2} \\ &= G(z^{1/2}) \\ &= \frac{1}{2}X(z^{1/2}) + \frac{1}{2}X(-z^{1/2}). \end{aligned}$$

10.45. In each part of this problem, we assume that the signal obtained by taking the inverse z-transform is called $x[n]$.

(a) Yes. The order of the numerator is equal to the order of the denominator in the given z-transform. Therefore, we can perform long-division to expand the z-transform such that the highest power of z in the expansion is 0. This would make $x[n] = 0$ for $n < 0$.

(b) No. This z-transform can be obtained by multiplying the z-transform of the previous part by z. Hence, its inverse is the inverse of the previous part shifted by 1 to the left. This implies that the resultant signal is not zero at $n = -1$.

(c) Yes. We can perform long-division to expand the z-transform such that the highest power of z in the expansion is -1. This would make $x[n] = 0$ for $n \leq 0$.

(d) No. When long-division is used to expand the z-transform, the highest power of z in the expansion is 1. This would make $x[-1] \neq 0$.

10.46. **(a)** Taking the z-transform of both sides of the difference equation relating $x[n]$ and $s[n]$ and simplifying, we get

$$H_1(z) = \frac{X(z)}{S(z)} = 1 - z^{-8}e^{-8\alpha} = \frac{z^8 - e^{-8\alpha}}{z^8}.$$

The system has an 8th order pole at $z = 0$ and 8 zeros distributed around a circle of radius $e^{-\alpha}$. This is shown in Figure S10.46. The ROC is everywhere on the z-plane except at $z = 0$.

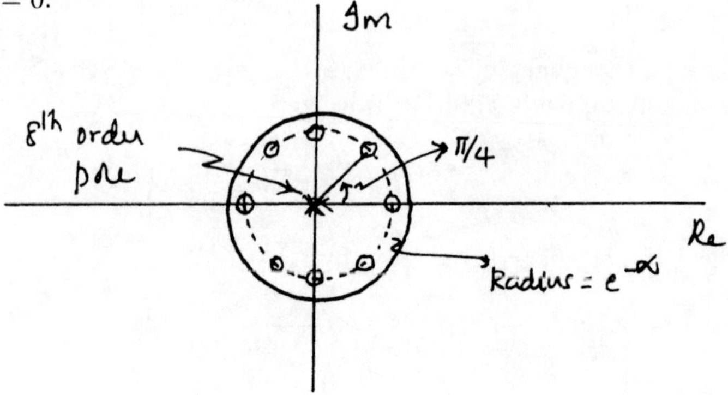

Figure S10.46

(b) We have

$$H_2(z) = \frac{Y(z)}{X(z)} = \frac{S(z)}{X(z)} = \frac{1}{H_1(z)}.$$

Therefore,

$$H_2(z) = \frac{1}{1 - z^{-8}e^{-8\alpha}} = \frac{z^8}{z^8 - e^{-8\alpha}}.$$

There are two possible ROCs for $H_2(z)$: $|z| < e^{-\alpha}$ or $|z| > e^{-\alpha}$. If the ROC is $|z| < e^{-\alpha}$, then the ROC does not include the unit circle. This in turn implies that the system would be unstable and anti-causal. If the ROC is $|z| > e^{-\alpha}$, then the ROC includes the unit circle. This in turn implies that the system would be stable and causal.

(c) We have

$$H_2(z) = \frac{1}{1 - z^{-8}e^{-8\alpha}}.$$

We need to choose the ROC to be $|z| > e^{-\alpha}$ in order to get s stable system. Now consider

$$P(z) = \frac{1}{1 - z^{-1}e^{-8\alpha}}$$

with ROC $|z| > e^{-\alpha}$. Taking the inverse z-transform, we get

$$p[n] = e^{-8\alpha n}u[n].$$

Now, note that

$$H_2(z) = P(z^8).$$

From Table 10.1 we know that

$$h_2[n] = \begin{cases} p[n/8] = e^{-\alpha n}, & n = 0, \pm 8, \pm 16, \cdots \\ 0, & \text{otherwise} \end{cases}$$

10.47. (a) From Clue 1, we have $H(-2) = 0$. From Clue 2, we know that when

$$X(z) = \frac{1}{1 - \frac{1}{2}z^{-1}}, \qquad |z| > \frac{1}{2}$$

we have

$$Y(z) = 1 + \frac{a}{1 - \frac{1}{4}z^{-1}}, \qquad |z| > \frac{1}{4}.$$

Therefore,

$$H(z) = \frac{Y(z)}{X(z)} = \frac{(1 + a - \frac{1}{4}z^{-1})(1 - \frac{1}{2}z^{-1})}{1 - \frac{1}{4}z^{-1}}, \qquad |z| > \frac{1}{4}.$$

Substituting $z = -2$ in the above equation and noting that $H(-2) = 0$, we get

$$a = -\frac{9}{8}.$$

(b) The response to the signal $x[n] = 1 = 1^n$ will be $y[n] = H(1)x[n]$. Therefore,

$$y[n] = H(1) = \frac{1}{4}.$$

10.48. From the pole-zero diagram, we may write

$$H_1(z) = A\frac{(z - \frac{3}{4}e^{j\pi/4})(z - \frac{3}{4}e^{-j\pi/4})}{(z - \frac{3}{4}e^{j3\pi/4})(z - \frac{3}{4}e^{-j3\pi/4})}$$

and

$$H_2(z) = B\frac{(z - \frac{1}{2}e^{j3\pi/4})(z - \frac{1}{2}e^{-j3\pi/4})}{(z - \frac{1}{2}e^{j\pi/4})(z - \frac{1}{2}e^{-j\pi/4})}$$

where A and B are constants. Now note that

$$H_2(z) = \frac{B}{A}H_1\left(\frac{3}{2}ze^{j\pi}\right) = \frac{B}{A}H_1\left(-\frac{3}{2}z\right).$$

Using the property 10.5.3 of the z-transform (see Table 10.1), we get

$$h_2[n] = \frac{B}{A}\left(-\frac{2}{3}\right)^n h_1[n].$$

We may rewrite this as

$$h_2[n] = g[n]h_1[n],$$

where $g[n] = (B/A)(-2/3)^n$. Note that since both $h_1[n]$ and $h_2[n]$ are causal, we may assume that $g[n] = 0$ for $n < 0$. Therefore,

$$g[n] = \frac{B}{A}\left(-\frac{2}{3}\right)^n u[n].$$

Now, clue 3 also states that $\sum_{k=0}^{\infty}|g[k]| = 3$. Therefore,

$$\sum_{k=0}^{\infty}\frac{B}{A}\left(-\frac{2}{3}\right)^k = 3$$

or

$$\frac{B}{A}\frac{1}{1-2/3} = 3 \quad \Rightarrow \quad \frac{B}{A} = 1.$$

Therefore,

$$g[n] = \left(-\frac{2}{3}\right)^n u[n].$$

10.49. (a) We may write the left side of eq. (P10.49-1) as

$$\sum_{n=N_1}^{\infty}|x[n]|r_1^{-n} = \sum_{n=N_1}^{\infty}|x[n]|\left(r_0\frac{r_1}{r_0}\right)^{-n} = \sum_{n=N_1}^{\infty}|x[n]|r_0^{-n}\left(\frac{r_1}{r_0}\right)^{-n}. \quad \text{(S10.49-1)}$$

Since $r_1 \geq r_0$, the sequence $(r_1/r_0)^{-n}$ decays with increasing n, i.e., as $n \to \infty$ $(r_1/r_0)^{-n} \to 0$. Therefore, $(r_1/r_0)^{-n} \leq (r_1/r_0)^{-N_1}$ for $n \geq N_1$. Substituting this in eq.(S10.49-1), we get

$$\sum_{n=N_1}^{\infty}|x[n]|r_1^{-n} = \sum_{n=N_1}^{\infty}|x[n]|r_0^{-n}\left(\frac{r_1}{r_0}\right)^{-n} \leq \left(\frac{r_1}{r_0}\right)^{-N_1}\sum_{n=N_1}^{\infty}|x[n]|r_0^{-n}.$$

Therefore, $A = (r_1/r_0)^{-N_1} = (r_0/r_1)^{N_1}$.

(b) The above inequality shows that if $X(z)$ has the finite bound B for $|z| = r_0$, then $X(z)$ has the finite bound $(r_0/r_1)^{N_1}B$ for $|z| = r_1 \geq r_0$. Thus, $X(z)$ converges for $|z| = r_1 \geq r_0$ and Property 4 of Section 10.2 follows.

(c) Consider a left-sided sequence $x[n]$ such that

$$x[n] = 0, \qquad n > N_2$$

and for which

$$\sum_{n=-\infty}^{\infty}|x[n]|r_0^{-n} = \sum_{n=-\infty}^{N_2}|x[n]|r_0^{-n}.$$

Then we need to show that if $r_1 \le r_0$,

$$\sum_{n=-\infty}^{N_2} |x[n]| r_1^{-n} \le P \sum_{n=-\infty}^{N_2} |x[n]| r_0^{-n}. \qquad \text{(S10.49–2)}$$

where P is a positive constant.

We may write the left side of eq. (S10.49-2) as

$$\sum_{n=-\infty}^{N_2} |x[n]| r_1^{-n} = \sum_{n=-\infty}^{N_2} |x[n]| \left(r_0 \frac{r_1}{r_0}\right)^{-n} = \sum_{n=-\infty}^{N_2} |x[n]| r_0^{-n} \left(\frac{r_1}{r_0}\right)^{-n}. \qquad \text{(S10.49–3)}$$

Since $r_1 \le r_0$, the sequence $(r_1/r_0)^{-n}$ decays with decreasing n, i.e., as $n \to -\infty$ $(r_1/r_0)^{-n} \to 0$. Therefore, $(r_1/r_0)^{-n} \le (r_1/r_0)^{-N_2}$ for $n \le N_2$. Substituting this in eq.(S10.49-3), we get

$$\sum_{n=-\infty}^{N_2} |x[n]| r_1^{-n} = \sum_{n=-\infty}^{N_2} |x[n]| r_0^{-n} \left(\frac{r_1}{r_0}\right)^{-n} \le \left(\frac{r_1}{r_0}\right)^{-N_2} \sum_{n=-\infty}^{N_2} |x[n]| r_0^{-n}.$$

Therefore, $P = (r_1/r_0)^{-N_2} = (r_0/r_1)^{N_2}$.

The above inequality shows that if $X(z)$ has the finite bound B for $|z| = r_0$, then $X(z)$ has the finite bound $(r_0/r_1)^{N_2} B$ for $|z| = r_1 \le r_0$. Thus, $X(z)$ converges for $|z| = r_1 \le r_0$ and Property 5 of Section 10.2 follows.

10.50. (a) From the given pole-zero plot, we get

$$H(z) = A \frac{z^{-1} - a}{1 - az^{-1}},$$

where A is some constant. Therefore,

$$H(e^{j\omega}) = A \frac{e^{-j\omega} - a}{1 - ae^{-j\omega}}$$

and

$$|H(e^{j\omega})|^2 = H(e^{j\omega}) H^* e^{j\omega} = |A|^2 \left[\frac{e^{-j\omega} - a}{1 - ae^{-j\omega}}\right] \left[\frac{e^{j\omega} - a}{1 - ae^{j\omega}}\right].$$

Therefore,

$$|H(e^{j\omega})|^2 = |A|^2 \frac{1 - ae^{-j\omega} - ae^{j\omega} + a^2}{1 - ae^{-j\omega} - ae^{j\omega} + a^2} = |A|^2.$$

This implies that $|H(e^{j\omega})| = |A|$ =constant.

(b) We get $|v_1|^2 = 1 + a^2 - 2a\cos(\omega)$.

(c) We get

$$|v_2|^2 = 1 + \frac{1}{a^2} - \frac{2}{a}\cos\omega = \frac{1}{a^2}[a^2 + 1 + 2a\cos\omega] = \frac{1}{a^2}|v_1|^2.$$

10.51. (a) We know that for a real sequence $x[n]$, $x[n] = x^*[n]$. Let us first find the z-transform of $y[n] = x^*[n]$ in terms of $X(z)$, the z-transform of $x[n]$. We have

$$\begin{aligned} Y(z) &= \sum_{n=-\infty}^{\infty} y[n]z^{-n} \\ &= \sum_{n=-\infty}^{\infty} x^*[n]z^{-n} \\ &= \left[\sum_{n=-\infty}^{\infty} x[n](z^*)^{-n}\right]^* \\ &= [X(z^*)]^* = X^*(z^*). \end{aligned}$$

Now, since $x[n] = x^*[n]$, we have $\mathcal{Z}\{x[n]\} = \mathcal{Z}\{x^*[n]\}$ which in turn implies that $X(z) = X^*(z^*)$.

(b) If $X(z)$ has a pole at $z = z_0$, then $1/X(z_0) = 0$. From the result of the previous part, we know that

$$\frac{1}{X^*(z_0^*)} = 0.$$

Conjugating both sides, we get $1/X(z_0^*) = 0$. This implies that $X(z)$ has a pole at z_0^*.

If $X(z)$ has a zero at $z = z_0$, then $X(z_0) = 0$. From the result of the previous part, we know that

$$X^*(z_0^*) = 0.$$

Conjugating both sides, we get $X(z_0^*) = 0$. This implies that $X(z)$ has a zero at z_0^*.

(c) (1) The z-transform of the given sequence is

$$X(z) = \frac{1}{1 - \frac{1}{2}z^{-1}} = \frac{z}{z - \frac{1}{2}}, \qquad |z| > 1/2.$$

Clearly, $X(z)$ has a pole at $z = 1/2$ and a zero at $z = 0$ and the property of part (b) holds.

(2) The z-transform of the given sequence is

$$X(z) = 1 - \frac{1}{2}z^1 + \frac{1}{4}z^{-2} = \frac{z^2 - (1/2)z + (1/4)}{z^2}, \qquad |z| > 0.$$

$X(z)$ has two zeros at $z = 1/2$ and two poles at $z = 0$. The property of part (b) still holds.

(d) Now, from part (b) of problem 10.43 we know that if $x[n]$ and $X(z)$ has a pole at $z_0 = \rho e^{j\theta}$, then $X(z)$ must have a pole at $(1/z_0) = (1/\rho)e^{-j\theta}$.

If $x[n]$ is real and $X(z)$ has a pole at $z_0 = \rho e^{j\theta}$, then from part (b) we know that $X(z)$ must have a pole at $z_0^* = \rho e^{-j\theta}$. Now, from part (b) of problem 10.43 we know that if $x[n]$ and $X(z)$ has a pole at $z_0^* = \rho e^{-j\theta}$, then $X(z)$ must have a pole at $(1/z_0^*) = (1/\rho)e^{j\theta}$.

A similar argument may be constructed for zeros.

10.52. We have

$$\begin{aligned}
X_2(z) &= \sum_{n=-\infty}^{\infty} x_2[n]z^{-n} \\
&= \sum_{n=-\infty}^{\infty} x_1[-n]z^{-n} \\
&= \sum_{n=-\infty}^{\infty} x_1[n]z^{n} \\
&= X_1(z^{-1}) = X_1(1/z).
\end{aligned}$$

Using an argument similar to the one used on part (b) of problem 10.43, we may argue that if $X_1(z)$ has a pole (or zero) at $z = z_0$, then $X_2(z)$ must have a pole (or zero) at $z = 1/z_0$.

10.53. Let us assume that $x[n]$ is a sequence with z-transform $X(z)$ which has the ROC $\alpha < |z| < \beta$.

(a) (1) The z-transform of the sequence $y[n] = x[n - n_0]$ is

$$\begin{aligned}
Y(z) &= \sum_{n=-\infty}^{\infty} y[n]z^{-n} \\
&= \sum_{n=-\infty}^{\infty} x[n-n_0]z^{-n}
\end{aligned}$$

Substituting $m = n - n_0$ in the above equation, we get

$$\begin{aligned}
Y(z) &= \sum_{m=-\infty}^{\infty} x[m]z^{-m-n_0} \\
&= z^{-n_0} \sum_{m=-\infty}^{\infty} x[m]z^{-m} \\
&= z^{-n_0} X(z).
\end{aligned}$$

Clearly, $Y(z)$ converges where $X(z)$ converges except for the addition or deletion of $z = 0$ because of the z^{-n_0} term. Therefore, the ROC of $Y(z)$ is $\alpha < |z| < \beta$ except for the possible addition or deletion of $z = 0$ in the ROC.

(2) The z-transform of the sequence $y[n] = z_0^n x[n]$ is

$$\begin{aligned}
Y(z) &= \sum_{n=-\infty}^{\infty} y[n]z^{-n} \\
&= \sum_{n=-\infty}^{\infty} z_0^n x[n]z^{-n} \\
&= \sum_{n=-\infty}^{\infty} x[n](z/z_0)^{-n} \\
&= X(z/z_0)
\end{aligned}$$

Since $X(z)$ converges for $\alpha < |z| < \beta$, $Y(z)$ converges for $\alpha < |z/z_0| < \beta$. Therefore, the ROC of $Y(z)$ is $|z_0|\alpha < |z| < |z_0|\beta$.

(3) The z-transform of the sequence $y[n] = x[-n]$ is

$$\begin{aligned} Y(z) &= \sum_{n=-\infty}^{\infty} y[n]z^{-n} \\ &= \sum_{n=-\infty}^{\infty} x[-n]z^{-n} \\ &= \sum_{n=-\infty}^{\infty} x[n]z^{n} \\ &= X(1/z) \end{aligned}$$

Since $X(z)$ converges for $\alpha < |z| < \beta$, $Y(z)$ converges for $\alpha < |1/z| < \beta$. Therefore, the ROC of $Y(z)$ is $(1/\beta) < |z| < (1/\alpha)$.

(b) (1) From Problem 10.51(a), we know that the z-transform of the sequence $y[n] = x^*[n]$ is $Y(z) = X^*(z^*)$. The ROC of $Y(z)$ is the same as the ROC of $X(z)$.

(2) Suppose that the ROC of $x[n]$ is $\alpha < |z| < \beta$. From subpart (2) of part (a), the z-transform of $y[n] = z_0^n x[n]$ is

$$Y(z) = X(z/z_0)$$

with ROC $|z_0|\alpha < |z| < |z_0|\beta$. Therefore, $R_y = |z_0|R_x$.

10.54. **(a)** Let $x[n] = 0$ for $n > 0$. Then,

$$\begin{aligned} X(z) &= \sum_{n=-\infty}^{\infty} x[n]z^{-n} \\ &= \sum_{n=-\infty}^{0} x[n]z^{-n} \\ &= x[0] + x[-1]z + x[-2]z^2 + \cdots \end{aligned}$$

Therefore,

$$\lim_{z\to 0} X(z) = x[0].$$

(b) Let $x[n] = 0$ for $n < 0$. Then,

$$\begin{aligned} X(z) &= \sum_{n=-\infty}^{\infty} x[n]z^{-n} \\ &= \sum_{n=0}^{\infty} x[n]z^{-n} \\ &= x[0] + x[1]z^{-1} + x[-2]z^{-2} + \cdots \end{aligned}$$

Therefore,

$$\lim_{z\to\infty} z(X(z) - x[0]) = \lim_{z\to\infty} z\{x[1]z^{-1} + x[-2]z^{-2} + \cdots\} = x[1].$$

10.55. **(a)** From the initial value theorem, we have

$$\lim_{z\to\infty} X(z) = x[0] = \text{non-zero and finite.}$$

Therefore, as $z \to \infty$, $X(z)$ tends to a finite non-zero value. This implies that $X(z)$ has neither poles nor zeros at infinity.

(b) A rational z-transform is made up of factors of the form $1/(z-a)$ and $(z-b)$. Note that the factor $1/(z-a)$ has a pole at $z=a$ and a zero at $z=\infty$. Also note that the factor $(z-b)$ has a zero at $z=b$ and a pole at $z=\infty$. From the results of part (a), we know that a causal sequence has no poles or zero at infinity. Therefore, all zeros at infinity contributed by factors of the form $1/(z-a)$ must be cancelled out by the poles at infinity contributed by factors of the form $(z-b)$ This implies that the number of factors of the form $(z-b)$ equals the number of factors of the form $1/(z-a)$. Consequently, the number of zeros in the finite z-plane must equal the number of poles in the finite z-plane.

10.56. **(a)** The z-transform of $x_3[n]$ is

$$\begin{aligned}
X_3(z) &= \sum_{n=-\infty}^{\infty} x_3[n]z^{-n} \\
&= \sum_{n=-\infty}^{\infty} \left[\sum_{k=-\infty}^{\infty} x_1[k]x_2[n-k]\right] z^{-n} \\
&= \sum_{k=-\infty}^{\infty} x_1[k] \left[\sum_{n=-\infty}^{\infty} x_2[n-k]z^{-n}\right] \\
&= \sum_{k=-\infty}^{\infty} x_1[k]\mathcal{Z}\{x_2[n-k]\} \\
&= \sum_{k=-\infty}^{\infty} x_1[k]\hat{X}_2(z)
\end{aligned}$$

(b) Using the time shifting property (10.5.2), we get

$$\hat{X}_2(z) = \mathcal{Z}\{x_2[n-k]\} = z^{-k}X_2(z),$$

where $X_2(z)$ is the z-transform of $x_2[n]$. Substituting in the result of part (a), we get

$$X_3(z) = X_2(z) \sum_{k=-\infty}^{\infty} x_1[k]z^{-k}.$$

(c) Noting that the z-transform of $x_1[n]$ may be written as

$$X_1(z) = \sum_{k=-\infty}^{\infty} x_1[k]z^{-k},$$

we may rewrite the result of part (b) as

$$X_3(z) = X_1(z)X_2(z).$$

10.57. **(a)** $X_1(z)$ is a polynomial of order N_1 in z^{-1}. $X_2(z)$ is a polynomial of order N_2 in z^{-1}. Therefore, $Y(z) = X_1(z)X_2(z)$ is a polynomial of order $N_1 + N_2$ in z^{-1}. This implies that $M = N_1 + N_2$.

(b) By noting that $y[0]$ is the coefficient of the z^0 term in $Y(z)$, $y[1]$ is the coefficient of the z^{-1} term in $Y(z)$, and $y[2]$ is the coefficient of the z^{-2} term in $Y(z)$, we get

$$\begin{aligned} y[0] &= x_1[0]x_2[0], \\ y[1] &= x_1[0]x_2[1] + x_1[1]x_2[0], \\ y[2] &= x_1[0]x_2[2] + x_1[1]x_2[1] + x_1[2]x_2[0]. \end{aligned}$$

(c) We note the pattern that emerges from part (b). The k-th point in the sequence $y[n]$ is the coefficient of z^{-k} in $Y(z)$. The z^{-k} term of $Y(z)$ is formed by the following sum: (the product of the z^0 term of $X_1(z)$ with the z^{-k} term of $X_2(z)$) + (the product of the z^{-1} term of $X_1(z)$ with the z^{-k+1} term of $X_2(z)$) + (the product of the z^{-2} term of $X_1(z)$ with the z^{-k+2} term of $X_2(z)$) + ... + the (product of the z^{-N_1} term of $X_1(z)$ with the z^{-k+N_1} term of $X_2(z)$).
Therefore,

$$y[k] = \sum_{m=0}^{N_1} x_1[m]x_2[k-m].$$

Since $x_1[m] = 0$ for $m > N_1$ and $m < 0$, we may rewrite this as

$$y[k] = \sum_{m=-\infty}^{\infty} x_1[m]x_2[k-m].$$

10.58. Consider a causal and stable system with system function $H(z)$. Let its inverse system have the system function $H_i(z)$. The poles of $H(z)$ are the zeros of $H_i(z)$ and the zeros of $H(z)$ are the poles of $H_i(z)$.

For $H(z)$ to correspond to a be causal and stable system, all its poles must be within the unit circle. Similarly, for $H_i(z)$ to correspond to a be causal and stable system, all its poles must be within the unit circle. Since the poles of $H_i(z)$ are the zeros of $H(z)$, the previous statement implies that the zeros of $H(z)$ must be within the unit circle. Therefore, all poles and zeros of a minimum-phase system must lie within the unit circle.

10.59. **(a)** From Figure S10.59, we have

$$W_1(z) = X(z) - \frac{k}{3}z^{-1}W_1(z) \quad \Rightarrow \quad W_1(z) = X(z)\frac{1}{1+\frac{k}{3}z^{-1}}.$$

Also,

$$W_2(z) = -\frac{k}{4}z^{-1}W_1(z) = -X(z)\frac{\frac{k}{4}z^{-1}}{1+\frac{k}{3}z^{-1}}.$$

Therefore, $Y(z) = W_1(z) + W_2(z)$ will be

$$Y(z) = X(z)\frac{1}{1+\frac{k}{3}z^{-1}} - X(z)\frac{\frac{k}{4}z^{-1}}{1+\frac{k}{3}z^{-1}}.$$

Finally,

$$H(z) = \frac{Y(z)}{X(z)} = \frac{1-\frac{k}{4}z^{-1}}{1+\frac{k}{3}z^{-1}}.$$

Since $H(z)$ corresponds to a causal filter, the ROC will be $|z| > |k|/3$.

(b) For the system to be stable, the ROC of $H(z)$ must include the unit circle. This is possible only if $|k|/3 < 1$. This implies that $|k|$ has to be less than 3.

(c) If $k = 1$, then

$$H(z) = \frac{1-\frac{1}{4}z^{-1}}{1+\frac{1}{3}z^{-1}}.$$

The response to $x[n] = (2/3)^n$ will be of the form

$$y[n] = x[n]H(2/3) = \frac{5}{12}(2/3)^n.$$

10.60. The unilateral z-transform of $y[n] = x[n+1]$ is

$$\begin{aligned}
\mathcal{Y}(z) &= \sum_{n=0}^{\infty} y[n]z^{-n} \\
&= y[0] + y[1]z^{-1} + y[2]z^{-2} + \cdots \\
&= x[1] + x[2]z^{-1} + x[3]z^{-2} + \cdots \\
&= z\{x[0] + x[1]z^{-1} + x[2]z^{-2} + x[3]z^{-3} + \cdots\} - zx[0] \\
&= z\mathcal{X}(z) - zx[0].
\end{aligned}$$

10.61. **(a)** The unilateral z-transform of $y[n] = x[n+3]$ is

$$\begin{aligned}
\mathcal{Y}(z) &= \sum_{n=0}^{\infty} y[n]z^{-n} \\
&= \sum_{n=0}^{\infty} x[n+3]z^{-n} \\
&= \sum_{n=-3}^{\infty} x[n+3]z^{-n} - x[0]z^3 - x[1]z^2 - x[2]z \\
&= \sum_{n=0}^{\infty} x[n]z^{-n+3} - x[0]z^3 - x[1]z^2 - x[2]z \\
&= z^3\sum_{n=0}^{\infty} x[n]z^{-n} - x[0]z^3 - x[1]z^2 - x[2]z \\
&= z^3\mathcal{X}(z) - x[0]z^3 - x[1]z^2 - x[2]z
\end{aligned}$$

(b) The unilateral z-transform of $y[n] = x[n-3]$ is

$$\begin{aligned}
\mathcal{Y}(z) &= \sum_{n=0}^{\infty} y[n]z^{-n} \\
&= \sum_{n=0}^{\infty} x[n-3]z^{-n} \\
&= \sum_{n=3}^{\infty} x[n-3]z^{-n} + x[-1]z^{-2} + x[-2]z^{-1} + x[-3] \\
&= \sum_{n=0}^{\infty} x[n]z^{-n-3} + x[-1]z^{-2} + x[-2]z^{-1} + x[-3] \\
&= z^{-3}\sum_{n=0}^{\infty} x[n]z^{-n} + x[-1]z^{-2} + x[-2]z^{-1} + x[-3] \\
&= z^{-3}\mathcal{X}(z) + x[-1]z^{-2} + x[-2]z^{-1} + x[-3]
\end{aligned}$$

(c) We have

$$y[n] = \sum_{k=-\infty}^{n} x[k] = \sum_{m=0}^{\infty} x[n-m].$$

Therefore,

$$\begin{aligned}
\mathcal{Y}(z) &= \sum_{m=0}^{\infty} z^{-m}\mathcal{X}(z) + \sum_{m=1}^{\infty} z^{-m}\sum_{l=1}^{m} x[-l]z^{l} \\
&= \frac{\mathcal{X}(z)}{1-z^{-1}} + \sum_{m=1}^{\infty} z^{-m}\sum_{l=1}^{m} x[-l]z^{l}
\end{aligned}$$

10.62. Note that

$$\phi_{xx}[n] = \sum_{k=-\infty}^{\infty} x[k]x[n+k] = x[n] * x[-n].$$

Now, applying the convolution property, the z-transform of $\phi_{xx}[n]$ is

$$\Phi_{xx}(z) = X(z)\mathcal{Z}\{x[-n]\}.$$

From the time-reversal property we know that the z-transform of $x[-n]$ is $X(1/z)$. Therefore,

$$\Phi_{xx}(z) = X(z)X(1/z).$$

10.63. **(a)** Since the ROC is $|z| < 1/2$, the sequence is left-sided. Using the power-series expansion, we get

$$\log(1-2z) = -\sum_{n=1}^{\infty}\frac{2^n z^n}{n} = -\sum_{n=-\infty}^{-1} - \frac{2^{-n}z^{-n}}{n}.$$

Therefore,

$$x[n] = \frac{2^{-n}}{n}u[-n-1].$$

(b) Since the ROC is $|z| > 1/2$, the sequence is right-sided. Using the power-series expansion, we get

$$\log(1-(1/2)z^{-1}) = -\sum_{n=1}^{\infty}\frac{(1/2)^n z^{-n}}{n}.$$

Therefore,

$$x[n] = -\frac{2^{-n}}{n}u[n-1].$$

10.64. Let us define $Y(z)$ to be

$$Y(z) = -z\frac{d}{dz}X(z).$$

Then using the differentiation property of the z-transform, we get

$$y[n] = nx[n].$$

(a) Now,

$$Y(z) = -z\frac{d}{dz}X(z) = z\frac{2}{1-2z} = -\frac{1}{1-\frac{1}{2}z^{-1}}.$$

Noting that the ROC of $Y(z)$ is $|z| < (1/2)$ (the same as the ROC of $X(z)$), we get

$$y[n] = \left(\frac{1}{2}\right)^n u[-n-1].$$

Therefore,

$$x[n] = \frac{1}{n}\left(\frac{1}{2}\right)^n u[-n-1] = \frac{2^{-n}}{n}u[-n-1].$$

This is same as the answer obtained for Problem 10.63(a).

(b) In this part,

$$Y(z) = -z\frac{d}{dz}X(z) = \frac{\frac{1}{2}z^{-1}}{1-\frac{1}{2}z^{-1}}.$$

Noting that the ROC of $Y(z)$ is $|z| > (1/2)$ (the same as the ROC of $X(z)$), we get

$$y[n] = -\frac{1}{2}\left(\frac{1}{2}\right)^{n-1} u[n-1].$$

Therefore,

$$x[n] = -\frac{1}{n}\left(\frac{1}{2}\right)^{n} u[n-1] = -\frac{2^{-n}}{n}u[n-1].$$

This is same as the answer obtained for Problem 10.63(b).

10.65. (a) From the given $H_c(s)$, we get

$$|H_c(j\omega)| = \frac{|a-j\omega|}{|a+j\omega|} = \frac{\sqrt{a^2+\omega^2}}{\sqrt{a^2+\omega^2}} = 1.$$

(b) Applying the bilinear transformation, we get

$$H_d(z) = \frac{a - \frac{1-z^{-1}}{1+z^{-1}}}{a + \frac{1-z^{-1}}{1+z^{-1}}} = \frac{a-1}{a+1}\left[\frac{1+z^{-1}\frac{a+1}{a-1}}{1+z^{-1}\frac{a-1}{a+1}}\right].$$

Therefore, $H_d(z)$ has a pole at $z = (a-1)/(a+1)$ and a zero at $z = (a+1)/(a-1)$. Since a is real and positive,

$$\left|\frac{a-1}{a+1}\right| \leq 1 \quad \text{and} \quad \left|\frac{a+1}{a-1}\right| \geq 1.$$

Therefore, the pole of $H_d(z)$ lies inside the unit circle and the zero of $H_d(z)$ lies outside the unit circle.

(c) $H_d(z)$ may be rewritten as

$$H(z) = \frac{(a-1)+z^{-1}(a+1)}{(a+1)+z^{-1}(a-1)}.$$

Therefore,

$$|H(e^{j\omega})| = \frac{|(a-1)+e^{-j\omega}(a+1)|}{|(a+1)+e^{-j\omega}(a-1)|} = \frac{|(a-1)+(\cos\omega - j\sin\omega)(a+1)|}{|(a+1)+(\cos\omega - j\sin\omega)(a-1)|}.$$

This may be written as

$$\begin{aligned}|H(e^{j\omega})| &= \frac{\sqrt{(a-1)^2+\cos^2\omega(a+1)^2+2(a+1)(a-1)\cos\omega+(a+1)^2\sin^2\omega}}{\sqrt{(a+1)^2+\cos^2\omega(a-1)^2+2(a+1)(a-1)\cos\omega+(a-1)^2\sin^2\omega}} \\ &= \frac{\sqrt{(a-1)^2+(a+1)^2+2(a+1)(a-1)\cos\omega}}{\sqrt{(a+1)^2+(a-1)^2+2(a+1)(a-1)\cos\omega}} = 1\end{aligned}.$$

10.66. (a) We are given that

$$H_d(z) = H_c\left(\frac{1-z^{-1}}{1+z^{-1}}\right).$$

Therefore,

$$H_d(e^{j\omega}) = H_c\left(\frac{1-e^{-j\omega}}{1+e^{-j\omega}}\right) = H_c\left(\frac{e^{j\omega/2}-e^{-j\omega/2}}{e^{j\omega/2}+e^{-j\omega/2}}\right) = H_c\left(j\tan\frac{\omega}{2}\right).$$

(b) From the given $H_c(s)$, we get

$$H_c(0) = \frac{1}{(e^{j\pi/4})(e^{-j\pi/4})} = 1$$

and

$$H_c(\infty) = \frac{1}{\lim_{s\to\infty}(s+e^{j\pi/4})(s+e^{-j\pi/4})} = 0.$$

Now,

$$\begin{aligned} |H_c(j\omega)| &= \frac{1}{|(j\omega+e^{j\pi/4})(j\omega+e^{-j\pi/4}|} = \frac{1}{|-\omega^2+2\cos(\pi/4)j\omega+1|} \\ &= \frac{1}{\sqrt{(1-\omega^2)^2+4\omega\cos^2(\pi/4)\omega^2}}. \end{aligned}$$

Clearly, $|H_c(j\omega)|$ decreases monotonically with increasing ω.

(c) (1) We are given that

$$H_d(z) = H_c\left(\frac{1-z^{-1}}{1+z^{-1}}\right).$$

Therefore,

$$H_d(z) = \frac{1}{(\frac{1-z^{-1}}{1+z^{-1}}+e^{j\pi/4})(\frac{1-z^{-1}}{1+z^{-1}}+e^{-j\pi/4})}.$$

This may be rewritten as

$$H_d(z) = \frac{1}{(1+e^{j\pi/4})(1+e^{-j\pi/4})}\frac{(1+z^{-1})^2}{[1-z^{-1}\frac{1+e^{j\pi/4}}{1-e^{j\pi/4}}][1-z^{-1}\frac{1+e^{-j\pi/4}}{1-e^{-j\pi/4}}]}.$$

Therefore, $H_d(z)$ has exactly two poles which lie at $z = -(1+e^{j\pi/4})/(1-e^{j\pi/4})$ and $z = -(1+e^{-j\pi/4})/(1-e^{-j\pi/4})$. It can be easily shown that both these poles lie inside the unit circle.

(2) From the result of part (a), we have

$$H_d(e^{j0}) = H_c(j\tan 0) = H_c(j0) = 1.$$

(3) We have

$$\begin{aligned} |H_d(e^{j\omega})| &= \left|H_c\left(j\tan\tfrac{\omega}{2}\right)\right| = \frac{1}{|1-\tan^2(\omega/2)+\sqrt{2}j\tan(\omega/2)|} \\ &= \frac{1}{\sqrt{(1-\tan^2(\omega/2))^2+2\tan^2(\omega/2)}} = \frac{1}{1+\tan^4(\omega/2)}. \end{aligned}$$

As ω increases from 0 to pi, $\tan(\omega/2)$ increases monotonically from 0 to ∞. Therefore, $|H_d(e^{j\omega})|$ decreases monotonically from 1 to 0.

(4) The half-power frequency ω_d satisfies the relationship

$$|H_d(e^{j\omega_d})|^2 = \frac{1}{2} = \left|H_c\left(j\tan\frac{\omega_d}{2}\right)\right|^2.$$

We know that $|H_c(j)|^2 = 1/2$. Therefore,

$$j\tan\frac{\omega_d}{2} = j \quad \Rightarrow \quad \omega_d = \pi/2.$$

Chapter 11 Answers

11.1. The system shown in Figure P11.1 may be looked at as a parallel interconnection of the a system with system function $H_0(z)$ with a feedback system of the form shown in Figure 11.3(b). From Section 10.8.1 we know that the feedback system has a closed-loop system function $Q(z)$ given by

$$Q(z) = \frac{H_1(z)}{1 + G(z)H_1(z)}.$$

Therefore, the system function of the parallel interconnection is

$$Q_1(z) = Q(z) + H_0(z) = H_0(z) + \frac{H_1(z)}{1 + G(z)H_1(z)}.$$

11.2. The system shown in Figure P11.2 may be redrawn as shown in the left-hand side sketch of Figure S11.2. Here,

$$Q(s) = \frac{H_1(s)}{1 + G_1(s)H_1(s)}.$$

The system may be further simplified as shown in the right-hand side sketch of Figure S11.2.

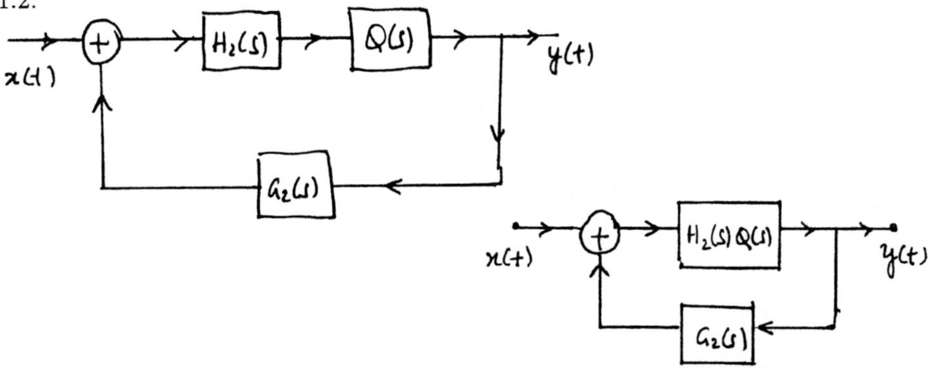

Figure S11.2

From the above figures it is clear that the overall system function is

$$Q_1(s) = \frac{\frac{H_2(s)H_1(s)}{1+G_1(s)H_1(s)}}{1 + \frac{G_2(s)H_2(s)H_1(s)}{1+G_1(s)H_1(s)}} = \frac{H_1(s)H_2(s)}{1 + G_1(s)H_1(s) + G_2(s)H_1(s)H_2(s)}.$$

11.3. From Section 11.1 we know that the overall system function of this feedback system is

$$Q(s) = \frac{\frac{1}{s-1}}{1 + \frac{s-b}{s-1}} = \frac{1}{2s - b - 1}.$$

$Q(s)$ has a pole at $(b+1)/2$. For this system to be causal and stable, the pole has to be in the left-half of the s-plane. Therefore, we require that

$$\frac{b+1}{2} < 0 \quad \Rightarrow \quad b < -1.$$

11.4. Taking the Laplace transform of the given differential equation we obtain

$$Q(s) = \frac{Y(s)}{X(s)} = \frac{s}{s^2+s+1}.$$

We also know from Section 11.1 that

$$Q(s) = \frac{H(s)}{1+G(s)H(s)}.$$

Since it is given that $H(s) = 1/(s+1)$, we obtain

$$Q(s) = \frac{1}{s+1+G(s)}.$$

Comparing with the first equation, we obtain

$$G(s) = \frac{1}{s}.$$

11.5. From Section 11.1 we know that the overall system function of this feedback system is

$$Q(z) = \frac{\frac{1}{1-(1/2)z^{-1}}}{1+\frac{1-bz^{-1}}{1-(1/2)z^{-1}}} = \frac{1}{2-(b+1/2)z^{-1}}.$$

$Q(z)$ has a pole at $(2b+1)/4$. For this system to be causal and stable, the pole has to be inside the unit circle. Therefore, we require that

$$\left|\frac{2b+1}{4}\right| < 1 \quad \Rightarrow \quad -\frac{5}{2} < b < \frac{3}{2}.$$

11.6. From Section 11.1 we know that the overall system function of this feedback system is

$$Q(z) = \frac{1-z^{-N}}{1-z^{-1}} = 1 + z^{-1} + z^{-2} + \cdots + z^{N-1}.$$

Comparing with the definition of the z-transform given in eq. (10.3), we know that the inverse z-transform of $Q(z)$ is

$$q[n] = \begin{cases} 1, & 0 \le n \le N-1 \\ 0, & \text{otherwise} \end{cases}.$$

Clearly, the system is FIR.

11.7. Using the techniques outlined in Section 11.2, we may draw the root locus for the given system. In Figure 11.7, we show the root loci for $K > 0$ and $K < 0$.

For $K > 0$, the root locus never crosses over to the right-half of the s-plane. Therefore, the system is stable for all positive values of K. But for $K < 0$, the instability occurs when the root locus crosses over to the right-half of the s-plane through the point $s = 0$. Therefore, the value of K at which instability begins to occur is obtained from eq. (11.52) to be

$$K = -\frac{1}{|G(0)H(0)|} = -6.$$

Therefore, the system is stable for $K > -6$.

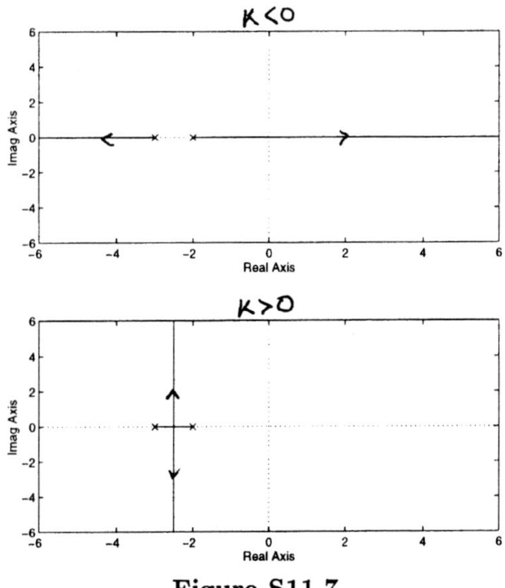

Figure S11.7

11.8. The root locus for $K < 0$ is as shown in Figure S11.8. Clearly, the poles cross over to the right-half of the s-plane at some $s = \pm\omega_0$. We also know that at this point, the poles satisfy:

$$\frac{j\omega_0 - 1}{-\omega_0^2 + 3j\omega + 2} = -\frac{1}{K}.$$

Equating real and imaginary parts on both sides, we get $K = -3$ and $\omega_0 = \sqrt{5}$. Therefore, the system is stable for $K > -3$.

11.9. Using the techniques outlined in Section 11.2, we may draw the root locus for the given system. In Figure S11.9, we show the root loci for $K > 0$ and $K < 0$.

From these figures, it is clear that the root locus always lies on the real axis for all values of K. Therefore, the feedback system can have closed-loop poles *only* on the real axis. In order for a system to have an oscillatory impulse response, it must have poles which have a nonzero imaginary part. Clearly the given system does not have an oscillatory impulse response.

11.10. Since two branches of the root locus for $K > 0$ starts at $s = -1$ and ends at $s = 1$, we can conclude from properties 1 and 2 in Section 11.3.5 that there are two poles at $s = -1$ and at least one zero at $s = 1$. Since the entire real axis is a part of some branch of the root locus for $K < 0$, we may conclude from property 3 in Section 11.3.5 we know that this is possible only if there are two zeros at $s = 1$.

11.11. Using the techniques developed in Section 11.3 we may draw the root locus $K > 0$. This is as shown in the Figure S11.11.

From this figure it is clear that the system becomes unstable when the root locus crosses

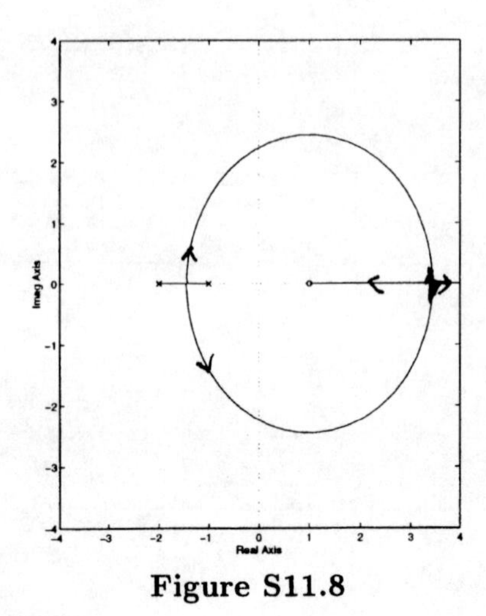

Figure S11.8

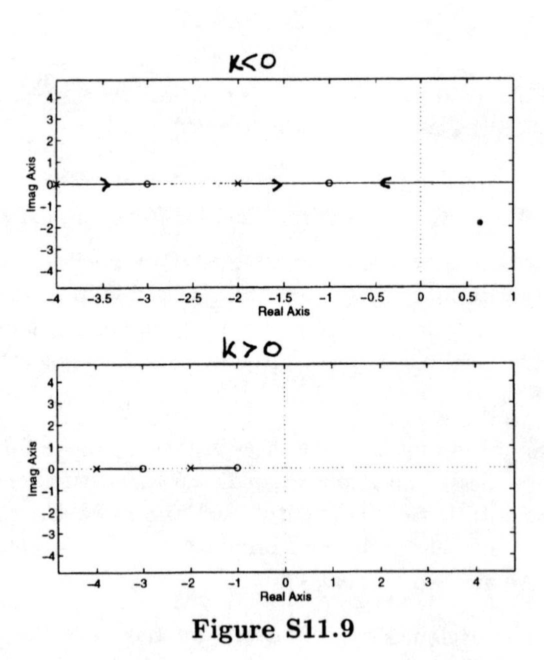

Figure S11.9

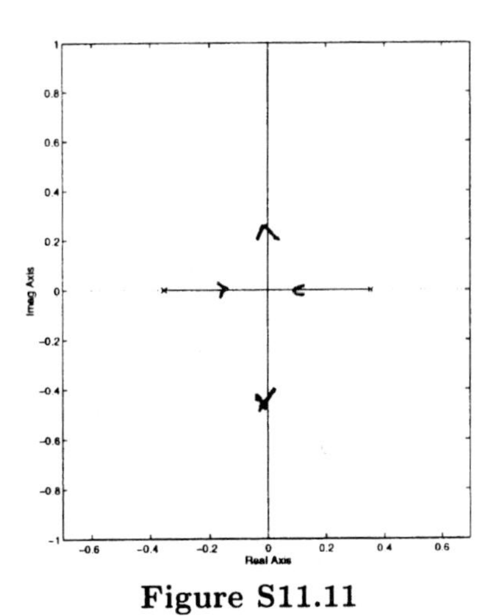

Figure S11.11

either $|z| = 1$. From eq. (11.52), we know that the corresponding value of K is

$$K = \frac{1}{H(j)G(j)} = \frac{1}{H(-j)G(-j)} = \frac{5}{4}.$$

Therefore, the system is stable for $0 < K < \frac{5}{4}$.

11.12. In Figure S11.12, we show the root loci for $K > 0$ and $K < 0$ for different positions of the poles and zeros. Clearly the root loci stay on the real axis only if the poles and zero alternate positions on the real axis.

11.13. Note that the system may be viewed as shown in Figure S11.13. We note now that

$$H(z) = \frac{z^2}{z^2 - z - 4}$$

and

$$G(z) = \frac{K}{z}.$$

Therefore, the closed loop system function is

$$Q(z) = \frac{H(z)}{1 + G(z)H(z)} = \frac{z^2}{z^2 + (K-1)z - 4}.$$

Therefore, the closed-loop poles of this system lie at

$$z = \frac{-(K-1) \pm \sqrt{(K-1)^2 + 16}}{2}.$$

It can be easily shown that the magnitude of at least one of the poles is always greater than 1. Therefore, the system is never stable.

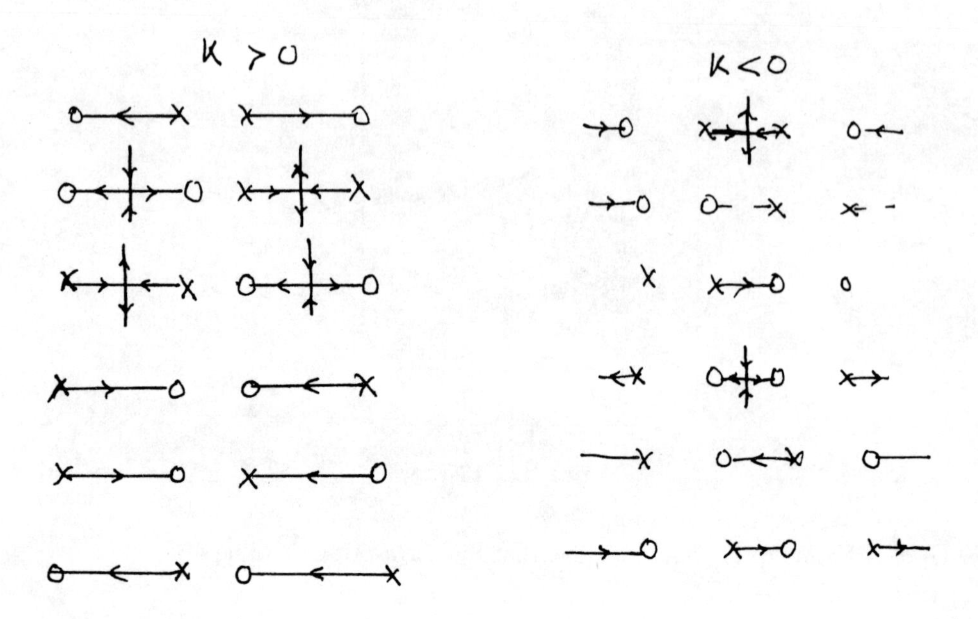

Figure S11.12

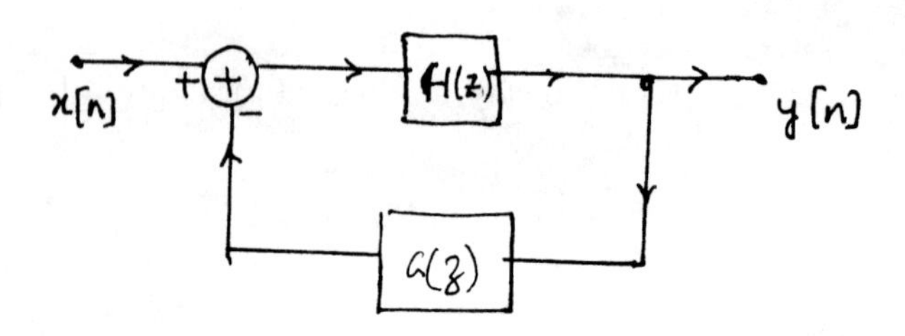

Figure S11.13

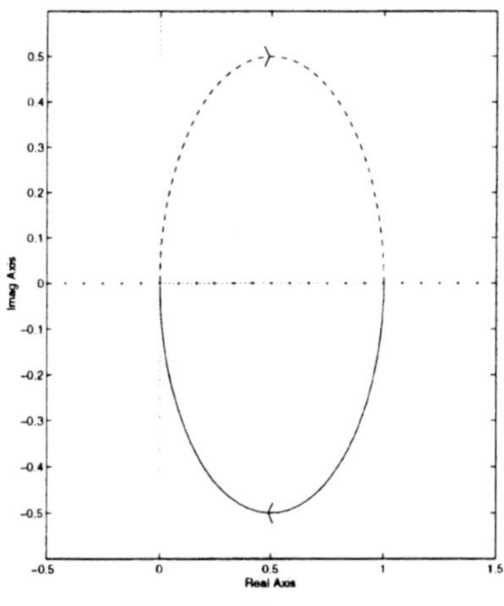

Figure S11.15

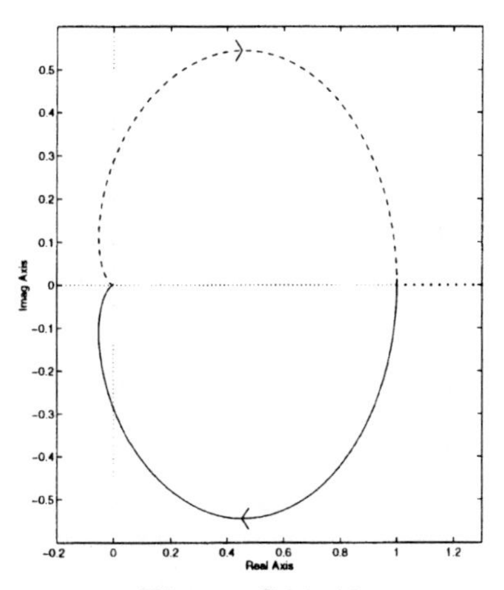

Figure S11.16

11.14. The number of clockwise encirclements is given by the difference between the number of zeros and number of poles within C.

(a) There is one pole and one zero within C. Therefore, $W(p)$ encircles the origin 0 times.

(b) There are 2 zeros and 1 pole within C. Therefore, $W(p)$ encircles the origin once.

11.15. The Nyquist plot for this system is shown in Figure S11.15. Since the system has zero right-half poles, the number of counterclockwise encirclements of the point $-1/K$ must be zero. From the figure, it is clear that this is possible for $-1/K > 1$ and $-1/K < 0$, that is, $K > -1$.

11.16. The Nyquist plot for this system is shown in Figure S11.16. Since the system has zero right-half poles, the number of counterclockwise encirclements of the point $-1/K$ must be zero. From the figure, it is clear that this is possible for $-1/K > 1$ and $-1/K < 0$, that is,

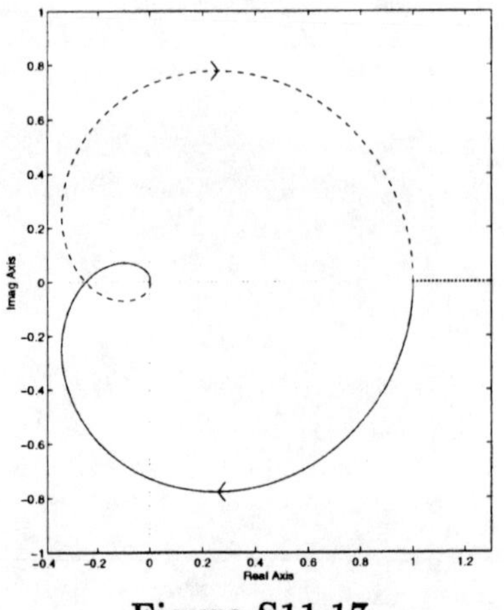

Figure S11.17

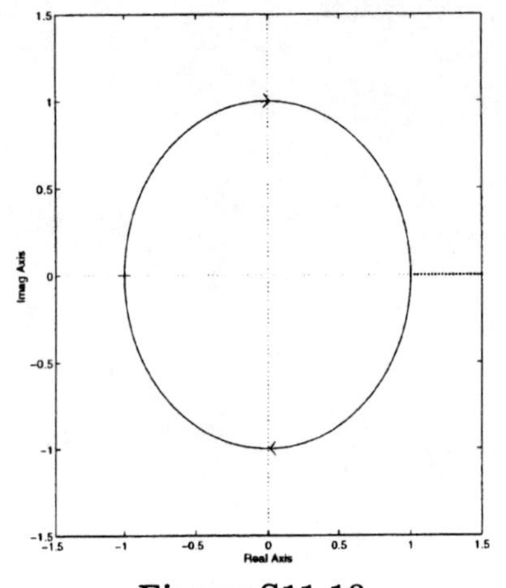

Figure S11.18

$K > -1$.

11.17. 11.17 The Nyquist plot for this system is shown in Figure S11.17. Since the system has zero right-half poles, the number of counterclockwise encirclements of the point $-1/K$ must be zero. From the figure, it is clear that this is possible for $-1/K > 1$ and $-1/K < -1/4$. Therefore, the system is stable for $-1 < K < 4$.

11.18. The Nyquist plot for this system is shown in Figure S11.18. Since the system has zero poles outside the unit circle, the number of counterclockwise encirclements of the point $-1/K$ must be zero. From the figure, it is clear that this is possible for $-1/K > 1$ and $-1/K < 1$. Therefore, the system is stable for $-1 < K < 1$.

11.19. Let us consider the continuous-time case. The Nyquist plot may be viewed as a plot of the function $G(j\omega)H(j\omega)$. Note that if this plot passes through a point $-1/K$, then

$G(j\omega_0)H(j\omega_0) = -1/K$ at this point. Therefore, the denominator of the overall system function $Q(s) = H(s)/[1 + KG(s)H(s)]$ evaluated at $j\omega_0$ will be zero. This implies the system $Q(s)$ has a pole at $j\omega_0$ which makes it unstable. A similar argument may be made for the discrete-time case.

11.20. Note that

$$H(j\omega) = \frac{1 + j\omega}{-\omega^2 + j\omega + 1}.$$

In order to determine the gain margin, we first need to find the ω at which $\sphericalangle H(j\omega) = -\pi$. Clearly, for this to happen, we need

$$1 + j\omega = -(-\omega^2 + j\omega + 1).$$

This is not possible for any value of ω. Therefore, the system has an infinite gain margin.

In order to determine the phase margin, we first need to find the ω at which $|H(j\omega)| = 1$. Noting that

$$|H(j\omega)| = \frac{1 + \omega^2}{(1 - \omega^2)^2 + \omega^2},$$

we find that $|H(j\omega)| = 1$ at $\omega_0 = \sqrt{2}$. Now,

$$\sphericalangle H(j\omega_0) = \tan^{-1}\sqrt{2} - \tan^{-1}\left(\frac{\sqrt{2}}{-1}\right) = -\pi + 2\tan^{-1}\sqrt{2}.$$

Therefore, the phase margin is $2\tan^{-1}\sqrt{2}$.

11.21. From Figure P11.21, we may obtain the overall system function to be

$$Q(s) = \frac{K(s + 100)}{(1 + K)s + (100 + K)}.$$

(i) When $K = 0.1$, the pole is at $s = -91$ and the zero is at $s = -100$.

(ii) When $K = 1$, the pole is at $s = -50.5$ and the zero is at $s = -100$.

(iii) When $K = 10$, the pole is at $s = -10$ and the zero is at $s = -100$.

(iv) When $K = 100$, the pole is at $s = -1.98$ and the zero is at $s = -100$.

11.22. The closed-loop system function is given by

$$Q(s) = \frac{H(s)}{1 + G(s)H(s)}.$$

(a) From the given $H(s)$ and $G(s)$, we obtain

$$Q(s) = \frac{1}{(s + 1)(s + 3) + 1} = \frac{1}{(s + 2)^2}.$$

Taking the inverse Laplace transform, we get

$$q(t) = te^{-2t}u(t).$$

(b) From the given $H(s)$ and $G(s)$, we obtain

$$Q(s) = \frac{s+1}{(s+1)(s+3)+1} = \frac{s+1}{(s+2)^2}.$$

The partial fraction expansion of $Q(s)$ is

$$Q(s) = \frac{1}{s+2} - \frac{1}{(s+2)^2}.$$

Taking the inverse Laplace transform, we get

$$q(t) = e^{-2t}u(t) - te^{-2t}u(t).$$

(c) From the given $H(s)$ and $G(s)$, we obtain

$$Q(s) = \frac{1/2}{1+(1/2)e^{-s/3}}.$$

Taking the inverse Laplace transform, we get

$$q(t) = \left(\frac{1}{2}\right)\left(-\frac{1}{2}\right)^t \sum_{k=0}^{\infty} \delta(t - k/3).$$

11.23. The closed-loop system function is given by

$$Q(z) = \frac{H(z)}{1+G(z)H(z)}.$$

(a) From the given $H(z)$ and $G(z)$, we obtain

$$Q(z) = \frac{z^{-1}}{1+\frac{1}{6}z^{-1} - \frac{1}{6}z^{-2}}.$$

The partial fraction expansion of $Q(z)$ is

$$Q(z) = \frac{6/5}{1-\frac{1}{3}z^{-1}} - \frac{6/5}{1+\frac{1}{2}z^{-1}}.$$

Taking the inverse z-transform, we get

$$q[n] = \frac{6}{5}\left(\frac{1}{3}\right)^n u[n] - \frac{6}{5}\left(-\frac{1}{2}\right)^n u[n].$$

(b) From the given $H(z)$ and $G(z)$, we obtain

$$Q(z) = \frac{\frac{2}{3} - \frac{1}{2}z^{-1} + \frac{1}{12}z^{-2}}{1+\frac{1}{6}z^{-1} - \frac{1}{6}z^{-2}}.$$

Now, the inverse z-transform is

$$q[n] = \frac{2}{3}q_a[n] - \frac{1}{2}q_a[n-1] + \frac{1}{12}q_a[n-2],$$

where $q_a[n]$ is the impulse response obtained in part (a).

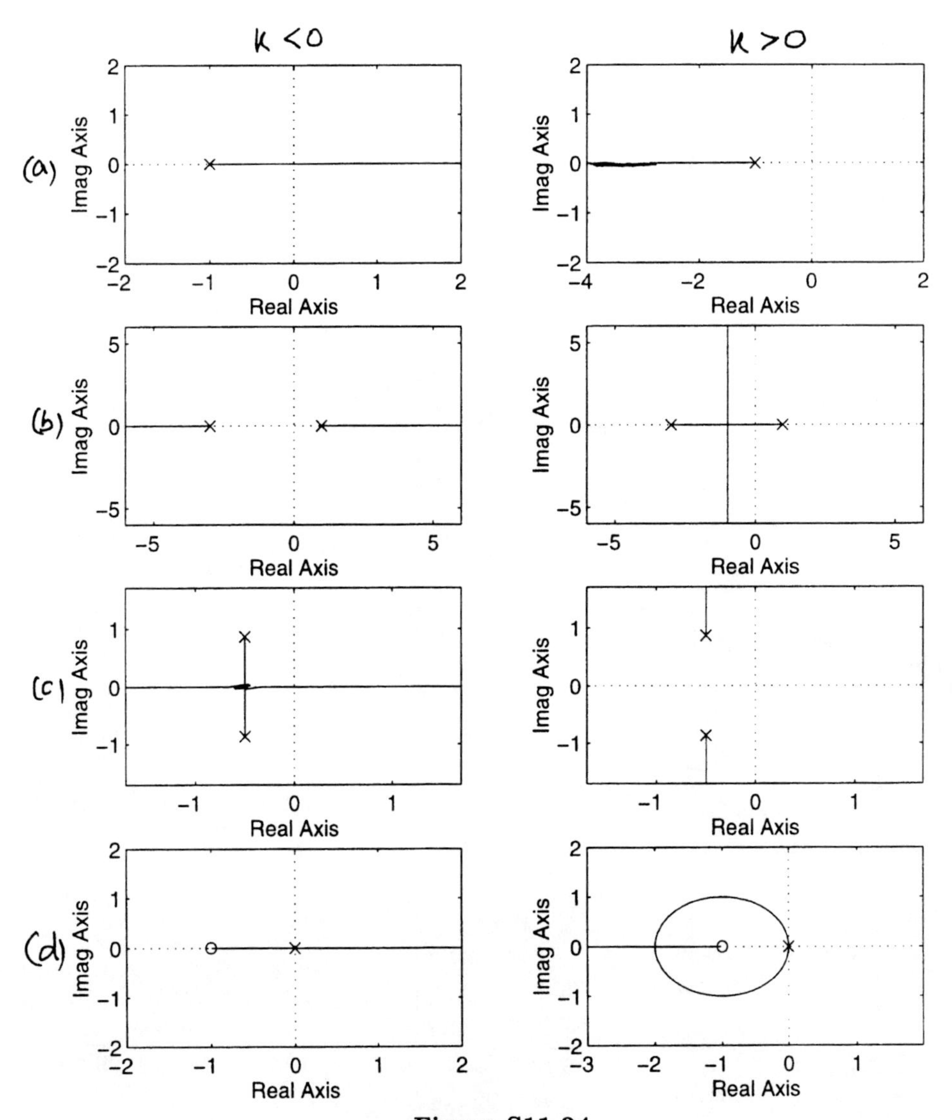

Figure S11.24

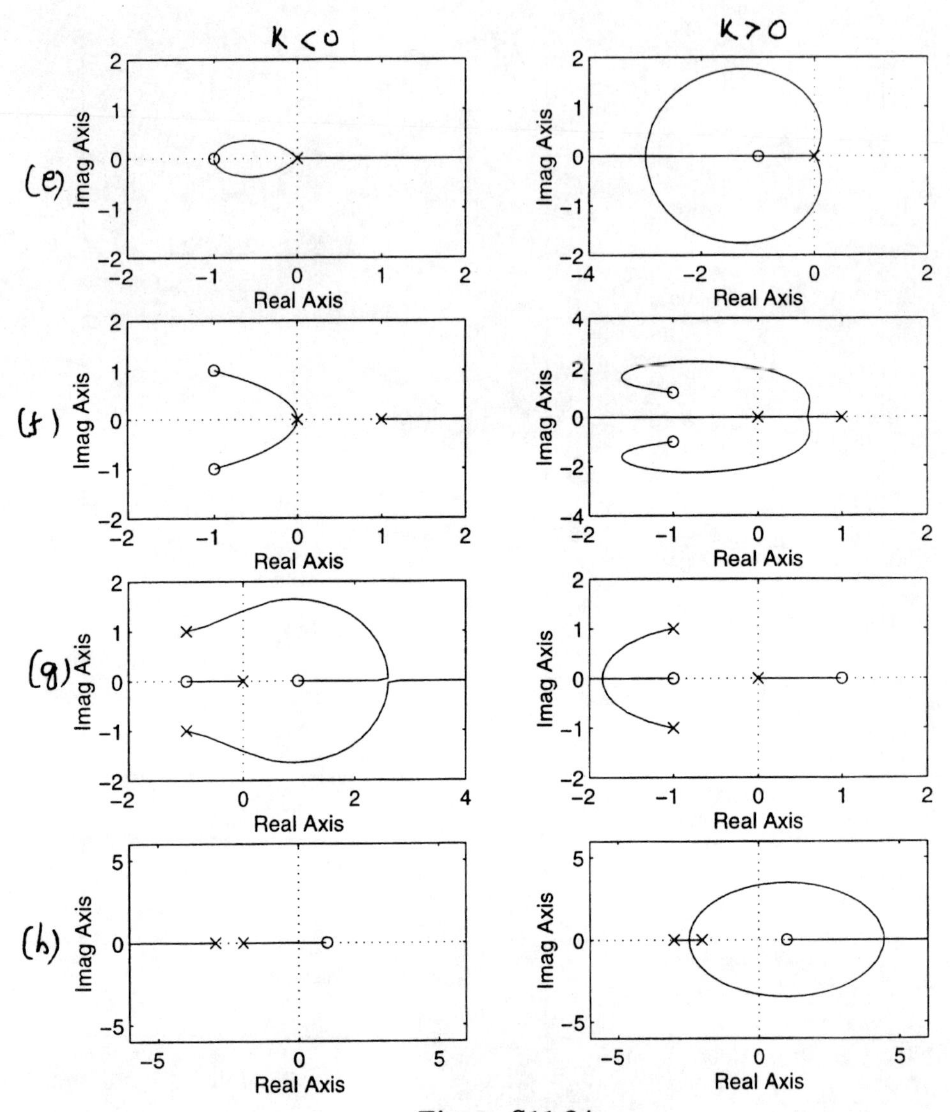

Figure S11.24

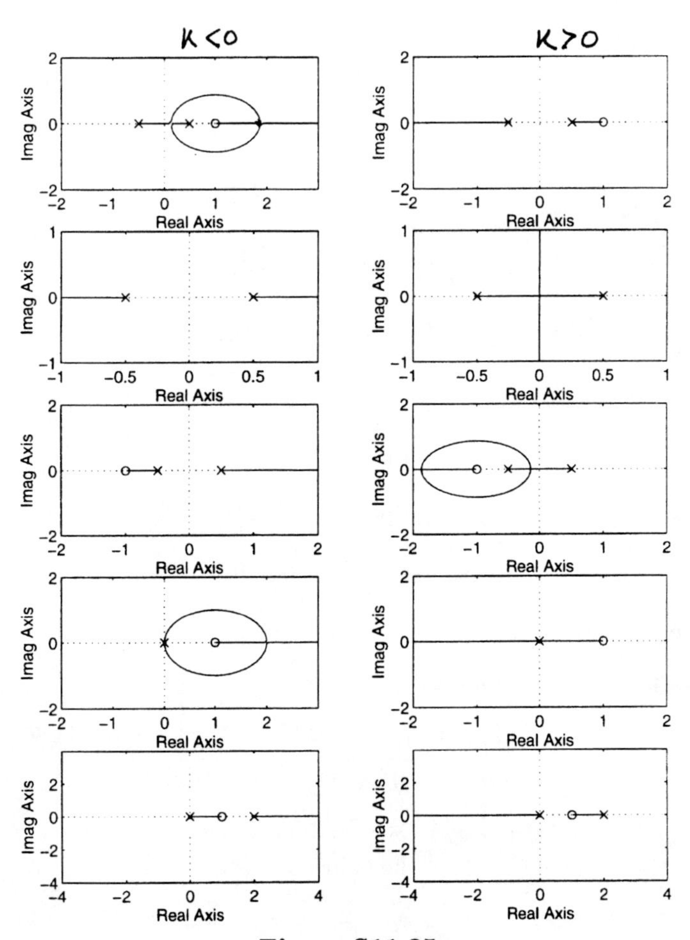

Figure S11.25

11.24. The root-loci are as shown in Figure S11.24.

11.25. The root-loci are as shown in Figure S11.25.

11.26. The root-loci are as shown in Figure S11.26.

11.27. **(a)** The root locus is as shown in Figure S11.27.

(b) The root locus is as shown in Figure S11.27.

(c) To have no oscillatory behavior, the closed-loop poles must lie on the real axis. We know that the closed loop poles must satisfy

$$G(s)H(s) = -1 \quad \Rightarrow \quad \frac{s+2}{s^2+2s+4} = -\frac{1}{K}.$$

Therefore, the closed-loop poles satisfy

$$s^2 + (2+K)s + (4+2K) = 0.$$

In order for these poles to be real, we require that

$$(2+K)^2 = 4(4+2k) \quad \Rightarrow \quad K = 6.$$

11.28. The plots are as shown in Figure S11.28

11.29. The plots are as shown in Figure S11.29.

11.30. The plots are as shown in Figure S11.30.

11.31. The plots are as shown in Figure S11.31.

11.32. **(a)** The closed-loop system function is

$$Q(s) = \frac{H(s)}{1+KG(s)H(s)} = \frac{D_2(s)N_1(s)}{D_1(s)D_2(s)+KN_1(s)N_2(s)}.$$

Clearly, $Q(s) = 0$ either when $D_2(s) = 0$ or when $N_1(s) = 0$. Therefore, the zeros of $Q(s)$ are the poles of $G(s)$ and the zeros of $H(s)$.

(b) When $K = 0$,

$$Q(s) = \frac{N_1(s)}{D_1(s)}.$$

Therefore, the poles of $Q(s)$ are the poles of $H(s)$, and the zeros of $Q(s)$ are the zeros of $H(s)$.

(c) We may write $Q(s)$ as

$$Q(s) = \frac{p^2(s)}{p(s)q(s)}\left[\frac{\frac{D_2(s)}{p(s)}\frac{N_1(s)}{p(s)}}{\frac{D_1(s)}{q(s)}\frac{D_2(s)}{p(s)} + K\frac{N_1(s)}{p(s)}\frac{N_2(s)}{q(s)}}\right] = \frac{p(s)}{q(s)}\left[\frac{\tilde{H}(s)}{1+K\tilde{G}(s)\tilde{H}(s)}\right].$$

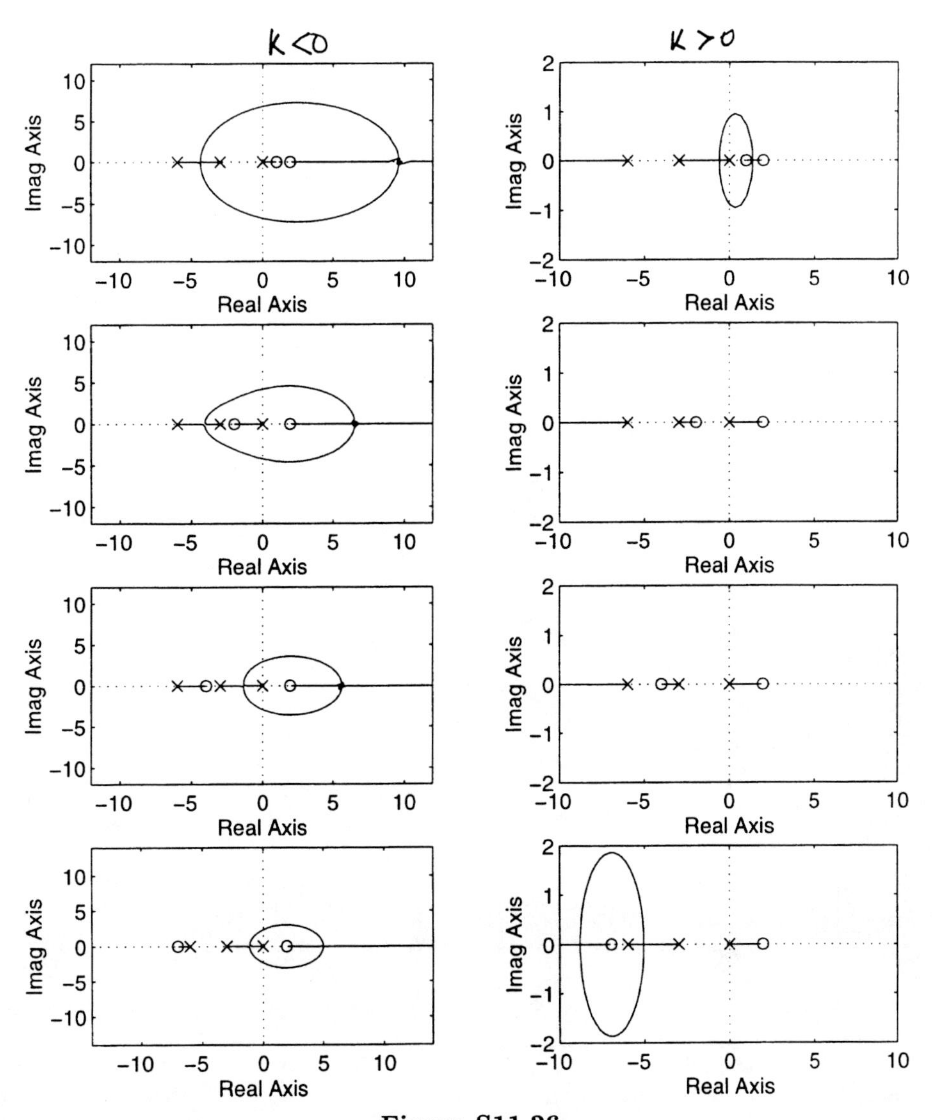

Figure S11.26

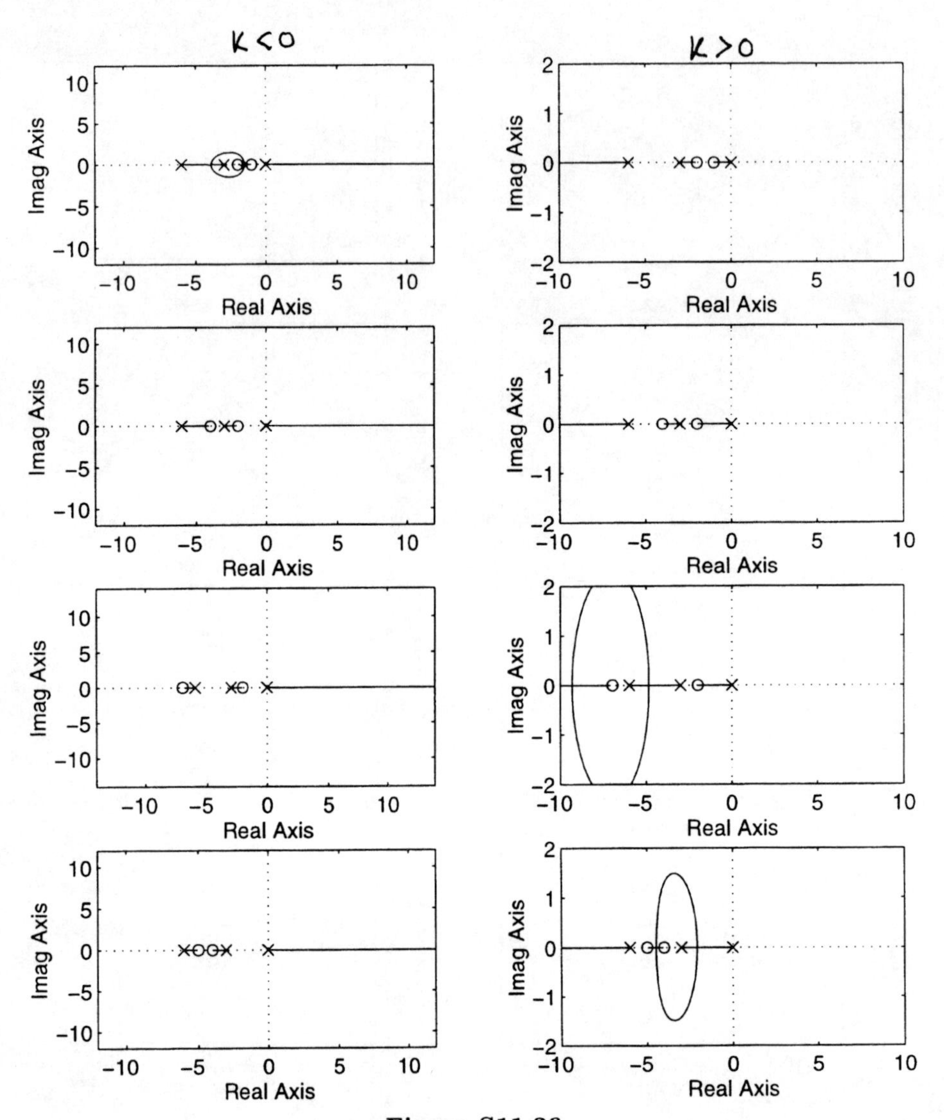

Figure S11.26

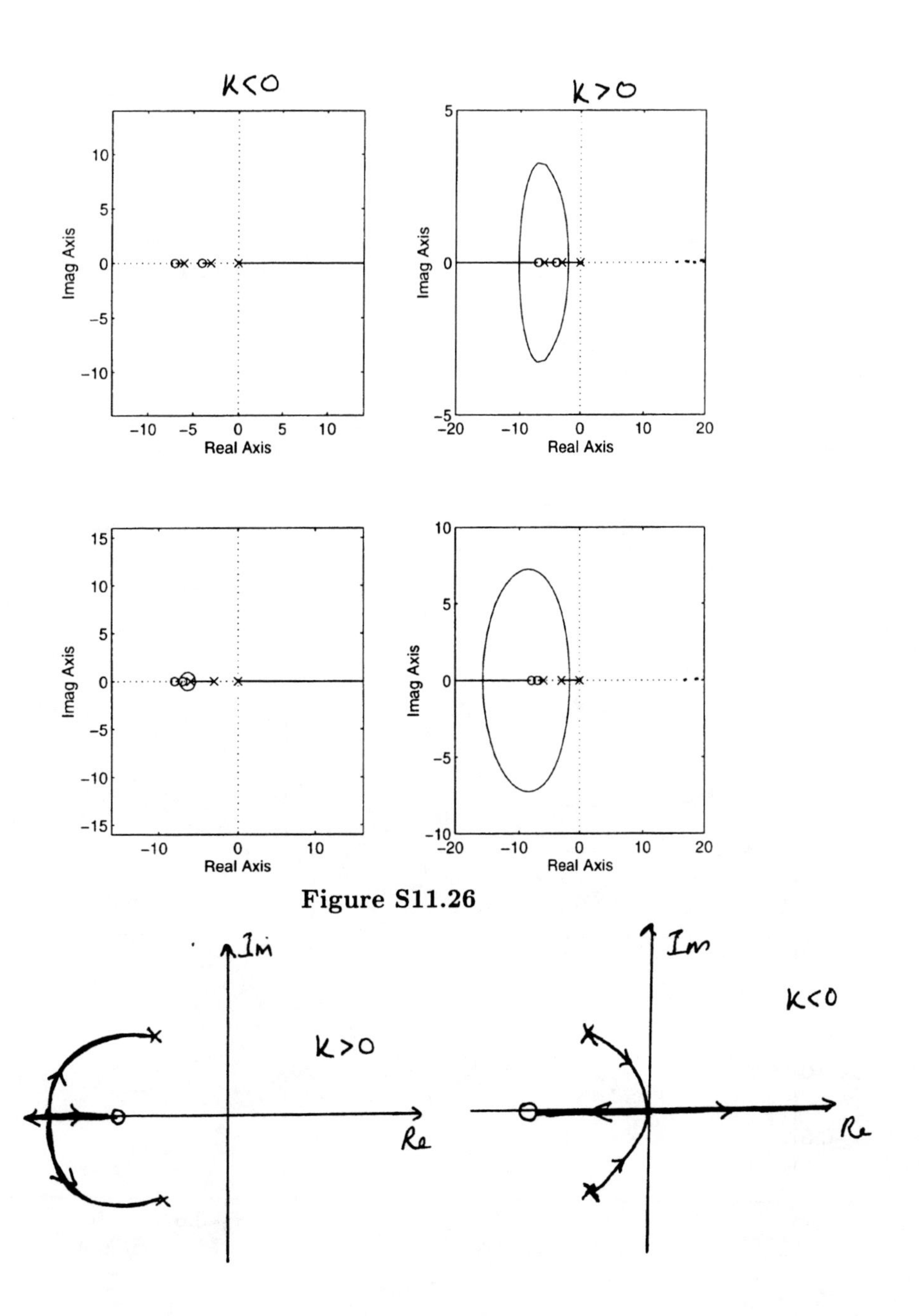

Figure S11.26

Figure S11.27

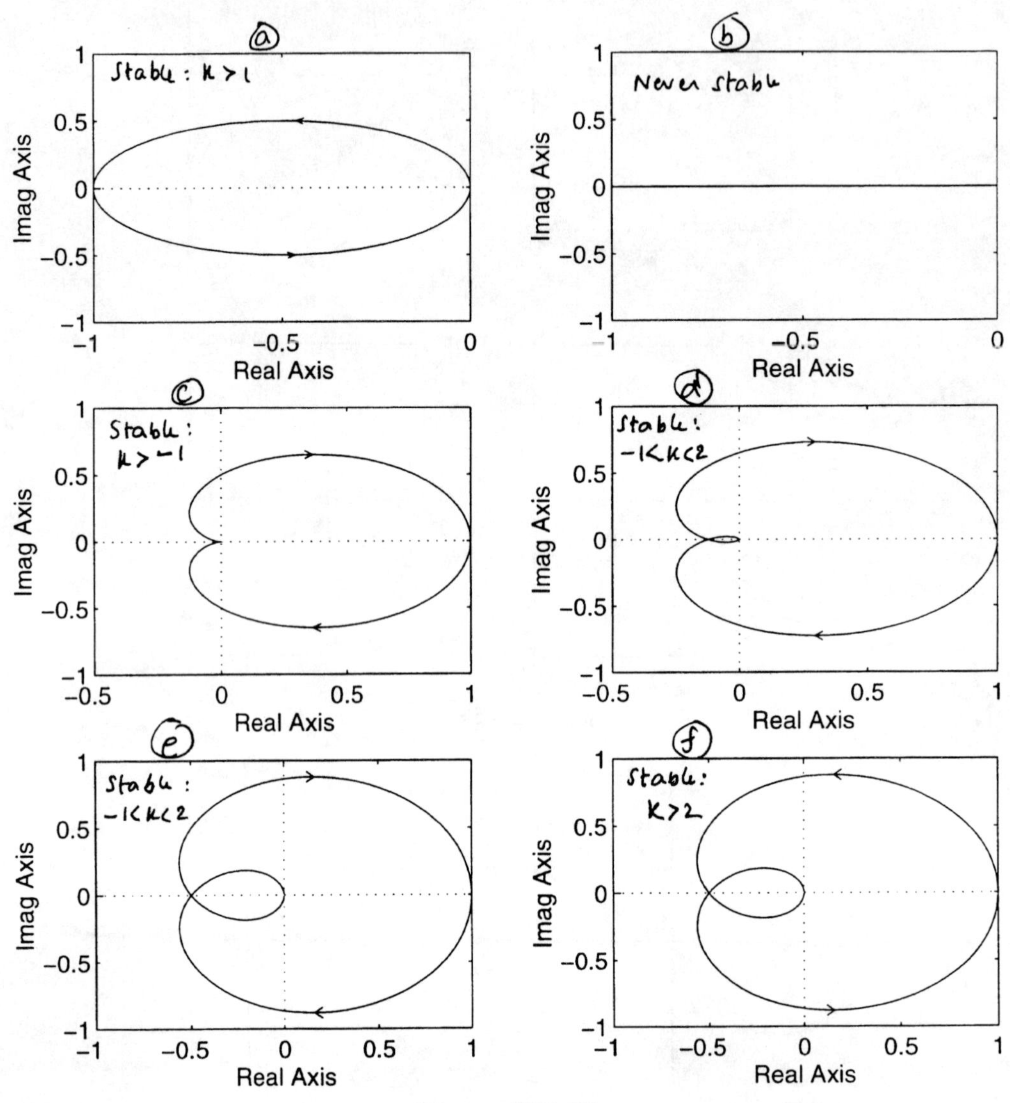

Figure S11.28

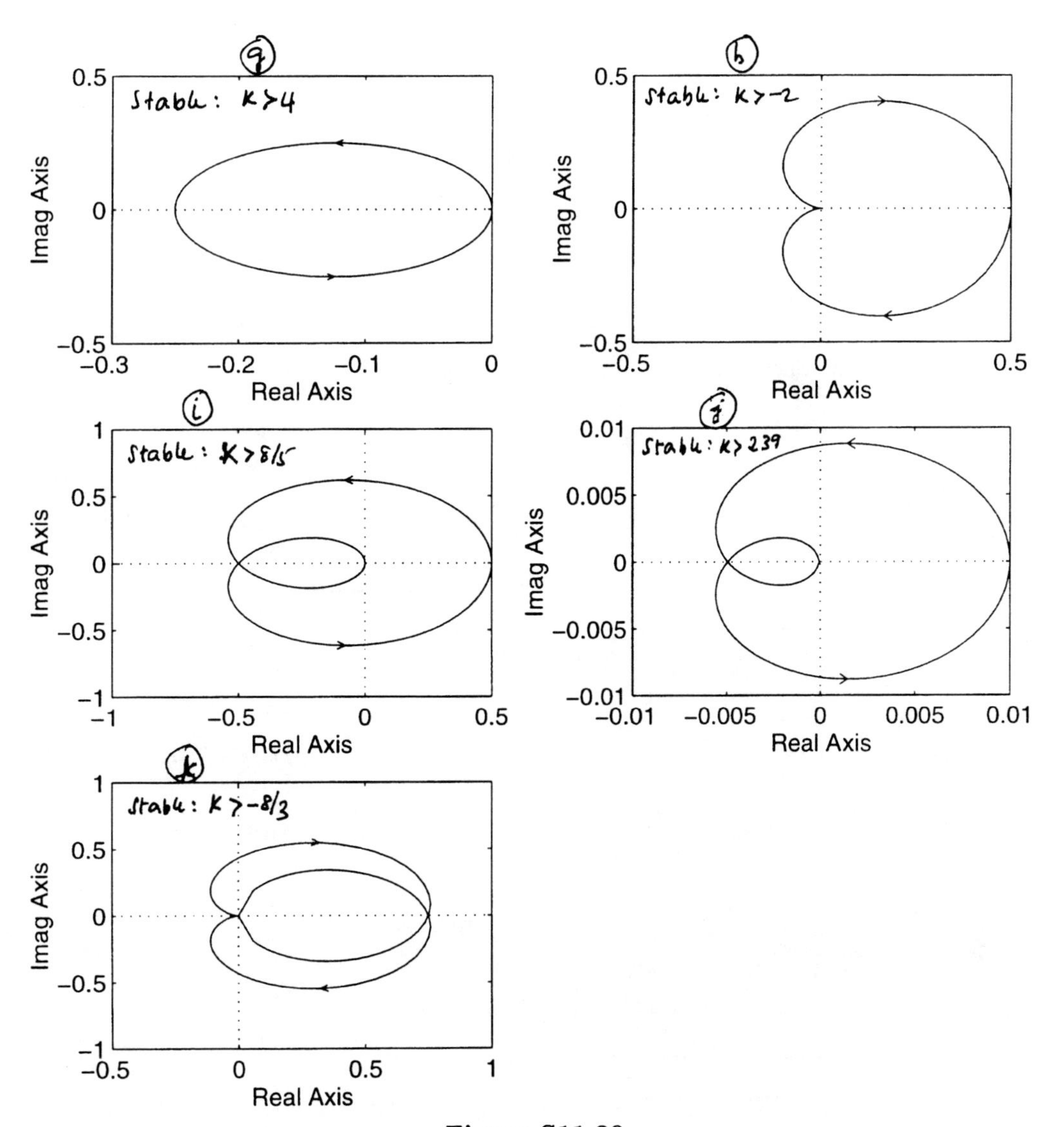

Figure S11.28

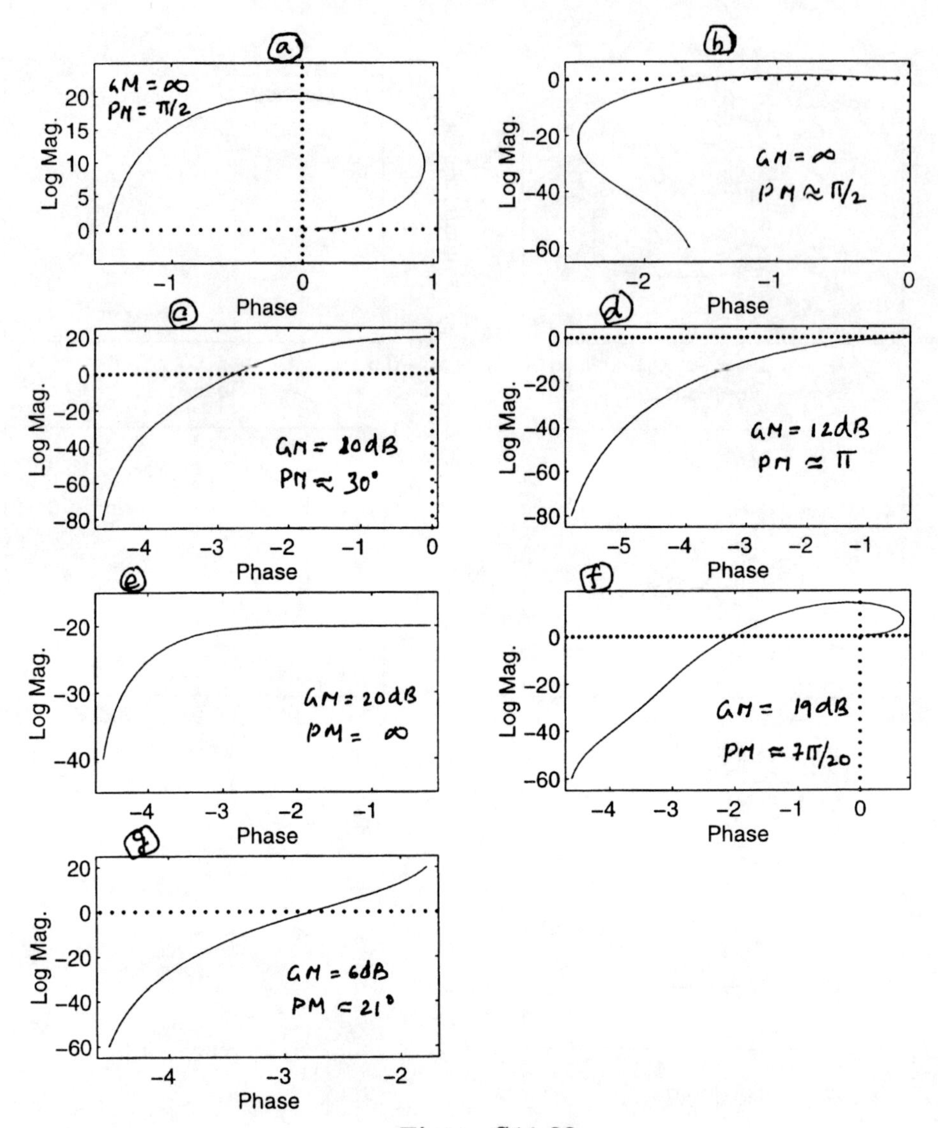

Figure S11.29

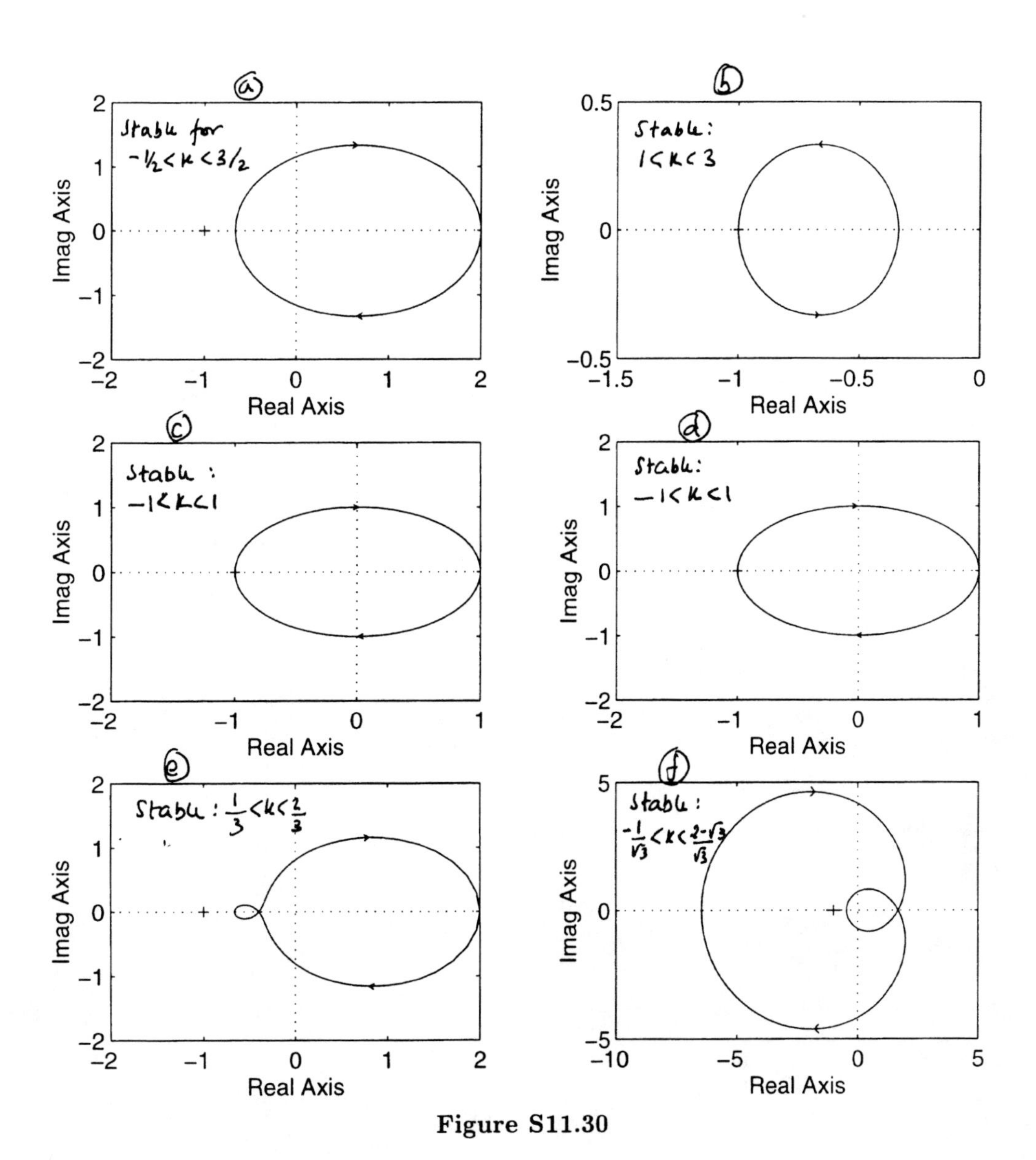

Figure S11.30

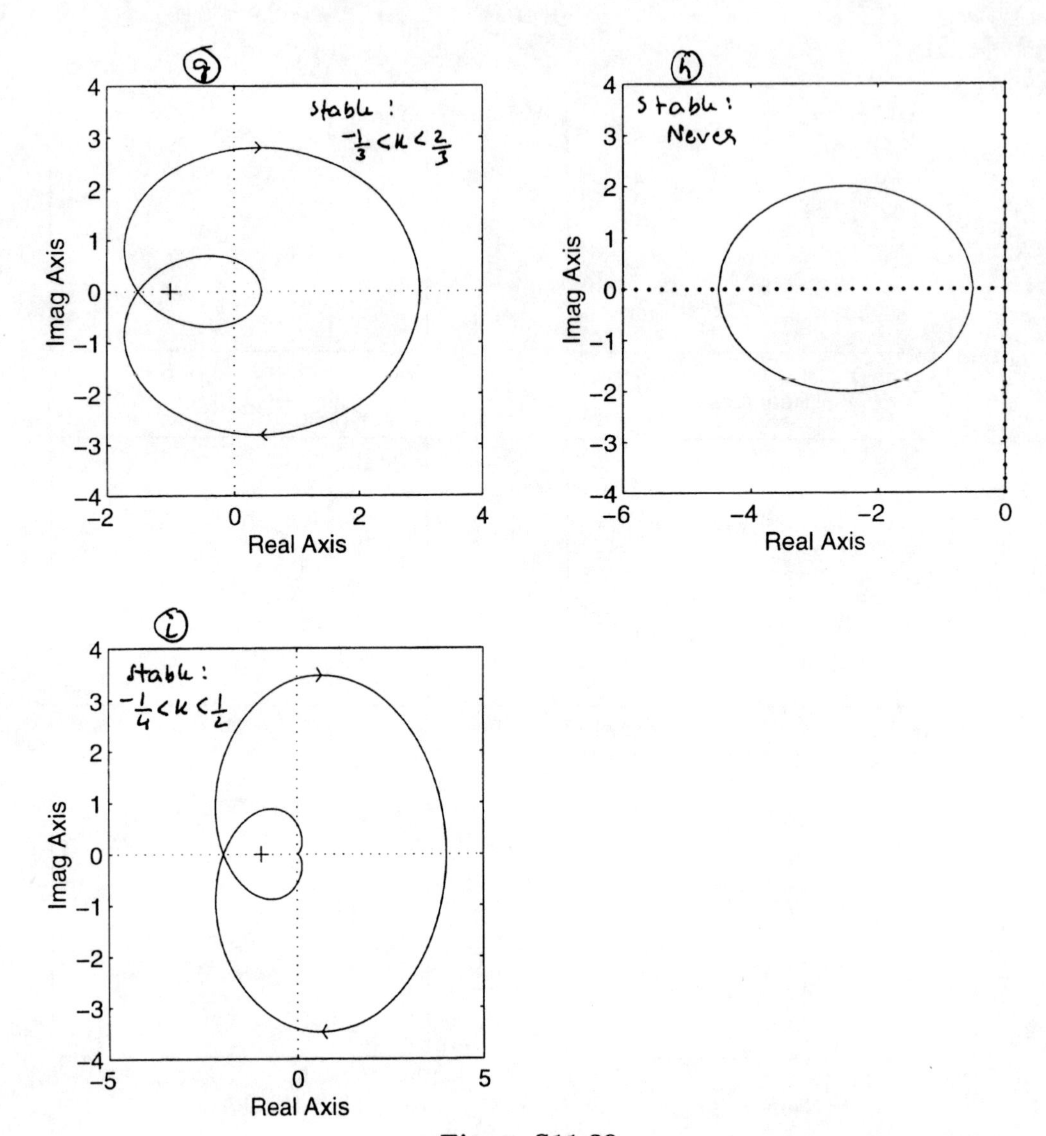

Figure S11.30

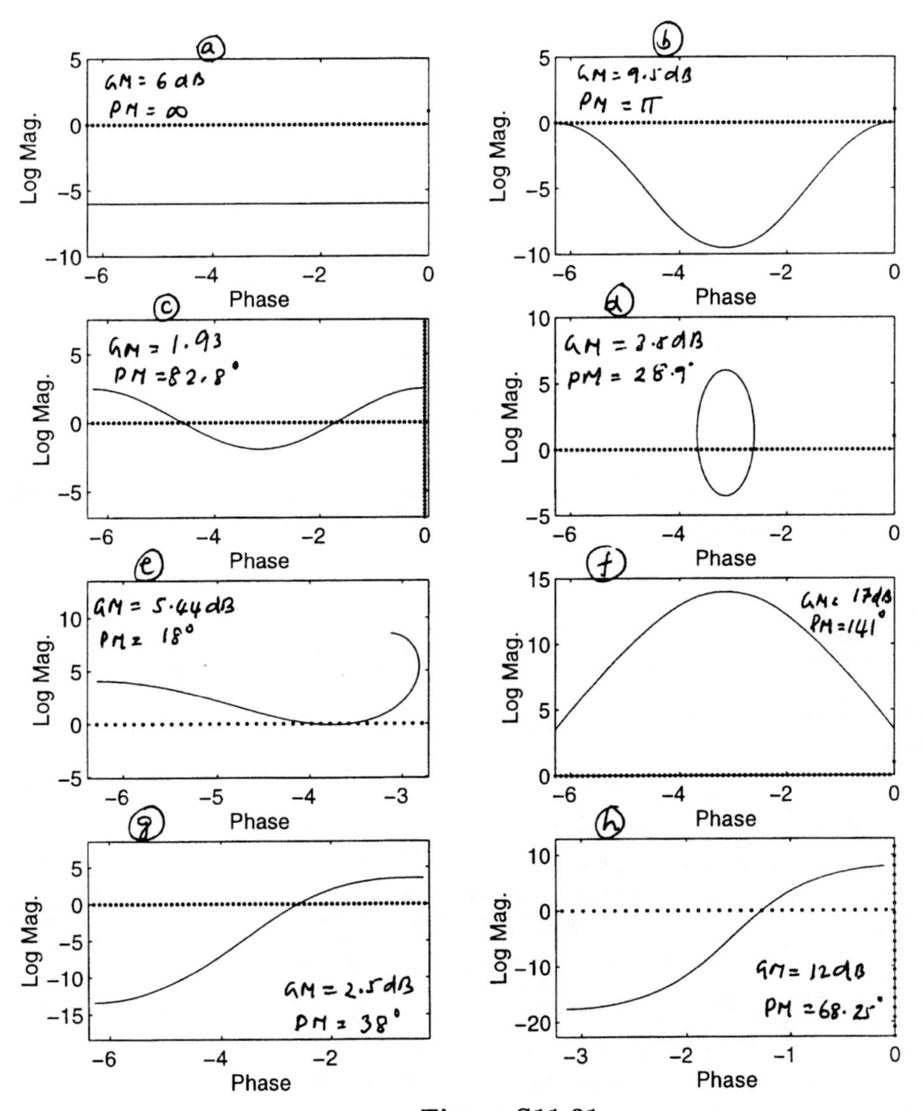

Figure S11.31

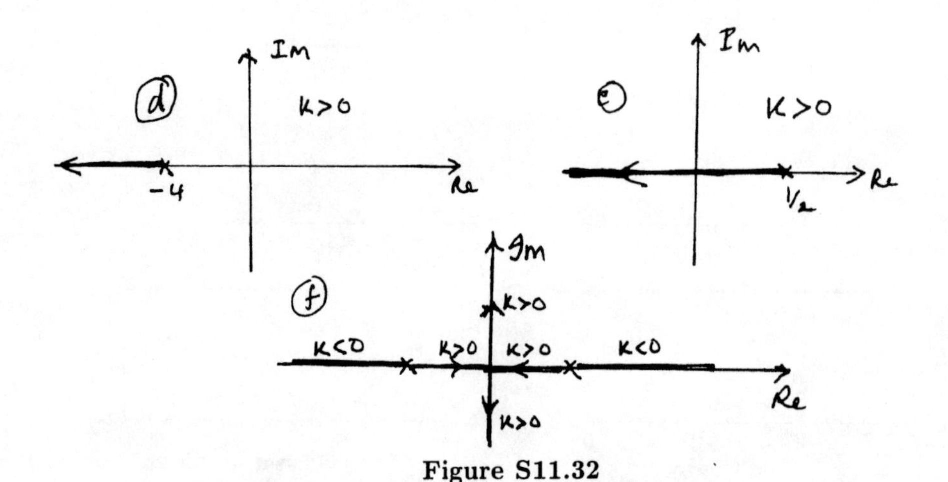

Figure S11.32

(d) In this case

$$Q(s) = \frac{s+1}{s+2}\left[\frac{1}{s+k+4}\right].$$

The zero which is independent of K is at $s = -1$. The pole which is independent of K is at $s = -2$. The root locus for the remaining closed-loop pole is as shown in Figure S11.32 for $K > 0$.

(e) In this case

$$Q(z) = (z+1)\left[\frac{1}{z+k-(1/2)}\right].$$

The zero which is independent of K is at $z = -1$. The root locus for the pole is as shown in Figure S11.32 for $K > 0$.

(f) (i) For this case, we have $G(z)H(z) = 1/[(z-2)(z+2)]$. The root-locus for $K > 0$ and $K < 0$ are shown in Figure S11.32.

(ii) The system is stable for when the closed-loop poles are within the unit circle. The closed-loop poles satisfy the condition

$$G(z)H(z) = -1/K.$$

Therefore, looking at the plots from before, it is clear that as K increases, the system becomes stable when $G(1)H(1) = -1/K$. That is, the system becomes stable when $K > 3$. As K continues to increase, the system again becomes unstable when $G(j)H(j) = -1/K$. That is, the system becomes unstable when $K > 5$. Therefore, the system is stable for $3 < K < 5$.

(iii) When $K = 4$, $Q(z) = 1$. Therefore, $q[n] = \delta[n]$.

11.33. The root loci are as shown in Figure S11.33.

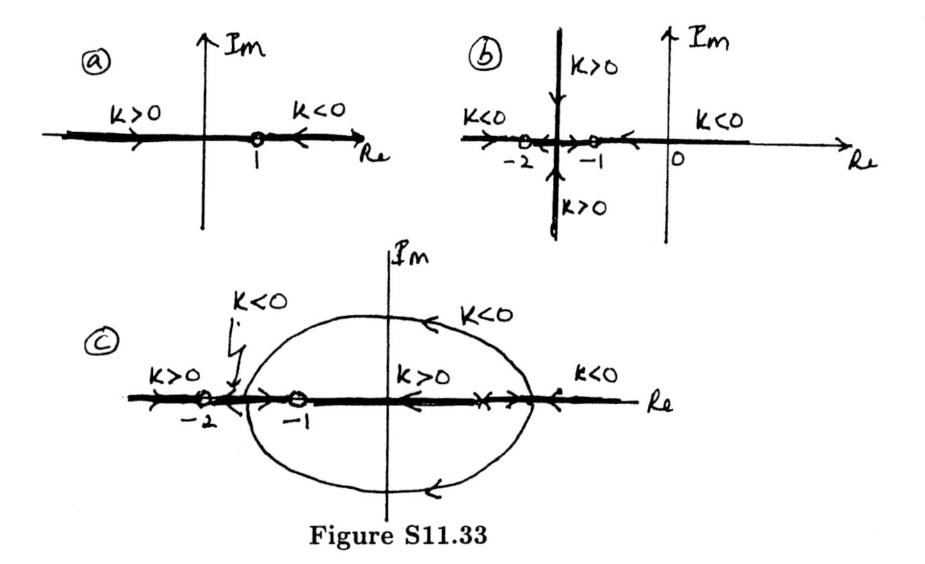

Figure S11.33

11.34. **(a)** For $K > 0$ and for large s, Figure P11.34 shows that the angle contributed by any pole of $G(s)H(s)$ is approximately equal to the angle contributed by any zero of $G(s)H(s)$. Therefore,

$$\sphericalangle\{G(s)H(s)\} = (m - n)\theta$$

where θ is the angle contributed by any zero. To be on the root-locus we require that (angle criterion)

$$(n - m)\theta = (2k + 1)\pi, \qquad k = 0, 1, 2, \cdots, (n - m - 1).$$

This implies that

$$\theta = \frac{(2k + 1)\pi}{n - m}, \qquad k = 0, 1, 2, \cdots, (n - m - 1).$$

Similarly, for $K < 0$,

$$(n - m)\theta = 2k\pi, \qquad k = 0, 1, 2, \cdots, (n - m - 1).$$

This implies that

$$\theta = \frac{2k\pi}{n - m}, \qquad k = 0, 1, 2, \cdots, (n - m - 1).$$

(b) (i) By expanding the right-hand side we get the s^{r-1} term to be

$$s^{r-1}(-\zeta_1 - \zeta_2 \cdots - \zeta_r).$$

Equating the coefficients of s^{r-1}, we get

$$f_{r-1} = -\sum_{i=1}^{r} \zeta_i.$$

(ii) Assuming that $a_n = b_m = 1$ and performing long division gives

$$\frac{1}{G(s)H(s)} = s^{n-m} + (a_{n-1} - b_{m-1})s^{n-m-1} + \cdots .$$

Therefore,

$$\gamma_{n-m-1} = a_{n-1} - b_{m-1}.$$

Using the result of part (b-i) with eq. (P11.34-2) gives us

$$a_{n-1} = -\sum_{k=1}^{n} \alpha_k, \qquad b_{m-1} = -\sum_{k=1}^{m} \beta_k.$$

Therefore,

$$\gamma_{n-m-1} = a_{n-1} - b_{m-1} = \sum_{k=1}^{m} \beta_k - \sum_{k=1}^{n} \alpha_k.$$

(iii) From eq. (P11.34-1), we have

$$\frac{1}{G(s)H(s)} + K = 0.$$

Now, for large s, eq .(P11.34-3) may be approximated as

$$\frac{1}{G(s)H(s)} \approx s^{n-m} + \gamma_{n-m-1}s^{n-m-1} + \cdots + \gamma_0.$$

Therefore,

$$s^{n-m} + \gamma_{n-m-1}s^{n-m-1} + \cdots + \gamma_0 + K = 0.$$

(iv) From (i), if ζ_i are the $n-m$ poles, then

$$\sum_{i=1}^{n-m} \zeta_i = -\gamma_{n-m-1} = b_{m-1} - a_{n-1}.$$

(c) (i) Here, $n - m = 3$. For $K > 0$, $\theta = (2k+1)\pi/3$ with $k = 0, 1, 2$. This implies that $\theta = \pi/3, \pi, -\pi/3$.

For $K < 0$, $\theta = 2k\pi/3$ with $k = 0, 1, 2$. This implies that $\theta = 0, 2\pi/3, -2\pi/3$.

(ii) Intersection point is $= \frac{\alpha_k}{3} = \frac{-1-3-5}{3} = -3$. Therefore, intersection point is $s = -3$.

(iii) The root locus is as shown in the Figure S11.34.

(d) Plots are shown in Figure S11.34.

(e) For $n - m > 3$, the poles go to ∞ at asymptotic angles which are $(2k+1)\pi/(n-m)$ for $K > 0$ and $2k\pi/(n-m)$ for $K < 0$. The smallest angles are $\pi/(n-m) \le \pi/3$ and 0. This implies that at least some of the poles will enter the right-half of the s-plane for large enough K. This causes instability for large K.

(f) The root locus is as shown in Figure S11.34.

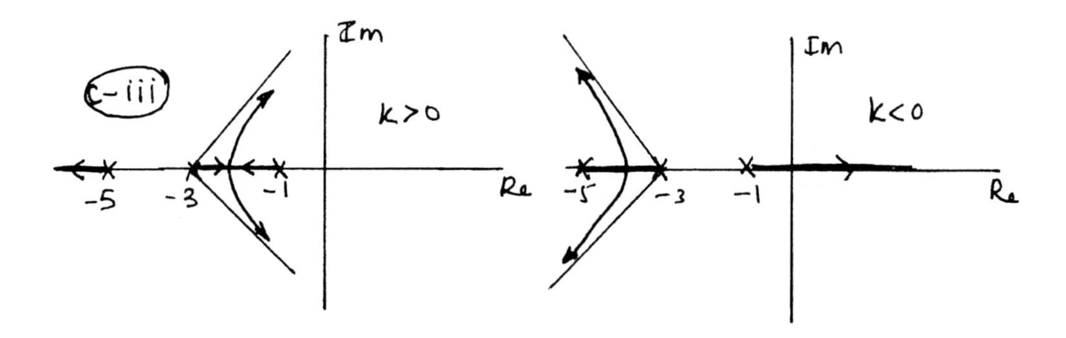

Figure S11.34−1

(g) If $n > m$, then at least 1 pole goes towards infinity for large K. This implies that there will be a pole outside the unit circle for large K. This in turn implies that the system would be unstable for large K.

11.35. (a) We have

$$G(j\omega)H(j\omega) = -\frac{1}{K}.$$

Therefore,

$$K(j\omega - 1) = \omega^2 - 3j\omega - 2.$$

Equating the real and imaginary parts on both sides, we get $K = -3$ and $\omega = \pm\sqrt{5}$. Therefore, the system is stable for $-3 < K < 2$.

(b) *Continuous-time systems:* For $|K|$ sufficiently large, one of the poles approaches the zero that is in the right-half of the s-plane. Therefore, the system will have a right-half pole. This makes it unstable.

Discrete-time systems: For $|K|$ sufficiently large, one of the poles approaches the zero that is outside the unit circle. Therefore, the system will have a pole outside the unit circle. This makes it unstable.

11.36. (a) The root locus shows three poles which move towards infinity. The asymptote angles are $-\pi/3$, $\pi/3$ and π for $K > 0$ and $0, -2\pi/3$, and $2\pi/3$ for $K < 0$. The asymptotes intersect at $s = -1$. Therefore, the root locus is as shown in Figure S11.36.

(b) We have

$$G(j\omega)H(j\omega) = -\frac{1}{K}.$$

Therefore,

$$-K = -j\omega^3 - 3\omega^2 + 2j\omega.$$

Equating the real and imaginary parts on both sides, we get $K = 6$ and $\omega = \pm\sqrt{2}$.

(c) From eq. (P11.36-1), we have

$$G(s)H(s) = \frac{1}{s^3 + 3s^2 + 2s}.$$

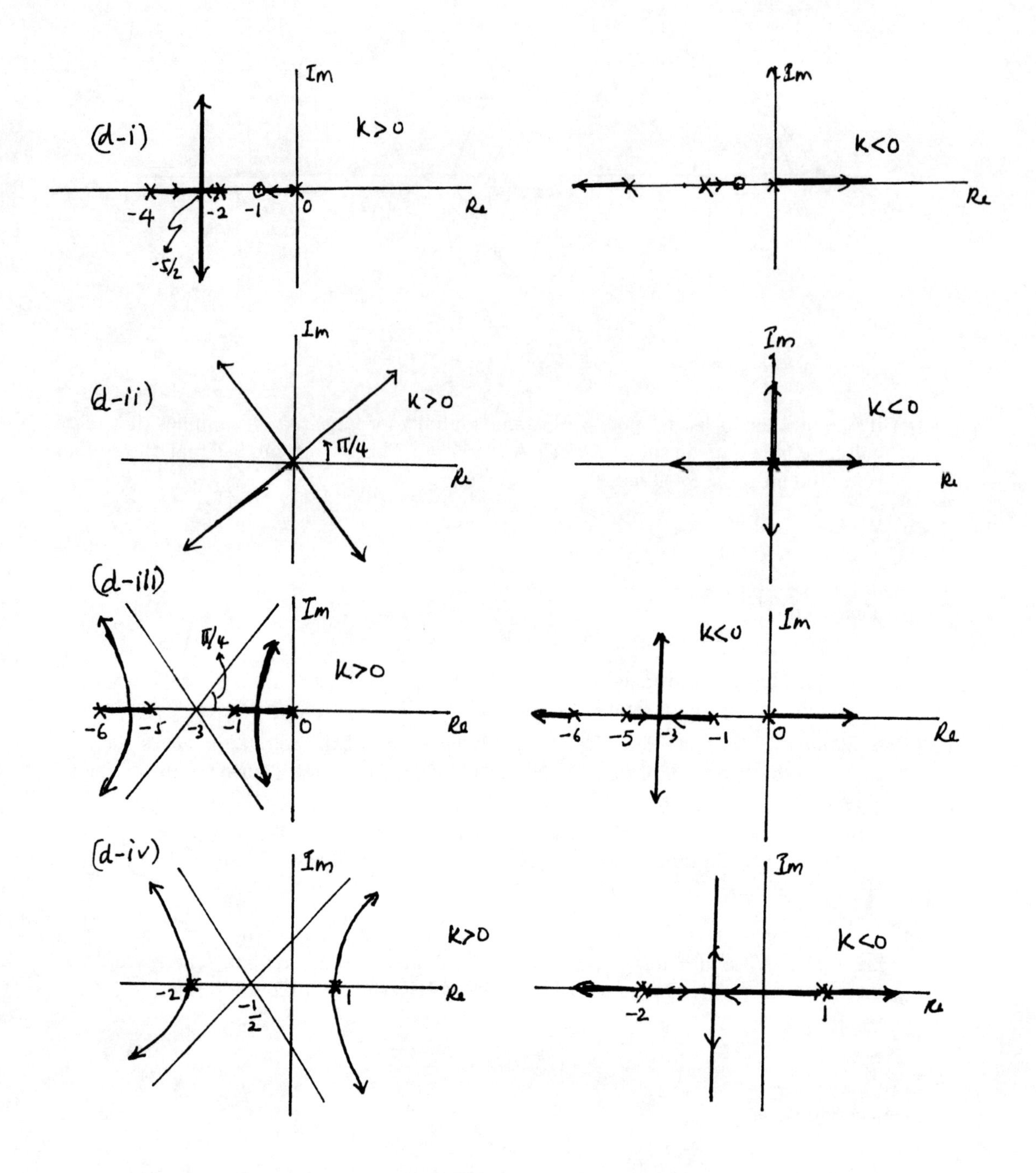

Figure S11.34 – 2

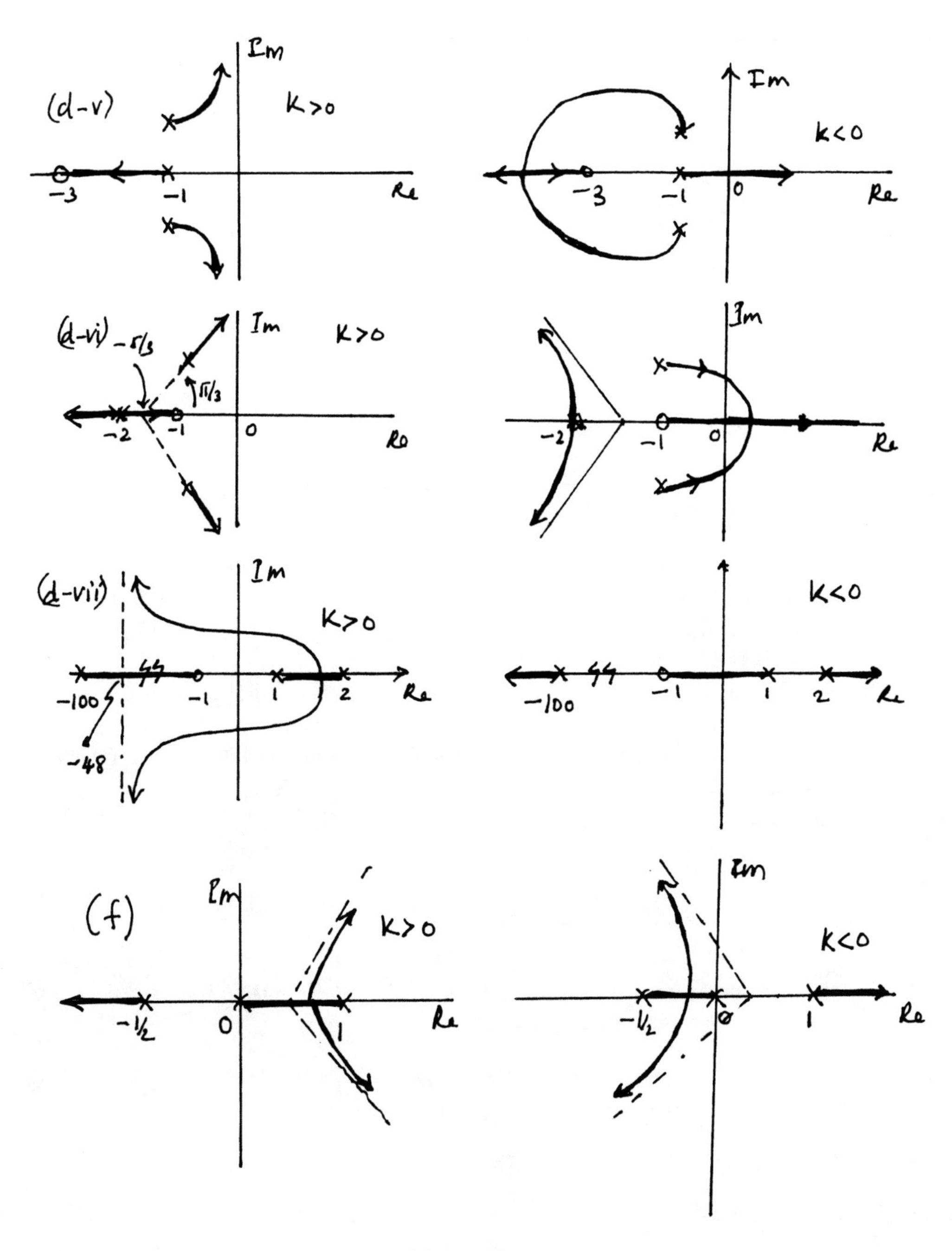

Figure S11.34 – 3

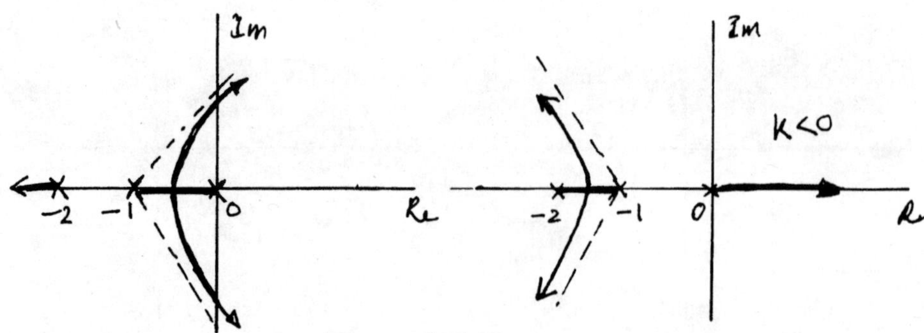

Figure S11.36

Now since $1/[G(s)H(s)] = -K$,

$$s^3 + 3s^2 + 2s = -K.$$

(i) For $K > 0$, $-1 \leq s \leq 0$ is on the root locus. Therefore, $p(s) = -K < 0$. If the breakaway point occurs at $K = K_0$, then the poles are no longer on the real axis. Therefore, the breakaway point occurs at the maximum value of K for which the poles are still real. Note that $p(s)$ in the range $-1 \leq s \leq 0$ will be minimum at this point. Therefore, $K_0 = -p(s_+)$.

(ii) For $K < 0$, $-2 \leq s \leq -1$ is on the root locus. Therefore, $p(s) = -K > 0$. If the breakaway point occurs at $K = K_1$, then the poles are no longer on the real axis. Therefore, the breakaway point occurs at the minimum value of K for which the poles are still real. Note that $p(s)$ in the range $-2 \leq s \leq -1$ will be maximum at this point. Therefore, $K_1 = -p(s_-)$.

(iii) Equating $dp(s)/ds$ to zero,

$$3s^2 + 6s + 2 = 0 \quad \Rightarrow \quad s = -1 \pm \frac{1}{\sqrt{3}}.$$

The breakaway points are $s_+ = -1 + 1/\sqrt{3}$ for $K > 0$ and $s_- = -1 - 1/\sqrt{3}$ for $K < 0$. The corresponding gains are

$$K_0 = -(s_+^3 + 3s_+^2 + 2s_+) = 0.385$$

and

$$K_1 = -(s_-^3 + 3s_-^2 + 2s_-) = -0.385$$

11.37. (a) We have $G(s)H(s) = K/[(s+10)(s-2)]$. The root locus for $K > 0$ is shown in Figure S11.37.

From the figure, it is clear that the system becomes stable when the pole crosses over to the left-half at $s = 0$. The corresponding gain is $K = 20$. Therefore, the system is stable for $K > 20$.

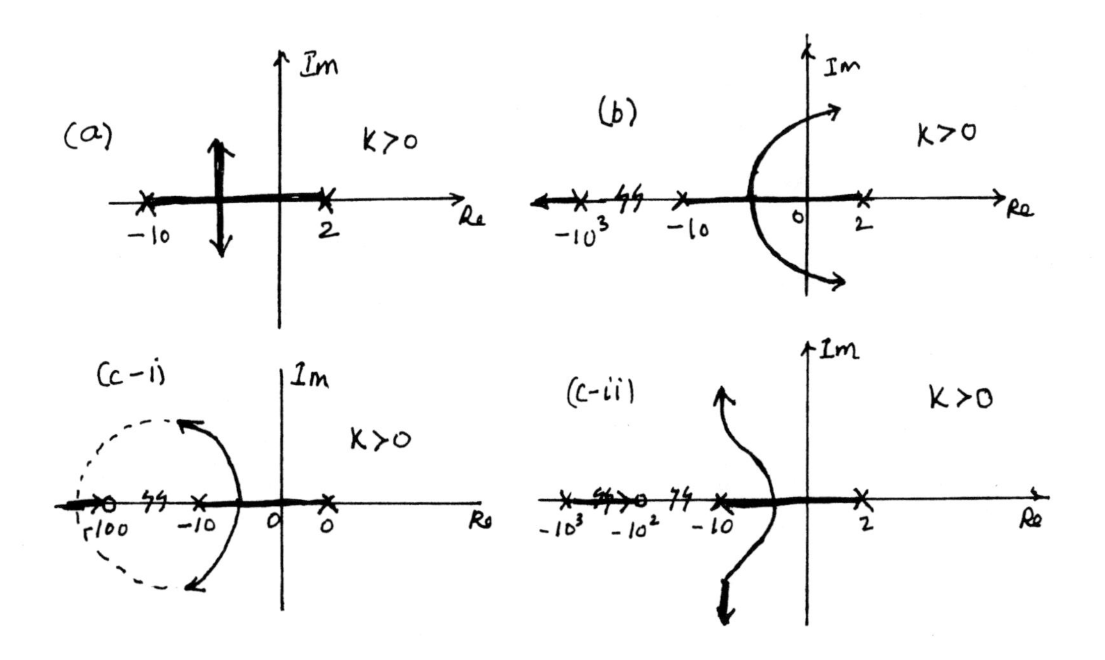

Figure S11.37

(b) In this case, for $K > 0$, poles go to infinity at angles of $-\pi/3, \pi/3$, and π. Therefore, as shown in the root locus in Figure S11.37, for sufficiently large values of K the poles cross over to the right half of the s-plane. This makes the system unstable for large K.

(c) (i) If $H(s)$ is given by eq. (P11.37-1), then

$$G(s)H(s) = \frac{K(s+100)}{(s+10)(s-2)}.$$

The root locus is drawn in Figure S11.37 for $K > 0$. The system is stable for large K.

(ii) If $H(s)$ is given by eq. (P11.37-1), then

$$G(s)H(s) = \frac{1000K(s+100)}{(s+10)(s-2)(s+1000)}.$$

The root locus is drawn in Figure S11.37 for $K > 0$. The system is stable for large K.

11.38. **(a)** When $a = 1/2$,

$$G(z)H(z) = \frac{K(z-1/2)}{z(z-1)}.$$

The root locus for $K > 0$ and $K < 0$ is shown in Figure S11.38.

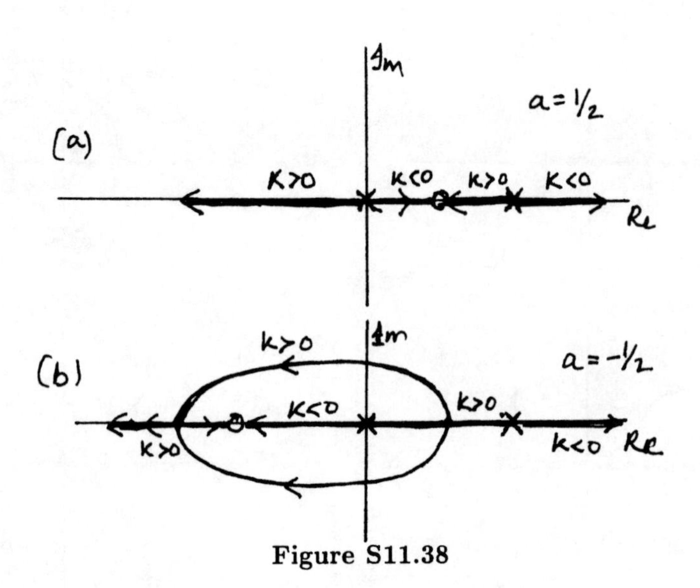

Figure S11.38

(b) When $a = -1/2$,

$$G(z)H(z) = \frac{K(z + 1/2)}{z(z-1)}.$$

The root locus for $K > 0$ and $K < 0$ is shown in Figure S11.38.

(c) If the closed-loop impulse response is $q[n] = (A + Bn)\alpha^n$, the denominator of the closed-loop system function

$$Q(z) = \frac{H(z)}{1 + G(z)H(z)}$$

must be of the form $(z - \alpha)^2$. This implies that

$$z^2 + (K-1)z + \frac{1}{2}K = (z - \alpha)^2. \qquad \text{(S11.39–1)}$$

Equating coefficients of different powers of z on both sides, we get

$$K = 2 \pm \sqrt{3}.$$

Since $|\alpha| < 1$, we know from eq. (S11.38-1) that $K/2 < 1$. Therefore, $K = 2 - \sqrt{3}$.

11.39. (a) The root locus is as shown in the Figure S11.39.

(b) The closed-loop system function is

$$Q(z) = \frac{z}{z(1+K) - 1/2}. \qquad \text{(S11.39–1)}$$

Clearly, as K is decreased from 0 to -1, the closed-loop pole given by $1/[2(1+K)]$ goes from $1/2$ to ∞. When K is decreased below -1, the pole moves from $-\infty$ towards 0. The root locus for $K < 0$ is as shown in Figure S11.39.

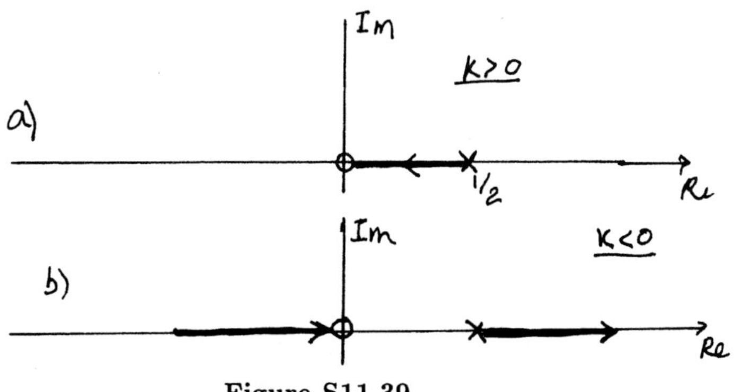

Figure S11.39

(c) From the figures above and the analysis of (b), it is clear that the system is stable for $K < -1$.

(d) Noting that $H(z) = Y(z)/E(z)$, The difference equation relating $y[n]$ and $e[n]$ is

$$y[n] - \frac{1}{2}y[n-1] = e[n].$$

Noting that $G(z) = R(z)/Y(z)$, we get

$$r[n] = Ky[n].$$

Also, note that $e[n] = x[n] - r[n]$. Therefore,

$$e[n] = x[n] - Ky[n].$$

Note that here, $e[n]$ depends on $y[n]$ and $y[n]$ in turn depends on $e[n]$.

If instead $G(z) = Kz^{-1}$, then

$$r[n] = Ky[n-1]$$

and

$$e[n] = x[n] - r[n] = x[n] - Ky[n-1].$$

Here, $e[n]$ depends on $y[n-1]$.

(e) Since the pole is not at $|z| = \infty$, we know that $K \neq -1$. From eq. (S11.39-1), we get

$$y[n] = \frac{1}{1+K}\left[x[n] + \frac{1}{2}y[n-1]\right].$$

Assuming $x[n] = 0$ for $n < n_0$, the above difference equation shows that $y[n] = 0$ for $n < n_0$. Therefore, the system is causal.

11.40. From the given $G(z)$ and $H(z)$, we have

$$G(z)H(z) = \frac{K(z-1/4)(z-1/2)}{z(z^2 - \frac{7\sqrt{2}}{8}z + \frac{49}{64})}.$$

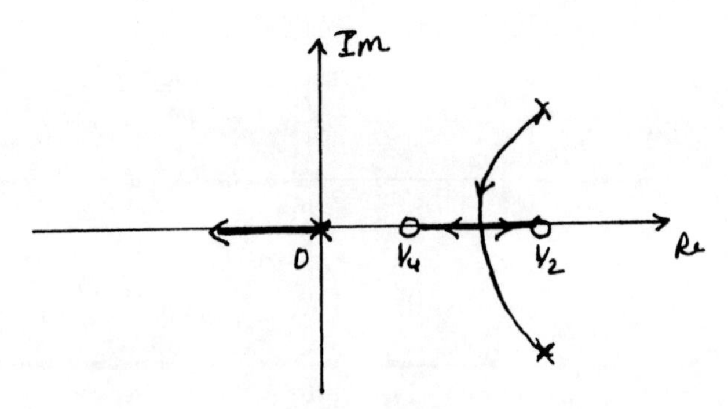

Figure S11.40

The root-locus for $K > 0$ is as shown in Figure S11.40.

Damping is improved when the closed-loop poles are not too close to the unit circle. If we choose $z = -1/2$ to be a closed-loop pole, then $K = 1.06$. This value of K provides improved damping.

11.41. (a) (i) The closed-loop system function is

$$Q(z) = \frac{H(z)}{1 + G(z)H(z)} = \frac{z^2 - 1}{z^3 + (K - 3/4)z + (K - 1/4)}.$$

(ii) Sum of the closed loop poles is the negative of the coefficient of z^2. Since this is zero, it is independent of K.

(b) The closed loop poles are the solutions of the following equation:

$$z^n + a_{n-1}z^{n-1} + \cdots + a_0 + K(z^m + b_{m-1}z^{m-1} + \cdots + b_0) = 0.$$

If $m < n - 2$, then the coefficient a_{n-1} of z^{n-1} is independent of K and the sum of the roots which is equal to $-a_{n-1}$ is independent of K.

11.42. (a) We know that

$$G(z)H(z) = \frac{z}{(z - 1/2)(z - 1/4)} = -\frac{1}{K}.$$

If $z = -1$, then from the above equation $K = 15/8$.

(b) If $z = 1$, then from the above equation $K = -3/8$.

(c) The system is stable for $(-3/8) < K < (15/8)$.

11.43. (a) The root-locus is sketched in Figure S11.43 for $K < 0$ and $K > 0$.

(b) Let the poles cross the unit circle at $z_0 = a + jb$ and $z_0* = a - jb$. Note that $|z_0| = |z_1| = |a + jb| = \sqrt{a^2 + b^2} = 1$. We also have

$$G(z_0)H(z_0) = -\frac{1}{K}.$$

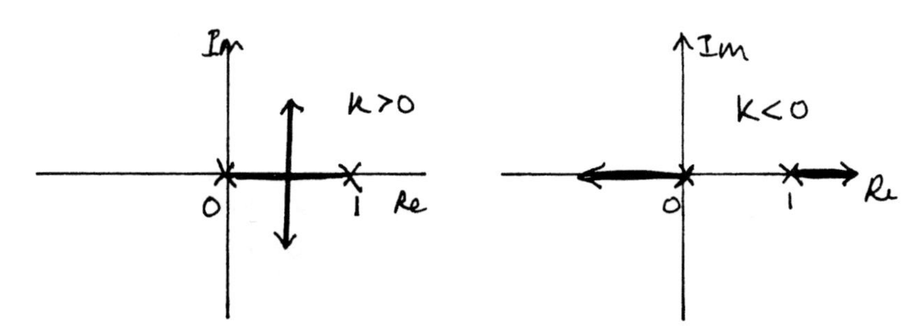

Figure S11.43

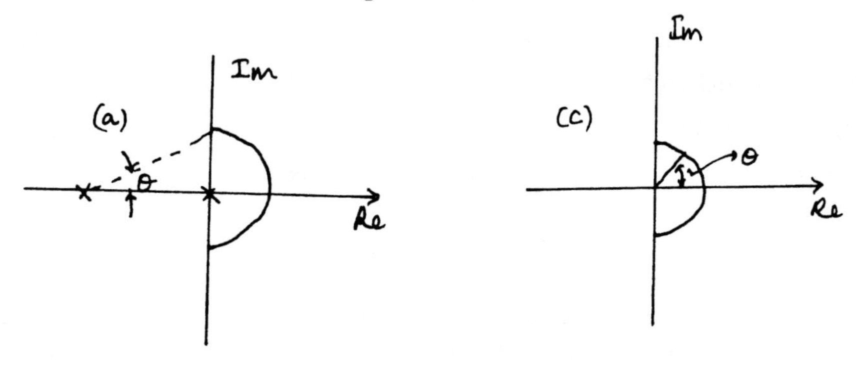

Figure S11.44 — 1

Therefore,

$$z_0(z_0 - 1) = -K.$$

This implies that

$$-(a^2 - b^2 - a - jb + 2jab) = K.$$

Equating the imaginary parts on both sides, we get $a = 1/2$ and $b = \sqrt{3}/2$. From this we get $K = 1$. The exit points are $(1/2) \pm j(\sqrt{3}/2)$.

11.44. (a) At the point $j0^+$ on the infinitesimal circle around the origin, the angle ϕ contributed to $\sphericalangle G(j0^+)H(j0^+)$ (see Figure S11.44) by the pole at $s = -1$ is approximately zero. Therefore, $\sphericalangle G(j0^+)H(j0^+)$ is only due to the pole at the origin. Therefore,

$$\sphericalangle G(j0^+)H(j0^+) = -\frac{\pi}{2}.$$

Similarly,

$$\sphericalangle G(j0^+)H(j0^+) = \frac{\pi}{2}.$$

(b) We have

$$G(j\omega)H(j\omega) = \frac{-\omega^2 - j\omega}{\omega^4 + \omega^2}.$$

It is easy to see that when $\omega = \pm\infty$, $G(j\omega)H(j\omega) = 0$. Now note that as ω increases from $-\infty$ to 0^-, $G(j\omega)H(j\omega)$ changes from ∞ to $-1 + j\infty$. Figure P11.44(b) shows this fact. Also, as ω decreases from ∞ to 0^+, $G(j\omega)H(j\omega)$ changes from ∞ to $-1 - j\infty$. This is again depicted in Figure P11.44(b).

(c) We may use the argument presented in part (a) to prove this statement. Note that for any point on the infinitesimal circle, the contribution to $\sphericalangle G(s)H(s)$ by the pole at $s = -1$ is negligible. This is shown in the figure below.

Clearly, the only contribution to $\sphericalangle G(s)H(s)$ comes from the pole at the origin. From Figure S11.44, we note that this contribution is θ. Therefore,

$$\sphericalangle G(s)H(s) = -\theta.$$

(d) The system is stable for $-(1/K) < 0$. This implies that the system is stable for $K > 0$.

(e) The Nyquist plots are as shown in Figure S11.44.

(f) The Nyquist plots are as shown in Figure S11.44.

11.45. (a) Note that the system function of the overall system is

$$P(s) = H(s)C(s) = \frac{1}{(s+1)(s+3)}.$$

Although it appears as if the overall system is now stable, this is still not a good way to achieve stabilization. The reason in that since $H(s)$ is unstable, the output of $H(s)$ will not be bounded for some inputs. In practical systems, this will lead to unpredictable outputs for some inputs. In such cases, $C(s)$ can in now way undo the damage caused by $H(s)$ since it follows $H(s)$.

(b) The Nyquist plot for this system is shown in Figure S11.45. Since the system has right-half pole, we require that the Nyquist plot encircle the $-1/K$ point once in the counter-clockwise direction. This is clearly not possible. Therefore, the system is not stable for any value of K.

(c) In this case, $G(s)C(s) = K(s+a)/[(s+1)(s-\mathbf{2})]$. The Nyquist plot for $0 < a < 1$ and for $a > 1$ is shown in Figure S11.45.

From these figures, it is clear that for $0 < a < 1$, the system is stable for $-1/K$ in the range $-a/2$ to 0. This implies that for $0 < a < 1$, the system is stable for $K > (2/a)$. It is also clear for the figure that for $a > 1$, the system is stable for $-1/K$ in the range -1 and 0. This implies that for $a > 1$, the system is stable for $K > 1$.

(d) The denominator of the closes-loop system function is

$$s^2 + s(K-1) + 2(K-1).$$

This must be equal to $s^2 + \omega_n s + \omega_n^2$. Therefore,

$$K - 1 = \omega_n \quad \text{and} \quad 2(K-1) = \omega_n^2.$$

Solving these two equations, we get $K = 1$ or $K = 3$. Since $\omega_n > 0$, K has to be 3.

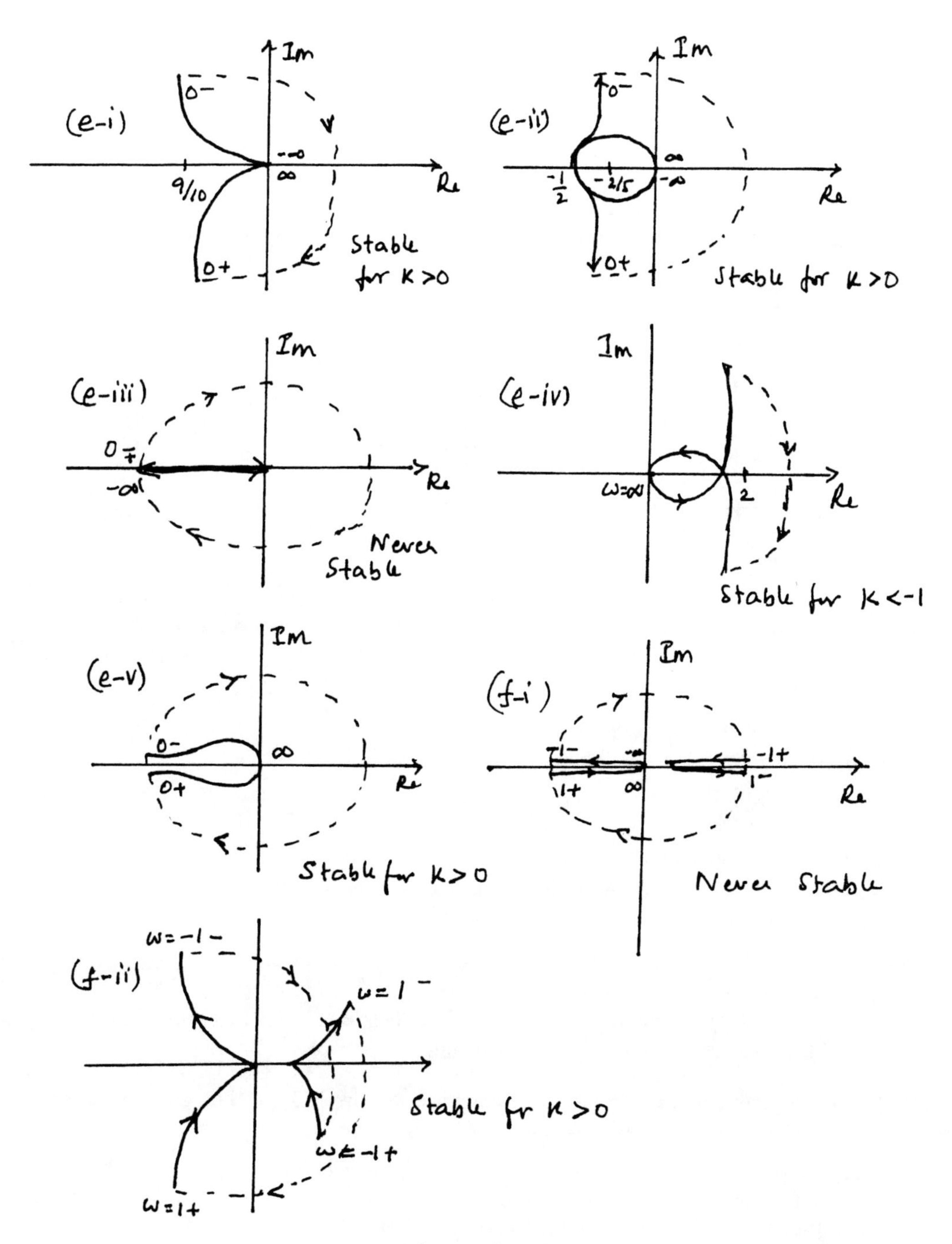

Figure S11.44

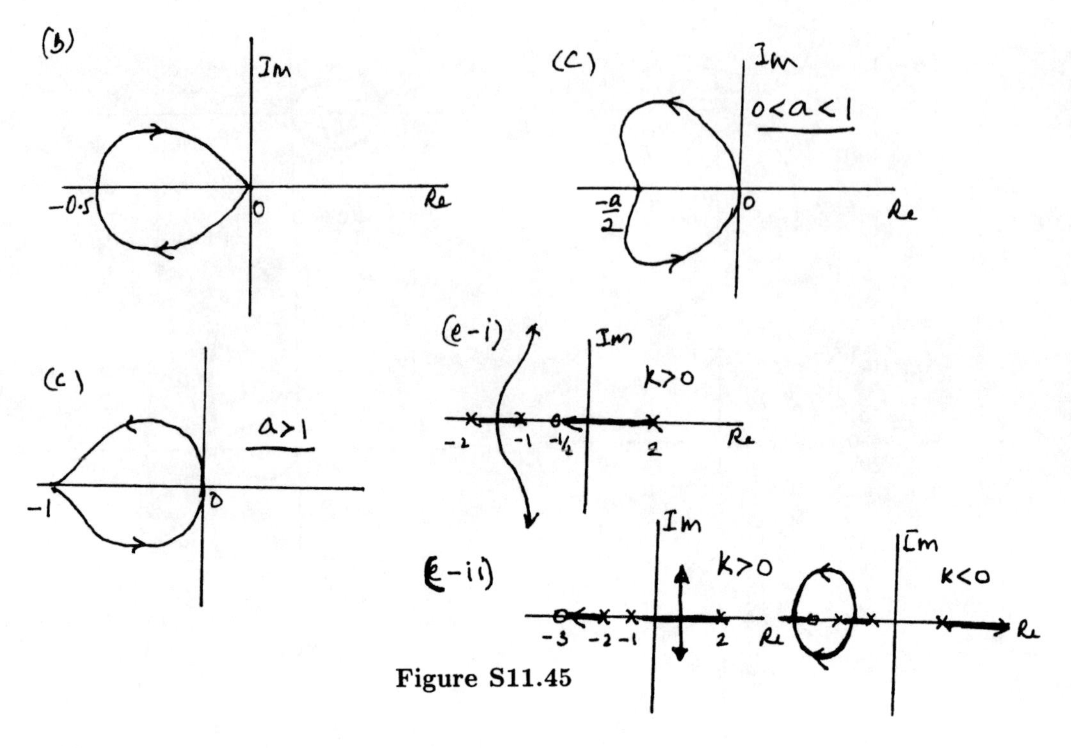

Figure S11.45

(e) (i) In this case,

$$G(s)H(s) = \frac{K(s+1/2)}{(s+2)(s+1)(s-2)}.$$

The root locus for this system for $K > 0$ is as shown in Figure S11.45. Clearly, for sufficiently large K, all poles are in the left-half of the s-plane.

(ii) In this case,

$$G(s)H(s) = \frac{K(s+3)}{(s+2)(s+1)(s-2)}.$$

The root locus for this system for $K > 0$ is as shown in Figure S11.45. Clearly, there are always poles in the right-half of the s-plane.

11.46. **(a)** The log magnitude-phase plot is as shown in Figure S11.46. The gain margin is 20 dB at $\omega = 100$ and the phase margin is $\pi/4$ at $\omega = 10$

(b) The extra phase added to the phase of the original system is $-\omega\tau$. Now, at $\omega = 10$, this phase cannot exceed the negative of the phase margin. Therefore, in order to ensure stability $10\tau < \pi/4$. This implies that $\tau < \pi/40$.

(c) For the phase margin, we first need to find ω_0 such that $|G(j\omega_0)H(j\omega_0)| = 1$. For this we need $\omega_0 = 12.66$. Now, $\sphericalangle G(j\omega_0)H(j\omega_0) = -133.7°$. Therefore the phase margin is $180 - 133.7 = 46.3°$.

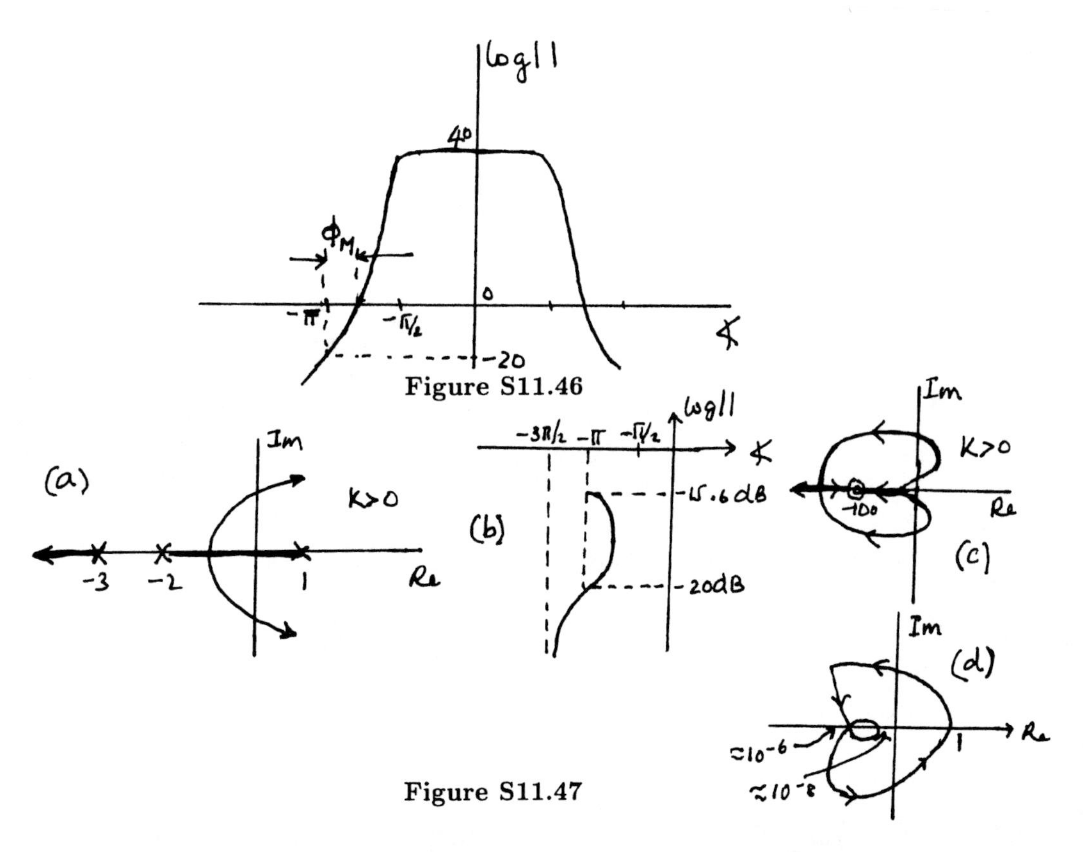

Figure S11.46

Figure S11.47

For the gain margin, we first need to find ω_0 such that $\sphericalangle G(j\omega_0)H(j\omega_0) = -\pi$. For this we need $\omega_0 = 93$. Now, $|G(j\omega_0)H(j\omega_0)| = -24.7$ dB. Therefore the gain margin is 24.7 dB.

11.47. **(a)** The root locus is drawn below for $K > 0$. The system is stable for $6 < K < 10$.

(b) For $K = 7$,

$$Q(s) = \frac{(s-1)(s+2)(s+3)}{s^3 + 4s^2 + s + 1}.$$

Using the Routh criterion, we can argue that all poles have negative real parts since all denominator coefficients are positive. Therefore, the system is stable. The log magnitude-phase plot is as shown in Figure S11.47.

We see that when $\omega = 0^+$, $\sphericalangle G(j\omega)H(j\omega) = -\pi$ and $20\log_{10}|G(j\omega)H(j\omega)| = -15.5$ dB. For this case, $7|G(j\omega)H(j\omega)| = 7/6 > 1$. Also, when $\omega = 1$, $\sphericalangle G(j\omega)H(j\omega) = -\pi$ and $20\log_{10}|G(j\omega)H(j\omega)| = 20$ dB. For this case, $7|G(j\omega)H(j\omega)| = 0.7 < 1$.

(c) The root locus is as shown in Figure S11.47 for $K > 0$. There are poles on the $j\omega$-axis when $K = 8.51$ and $K = 4.69 \times 10^5$.

(d) The Nyquist plot is as shown in Figure S11.47. The system is stable when $K < 8$ and for $K \gtrsim 10^5$.

11.48. **(a)** As $\epsilon \to 0$, then the contribution to $\sphericalangle G(e^{j0^+})H(e^{j0^+})$ is from only the pole on the unit circle. Therefore,

$$\sphericalangle G(e^{j0^+})H(e^{j0^+}) = -\frac{\pi}{2}.$$

Similarly,

$$\sphericalangle G(e^{j0^-})H(e^{j0^-}) = \frac{\pi}{2}.$$

(b) We have

$$G(e^{j\omega})H(e^{j\omega}) = \frac{(\cos 2\theta - \cos\theta) - j(\sin 2\theta - \sin\theta)}{2(1-\cos\theta)}.$$

It can be easily verified that if this is plotted for various values of ω, we would get a plot looking like Figure P11.48(b). Furthermore, from part (a) we know that when $\omega = 0^+$, $\sphericalangle G(e^{j0^+})H(e^{j0^+}) = -\frac{\pi}{2}$. We see that this is indeed true. Also, from part (a) we know that when $\omega = 0^- = 2\pi^-$, $\sphericalangle G(e^{j0^-})H(e^{j0^-}) = \frac{\pi}{2}$. We see that this is also true.

(c) When $\omega = \pi/3$, $\sphericalangle G(e^{j\omega})H(e^{j\omega}) = -\pi$. Now, at this point

$$|G(e^{j\omega})H(e^{j\omega})| = \frac{(-1)^2 + 0^2}{2(1-1/2)} = 1.$$

(d) Using an argument similar to the one used in Problem 11.44(c), we may conclude that

$$\sphericalangle G(z)H(z) = -\theta.$$

(e) The system is stable for $-\infty < -(1/K) < -1$. That is the system is stable for $0 < K < 1$.

(f) The Nyquist plots are as shown in the figures below.

11.49. **(a)** The dc gain is $H(j0) = G$.

(b) The time constant is $1/a$.

(c) The frequency response of the system is

$$H(j\omega) = \frac{Ga}{j\omega + a}.$$

At $\omega = a$, $|H(ja)| = G/\sqrt{2} = H(j0)/\sqrt{2}$. Therefore, the bandwidth of the amplifier is a.

(d) The system function of the closes loop system is

$$Q(s) = \frac{Ga}{s + a + KGa}.$$

The dc gain is $Q(j0) = G/(1+KG)$. The time constant is $1/[a(1+KG)]$. The bandwidth is $a(1+KG)$.

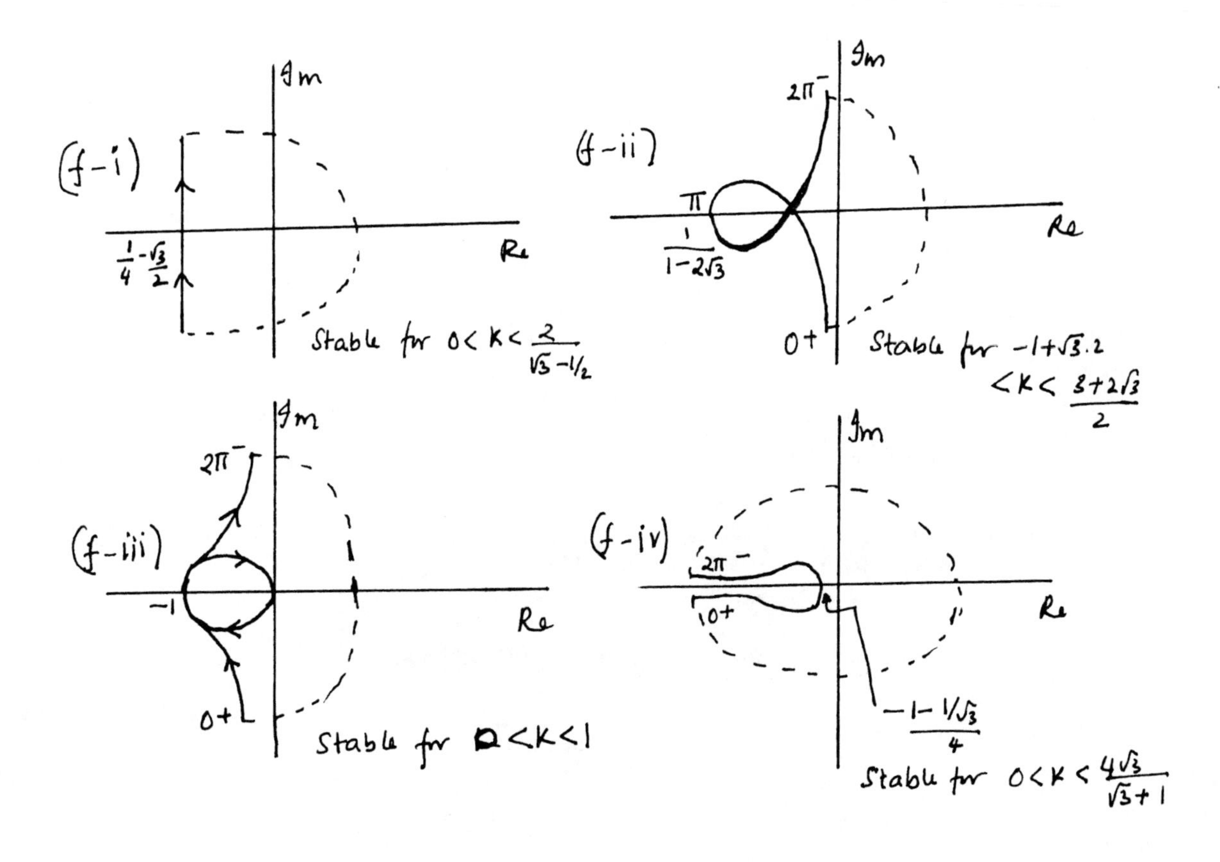

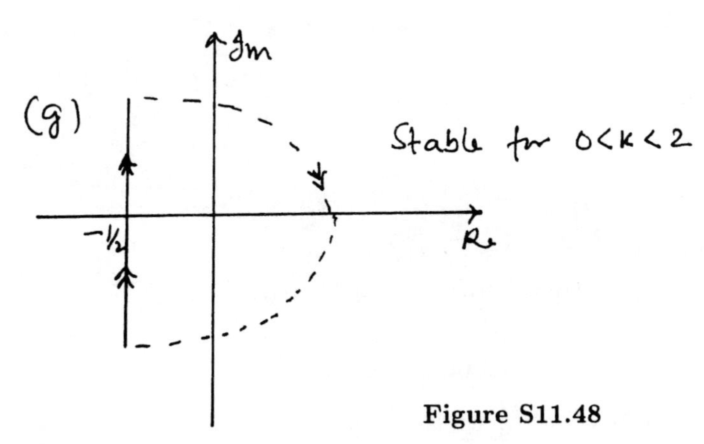

Figure S11.48

(e) The new bandwidth $a(1 + KG)$ is twice the old bandwidth a when $K = 1/G$. The corresponding (i) time-constant is $1/2a$ and (ii) dc gain is $G/2$.

11.50. (a) For both Figures P11.50(b) and P11.50(c), we have

$$H(s) = \frac{V_o(s)}{V_i(s)} = \frac{-KZ_2(s)}{KZ_1(s) + Z_1(s) + Z_2(s)}.$$

(b) For $K >> 1$, $KZ_1(s) >> Z_1(s) + Z_2(s)$. Therefore,

$$H(s) \approx \frac{-KZ_2(s)}{KZ_1(s)} = -\frac{Z_2(s)}{Z_1(s)}.$$

11.51. (a) When $K = 10^6$ and $R_2/R_1 = 1$, then $H(s) = -1$. This is the same as $-R_2/R_1$. When $K = 10^6$ and $R_2/R_1 = 10^3$, then $H(s) = -999$. This is approximately same as $-R_2/R_1$.

(b) (i) When $R_2/R_1 = 100$ and $K = 10^6$, $H(s) = -99.989$. When $R_2/R_1 = 100$ and $K = 5 \times 10^5$, $H(s) = -99.9798$. The percentage change is 0.01%.

(ii) K should be 9898.

11.52. In this case, using the result of Problem 11.50(a) we have

$$H(s) = \frac{-KC}{s(K+1)R + C}.$$

When $K >> 1$,

$$H(s) = \frac{-C}{Rs}.$$

Clearly, $H(s)$ performs the job of an integrator. The approximation breaks down when

$$s(K+1)R \approx C \quad \Rightarrow \quad \omega \approx \frac{C}{(K+1)R}.$$

11.53. (a) Using Kirchoff's law,

$$v_o(t) = v_d(t) + i_d(t)R + v_i(t).$$

Also,

$$v_-(t) = v_o(t) - v_d(t) = v_i(t) + i_d(t)R.$$

This implies that

$$v_o(t) = -K[v_o(t) - v_d(t)].$$

(b) For large K, eq. (P11.53-3) may be written as

$$v_o(t)[1 + K] = Kv_d(t) \quad \Rightarrow \quad v_o(t) \approx v_d(t).$$

This implies that

$$v_o(t) \approx \left(\frac{1+K}{K}\right) v_o(t) + RMe^{qv_o(t)/kT} + v_i(t).$$

This in turn may be approximates as

$$v_o(t) \approx -K[v_i(t) + RMe^{qv_o(t)/kT}].$$

Now note that from Figure P11.53(b),

$$v_o(t) = -K[v_i(t) + RMe^{qv_o(t)/kT}].$$

(c) We may rewrite the equation obtained in the previous part as

$$RMe^{qv_o(t)/kT} = \frac{v_o(t) + Kv_i(t)}{-K}.$$

For large K, this becomes

$$RMe^{qv_o(t)/kT} = -v_i(t).$$

Taking the log of both sides and simplifying, we get

$$v_o(t) = \frac{kT}{q} \ln\left[-\frac{v_i(t)}{RM}\right].$$

11.54. (a) Since $X_f(s) = -X_i(s)H(s)G(s)$, and $G(s)H(s) = -1$, we may conclude that $X_f(s) = X_i(s)$ and $x_f(t) = x_i(t)$.

(b) The closed-loop gain will be infinite.

(c) (i) By simple algebraic manipulation, the required relation may be proved.

(ii) Substituting for $Z_1(s)$, $Z_2(s)$, and $Z_3(s)$ in the equation for $G(s)H(s)$, we get

$$G(j\omega)H(j\omega) = \frac{-AX_1(j\omega)X_2(j\omega)}{jR_0(X_1(j\omega) + X_2(j\omega) + X_3(j\omega)) - X_2(j\omega)(X_1(j\omega) + X_3(j\omega))}.$$

If we want to produce oscillations, then we need $G(j\omega)H(j\omega) = -1$. Therefore,

$$\frac{-AX_1(j\omega)X_2(j\omega)}{jR_0(X_1(j\omega) + X_2(j\omega) + X_3(j\omega)) - X_2(j\omega)(X_1(j\omega) + X_3(j\omega))} = -1. \qquad \text{(S11.54-1)}$$

Cross-multiplying and equating imaginary parts on both sides, we get

$$R_0[X_1(j\omega) + X_2(j\omega) + X_3(j\omega)] = 0.$$

Assuming that R_0 is not zero, we get

$$X_1(j\omega) + X_2(j\omega) + X_3(j\omega) = 0.$$

(iii) Equating the real parts on both sided of eq. (S11.54-1), we get

$$A = -\frac{X_1(j\omega) + X_3(j\omega)}{X_1(j\omega)}.$$

Using the result of part (i), we get

$$A = \frac{X_2(j\omega)}{X_1(j\omega)}.$$

(iv) From the result of part (ii), we know that at the oscillation frequency ω_0

$$\omega_0 L + \omega_0 L - \frac{1}{\omega_0 C} = 0.$$

Therefore,

$$\omega_0 = \frac{1}{\sqrt{2LC}}.$$

11.55. (a) From Figure P11.55(a), we get

$$H(z) = \sum_{i=1}^{N} c_i z^{-i}.$$

The closed-loop system function for the system of Figure P11.55(b) is

$$Q(z) = \frac{K}{1 + KH(z)} = \frac{K}{1 + K\sum_{i=1}^{N} c_i z^{-i}}.$$

(b) In this case

$$Q(z) = \frac{K}{1 + KH(z)} = \frac{K\sum_{i=0}^{N} d_i z^{-i}}{\sum_{i=0}^{N} d_i z^{-i} + K\sum_{i=1}^{N} c_i z^{-i}}.$$

We need $Q(z)$ to be of the form

$$Q(z) = \frac{\sum_{i=0}^{N} b_i z^{-i}}{\sum_{i=0}^{N} a_i z^{-i}}$$

Equating coefficients in eqs. (S11.55-1) and (S11.55-2), we get

$$a_0 = d_0, \qquad K = \frac{b_0}{a_0}, \qquad d_i = \frac{1}{K} b_i, \qquad c_i = \frac{1}{K}[a_i - d_i].$$

11.56. (a) Using the given approximations and the fact that $a(t) = 0$, we get

$$L\frac{d^2\theta(t)}{dt^2} = g\theta(t) + Lx(t).$$

Taking the Laplace transform of both sides of the above equation and simplifying, we get

$$H(s) = \frac{\theta(s)}{X(s)} = \frac{1}{s^2 - g/L}.$$

The system has poles at $s = \sqrt{g/L}$ and $s = -\sqrt{g/L}$. Clearly, one of these poles is always in the right-hand side of the s-plane. This implies that the system is unstable.

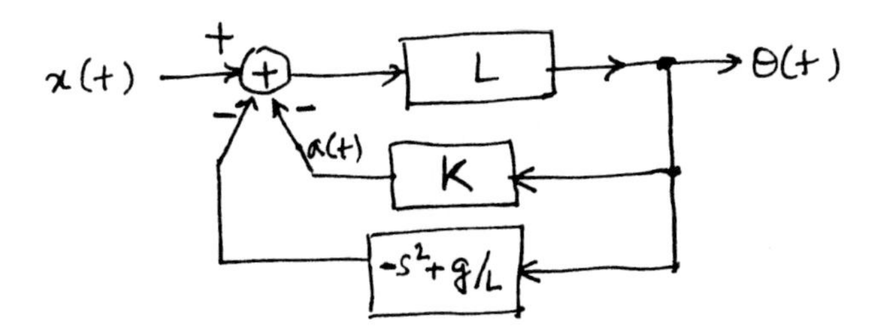

Figure S11.56

(b) The block-diagram of the linearized system is as shown in Figure S11.56.

The closed loop system function is

$$Q(s) = \frac{1}{s^2 - (g/L) + (K/L)}.$$

The poles of $Q(s)$ are at $s = \sqrt{(g-K)/L}$ and $s = -\sqrt{(g-K)/L}$. Clearly, one of the poles is always in the right-hand side of the s-plane. This implies that the system is unstable. If now $K = 2g$, then the poles are on the imaginary axis and the system has a purely oscillatory response.

(c) In this case,

$$L\frac{d^2\theta(t)}{dt^2} = g\theta(t) - K_1\theta(t) + K_2\frac{d\theta(t)}{dt} + Lx(t).$$

Therefore, the closed-loop system function is

$$Q(s) = \frac{\theta(s)}{X(s)} = \frac{L}{Ls^2 + K_2 s - g + K_1}.$$

The poles of $Q(s)$ are at

$$s = \frac{-K_2 \pm \sqrt{K_2^2 - 4L(K_1 - g)}}{2L}.$$

If $K_2 > 0$ and $K_1 > g$, the poles are in the left-hand side of the s-plane. This would make the system stable.

Since $\omega_n^2 = 9 = 2K_1 - 19.6$, $K_1 = 14.3\ rad/sec^2$. Also, $2K_2 = 2\zeta\omega_n$. This implies that $K_2 = 3$ rad/sec.

11.57. The closed-loop system function is

$$Q(s) = \frac{H_c(s)H_p(s)}{1 + H_c(s)H_p(s)}.$$

(a) Here, the closed-loop system function is

$$Q(s) = \frac{K\alpha}{s + \alpha(1+K)}.$$

We can always choose K such that the pole $s = -\alpha(1+K)$ is in the left-half of the s-plane.

Taking the inverse Laplace transform of $Q(s)$, we get

$$q(t) = K\alpha e^{-(\alpha+K\alpha)t}u(t).$$

Now, if $x(t) = \delta(t)$, then

$$y(t) = q(t) = K\alpha e^{-(\alpha+K\alpha)t}u(t).$$

Therefore, the error is

$$e_0(t) = \delta(t) - q(t) = \delta(t) - K\alpha e^{-(\alpha+K\alpha)t}u(t).$$

If $q(t)$ corresponds to a stable system, then $q(t)$ will be a decaying exponential. Therefore, $e(t)$ also decays with time.

Now, if $x(t) = u(t)$, then $y(t) = u(t) * q(t)$ and the error is

$$e_1(t) = u(t) - u(t) * q(t) = u(t) * [\delta(t) - q(t)] = u(t) * e_0(t).$$

We know that $e_0(t)$ is a function that decays with time. Obviously, $e_1(t) = u(t) * e_0(t)$ cannot be a function that decays with time.

(b) Here, the closed-loop system function is

$$Q(s) = \frac{H_c(s)H_p(s)}{1 + H_c(s)H_p(s)} = \frac{\alpha(K_1 s + K_1)}{s^2 + s(\alpha + K_1\alpha) + \alpha K_2}.$$

If $\alpha + K_1\alpha > 0$ and $\alpha K_2 > 0$, then the system will be stable.

In this case,

$$d(t) = K_1 e(t) + K_2 \int_{-\infty}^{t} e(t)dt.$$

This implies that $e(t)$ can go to zero and still result in $d(t) =$ constant. This implies that the system can track a step.

(c) If we use the PI controller, the closed-loop system function is

$$Q(s) = \frac{K_1 s + K_2}{s^3 - 2s^2 + (K_1+1)s + K_2}.$$

By the Routh criterion, all coefficients in the denominator have to be positive to get all roots to lie in the left-half of the s-plane. Clearly, this is not true for $Q(s)$. Therefore, the system is unstable.

For the PID controller,

$$Q(s) = \frac{K_3 s^2 + K_1 s + K_2}{s^3 + s^2(K_3 - 2) + 2(K_1+1)s + K_2}.$$

The system is stable if $K_3 > 2$ and $K_1, K_2 > 0$.

11.58. **(a)** We have $X(s) = 1/s$ and

$$E(s) = \frac{1}{s}\left[\frac{1}{1+H(s)}\right].$$

Using the final-value theorem, we have

$$e(\infty) = \lim_{s\to 0} sE(s) = \frac{1}{1+H(0)}.$$

For $l = 1$, $H(0) = \infty$. Therefore, $e(\infty) = 0$. This implies that the system can track a step.

(b) In this case, $X(s) = 1/s^2$ and

$$E(s) = \frac{1}{s^2}\left[\frac{1}{1+H(s)}\right].$$

Using the final-value theorem, we have

$$e(\infty) = \lim_{s\to 0} sE(s) = \lim_{s\to 0}\frac{1}{s(1+H(s))}.$$

Here $e(\infty)$ is a constant(using the equation for $H(s)$ given in the problem). Therefore, the system cannot track a ramp.

(c) In this case, $X(s) = 1/s^k$ and

$$E(s) = \frac{1}{s^k}\left[\frac{1}{1+H(s)}\right].$$

Using the final-value theorem, we have

$$e(\infty) = \lim_{s\to 0} sE(s) = \lim_{s\to 0}\frac{1}{s^{k-1}(1+H(s))}.$$

Here $e(\infty)$ is unbounded for $k > 2$.

(d) If $x(t) = u_{-k}(t)$, then $X(s) = 1/s^k$ and

$$E(s) = \frac{1}{s^k}\left[\frac{1}{1+H(s)}\right] = \frac{s^l}{s^k(s^l + G(s))},$$

where

$$G(s) = \frac{K\prod_{k=1}^{m}(s-\beta_k)}{\prod_{k=1}^{n-1}(s-\alpha_k)}.$$

Using the final-value theorem, we have

$$e(\infty) = \lim_{s\to 0} sE(s) = \lim_{s\to 0}\frac{s^l}{s^{k-1}(s^l + G(s))}. \qquad \text{(S11.58–1)}$$

(i) From eq. (S11.58-1), $e(\infty) = 0$ for $l \geq k$.

(ii) From eq. (S11.58-1), $e(\infty)$ =constant for $l = k - 1$.
(iii) From eq. (S11.58-1), $e(\infty)$ is unbounded for $l < k - 1$.

11.59. **(a)** If $x[n] = u[n]$, then

$$X(z) = \frac{1}{1 - z^{-1}} = \frac{z}{z - 1}, \qquad |z| > 1.$$

Therefore,

$$E(z) = X(z)\left[\frac{1}{1 + H(z)}\right] = \frac{z(z + 1/2)}{(z - 1)(z + 1/2) + 1}.$$

It may be easily verified that partial fraction expansion may be used to write $E(z)$ as

$$E(z) = 1 + \frac{P}{z - a} + \frac{Q}{z - b},$$

where $|a| < 1$ and $|b| < 1$ and P and Q are constants. Clearly, $e[n] = \delta[n]+$ sum of to damped exponentials. Therefore, $\lim_{n \to \infty} e[n] = 0$.

(b) Here,

$$H(z) = \frac{A(z)}{(z - 1)B(z)}.$$

where $A(z)$ and $B(z)$ are polynomials in z. Then,

$$E(z) = \frac{zB(z)}{(z - 1)B(z) + A(z)}.$$

The closed-loop system function is

$$Q(z) = \frac{(z - 1)B(z)}{(z - 1)B(z) + A(z)}.$$

Since it is given that $Q(z)$ is stable, we know that $Q(z)$ has no poles outside the unit circle. Also note that the $(z - 1)$ factor in the numerator cancels out the pole at $z = 1$ introduced by the step.

Since $E(z)$ and $Q(z)$ share the same denominators, we can also conclude that $E(z)$ also has no poles inside the unit circle. This implies that $e[n]$ is a stable signal and $\lim_{n \to \infty} e[n] = 0$.

(c) Here, $H(z) = 1/(z - 1)$. Therefore,

$$E(z) = X(z)\frac{1}{1 + H(z)} = 1.$$

This implies that $e[n] = \delta[n]$ and $e[n] = 0$ for $n \geq 1$.

(d) Here,

$$E(z) = X(z)\frac{1}{1 + H(z)} = 1 + \frac{1}{4}z^{-1}.$$

This implies that $e[n] = \delta[n] + (1/4)\delta[n - 1]$ and $e[n] = 0$ for $n \geq 2$.

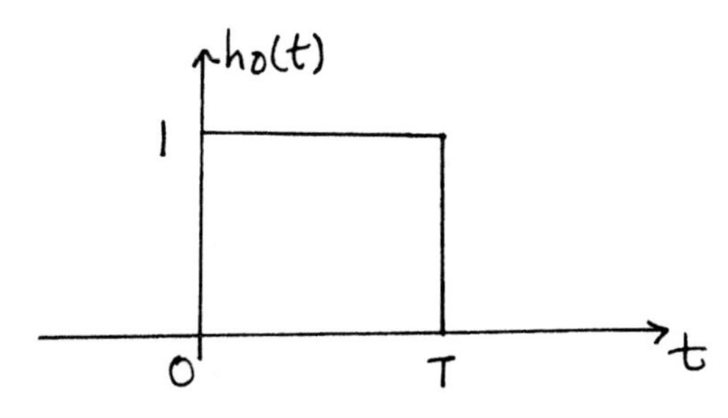

Figure S11.60

(e) Here,

$$E(z) = \sum_{k=0}^{N-1} a_k z^{-k} \quad \text{and } X(z) = \frac{z}{z-1}.$$

Since we know that

$$E(z) = X(z)\frac{1}{1+H(z)},$$

we get

$$H(z) = \frac{z - (z-1)\sum_{k=0}^{N-1} a_k z^{-k}}{(z-1)\sum_{k=0}^{N-1} a_k z^{-k}}.$$

(f) Here,

$$E(z) = X(z)\frac{1}{1+H(z)} = 1 + z^{-1}.$$

This implies that $e[n] = \delta[n] + \delta[n-1]$ and $e[n] = 0$ for $n \geq 2$.

11.60. (a) The output of the zero-order hold can be considered to be of the form $\left[\sum_{k=-\infty}^{\infty} e[k]\delta(t-kT)\right] *$ $h_0(t)$, where $h_0(t)$ is as shown in Figure S11.60. Now, the output of $H(s)$ will be of the form

$$\begin{aligned}
p(t) &= \left[\sum_{k=-\infty}^{\infty} e[k]\delta(t-kT)\right] * h_0(t) * h(t) \\
&= \left[\sum_{k=-\infty}^{\infty} e[k]\delta(t-kT)\right] * f(t) \\
&= \left[\sum_{k=-\infty}^{\infty} e[k]f(t-kT)\right]
\end{aligned}$$

where $f(t) = h_0(t) * h(t)$. Now, the output of the C/D system will be of the form

$$p[n] = p(nT) = \left[\sum_{k=-\infty}^{\infty} e[k]f(nT - kT)\right].$$

Since $f[n] = f(nT)$, we may write the above equation as

$$p[n] = \sum_{k=-\infty}^{\infty} e[k]f[n-k].$$

Since the overall system obviously obeys the convolution sum, we may conclude that it is LTI.

(b) Now, let $e[n] = u[n]$. Then, $\sum_{k=-\infty}^{\infty} \delta(t-kT)e[k] = \sum_{k=0}^{\infty} \delta(t-kT)e[k]$.. Therefore, the output of the zero-rder hold will be $u(t)$. Therefore, $p(t) = s(t)$ will be the step response of $H(s)$. Therefore, $p[n] = p(nT) = s(nT)$. Noting that when the input to $F(z)$ was $u[n]$ the output was $p[n] = s(nT)$, we conclude that the step response of the system is $q[n] = s(nT)$.

(c) Given $H(s)$, we know that

$$S(s) = \left[\frac{1}{s-1}\right]\frac{1}{s} = \frac{1}{s+1} + \frac{1}{s}.$$

Now since $q[n] = s(nT)$,

$$Q(z) = \frac{1}{1 - e^T z^{-1}} - \frac{1}{1 - z^{-1}} = \frac{z^{-1}(e^T - 1)}{(1 - z^{-1})(1 - e^T z^{-1})}.$$

Since $u[n] \stackrel{Z}{\longleftrightarrow} 1/(1 - z^{-1})$, we may conclude that

$$F(z) = (1 - z^{-1})Q(z) = \frac{z^{-1}(e^T - 1)}{(1 - z^{-1})(1 - e^T z^{-1})}, \qquad |z| > e^T.$$

(d) The root locus for

$$G(z)F(z) = \frac{K(e^t - 1)}{z - e^T}$$

is as shown in Figure S11.60. From this it is clear that the system becomes unstable when the roots just cross the unit circle at $z = \pm 1$. From this, we may find that the system is stable for

$$1 < K < \frac{e^T + 1}{e^T - 1}.$$

(e) Here,

$$G(z)F(z) = \frac{Kz(e^T - 1)}{(z + 1/2)(z - e^T)}.$$

The root locus is as shown in Figure S11.60. When $z = 1$ is on the root locus, then $K = 3/2$. The second pole is at $z = -(1/2)e^T$. When $z = -1$, then $K = (1/2)(1 + e^T)/(e^T - 1)$. If this has to be greater than 3/2, then $T < \ln 2$. Choosing T to be $\ln(3/2)$, we get $K = 2$. The poles are then at $\pm\sqrt{3}/2$. This is a stable system.